W9-ABM-711

Lecture Notes in Computer Science

Edited by G. Goos and J. Hartmanis
Series: I.F.I.P. TC7 Optimization Conferences

27

Optimization Techniques IFIP Technical Conference

Novosibirsk, July 1–7, 1974

Edited by G. I. Marchuk

Springer-Verlag
Berlin · Heidelberg · New York 1975

Prof. Dr. G. I. Marchuk
Computer Center
Novosibirsk 630090/USSR

AMS Subject Classifications (1970): 00 A 10, 49 B 10, 49 B 25, 65 K 05, 90 A 15, 90 C 10, 90 C 30, 90 D 25, 93 B 30, 93 C 20
CR Subject Classifications (1974): 5.17. 5.41, 8.1

ISBN 3-540-07165-2 Springer-Verlag Berlin · Heidelberg · New York
ISBN 0-387-07165-2 Springer-Verlag New York · Heidelberg · Berlin

Offsetdruck: Julius Beltz, Hemsbach/Bergstr.

PREFACE

The Proceedings are based on the papers presented at the IFIP Technical Conference on Optimization Techniques held in Novosibirsk, July 1-7, 1974. The Conference was organized by the IFIP Technical Committee on Optimization (TC-7) and by the Academy of Sciences of the USSR.

The Conference was devoted to a discussion of the following topics: Theory of Differential Games, Mathematical Programming and Numerical Algorithms, Optimal Control, System Modelling and Identification.

The papers given at the Conference reflected an active progress that had been made in the above mentioned areas and an increasing interest of specialists in optimization in the problems of economy and the environment.

The International Program Committee consisted of:
L. S. Pontrjagin, G. I. Marchuk (USSR), Chairman, A. V. Balakrishnan (USA), J. L. Lions (France), J. Stoer (West Germany), P. Conti (Italy), K. Yajima (Japan), and S. Olech (Poland).

TABLE OF CONTENTS

SYSTEM MODELLING AND IDENTIFICATION

OPTIMAL CONTROL

MATHEMATICAL PROGRAMMING AND NUMERICAL ALGORITHMS

THEORY OF GAMES

IDENTIFICATION — INVERSE PROBLEMS FOR PARTIAL DIFFERENTIAL EQUATIONS: A STOCHASTIC FORMULATION

A.V. Balakrishnan*
System Science Department; UCLA, USA

1. Introduction

This paper presents a stochastic formulation of a class of identification problems for partial differential equations, known as 'inverse' problems in the mathematical-physics literature. By introducing stochastic processes to model errors in observation as well as 'disturbance' we can provide a precise formulation to interpret what appear to be 'ad hoc' techniques, especially in the treatment of 'inverse' problems. More importantly, we can model unknown sources as stochastic disturbances leading to more general 'inverse' problems than considered hitherto.

Important as inverse problems are in the area of Geophysical applications, where they were studied initially, they would appear to be equally if not more important, in the newer application areas such as modelling and optimization of Environmental Systems, particularly in Water Resources. Diffusion equations arise for instance in stream pollution problems as well as in underground water flow problems, and these models would appear to be well enough founded to attempt system identification. In fact with ever increasing feasibility in high speed and low cost of digital and hybrid computation, the scope of such inverse problems in modelling and simulation is bound to widen even further.

We shall only deal with Cauchy problems for partial differential equations with <u>continuous</u> time observation (as opposed to 'discrete' time). A crucial point then is that the familiar 'Wiener process' modelling of observation errors is unrealistic, as pointed out in [1]. Instead we shall employ a 'white-noise' theory throughout, the relevant notion being explained in Section 2. This involves handling finitely-additive cylinder measures but for linear systems it actually provides a simpler set-up in dealing with partial differential equations. A second feature of our approach is the use of semigroup theory because within the restriction to time-invariant systems (a natural assumption for identification problems) it separates the more general and abstract structural aspects of the problems from the technicalities of the particular partial differential equation set-up involved,

*Research supported in part under AFOSR Grant No. 73-2492, Applied Math Division, USAF.

and retains the similarity to the familiar finite-dimensional formulation as much as possible.

2. The White Noise Process:

Let H denote a real separable Hilbert space and let

$$W = L_2((0,T);H)$$

so that W is a real separable Hilbert space also. Introduce the Gauss measure μ on W on the ring of cylinder sets. This measure is completely described by specifying that for any h in W, with ω denoting 'points' in W:

$$\int_H \exp i[\omega,h] \quad d\mu \quad = \quad \exp(-||h||^2/2)$$

For each ω, let

$$n(t,\omega) \quad = \quad \omega(t)$$

where of course $\omega(t)$ is only defined a.e. We call the 'process' $n(t,\omega)$ 'white noise'. More generally we shall continue to call

$$L\omega$$

white noise again, if L is a linear bounded transformation mapping W into another Hilbert space, and LL^* has a bounded inverse. Not every Borel measurable function $f(\omega)$ (mapping W into a Hilbert space H_1) is a random variable in the usual sense since μ being only finitely additive, need not be extendable as a countably additive measure on every sub-sigma-algebra of Borel sets. We shall mention only two classes of function where this is possible, (and is of significance to us); referring to the work of Gross [2], and [3] for further information on the subject. The first function is a linear transformation: $f(\omega) = L\omega$. This is a random variable in the sense that μ can be extended to be countably additive on the inverse images of Borel sets, if and only if L is Hilbert-Schmidt, and in that case

$$E(||L\omega||^2) = \text{Tr. } LL^* < \infty$$

The second functional is the homogeneous polynomial of degree two, scalar valued:

$$f(\omega) = [L\omega,\omega]$$

This is a random variable if and only if $(L+L^*)$ is nuclear. See [3]. We shall also need to use the theory of estimation involving linear random variables and stochastic semigroup equations; see [3], [7] for both of these topics.

3. A Class of Inverse Problems

Our point of departure will be a particular class of inverse problems for linear p.d.e., studied in depth by Lavrentiev et al [5]. We quote it with a slight modification in notation: Given the telegraphist's equation:

$$\left.\begin{aligned} &\frac{\partial^2 f}{\partial t^2} = \Delta f + a(\cdot)f + u(t,.) \text{ in}\Omega;\ t > 0 \\ &f(0,.) = 0 = \frac{\partial f}{\partial t}(0,.) \\ &\Omega = [x,y,z] \text{ with } z > 0 \\ &\frac{\partial f}{\partial z}(t,.) = 0\Big|_{z=0} \end{aligned}\right\} \tag{3.1}$$

The 'inverse' problem is to determine the unknown function $a(\cdot)$, given the solution of (3.1) $f(t,x_1,y_1,z_1)$, at the point (x_1,y_1,z_1), and the input $u(t,.)$. [Alternately, we may wish to determine a functional on $f(t,.)$ following Marchuk [6]. This can bring in further simplification.] Their technique (in outline, consult the paper for the details) is to first solve the Cauchy problem with the same initial and boundary conditions:

$$\frac{\partial f_o}{\partial t} = \Delta f_o + u(t,.) \tag{3.2}$$

and <u>linearize</u> (3.1) about this solution to obtain, setting

$$h = f - f_o \tag{3.3}$$

the linear equation (linear also in $a(\cdot)$):

$$\frac{\partial^2 h}{\partial t^2} = \Delta h + a(\cdot)\, f_o(t,.) \tag{3.4}$$

with the same initial and boundary conditions for h. This linearization technique is also basic in Marchuk's work [6]. We have thus the problem of determining $a(\cdot)$ from (3.4) and the given observation:

$$y(t) = h(t,x_1,y_1,z_1) = f(t,x_1,y_1,z_1) - f_o(t,x_1,y_1,y_1,z_1) \tag{3.5}$$

They proceed to show that this solution of this problem has the property of uniqueness, and obtain formulas for determining $a(\cdot)$ in special cases and otherwise introduce techniques based on integral geometry.

Our first step is to obtain an abstract formulation of this problem using the theory of semigroups of linear operators. Not only does such a formulation yield a useful measure of generality; it also enables us to see more clearly the relationship of this 'inverse' problem to the 'identification' problem as studied in the engineering literature. As is well known, we can recast (3.1) as an 'ordinary' differential equation in the Hilbert space $H = L_2(\Omega)$ as follows (see [7] for an elementary exposition):

$$\dot{x}(t) = Ax(t) + Bx(t) + u(t) \tag{3.6}$$

where A is the infinitesimal generator of a strongly continuous semigroup $S(t)$ (corresponding to the Cauchy problem with homogeneous boundary conditions) $u(t)$ denoting the function $u(t,.)$ as an element of H for each t; $x(0) = 0$, and finally, B is a linear bounded transformation of H into H that is to be determined given $u(t)$, $0 \leq t \leq T$ and the 'observation'

$$f(t;\, x_1, y_1, z_1) = Cx(t) \qquad 0 < t < T$$

In this example C is a linear transformation defined on the subdomain of continuous functions of H and is an unbounded, uncloseable operator, and hence our natural inclination to make C to be a bounded (continuous) operator has to be revised. On the other hand one can adduce 'physical' arguments for making C bounded; we may argue that any measuring device or instrument must correspond to a spatial smoothing and that 'pointwise' measurements are impossible. This view is shared by Marchuk [6]. Fortunately it is possible to construct a theory in which measurement at a point (or points) of Ω or its boundary can be permitted. Thus we assume that C is unbounded but that for each x in H.

$$S(t)x \in \cdot\, D(C) \qquad \text{for } t > 0 \text{ a.e.}$$

and that for $h(\cdot)$ in $L_2([0,T];H) = W$

$$\int_0^t S(t-\sigma)h(\sigma)d\sigma \in D(C) \qquad \text{a.e. } 0 < t < T$$

and that

$$C\int_0^t S(t-\sigma)h(\sigma)d\sigma = \int_0^t CS(t-\sigma)h(\sigma)d\sigma$$

and the righthand side defines a linear bounded transformation of W into

$$W_o = L_2([0,T];H_o)$$

where H_o is the range space of C and at least in theory can be a Hilbert space <u>not</u> necessarily finite dimensional. Needless to say, this assumption is satisfied in the present example. Let us next examine the method of solution of Lavrentiev [5]. We define $x_o(t)$ by:

$$\dot{x}_o(t) = Ax_o(t) + Bx_o(t) + u(t);\; x_o(0) = 0$$

Then defining

$$z(t) = x(t) - x_o(t)$$

we note that

$$\dot{z}(t) = Az(t) + Bx_o(t) + Bz(t)$$

Following the 'linearization' technique now means that we omit the third term $Bz(t)$ or redefine $z(t)$ by:

$$\dot{z}(t) = Az(t) + Bx_o(t); \; z(0) = 0 \tag{3.7}$$

and since $x_o(t) \in D(C)$ a.e., so does $z(t)$ and we obtain:

$$y(t) = Cz(t) \quad \text{a.e.} \tag{3.8}$$

The problem is that of determining B from (3.7) and (3.8). Thanks to the linearization we have:

$$y(t) = \int_o^t CS(t-\sigma)Bx_o(\sigma)d\sigma \quad \text{a.e.} \quad 0 < t < T \tag{3.9}$$

We can clearly see that this is expressible as:

$$y = LB$$

where L is a linear bounded transformation mapping the Banach space E(H), of linear bounded transformations on H into H, into W_o. The uniqueness theorem of Lavrentiev is then equivalent to saying that zero is not in the point spectrum of L. One method of determining B is the familiar least squares technique:

$$\text{Minimize} \quad ||y - LB||^2 \tag{3.10}$$

where B ranges over a known subspace of E(H). The important point here is that for any 'practical' algorithm the subspace has to be finite dimensional. Let M denote the closed subspace in which B is known to lie. Let L^* denote the adjoint of L with respect to M. Then denoting by B the minimizing solution (assuming that one exists in M) we can write the familiar 'pseudo-inverse'

$$B = (L^*L)^{-1} L^*y \tag{3.11}$$

An important point to note is the appearance of the adjoint operation, a feature emphasized (properly) by Marchuk [6].

Before we go on to the stochastic formulation let us briefly examine a second example of the inverse problem to indicate how to include non-homogeneous boundary conditions. This is also a problem considered by Lavrentiev [5, p. 39], except that we shall change the domain. Thus let Ω denote the sphere in R^3:

$$x^2 + y^2 + z^2 < a^2$$

and consider the heat conduction problem:

$$\frac{\partial f}{\partial t} = \Delta f + u(t)\, g(x,y,z) \qquad t > 0, \text{ in } \Omega$$
$$f(t,.)\Big|_{\Gamma} = h(t); \qquad f(0; x,y,z) = 0 \tag{3.12}$$

where Γ is the boundary of Ω and $h(t)$ is an element of $H_\Gamma = L_2(\Gamma)$, a.e. in t, and $h(\cdot)$ is an element of $W = L_2([0,T]; H_\Gamma)$. Let Γ_o denote the 2-sphere

$$x^2 + y^2 + z^2 = c^2 \qquad c^2 < a^2$$

Then the 'observation' or 'measurement' is the function:

$$f(t;x,y,z), \qquad (x,y,z)\varepsilon\Gamma_o$$

The inverse problem is to determine the function $g(x,y,z)$ given the scalar function $u(t)$ and the observation. This is a simpler problem in that it is actually 'linear' as we shall see. To obtain the abstract formulation we begin with the Cauchy problem corresponding to the homogeneous boundary value problem:

$$\frac{\partial f}{\partial t} = \Delta f \; ; \quad f(t,.)\Big|_{\Gamma} = 0 \; ; \quad f(0;.) \text{ given}$$

Let $S(t)$ denote the semigroup and A the generator with denoting $L_2(\Omega)$ as before. Next let us note that the domain is such that the Dirichlet problem:

$$\Delta f = 0 \; ; \quad f(\cdot)\Big|_{\Gamma} = h$$

where h is an element of H_Γ, has a unique solution given by

$$f = Dh$$

where D is a bounded linear operator mapping H_Γ into H. We shall call D the Dirichlet operator. The solution to (3.12) can now be expressed (see [4] for details):

$$x(t) = -\int_0^t AS(t-\sigma)D\, h(\sigma)d\sigma + \int_0^t S(t-\sigma)\, Bu(\sigma)d\sigma \qquad \text{a.e.} \tag{3.13}$$

where B denotes the element $g(\cdot)$ assumed to be in H, and we stress that $x(t)$ is defined only a.e. in general. Next let C denote the mapping corresponding to the observation. C is clearly defined on the subspace of continuous functions in H and maps this subspace into $H_o = L_2(\Gamma_o)$. Also C satisfies the conditions imposed earlier in the abstract version of the first problem. Thus we can write for the observation:

$$y(t) = -\int_0^t CAS(t-\sigma)\, Dh(\sigma)d\sigma + \int_0^t CS(t-\sigma)\, Bu(\sigma)d\sigma \qquad \text{a.e.,} \tag{3.14}$$

and each term on the right is an element of $W_o = L_2([0,T];H_o)$. We need to determine B from (3.14) knowing $h(\cdot)$ and $u(\cdot)$. No linearization is required in this problem since (3.14) is already linear in B. Moreover writing LB as before to denote the mapping given by

$$\int_0^t CS(t-\sigma)\ Bu(\sigma)d\sigma$$

we need to solve the linear equation

$$LB = r$$

where $r(\cdot)$ denotes the function:

$$r(t) = y(t) + \int_0^t CAS(t-\sigma)Dh(\sigma)d\sigma$$

and the problem is thus reduced to the one already considered except that no linearizing approximation is needed.

4. Stochastic Formulation

Having established a more general abstract setting using semi-group theory we proceed now to the stochastic formulation. This we shall do in two stages. In the first stage, stochastic aspects will arise primarily by modelling errors in the measurement (the observation) as an additive stochastic term. In practice of course other sources of error such as calibration errors, bias errors may be far more significant; but the point is that these 'systematic' errors are supposed to be known and even after they are corrected for there will always nevertheless remain a random error component which in an electrical instrument will be the shot or thermal noise. It is best modelled as a Gaussian process of large bandwidth; or in theory as 'white noise'.

Let H denote a separable Hilbert space and let $A(\theta)$ parametrised by θ denote a family of infinitesimal generators of semigroups (strongly continuous at the origin) $S(\theta;t)$. Our observation now takes the form:

$$v(t) = Cx(\theta_o;t) + n(t) \qquad 0 \leq t \leq T \tag{4.1}$$

where θ_o is the true parameter value (unknown to the experimenter or observer) and $n(t)$ is a white Gaussian process, and we need to estimate θ_o based on the observation of duration T. The observation $v(t)$ has its range in a Hilbert space H_o, and for each T the function $v(t)$ is such that

$$v(\cdot) \in W_o = L_2((0,T);H_o)$$

The noise process $n(\cdot)$ has its range also in W_o and to say that it is white Gaussian, it suffices to specify the characteristic function of the corresponding weak distribution by: (E denoting expectation, [,] denoting inner product in W_o):

$$E(\exp i[n,h]) = \exp -1/2(d[h,h]), \quad h \in W_o \tag{4.2}$$

where d is a positive number denoting the component-by-component error variance. C is a linear, possibly unbounded operator, but subject then to the assumptions placed in the previous section. Finally $x(\theta_o,t)$ satisfies

$$\dot{x}(\theta_o;t) = A(\theta_o)\, x(\theta_o;t) + B(\theta_o)\, u(t) \tag{4.3}$$

$$x(\theta_o;0) = 0$$

where $B(\theta)$ also parametrized by θ is a linear bounded operator, $u(\cdot)$ is the known source or input process

$$u(\cdot) \in W_i = L_2((0,T);H_i)$$

where H_i is a separable Hilbert space and $B(\theta)$ maps H_i into H. We also allow the parameter θ to have its range in a Hilbert space M or in any algorithmic theory in the neighborhood of a known value θ_1. We may also take the view that θ is Gaussian with mean θ_1 and variance (operator) Λ.

The next basic question concerns the notion of a good or optimal estimate. Since the observation is now stochastic any operation on the data will also be stochastic and we resort to the classical stastical estimation theory to settle this. Thus if we assume that θ is an unknown parameter, we shall accept asymptotically unbiased and asymptotically consistent estimates as optimal. As is well known, one way to obtain such estimates is to invoke the '(a posteriori) maximal likelihood' technique. The likelihood functional, or more properly, the Radon-Nikodym derivative of the measure induced by the process $v(\cdot)$ on W_o with respect to the Gauss measure (induced by $n(\cdot)$), is given by:

$$p(T;v/\theta) = \exp - 1/2(1/d)\ \{[Cx(\theta;.),Cx(\theta;.)] - 2[Cx(\theta;.),v]\} \ldots \tag{4.4}$$

for each fixed θ, and T, where $x(\theta;t)$ is the unique solution of:

$$\dot{x}(\theta;t) = A(\theta)x(\theta;t) + B(\theta)\, u(t); \quad x(\theta;0) = 0$$

Maximizing the 'likelihood' ratio is seen to be identical with the least squares technique. In practice we seek a root of

$$q(T;v/\theta) = \mathrm{Log}\ p(T;v/\theta)$$

in the neighborhood of the known parameter θ_1. Thus we seek a root of

$$[L_1\theta,\ v - Cx(\theta;.)] = 0 \tag{4.5}$$

where the operator L_1 is the Frechet derivative or gradient mapping:

$$L_1\theta = \frac{d}{d\lambda} Cx(\theta_1 + \lambda\theta;.)\Big|_{\lambda=0} \quad , \theta \in M \tag{4.6}$$

(where $Cx(\theta;.)$ is considered as an element of W_o). We note immediately that in the first example in section 3 the linearization technique of Lavrentiev is precisely this if we use the approximation:

$$Cx(\theta;.) \approx Cx(\theta_1;.) + L_1\theta \tag{4.7}$$

so that we are 'solving':

$$[L_1\theta, v - Cx(\theta_1;.) - L_1\theta] = 0$$

or

$$L_1^* L_1\theta = L_1^*(v - Cx(\theta_1;.))$$

or alternately we are minimizing:

$$[L_1\theta, L_1\theta] - 2 [L_1\theta, v - Cx(\theta_1;.)]$$

The uniqueness now is guaranteed by the 'non-singularity of $L_1^* L_1$ and we may view the Lavrentiev result as establishing this for the particular problem considered.

The stochastic formulation provides a lot more than merely a justification of the linearization procedures of Lavrentiev and Marchuk. In the first place as also noted by Marchuk independently [6], we can repeat the linearization or more properly, give an iteration technique for fixed T as follows:

$$\theta_{n+1} = \theta_n + (L_n^* L_n)^{-1} (v - Cx(\theta_n;.)) \tag{4.8}$$

where L_n is the gradient operator at $\theta = \theta_n$. This is a Newton-Raphson type algorithm which is not limited to linear systems. Moreover we can show the role played by T, the duration of the data. Thus we get first an 'identifiability' criterion:

$$\lim_{T \to \infty} (1/T) (L_o^* L_o) \quad \text{is non-singular} \tag{4.9}$$

where L_o is the gradient at $\theta = \theta_o$ for fixed T. Under this condition we can show that for sufficiently large T a root exists in a suitably small neighborhood of the unknown θ_o, and that this root (approximated by the Newton-Raphson algorithm (4.8)) satisfies the conditions of asymptotic unbiassedness and consistency. These considerations are made much simpler in example two of section 3 because of the linear dependence of $x(\theta;.)$ on θ in that case. In other words the gradient L does NOT depend on θ. Moreover in both examples it is possible to obtain differential equations (updating equations) for the estimate as a function of T and actually a degenerate linear Kalman-Bucy type equation in the case of example 2. See [8].

5. Stochastic Formulation Continued: Source Noise

Not only does the stochastic formulation of inverse problems lead to a precise meaning of the algorithms used; but it also allows us to consider a situation which has so far not been considered. This is the inclusion of errors in modelling the source or forcing function or input; or, alternately, the inclusion of sources not accounted for by the experimenter. The lack of introduction of unknown sources is not surprising since as we shall see the estimation algorithms then cannot be given any physical explanation such as 'least squares fit' or other 'wave-form matching' criterion.

We shall present only the mathematical theory here exploiting the semigroup theoretic formulation already developed. Thus the observation has the same form as before:

$$v(t) = Cx(\theta_o;t) + n(t) \tag{5.1}$$

where we use the same notation as section 4 but $x(\theta;t)$ is itself stochastic now defined by the stochastic equation:

$$\dot{x}(\theta_o;t) = A(\theta_o)\, x(\theta_o;t) + B(\theta_o)\, u(t) + F(\theta_o)\, n_s(t) \tag{5.2}$$

where the difference from (4.3) is the third term, which is a Gaussian stochastic process with (for each T)

$$n_s(\cdot) \in W_s = L_2([0,T];H_s)$$

where H_s is a separable Hilbert space and $F(\theta)$ for each θ is a linear bounded operator mapping H_s into H. The process $n_s(\cdot)$ is white Gaussian with

$$E(\exp i[n_s,h]) = \exp - 1/2[h,h]; \qquad h \in W_s$$

and is independent of the observation noise process $n(\cdot)$.

We need to make a basic assumption; that for each θ of interest the process

$$Cx(\theta;.)$$

where $x(\theta;.)$ is defined by

$$\dot{x}(\theta;t) = A(\theta)\, x(\theta;t) + B(\theta)u(t) + F(\theta)\, n_s(t);\ x(\theta;0) = 0 \tag{5.3}$$

is such that the corresponding covariance operator $R(\theta)$ defined by

$$E(\exp i[h,Cx(\theta;.)]) = \exp - 1/2\ [R(\theta)h,h] \tag{5.4}$$

is trace-class (nuclear). In that case we can again invoke the method of maximum likelihood to obtain asymptotically unbiased, consistent estimates.

First of all under the assumption of nuclearity on $R(\theta)$ the (finitely additive) measure induced by the process $v(\cdot)$ (for each assumed θ) is absolutely continuous with respect to the Gauss measure induced by $n(\cdot)$. Hence we may invoke the criterion of maximum likelihood to obtain optimal estimates. Thus the likelihood functional (R-N derivative) is given by:

$$p(T;v/\theta) = \exp - 1/2d \quad [C\hat{x}(\theta;.), C\hat{x}(\theta;.)] - 2[C\hat{x}(\theta;.), v(\cdot)] + 2 \text{ Tr.} \int_0^T C(CP(t))^* dt \qquad \ldots (5.5)$$

where $\hat{x}(\theta;.)$ is the solution of:

$$\dot{\hat{x}}(\theta;t) = A(\theta)\hat{x}(\theta;t) + B(\theta)u(t) + (CP(t))^* (1/d)(v(t) - Cx(\theta;t)) \quad x(\theta;0) = 0$$

$$[\dot{P}(t)x,y] = [P(t)A(\theta)^*x,y] + [P(t)x,A(\theta)^*y] + [Fx,Fy] - (1/d)[CP(t)x,CP(t)y]$$

$P(0) = 0$; x, y in the domain of $A(\theta)^*$

For each T our estimate is the root of the gradient of

$$q(T;v/\theta) = \text{Log } p(T;v/\theta) \qquad (5.6)$$

in a neighborhood of θ_1. The root will exist and be unique in a small enough neighborhood of θ_o under the identifiability condition which is now that

$$\lim_{T \to \infty} (1/T)(L_o^* L_o) \text{ be non-singular,}$$

where L_o is the gradient operator (Frechet derivative):

$$L_o\theta = \frac{d}{d\lambda} Cx(\theta_o + \lambda\theta)\Big|_{\lambda=0} ; \quad \theta \in M$$

Interpreting L_n as the similar derivative at $\theta = \theta_n$, the algorithm for estimation for each T now becomes:

$$\theta_{n+1} = \theta_n + (L_n^* L_n)^{-1} (L_n^*(v - C\hat{x}(\theta_n;.) - K_n)$$

where K_n is the Frechet derivative of

$$\text{Tr} \int_0^T C(CP(t))^* \, dt$$

with respect to in M at $\theta = \theta_n$. For a treatment of the case of noise on the boundary, See [4]. Note that in (5.5) the term non-linear in the data:

$$[Cx(\theta;.), v(\cdot)]$$

has the form

$$[L\omega,\omega]$$

where L is trace-class, and is a random variable as noted in Section 2.

REFERENCES

1. A.V. Balakrishnan: "On the approximation of Ito integrals by band-limited processes" SIAM Journal on Control 1974.
2. L. Gross: Harmonic Analysis in a Hilbert Space, American Mathematical Society Memories No. 46, 1963.
3. A.V. Balakrishnan: Stochastic Optimization Theory in Hilbert Spaces International Journal on Applied Mathematics and Optimization, Vol. 1, No. 2, 1974.
4. A.B. Balakrishnan: Identification and Stochastic Control of a Class of Distributed Systems with Boundary Noise, IRIA Conference on Stochastic Control and Computer Modelling, June, 1974. To be published by Springer-Verlag.
5. Lavrentiev, Romanov, Vasiliev: Multidimensional Inverse Problems for differential Equations, Lecture Notes in Mathematics, No. 167, Springer Verlag, 1970.
6. G. Marchuk: Perturbation Theory and the Statement of Inverse Problems, Lecture Notes in Computer Science, Vol. 4, Part II, Springer Verlag, 1973.
7. A.V. Balakrishnan: Introduction to Optimization Theory in a Hilbert space Lecture Notes in OR and Mathematical Systems, Springer-Verlag, Russian version, MIR, Moscow, 1974.
8. A.V. Balakrishnan: "Stochastic System Identification Techniques", in Stochastic Optimization and Control, edited by H.F. Karreman, John Wiley & Sons, 1968.

KEY PROBLEMS IN THE THEORY OF CENTRALIZED INTERACTION OF ECONOMIC SYSTEMS

K.A. Bagrinovsky

Institute of Economics & Industrial Engineering
Siberian Department of the USSR Academy of Sciences
Novosibirsk, USSR

The national economy of the USSR is a good example of a centralized hierarchical system. The main peculiarity of the system of planning and management in the USSR is the combination of productive and territorial principles.

The latter means that the development of plans and managing effects is carried out in two directions.

In terms of production these operations are carried out according to the hierarchy of governing bodies: the top bodies of planning and control - all-union (union-republic) ministries and departments - productive amalgamations and enterprises.

In terms of territory, plans and decisions are agreed upon with regional planning and control bodies (republic ministries and planning commissions of republics, regions and provinces).

Both directions are reflected in the system of models of territorial-production planning, that is being developed at the Institute of Economics of the Siberian Department of the USSR Academy of Sciences.

In the aspect of production, the system of planning the national economy of the USSR is a three-link one, and each link is associated with a certain problem stemming from the interaction of subsystems of different levels.

The aim of planning and control at the top level is to obtain consistent planned decisions in separate industries and in agriculture for a certain period of time. Each decision at the level of an industry should be stated so as to provide all related industries for material resources produced by this industry, as well as to meet the demands of the social sphere.

The production of fund-forming industries (machine-building, construction and others) should be sufficient to provide for the planned rates of growth in the national economy. Apart from this, the balance relations regulating the use of non-reproducible natural resources and, which is even more important, the utilization and reproduction of labour force must be satisfied. <u>This problem of production consistency</u> in decisions made at industrial level allows various model interpretations, some of which will be discussed later; but in each model essential are relations indicating what labour, financial and main productive funds and in what quantities are needed by a given industry to fulfil a given planned task. These relations are called sectoral cost functions and are formed during solving the problem of the next link, i.e. <u>the problem of working out a sectoral decision.</u>

In the process of solving this problem one should bring into agreement planned tasks developed for the sector on the whole during the period of solving the previous problem and planned-productive activity of units that are run on a self-supporting basis (productive amalgamations and enterprises). An aggregated planned task for an industry is transformed into a concrete plan of the development of the existing enterprises and of the creation of new enterprises and amalgamations. To fulfil successfully this task, various mathematical models are used, but it is always helpful to have the largest possible number of plan fulfilment versions for each economic unit. This need is met by solving <u>the problem of forming versions of the development of a certain unit</u>.

The solution of each problem mentioned above is carried out in the direct connection with the solution of the nearest link problem; the adjacent levels of economic control are exchanging certain portions of information. This exchange of information and the accepted suquence of procedures to transfer it constitute the process of "big iteration" whose successful realization is to lead to a plan agreed upon at all the levels of economic management.

The process of "big iteration" begins with the solution of the problem of productive reconcilement under some hypothesis about the functions of sectoral costs. The task given to an industry and the initial set of versions of the productive unit growth are the initial data for a sectoral problem. The solution of this problem is as follows: planned tasks for amalgamations are inputs in working out the versions of their development, and the estimated sectoral costs serve for the recalculation of planned tasks in the top planning link.

In the self-supporting link the set of versions is formed and the iteration cycle is repeated again.

As was mentioned above, the solution of each of the listed main problems is performed with the help of one or several mathematical models. The economic systems themselves being interconnected, the problems also appear interconnected, which entails special requirements also to models having to provide the "big iteration" process.

These models should be interconnected informationally so that the provision of reasonable inputs to each model would give a possibility to obtain a reasonable solution. For example, in a model for the problem of production reconcilement positive planned tasks for all the industries of the national economy should be guaranteed. The interconnections of models and sequences of procedures should be built up so as to provide for the convergence of the "big iteration" process. Any mathematical model which is used in solving the problem of production reconcilement is based on a balance system of equations that expresses the fact that the production output should be sufficient to cover all the material and capital costs of the national economy industries and to meet the demands of the social sphere, providing it with the necessary volume of non-productive (personal and public) consumption.

In the simplest case the basic balance system is as follows:

$$x_t = P(x_t) + Q(x_t, x_{t-1}) + u_t \tag{1}$$

Here x_t vector whose components are amounts of the production output by industries,

$P(x_t)$ vector of material costs whose components are summary material costs of industries,

Q a similar vector of capital costs,

u vector of non-productive consumption.

For this system the problem of the existence of non-negative solutions ($x_t \geqslant 0$) and of constructing a method for finding these solutionsis considered. The economic contents of the problem allow to use only monotony of operators P and Q.

At present for system (1) a number of conditions for existence of non-negative solution have been determined and iteration methods for obtaining solutions worked out.

A number of problems for more complicated versions of models have been solved as well.

Typical models for working out sectoral decisions are those in which a sought for decision is found by a procedure (a rule) of choice on a given set of versions.

Optimization problems of choice, that are well adjusted and used in practice, play here an essential role.

In the case of optimization by minimum cost criterion, functions of sectoral costs are simply determined in the process of developing the industry plan with the help of cost structure coefficients.

In this case a problem of choice for an industry is stated as follows:

$$\sum_{\kappa} A_{\kappa j}\, z_{\kappa j} \geqslant x_{jt} \qquad (2)$$

$$\sum_{\kappa} R_{\kappa j}\, z_{\kappa j} \leqslant R_j\,, \quad z_{\kappa j} \geqslant 0\,. \qquad (3)$$

Then the sought $\min\,(c, z) = g_j\,(x_j)$

and sectoral costs are:

$$p_{ij}(\kappa_j) = \alpha_{ij}\, g_j(x_j) \qquad (4)$$

where α_{ij} cost structure coefficients.

It is obvious, that functions of sectoral costs thus obtained will lead to a special type of operators P and Q of the previous problem. According to this, one should investigate into the relationship between the solutions of both problems and into an inverse problem about the properties of the solutions to sectoral problems.

To solve the problem of forming versions of the development of an economic unit, models describing the operation of the economic mechanism, including the mechanism of the motivation of workers, are used.

A typical model of this class has the following form:

$$y_l = \mu\,(x_l\,,\ \eta_l\,,\ \xi_l\,), \qquad (5)$$

Here y_l is vector whose components are characteristics of the state of a self-supporting unit with number l (the cost of basic productive funds, cost coefficients, the funds of economic incentives, etc.), η_l - vector of the parameters controlled by ministries (departments), ξ_l - vector of the parameters controlled by the management of the given amalgamation (enterprise), x - planned task for a given self-supporting unit, obtained as a result of solving the sectoral problem. By selecting different values of η_l and ξ_l with the same y_l different versions of state are obtained which are used to form the set $G(x_j)$ of the previous problem. The development of this type of models which are in general of simulation character is given.

The main line in solving the problem of territorial reconcilement of planned decisions is the use of inter-regional intersectoral optimization model. This model like that of intersectoral ties is a model of the top level. The problem of coordinating the solutions of both models have been examined and the conditions,on satisfying which certain aggregated values coincide, obtained. The problems of territorial reconcilement of sectoral plans down to the location of plants under construction are analyzed with the help of models of all-round growth of economic regions.

The appropriate problems are as a rule of multi-extremal nature and their solution is very difficult.

SOME STATEMENTS AND WAYS OF SOLVING DYNAMIC OPTIMIZATION PROBLEMS UNDER UNCERTAINTY

L.S.Belyaev
Siberian Power Institute
Ac.Sc., Irkutsk, USSR

Very many optimization problems, as related with system development, have a dynamic character and they are to be solved under uncertainty when for the part of initial information the probable description is neither known exactly nor available at all.

As is known [I], under uncertainty by a formalized solution of the problem it is generally impossible to find a single optimal variant. One may get but several rational variants that are the best on diverse criteria (Wald, Laplace, Savage et al.). But the final choice among them is made by experts based on their experience and intuition. Methods for solving such problems are insufficiently elaborated and they are investigated at the Siberian Power Institute, Ac.Sc. [2,3 et al.].

Under uncertainty it is expedient to make final decisions only for those actions (variants, objects, etc.) which are urgent and should be realised immediately. These actions concern usually the nearest time period called further "the first step". The other part of the studied period (period of "afteraction") is to be taken for the account of the consequences of these immediate actions, but the final decisions of the system development for this period can be made later.

At such a statement the question of the variant formation way (trajectory) of system development for the period of "afteraction" is complex enough. Apparently, such variants have to correspond on the one hand to actions of "the first step" (the following system development depends upon the choice of immediate objects) and on the other hand to impending conditions of system development (later decisions are made upon concrete situations). However taking into account that at making subsequent decisions uncertainty of information is also available and so the final choice is made by men (intuitively, informalized), it is impossible to foresee these decisions and definitely select the variants of system development for the period of "afteraction". Therefore here can be proposed diverse approximate ways [3]. Some possible statements will be considered below. The economical effect in the mentioned problems can be estimated by the functional:

$$З = \sum_{t=1}^{T} З_t (x_{t-1}, x_t, y_t), \qquad (I)$$

where: T = total number of time intervals ("steps");
y_t = vector of indefinite variables that characterizes "nature states" at time interval t;
x_{t-1} and x_t = vectors of parameters to optimize at the beginning and at the end of time interval t.

The <u>first</u> statement, the most simple (and the most rough) is to make the dynamic problem a statical one. This is simply enough when the nature state y_s (S = number of nature state) is taken as a concrete realisation of vector of random (indefinite) variables for the whole studied period T:

$$y_s = (y_{1S}, \ldots, y_{tS}, \ldots, y_{TS}) \qquad (2)$$

and the possible action of man x_i (i = number of action or variant) is the choice of a single-valued trajectory of the system development for the whole period T :

$$x_i = (x_{1i}, \dots, x_{ti}, \dots, x_{Ti}) . \quad (3)$$

At such a statement of the problem the general sequence of its solution can be the same as for the statical problems. Some realisations (2) and several possible actions (3) are chosen. This is done by the formal methods [3] or heuristic means.

For each action x_i and nature state we estimate the expenditures $З_{is}$ by functional (I). This results in a "pay matrix" $\| З_{is} \|$ on the basis of which a variant for realisation is chosen.

The shortcomings of this statement are obvious.

The <u>second</u> statement (and all following ones) supposes that the aim of the dynamic problem solution is the choice of an expedient action for the nearest time period ("first step") only. One of the possible ways of solving such a problem is [2] as follows:

(a) several possible actions are planned at the first step

$$x_i \equiv x_{1i} ;$$

(b) a series of realisations (2) for the whole studied period are chosen as before;

(c) for each planned action at the first step and the chosen nature state determinate optimization calculations are carried out for all steps beginning with the second one; this gives a pay matrix $\| З_{is} \|$,where $З_{is}$ = value of objective functional of the i-th action and the s-th state of nature is determined as

$$З_{is} = З_1 (x_i, y_{1s}) + \min_{x_t} \sum_{t=2}^{T} З_t (x_{t-1}, x_t, y_{ts}) . \quad (4)$$

The first summand in (4) is equal to the functional value in the first interval at fixed x_i and y_{1s} ; the second summand is an extreme value of the functional at the period of afteraction;

(d) rational actions at the "first step" are chosen on the basis of this matrix using special methods and criteria of solutions theory.

The merit of such a statement is the flexible and rather logical adaptation of various actions to different states of nature.

But it is laborious and doesn't quite correspond to the real situations. The "suboptimisation" of system development at the second and following steps at these or those conditions agrees with the assumption that further on (after the first step) we shall precisely know the impending conditions and shall act therefore optimally. In reality, making decisions at the next steps we shall be again under uncertainty and so cannot act optimally.

Taking into account the second circumstance we may not sometimes demand such a strict "suboptimisation" of system development at the period of "afteraction" and may take a simplified <u>third</u> statement. Here some possible actions at the first step and several nature states (2) for the whole studied period are also planned.

The difference is in the account method of the afteraction period and pay matrix calculation. For each action at the first step several (two-five) variants of system development at the period of afteraction are planned (numbers of these variants are designated by j):

$$(x_{2ij}, \dots, x_{tij}, \dots, x_{Tij}) \quad j = 1, 2, \dots \quad (5)$$

Further, for all variants of j functional (I) is calculated at each chosen nature state. These calculations become not optimization but only "value" (at fixed values x_{tij}) calculations not as in the above-mentioned second statement. Such a calculation determines the expenditure value:

$$З_{isj} = З_1(x_i, y_{1s}) + \sum_{t=2}^{T} З_t(x_{t-1\,ij}, x_{tij}, y_{ts}) . \qquad (6)$$

Now we suggest taking into account the adaptation of system development at the period of afteraction at diverse nature states by the choice of such a variant (from the mentioned number) for which the expenditures at the given nature state will be minimal:

$$З_{is} = \min_j З_{isj} . \qquad (7)$$

This agrees with the assumption that at the second and following steps we shall choose this or that variant of the further system development dependent upon the factual conditions, but this circumstance is accounted here in the simplified form. The values $З_{is}$ obtained by relation (7) are used for filling the pay matrix $\| З_{is} \|$. Its following analysis and choice of rational actions is going on in the usual order.

The fourth statement differs from the above ones as follows: each rational action at the first step (optimal on the corresponding criterion) is determined at the assumption that at all steps during the afteraction period the choice is implemented by this criterion. In other words, we choose a certain criterion, for instance that of Wald, and optimize the system development for the whole studied period. Optimal action obtained for the first step belongs to the rational actions. Then we optimize the system development (also for the whole studied period) by another criterion (for example the Savage criterion) and we get one more rational action at the first step, etc.

Having fixed the criterion of optimality K we come to the problem similar to stochastic dynamic problems (at the known probable description) when the extremum of the mathematical expectation of the functional is searched for. It can be solved using ideas and methods of dynamic programming. For instance, if we take the Wald criterion (minimax expenditures) then for each t-th time interval the following functional equation must be solved (starting from the end):

$$\overset{\circ}{K}{}_t^T(x_{t-1}) = \min_{x_t} \max_{y_t} \left[З_t(x_{t-1}, x_t, y_t) + \overset{\circ}{K}{}_{t+1}^T(x_t) \right] , \qquad (8)$$

where:

$\overset{\circ}{K}{}_t^T$ = minimal possible criterion value for the period from the beginning of the t-th interval to the end of the studied period (it depends upon the vector value x_{t-1}).

The solution of such a problem at the continuous values of vectors x and y , especially at their great dimension goes with great (possibly even unsolvable) computational difficulties. But, if a finite and not too large amount of vector values x_t is taken, characterising diverse system states and also some limited number of nature states (making the problem discrete), then the solution becomes practically possible. In [3] there is the algorithm of problem solution in such a statement on the base of which the computer program is worked out.

The expedient usage of this or that mentioned statements depends upon the peculiarities of the problem: its general laboriousness (accounting for the computers used), the possible dates and the solution periodicity, etc. For each problem this question has to be specified. The second statement is the most widely used now.

In [3] there are examples of problem solutions for power systems accounting for the uncertainty of initial information.

References.

I. Льюс Р.Д., Райфа Х. Игры и решения. М., изд-во иностр. лит., 1961.

2. Беляев Л.С. Вопросы оптимизации больших систем в вероятностных ситуациях. "Экономика и математические методы", т.3,вып.6, 1967.

3. Сборник "Вопросы построения автоматизированных информационных систем управления развитием электроэнергетических систем", вып.I. Учет неопределенности исходной информации, СЭИ СО АН СССР,Иркутск, 1973.

A NEW ALGORITHM FOR GAUSS MARKOV IDENTIFICATION

P. BERNHARD (*)

A. BENVENISTE (**) G. COHEN (**) J. CHATELON (***)

SUMMARY.

Based on the theory of Gauss Markov representation established by FAURRE [4], and on a new theorem about the inverse of the spectrum of the signal process, a new algorithm is proposed, using frequency methods, but which does not appeal to classical spectral factorization, nor to any algebraic Riccati equation.

(*) Centre d'Automatique de l'Ecole Nationale Supérieure des Mines de Paris, Fontainebleau, and IRIA, Rocquencourt, France

(**) Centre d'Automatique de l'Ecole Nationale Supérieure des Mines de Paris, Fontainebleau, France.

(***) Centre d'Automatique de l'Ecole Nationale Supérieure des Mines de Paris, Fontainebleau, France, and ECAM France.

1. Introduction.

The Kalman filtering theory starts from the following internal description of the signal $y(.)$ to be filtered :

$$x(k+1) = Fx(k) + v(k), \qquad (1)$$

$$y(k) = Hx(k) + w(k). \qquad (2)$$

Here, $x(k) \in R^n$, $y(k) \in R^m$, (v', w')[1] is a white noise of appropriate dimension, characterized by

$$E\begin{pmatrix} v(k) \\ w(k) \end{pmatrix} = \begin{pmatrix} 0 \\ 0 \end{pmatrix}, \quad E\begin{pmatrix} v(k) \\ w(k) \end{pmatrix}(v'(k)w'(l)) = \begin{pmatrix} Q & S \\ S' & R \end{pmatrix}\delta_{kl}.$$

F, H, Q, S and R are constant matrices, F is asymptotically stable, $Q = LL'$ with (F, L) completely controllable, and R is assumed positive definite (regular case).

On the other hand, Wiener filtering theory starts from the following external description of the process :

$$\Lambda_k = E[y(k+l)y'(l)] \qquad (3)$$

The Λ_k can be directly measured from a realization of $y(.)$ provided the signal process is ergodic.

The aim of Gauss Markov identification is to fill the gap between the two approaches. More precisely, the object is to be able to construct the Kalman filter of a process knowing only its external description. One approach whould be to use classical Wiener filtering techniques, and then find an internal description of the filter obtained. The obstacle there is the spectral factorization which turns out to be difficult to perform for a matrix spectrum.

The alternate approach is to find an internal representation of the process, and then compute the Kalman filter via classical means. The drawback of this method was that two algebraic matrix Riccati equations

(1) a prime on a vector or matrix means "transposed".

had to be solved : one to find the internal description, one to compute the filter gain. FAURRE [4] has shown that this number can be reduced to one by finding precisely the filter as one particular internal representation of y. As a matter of fact, it was pointed out by MEHRA [7] that the filter can be written as

$$\hat{x}(k+1) = F\hat{x}(k) + T\eta(k) \tag{4}$$

$$y(k) = H\hat{x}(k) + \eta(k) \tag{5}$$

and it is known that the innovation process $\eta(k)$ is white. What is often sought is the Wiener estimate $\hat{y} = H\hat{x}$.

In this paper, we give a new way of finding the filter from the Λ_k's, or more precisely from the spectrum of y :

$$\Gamma(z) = \sum_{k=-\infty}^{+\infty} \Lambda_k z^{-k} \tag{6}$$

We carry out all the analysis in discrete time as it is the formulation which is computationally usefull, but it carries over to the continuous case (it is only simpler). Let us point out also that by duality it applies to the linear quadratic stable regulator problem. See [1], [8].

2. Spectrum of $y(.)$.

It is well known that the link between (1) (2) on one hand and (3) on the other, is given, using

$$P = E[x(k)x'(k)], \tag{7}$$

which is computed through (9), and

$$G = E[x(k+1)y'(k)], \tag{8}$$

which is usually defined by (10), by the equations :

$$P - FPF' = Q, \tag{9}$$

$$G - FPH' = S, \tag{10}$$

$$\Lambda_0 - HPH' = R, \tag{11}$$

$$\Lambda_k - HF^{k-1}G = 0 \ , \ k \geq 1. \tag{12}$$

Λ_k depends only on H, F and G. Therefore, for fixed Λ_0, H, F, G, placing various non negative definite $P's$ in (9), (10), (11), we obtain various triples (Q, S, R) that all give the same external description. All these models will be termed "equivalent".
Such a model has a meaning only if P is such that the resulting noise covariance matrix built with (Q, S, R) is non negative definite. However, one can show that the filter can be computed from any of these models (meaningfull or not), and that all have the same filter. (i.e. the same gain T in (4)).

From (12), it is clear that the spectrum $\Gamma(z)$ of y can be written

$$\Gamma(z) = \Lambda_0 + H(zI - F)^{-1}G + G'(\tfrac{1}{z}I - F')^{-1}H' \tag{13}$$

<u>DEFINITION</u>. A triple (H, F, G) satisfying (13) is called an additive realization of Γ. It is said minimal if the dimension of F is minimal among all such realizations.

F being, by hypothesis, asymptotically stable, the additive decomposition of Γ is unique, with the first term constant, the second realizable and the third antirealizable. Then, it follows from classical realization theory [6] that all minimal realizations are identical up to a change of basis in the state space.

The result on which the new algorithm is based is the following :

<u>THEOREM</u>. <u>There exists a minimal additive realization</u> $(\bar{H}, \bar{F}, \bar{G})$ of $\Gamma'^{-1}(z)$ with

$$\bar{H} = T'$$

$$\bar{F} = (F - TH)'$$

<u>COROLLARY</u>. <u>The matrix</u> F - TH <u>is asymptotically stable iff the inverse of the spectrum has no pole on the unit circle</u>.

For a proof of the theorem, see [2] which gives a more complete account of the present theory and [1]. The corollary follows immediately.

3. Algorithms

All algorithms first involve evaluation of the $\Lambda_k's$. An efficient way of doing so is via the spectrum $\Gamma(e^{iT\omega})$ for discrete values of ω. (And using fast FOURIER transforms : FFT's). Then, from the Λ_k's, a standard realization algorithm yields H, F and G (see [5], [3].

The algorithm we propose is to invert $\Gamma(e^{iT\omega})$ for all values of ω. Then, do a FFT, obtain a sequence V_k, and a realization algorithm as previously. Obtain matrices $\bar{H}$ and $\bar{F}$. If a filter for x were sought this would suffice :

$$\hat{x}(k+1) = \bar{F}'x(k) + \bar{H}'y(k)$$

However, x has no meaning of its own, and is defined up to a change of basis. Thus we want $\hat{y} = H\hat{x}$. But if we have realized both the sequence Λ_k and V_k, the matrices obtained are usually not in the same basis. We must therefore match these bases. Having obtained the two realizations, there exists a non singular matrix M such that

$$\bar{F}' = M(F - TH)M^{-1} \tag{14}$$

$$\bar{H}' = MT \tag{15}$$

Hence knowing H, F, $\bar{H}$ and $\bar{F}$, we can compute M :

$$F'M - MF = -\bar{H}'H \tag{16}$$

This is a linear equation for M. Then T can be obtained from (15). Germain, of IRIA, pointed out to us that when y is scalar, we can dispense with this equation by realizing both the V_k's and the Λ_k's in companion form (which Rissanen's algorithm does). Then, the coefficie of T is just the difference of those in F and $\bar{F}'$. (It might be possible to do the same thing for the multivariable case, using an appropriate canonical form).

4. Conclusion.

The main contribution of this paper, as we see it, is to show that the algebraic Riccati equation, or a substitute, is by no means a necessary tool apart from standard spectral factorization.

Now, it is interesting to discuss the relative merits of this algorithm as compared to FAURRE's [4], which is discussed in parallel in [2].

Its drawbacks are the need for three FFT's instead of two, and two realization algorithms instead of one. However these are extremely performing algorithms, and it is not a very serious problem. Another drawback is the necessity to match the bases. We have seen that this could be improved upon.

The good point is replacing the Riccati equation by a sequence of matrix inversions. Notice that the Riccati equation involves a matrix inversion of the same dimension at each step. But this is nothing as compared to the real numerical difficulties associated with this equation, as is well known in the literature. And it is of dimension n, larger that the dimension m of Γ.

Therefore, one should expect that the new algorithm is more efficient every time the dimension of the state vector is large as compared to that of the signal.

BIBLIOGRAPHY.

[1] P. BERNHARD et G. COHEN : "Etude d'une fonction frequentielle intervenant en commande optimale avec une application à la réduction de la taille de ce problème". R.A.I.R.O, yellow series, J-2., 1973.

[2] P. BERNHARD et al. "Same title as this paper". Internal report A/54 of Centre d'Automatique de l'ENSMP, Fontainebleau, France 1974.

[3] P. FAURRE et CHATAIGNER : "Identification en temps réel et en temps différé par factorisation de matrices de Hankel" proceeding, French-Swedish symposium on process control, IRIA, 1971.

[4] P. FAURRE : "Representation markovienne de processus aléatoires". IRIA report n° 13, IRIA, France, 1973.

[5] B. L. HO, RE. KALMAN : "Effective construction of linear state variable models from input/output functions". Proc. 3rd Allerton Conference, pp 449-459, 1965.

[6] R.E. KALMAN. "Modern theory of realization" in Kalman, FALB and ARBIB : Topics in Mathematical System Theory (Chapt 10). McGraw Hill, 1969.

[7] MEHRA : "On the Identification of variances and adaptive Kalman filtering". IEEE transactions on A.C., AC-15, n° 2, pp 175-184, 1970.

[8] J.C. WILLEMS. : "least squares stationary optimal control and the algebraic Riccati equation", IEEE transactions on A.C., AC-16, pp 621-634, 1971.

ON OPTIMALITY CRITERIA IN IDENTIFICATION PROBLEMS

I.S. Durgarian N.S. Rajbman
Institute of Control Sciences
Moscow, USSR

The identification problem is largely formulated as a problem of finding a plant operator estimate optimal in terms of a specified criterion. The resultant solution permits estimation of the model adequacy. The criterion and the model structure are selected individually for each model. For complex systems the large number of plant parameter interrelations and of external disturbances make a completely adequate model non-feasible or undesirable. True, a more accurate model and a better system description is supposed to give a better forecast and control of the system but studying each input and its responses takes more time, funds and material resources. Hence it is clear that identification of large systems requires quantitative assessment of the effect of each system input on its output variable and of the accuracy with which a model where these inputs are integrated simulates the actual processes; a decision should be made, which system variables should be represented in the model. Identification is also impossible without systems approach, studying the hierarchical structure and knowledge of the entire system functioning.

We will describe optimal selection of plant information indices with the techno-economic criterion. The results related to the model performance act as constraints.

1. The desired product accuracy for a complex plant or especially for a set of plants can be obtained in a number of ways, for instance, by varying the input or state variables or both. Calculation of optimal characteristics requires establishment of input and state variable indices such that would characterize the internal state of plants and ensure the desired output quality in a "best" way, or by a specified criterion, or objective function. Model referenced solution generally relies on numerical characteristics closely related to specified requirements. The mean output variable characterizes the nominal value of a qualitative index (centers middle of the tolerance field, nominal size, etc.); variance, the

admissible deflection of the output variable (tolerance field); entropy, the output variable scatter. Consequently, the control should ensure the desired values of output and state numerical indices. Quite naturally, methods of ensuring the desired quality vary with optimality criteria and a control optimal in terms of one criterion can be far from the optimum in terms of the other. Let us consider certain details of techno-economic optimal control and try meeting the requirements of a certain comprehensive index integrating several techno-economic indices. Selection of optimality criteria is a major complex affair untractable by purely mathematical tools. Unlike the statistically optimal systems which, if optimal in terms of one criterion, are near-optimal in terms of others, these systems are not optimal in terms of another criterion.Therefore selection for a complex industrial process is heavily dependent on specific conditions and problems posed in design of new/or automation of existing processes. In many cases the definition of general design problems includes optimality criteria or the necessity of comparative analysis of data obtained with different criteria is indicated.

This section will describe the technique of calculating the basic indices of a technological line so as to ensure the desired production at minimal costs.

Let the technological chain be modelled as multiple regression of the output variable with respect to input variables

$$M\{y(t)/x_1(t),\dots,x_n(t)\} = \sum_{i=1}^{n} b_i x_i(t). \tag{2.1}$$

For dynamic plants the regression integrates the lag τ with respect to $x_i(t-\tau_i)$. In further discussion the arguments are not written so as not to encumber the equations.

The variance at the output can be given as a sum of two variables

$$\mathcal{D}\{y\} = \mathcal{D}\{M\{y/x_1,\dots,x_n\}\} + M\{\mathcal{D}\{y/x_1,\dots,x_n\}\}, \tag{2.2}$$

where $\mathcal{D}\{M\{y/x_1,\dots,x_n\}\}$ characterizes that part of the overall variance for the output variable y which is caused by the inputs and $M\{\mathcal{D}\{y/x_1,\dots,x_n\}\}$ is that part which is caused the by other factors except $x_1,\dots,x_n$.

The following formula for $\mathcal{D}\{M\{y/x_1,\dots,x_n\}\}$ is convenient for practical computation

$$\mathcal{D}\{M\{y/x_1,\dots,x_n\}\} = \sum_{i=1}^{n} b_i \operatorname{cov}(x_i y), \tag{2.3}$$

where $\operatorname{cov}(x_i y) = M\{[x_i - M\{x_i\}][y - M\{y\}]\} = r_{x_i y}\sigma_i \sigma_y$.

The output variable found through (2.1) should ensure the conditions

$$\mathcal{D}\{M\{y/x_1,\dots,x_n\}\} \leq \mathcal{D}_3\{y\}, \tag{2.4}$$

where $\mathcal{D}_3\{y\}$ is the specified value of the output variable variance.

For an automatic line of n jobs the total cost of an article or part C_Λ is composed of the input quality cost and the cost of each of the operations C_i

$$C_\Lambda = \sum_{i=0}^{n} C_i . \tag{2.5}$$

The costs C_i (in this case techno-economic indices are used as optimality criteria) can be represented in the form

$$C_i = A_i + f(\delta_i), \tag{2.6}$$

where A_i are constant values of the index elements independent of the accuracy δ_i for the output product and $f(\delta_i)$ are variable values of index elements dependent on δ_i.

The shape of the dependence $f(\delta_i)$ can normally be determined for a specific plant or line on the knowledge of normal operation data.

If the accuracy is characterized in terms of the r.m.s. error σ then formula (2.6) can be written in the form

$$C_i = A_i + \frac{k_i B_i}{6\sigma_i} \tag{2.7}$$

where the values A_i and B_i are determined for each job and k_i is the coefficient of relative scatter whose value depends on the job error distribution law. In more precise terms the derivation of (2.7) is given in Ref. 1. That formula represents the trade-offs of the job cost C_i and the job output accuracy characteristic σ_i.

Knowing the line characteristics and their relations to techno-economic indices one can compute the optimal line by mathematical programming, resolving multipliers or Lagrange conditional multiplier techniques.

Let us formulate the following nonlinear programming problem : Find the minimal value of the overall machining cost

$$\gamma = \sum_{i=0}^{n} \frac{k_i B_i}{6\sigma_i} \tag{2.8}$$

defined as a function of n variables $\sigma_1, \dots, \sigma_n$ which should satisfy the constraints

$$\sum_{i=1}^{n} b_i \tau_{x_i y} \sigma_i \sigma_y \leq \mathcal{D}_3 \{y\}$$
$$\sigma_{i\,max} > \sigma_i > \sigma_{i\,min} \tag{2.9}$$

If the trade-offs of the costs and the r.m.s. deflection are linear and (2.8) can be given in the form

$$k = \sum_{i=1}^{n} p_i \sigma_i , \tag{2.10}$$

where p_i are certain coefficients then we arrive at a linear programming problem with the same constraints (2.9).

In the case when the linear function (2.10) is maximal inside the definition region and the constraint (2.4) is given as the equality

$$\sum_{i=1}^{n} c_i^2 \sigma_i^2 + 2 \sum_{i=1}^{n-1} \sum_{j=i+1}^{n} c_i c_j \tau_{x_i x_j} \sigma_i \sigma_j = \mathcal{D}_3 \tag{2.11}$$

finding the conditional extremum of the function is reduced to finding the extremum of the function

$$\Phi = K + f\beta , \tag{2.12}$$

where f is a Lagrangean multiplier and

$$\beta = \sum_{i=1}^{n} c_i^2 \sigma_i^2 + 2 \sum_{i=1}^{n-1} \sum_{j=i+1}^{n} c_i c_j \tau_{x_i x_j} \sigma_i \sigma_j - \mathcal{D}_3 = 0$$

The vanishing of the partial derivatives leads to equations for unknown σ_i . For the case under consideration we have

$$\sigma_i^2 = \frac{p_i^2 \, \mathcal{D}\{M\{y/x_1,\dots,x_n\}\}}{4c_i^2\left[\frac{1}{4}\sum_{i=1}^{n}\frac{p_i^2}{c_i^2} + \frac{1}{2}\sum_{i=1}^{n}\sum_{j=i+1}^{n} r_{x_i x_j}\frac{p_i p_j}{c_i c_j}\right]} \tag{2.13}$$

If the value of γ in eq. (2.8) is to be minimized with

$$\varphi = \sum_{i=1}^{n} b_i r_{x_i y}\sigma_i\sigma_y - \mathcal{D}_3 = 0,$$

the Lagrange method leads to the expression

$$\sigma_i = \frac{\mathcal{D}\{M\{y/x_1,\dots,x_n\}\}\left(k_i B_i/6\sigma_y b_i r_{x_i y}\right)^{1/2}}{2{,}45\sum_{i=1}^{n}\left(k_i B_i \sigma_y b_i r_{x_i y}\right)^{1/2}} \tag{2.14}$$

Similarly the problem of finding the plant optimal entropic characteristic of is solved when the accuracy of plant functioning depends on the entropy of its output variable [2].

Note that in equality (2.2) the first sum represents the model variance, i.e. $\mathcal{D}\{y^*\}$. Similarly, represent the entropy of the plant output in the form:

$$H\{y\} = H\{y^*\} + \Delta H \tag{2.15}$$

where ΔH represents an error occurring when the plant is replaced by its model.

The linear model of the plant with n inputs and n outputs can be described in the form

$$y^* = AX$$

where $X = (X_1,\dots,X_n)$ is the vector of model inputs,
$y^* = (y_1^*,\dots,y_n^*)$ is the vector of model outputs, and

$$A = \begin{pmatrix} a_{11} & \cdots & a_{1n} \\ \vdots & & \vdots \\ a_{n1} & \cdots & a_{nn} \end{pmatrix}$$ is a

matrix of regression coefficients.

Then the entropy $H\{y^*\}$ can be expressed in terms of the entropy $H\{X\}$ in the following way [2]:

$$H\{y^*\} = H\{X\} + \log \frac{l_x}{l_{y^*}} |\det A| \quad , \tag{2.16}$$

where l_x and l_{y^*} are degrees of accuracies of measuring the values X and y^*, and $|\det A|$ is the magnitude of the determinant of the matrix A.

Formula (2.16) is correct in the case of a linear model of the plant. However in a more general case, where for vectorial X and y^* $y^* = \varphi(X)$ in this case there exists the unambiguous inverse transformation $X = \psi(y^*)$ we have

$$H\{y^*\} = H\{X\} + M[\log|\mathcal{J}(X)|] + \log \frac{l_x}{l_{y^*}} , \tag{2.17}$$

where $\mathcal{J}(X)$ is the Jacobian of the function φ, turning into its derivative in the case of the scalar X and y^*. It is obvious that in the linear case

$$\mathcal{J}(X) = \det A$$

we shall also arrive at the expression (2.16).

In the general case the entropy of the vector X can be expressed in terms of entropies of components as

$$H\{X\} = \sum_{i=1}^{n} H\{X_i\} - \sum_{i=2}^{n} I(X_1, \dots, X_{i-1} ; X_i), \tag{2.18}$$

where $I(X_1, \dots, X_{i-1} ; X_i)$ is the amount of data on the value X_i contained in the vector $(X_1, \dots, X_{i-1})$.

In the simplest case where the joint distribution of the values $X_1, \dots, X_n$ is expressed by the normal law (note that this does not testify to the normality of the joint distribution (X, y), or linearity of the regression), (2.18) can be written in the form

$$H\{X\} = \sum_{i=1}^{n} H\{X_i\} + \sum_{i=2}^{n} \log \left(1 - R^2(X_i ; X_1, \dots, X_{i-1})\right)^{1/2} \tag{2.19}$$

where $R(X_i; X_1, \ldots, X_{i-1})$ is a multiple coefficient of correlation X_i with $(X_1, \ldots, X_{i-1})$.

Substituting (2.19) into (2.17) we have

$$H\{y^*\} = \sum_{i=1}^{n} H\{X_i\} + \sum_{i=2}^{n} \log(1 - R^2(X_i; X_1, \ldots, X_{i-1}))^{1/2} + \\ + M[\log|J(X)|] + \log \frac{l_x}{l_{y^*}} \tag{2.20}$$

Assume that the entropy $H\{y^*\}$ should be maintained at a certain given level. The necessary entropy $H\{X\}$ of the input can be obtained immediately by equations (2.17) or (2.20). In more complicated cases it is necessary either to investigate the interrelations between the entropies of input values or to introduce additional constraints. Determine what the values $H\{X_1\}, \ldots, H\{X_n\}$ should be to provide the necessary value $H\{y^*\}$.

With some assumptions following from the physical nature of the plants the following dependence of the C_i on the input entropy H_i can be assumed for the i-th input:

$$C_i = \frac{1}{k_i} \log \frac{H_{oi}}{H_i},$$

where H_{oi} is the initial entropy of the i-th input, k_i is a coefficient. The values H_{oi} and k_i are considered to be given. For simplicity assume $k_1 = \ldots = k_n = k$.

Formulate now the following problem. It is desired to find the values $H_1, \ldots, H_n$ satisfying the equation:

$$\sum_{i=1}^{n} H_i - H\{y^*\} + \sum_{i=2}^{n} \log(1 - R^2(X_1, \ldots, X_{i-1}; X_i))^{1/2} + \\ + M[\log|J(X)|] + \log \frac{l_x}{l_{y^*}} = 0, \tag{2.21}$$

so that the total costs

$$\Phi = \sum_{i=1}^{n} C_i = \frac{1}{k} \sum_{i=1}^{n} \log H_{oi} - \frac{1}{k} \sum_{i=1}^{n} \log H_i \tag{2.22}$$

should be minimal.

The conditional extremum of the expression Φ is obtained when the values of the input entropies are equal, i.e. when

$$H\{X_1\} = \ldots = H\{X_n\} = \frac{1}{n} \sum_{i=1}^{n} H\{X_i\}.$$

If we do not restrict ourselves to the case of equality of all k_i then we obtain the extremum when the values are related as follows:

$$H\{X_1\} \cdot k_1 = H\{X_2\} \cdot k_2 = \ldots = H\{X_n\} \cdot k_n .$$

3. A most complete solution of the optimization problem with a techno-economic criterion would require several stages: 1) data acquisition for control; 2) data processing and transfer; 3) decisions-making using the results of data processing and forming control in accordance with the decision made.

Usually the predicted accuracy of plant functioning can be maintained by means of various alternative sequences of actions at each stage. Completion of each stage of work involves certain expenditures. Denote rosts for data acquisition as a; decision - making and control generation costs as c ; profit (as a result of increasing the accuracy of plant functioning) as d .

It is obvious that the expenditures a, b, c and d are functions of the input X chosen in identification (in a general case of the vectorial input.). Moreover, assume a, b, c and d are functions of some parameters α, β, γ . These can be associated with different methods and techniques of data acquisition and processing and ways of their utilization etc.

In the light of the above, the maximal profit in the case of utilization of our model can be found by the formula

$$p = \max_{\alpha,\beta,\gamma} \{ d(X,\alpha,\beta,\gamma) - a(X,\alpha) - b(X,\beta) - c(X,\gamma) \} = \\ = y(X) - z(X) . \tag{3.1}$$

The physical sense of (3.1) is obvious. The profit is a function of the income $y(X)$, obtained in control, and costs $z(X)$ of this control.

It can be assumed that the values y and z are related as follows:

$$y = y_m (1 - c e^{-kz}) , \tag{3.2}$$

where y_m is income obtained in the case of "ideal" control, i.e. when the realizable value is maintained exactly at a predicted level; c and k are coefficients the values of which are determined for process (a plant). Considering (3.2), formula (3.1) can be rewritten in the form

$$p = y(z) - z = y_m(1 - ce^{-kz}) - z = p(z). \qquad (3.3)$$

The value of the expenditure z_0 at which the maximum is achieved in the expression (3.3) represents those costs at which the maximal profit will be provided equal to

$$Q = \max_X \left\{ \max_{\alpha,\beta,\gamma} \left\{ d(X,\alpha,\beta,\gamma) - a(X,\alpha) - b(X,\beta) - c(X,\gamma) \right\} \right\} = \max_z p(z) = p(z_0). \qquad (3.4)$$

For finding z_0 it is sufficient to differentiate (3.3) over z and solve the equation thus obtained

$$y_m kc e^{-kz} - 1 = 0. \qquad (3.5)$$

It is obvious that the value $p(z)$ can be used as a criterion of identification and control performance. In the best variant the value $p(z)$ is maximal and equal to (3.4).

References

I. Райбман Н.С. Корреляционные методы определения приближенных характеристик автоматических линий. Изд-во АН СССР, "Энергетика и автоматика", №I, I96I.

2. Пугачев В.С. Теория случайных функций и ее применение к задачам автоматического управления. Физматгиз, I962.

NONSTATIONARY PROCESSES FOR MATHEMATICAL PROGRAMMING PROBLEMS UNDER THE CONDITIONS OF POORLY FORMALIZED CONSTRAINTS AND INCOMPLETE DEFINING INFORMATION

I.I. Eremin, Vl.D. Mazurov
Institute of Mathematics and Mechanics, Ural Scientific Center of the Academy of Sciences of the USSR
S.Kovalevskaja st. 16, Sverdlovsk, USSR

1. Introduction.

The methods of Mathematical Programming are an effective means for the analysis of complicated systems: technical, economical, some natural systems, etc. However, the following conditions are the reason of the necessity of adaptation of these methods:

- nonstationary and poorly defined processes;
- the relations between the components of the system are poorly formalized;
- the degree of determinancy of the available information can be arbitrary.

It seems, that this problem of adaptation can be solved by means of generalization and development of iterative methods, taking into consideration from step to step a more accurate definition of model and basic information. For example, the poorly formalized relations can be defined by means of methods of Pattern Recognition.

The characteristical properties of iterative procedures are the following:

1). The solution has a form of consequence.

2). Simplicity of iteration.

3). Stability in spite of random noise.

4). Non-sensibility of iterative operator to the change of basic information.

5). The slow convergence (it follows from 1).- 4).).

It is possible to assume that the iterative sequence $\{x_k\}$ is given in general by means of a multivalued transformation F of type

$$x_{k+1} \in F(x_k)$$

rather than by an iterative operator.

In model $\mathcal{M}$ (of plant $\mathfrak{M}$) of Mathematical Programming we point out its ingredients:

x is the state space vector;

$f(x)$ is the criterion function;

s is the defining information;

σ is the system of restrictions.

Thus, if, e.g., $\mathcal{M}$ is the Linear Programming model of type

$$max \{\langle c, x\rangle : Ax \leqq b, \ x \geqq 0\} \qquad (1.1)$$

then $s = (c, A, b)$, $f(x) = \langle c, x\rangle$, σ is the set of relations in between s and x.

Let F be the transformation (iterative operator):

$$\left[x_{t+1} \in F(x_t) \ (\forall t)\right] \Longrightarrow \left[\{x_t\}' \subset Arg\, Z\right] \qquad (1.2)$$

where Z is the Mathematical Programming Problem. In the transformation F we extract in explicit form the information ingredient s of model $\mathcal{M}$: $F = F[s]$.

If $s = s_t$ is the sequence then we get the process

$$x_{t+1} \in F[s_t](x_t) ; \qquad (1.3)$$

s_t is the parameter of F.

We give a further generalization of process (1.3):

$$x_{t+1} \in \Phi[s_t, u_t](x_t), \qquad (1.4)$$

where u_t is the controlling vector. This vector can be the formal (vector, algorithm, etc.) or non-formal (expert) object.

2. Iterative transformation including the block of pattern recognition for taking into account the poorly formalized restrictions.

We give a further generalization of process (1.4) which presupposes the possibility of a step by step innovation procedure for the poorly formalized constraints. Here we consider, for example, the Mathematical Programming Problem

$$max \{f(x) : x \in \mathcal{M}, \quad g(x) \geqslant 0\}. \qquad (2.1)$$

Let the restriction $g(x) \leq 0$ be poorly formalized: the function $g(x)$ is not given. The expert $(\equiv [*])$ gives the information about $g(x)$ for each $x \in R^n$:

$$[*](x) = \begin{cases} +1 & \text{if} \quad g(x) \leq 0, \\ -1 & \text{if} \quad g(x) > 0. \end{cases}$$

If $g_t(x)$ is the approximation of $g(x)$, $g_t(x) \in G$, then s_t is the informational ingredient of the model

$$max \{ f(x) : x \in \mathcal{M}, \quad g_t(x) \leq 0 \}. \qquad (2.2)$$

Let the elements x_t, $g_t(x) \in G$ and sets $\mathcal{M}_t$, $\mathcal{N}_t$ be given:

$$g_t(x) \leq 0 \quad (\forall x, \, x \in \mathcal{M}_t), \quad g_t(x) > 0 \quad (\forall x, x \in \mathcal{N}_t).$$

We find x_{t+1}, $\mathcal{M}_{t+1}$, $\mathcal{N}_{t+1}$, $g_{t+1}(x)$, s_{t+1} :

$$x_{t+1} \in \Phi[s_t, u_t](x_t),$$

$$\mathcal{M}_{t+1} = \begin{cases} \mathcal{M}_t \cup \{x_{t+1}\} & \text{if} \quad [*](x_{t+1}) = +1 \; ; \\ \mathcal{M}_t & \text{if} \quad [*](x_{t+1}) = -1 \; ; \end{cases}$$

$$\mathcal{N}_{t+1} = \begin{cases} \mathcal{N}_t & \text{if} \quad [*](x_{t+1}) = +1 \; ; \\ \mathcal{N}_t \cup \{x_{t+1}\} & \text{if} \quad [*](x_{t+1}) = -1 \; ; \end{cases}$$

$g_{t+1}(x) \in G$, $g_{t+1}(x)$ is a discriminant function for $\mathcal{M}_{t+1}$ and $\mathcal{N}_{t+1}$;

s_{t+1} is the informational ingredient of (2.2) by $t \longrightarrow t+1$.

We get the tracking procedure

$$x_{t+1} \in \Phi[s_t, u_t](x_t).$$

3. Examples.

Here we consider the convex programming problem

$$max \{ f[s](x) : f_j[s](x) \leqq 0 \quad (j \in \overline{1,m}) \} \equiv \tilde{f}[s](x); \quad (3.1)$$

$-f$, f_j are convex functions $(\forall s)$ and continuous functions concerning $z = (s; x)$.

3.1. Normal processes.

The nature of nonstationarity of s in (3.1) may be of various origin. Here we consider the case, when s is characterized by an uncertainty which may be gradually decreased and which hence allows an innovation of the estimate for vector s itself. Formally, $s_t \in S_t$, $S_t \supset S_{t+1}$ $(\forall t)$. If $s_t = arg\ min \{ \| s \| : s \in S_t \}$ then we call $\{ s_t \}$ a normal sequence.

Let S_t be convex and closed $(\forall t)$, $\bigcap_t S_t = S \neq \emptyset$, $\{ s_t \}$ a normal sequence. Then $s_t \longrightarrow \bar{s} \in S$. We define

$$\mathcal{M}(s) = \{ x : f_j[s](x) \leqslant 0 \quad (j \in \overline{1,m}) \}.$$

Theorem 3.1. Let $F[s](x)$ be a closed multivalued $\mathcal{M}(s)$ - Feyer - transformation of R^n into set of all subsets of R^n; the system $f_j[\bar{s}](x) \leqslant 0$ $(j \in \overline{1,m})$ satisfies the Slater constraint qualification; let us assume that $\{ s_t \}$ is a normal sequence, $\bar{s} = lim\ s_t$. Then for the sequence

$$x_{t+1} \in F[s_t](x_t) \quad (3.2)$$

exists the element $\bar{x} \in \mathcal{M}(s)$ such as $lim\ x_t = \bar{x}$.

Theorem 3.2. We assume the conditions of theorem 3.1 and (3.1) is solvable for $s = \bar{s}$, $\bar{s} = lim\ s_t$, s_t is a normal sequence, the function $f[\bar{s}](x)$ is strongly convex. Define

$$\begin{aligned} &\varphi[s,u] = Arg \max_x \{ f[s](x) - \sum_{j=1}^{m} u_j f_j^{+2}[s](x) \}, \\ &u = (u_1, \ldots, u_m) \geqq 0, \\ &x_{t+1} \in \varphi[s_t, u_t], \quad u_j^{t+1} = u_j^t + f_j^+[s_t](x_t), \\ &j = 1, \ldots, m. \end{aligned} \quad (3.3)$$

Then

$$x_t \longrightarrow \bar{x} = arg \max_x \{ f[\bar{s}](x) : x \in \mathcal{M}(\bar{s}) \}.$$

3.2. The use of Pattern Recognition methods.

We consider the process (2.3) for (2.1).

Theorem 3.3. Assume $\mathcal{M}$ is a limited closed convex set; $g(x)$ is an affine function; $\mathcal{M} \cap \{x: g(x) \leq 0\} \neq \emptyset$; G is the set of all affine functions; $g_{t+1}(x) \in G$;

$\{x: g_{t+1}(x) = 0\} = \mathcal{H}_{t+1}$; $\mathcal{H}_{t+1} = \arg\max_{\mathcal{H}} \{\rho:$ $\| \mathcal{H} - \mathcal{M}_{t+1} \| = \| \mathcal{H} - \mathcal{N}_{t+1} \| = \rho$, $\mathcal{H}$ is discriminant hyperplane for $\mathcal{M}_{t+1}, \mathcal{N}_{t+1}\}$; $x_t = \arg\,(2.2)$; $x_t \in \mathcal{H}_t$ $(\forall t)$. Then $\lim \| \mathrm{co}\,\mathcal{M}_t - \mathrm{co}\,\mathcal{N}_t \| = 0$.

3.3. Processes under incomplete information.

Various interpretations of essence may be given for the processes under incomplete defining information. These, for example, are the following situations:

- the conduct of the modelled system is determined, but the information on the system is incomplete;
- the nonstationary properties of the defining information are in exact correspondence with those of modelled system;
- the information nonstationarity is an artificial technique due to the computational algorithm, when the informational ingredient acts as the control for the algorithm;
- the nonstationary informational properties are due to poor formalizability of the restrictions.

The Pattern Recognition methods may be used to construct the process models in the form of "input-output" transformations (with no assumptions of the availability of any interior descriptions of the system whatsoever) if the number of different output of the system is finite and is sufficiently small in relation to the capacity of the set of all different outputs. As an example one may demonstrate the problems of estimation by Pattern Recognition methods of the parameters for substance mixtures.

References

I. И.И. Ерёмин, Вл.Д. Мазуров.О нестационарных процессах математического программирования.-In:"Нестационарные процессы математического программирования", Works of Institute of Mathematics and Mechanics, N 14, Sverdlovsk, 1974.

DYNAMIC MODELS OF TECHNOLOGICAL CHANGES

L.V. Kantorovich, V.I. Zhjanov

Moscow

The report is devoted to one of the problems of mathematical modelling of economy - the problem of modelling technological change.

Alongside with multi-product models such as the input-output model, the Von Neumann model of expanding economy, models of linear programming, there are macroeconomic models of national economy on the basis of such aggregate indices as national income, aggregate demand, etc., which are widely applied for qualitative and quantitative analysis of economy.

Such models have been developed since the thirties: Feldman's Growth Model, research by the Keynes school, especially works by Harod, Domar, Samuelson, Hicks, Solow and other authors. It is well known that these models have been given practical application in government regulation of economy, in measures for reducing crises, unemployment, though with partial success.

The simplest one-product models of the development of economy can be used for global analysis of development of the economic system and consideration of the effect of technological change on the dynamics of the economic system. On the basis of these models it is possible to study the effect of technological change on the most important economic characteristics.

Let us consider an economic system in which a single product is produced. One part of this product is allocated to consumption and the other part to accumulation of fixed and working stocks, whereas in this model no distinction is made between these two stocks.

Under the assumption of instantaneous transformation of capital stocks and consideration of technological change this model may be described by the following equation:

$$\frac{dK(t)}{dt} = P(t) - V(t) = e^{\delta t}\,\mathcal{U}[K(t), T(t)] - V[t, K(t), T(t), P(t)] \qquad (1)$$

where $P(t)$ is the net product or the national income;

$V(t)$ is the total consumption, given in unit time. The production function $\mathcal{U}[K(t), T(t)]$ characterizes the amount of net output which can be produced by capital stock $K(t)$ per unit time and labour resources $T(t)$, δ characterizes the rate of neutral technological change.

Within the framework of this model an important parameter of the economic system can be determined, i.e., norm of efficiency of investments $\eta_э$. For $\eta_э$ the following expression is derived

$$\eta_э = \frac{\frac{1}{P(t)}\frac{dP(t)}{dt} - \frac{1}{T(t)}\frac{dT(t)}{dt} - \delta}{1 - \frac{V(t)}{P(t)} - \frac{1}{T(t)}\frac{dT(t)}{dt}\frac{K(t)}{P(t)}} \qquad (2)$$

All the variables in this formula have a precise economic content. Calculation of the norm of efficiency according to this formula without consideration of technological change gives a value of 22%, and with consideration of technological change, lag, obsolescence of stocks this value is equal to 18%.

As has been pointed out, it is assumed in the model the hypothesis about the instantaneous transformation of stocks, i.e., the assumption that the stocks can always be transformed from one form to another and, in particular, due to this it is possible to change from one production structure (labour-capital ratio) to another one without loss. More practicable is a similar model without the assumption about the instantaneous transformation of stocks.

Let us consider the structure of the one-product model.

$T(t)$ - labour resources at instant t is a fixed function. We introduce $\lambda(t)$ - type (or structure) of new stocks created at instant of time t, which are characterized by the value (expressed in the product) of single stocks (stocks per unit of labour). It is assumed that the stocks created at each instant of time t are single-type ($\lambda(t)$ - single-valued function of t). $\varphi(t)$ denotes the intensity of creating stocks, i.e., $\varphi(t)\,dt$ is the number of new work positions created during the time $[t, t+dt]$, then $\lambda(t)\,\varphi(t)\,dt$ is the volume of newly created stocks in the same interval. The functions $\lambda(t)$ and $\varphi(t)$ in the model are subject to calculation.

It is assumed that potential modes of production are characterized by the production function $\mathcal{U}(x,y)$, which indicates the amount of net product created by labour y when using the fixed stocks x per unit time (at the initial instant). It is assumed that the function $\mathcal{U}(x,y)$ is positively homogeneous of the first degree

$$\mathcal{U}(\lambda x, \lambda y) = \lambda\, \mathcal{U}[x,y] \qquad \text{when} \qquad \lambda > 0$$

and based on optimal modes, which renders a natural assumption about convexity $\mathcal{U}[x,1]$.

Technological change is present in the model by the following method. It is assumed that the amount of net product, produced per unit at the given quantity of stocks and expenditures of labour, increases exponentially depending on the instant of creating stocks τ, i.e., it exceeds by $e^{\delta\tau}$ times (δ is fixed non-negative number) the amount of products produced by the stocks, created at the initial instant, under the same conditions.

It is also assumed that in the process of development of the economy the labour resources are removed from stocks of a lower structure, which have been created formerly. The labour resources, which have been freed from the removed stocks, are used on stocks created anew, the remaining stocks are not used subsequently. Under the assumption of continuous growth of capital organic composition (new stocks) the policy of removing stocks from production is characterized by the function $m(t)$, namely all stocks which have been created up to a certain instant of time $m(t)$ are freed to the instant of time t. The function $m(t)$ in the model is subject to determination.

The investments for increasing the fixed and working stocks are specified through their intensity so that $\partial\ell(t)\,dt$ is volume of investments at time interval $[t, t+dt]$. The function $\varkappa(t)$ is specified in the model, however it can be placed in dependence on the national income at instant t or other parameters of the model.

The system of equations that describe the model takes the form:

$$\varphi(t) = \frac{dT(t)}{dt} - \varphi[m(t)]\frac{dm(t)}{dt} \tag{3}$$

$$\varphi(t)\,\lambda(t) = \varkappa(t) \tag{4}$$

$$\mathcal{U}[\lambda(t),1]\,\varphi(t) - \frac{\partial \mathcal{U}[\lambda(t),1]}{\partial x}\varkappa(t) - e^{\delta[m(t)-t]}\varphi(t)\mathcal{U}[\lambda[m(t)],1] = 0 \tag{5}$$

The system is resolved for $t > t_0$ (t_0 is a fixed number). The initial conditions are specified as $m(t_0) = m_0$ $(m_0 < t_0)$,

$\lambda(t) = \lambda_0(t)$, $\varphi(t) = \varphi_0(t)$ when $t \in [m_0, t_0]$

where m_0 is a fixed number, and $\lambda_0(t)$ and $\varphi_0(t)$ are functions determining the initial distribution (with $t \in [m_0, t_0]$) of stocks and labour.

Equation (3) reflects the labour balance. Equation (4) reflects the stocks balance. Equation (5) is a condition of differential optimization. This condition denotes that the increase of net product at each instant of time should be maximal, in other words the functions $m(t)$, $\lambda(t)$, $\varphi(t)$ should be determined so that the function $dP(t)/dt$ is maximal at each instant of time t. Here $P(t)$ is the amount of net product (national income) produced at instant t per unit time. For $P(t)$ the following formula holds true:

$$P(t) = \int_{m(t)}^{t} e^{\delta\tau}\,\mathcal{U}[\lambda(\tau),1]\,\varphi(\tau)\,d\tau \tag{6}$$

For the norm of efficiency of investments the following formula has been derived

$$n(t) = \frac{1}{\varkappa(t)} \left\{ \frac{dP(t)}{dt} - e^{\delta m(t)} \mathcal{U}[\lambda[m(t)],1] \frac{dT(t)}{dt} \right\} \tag{7}$$

Let us specify the next form of the functions included in equations (3)-(5).

$\mathcal{U}[x,y] = x^{\alpha} y^{1-\alpha}$; $T(t) = T_0 e^{\rho t}$; $\varkappa(t) = \varkappa_0$ is a fixed positive number. Under the assumption of a possibility of expansion of equations (3)-(5) into infinite series and confining ourselves to linear terms of values δ and ρ we have

$$m(t) = \beta t \left[1 + \delta \frac{1-\beta}{2\alpha} t\right]$$

$$\varphi(t) = \frac{1}{t}\left[1 + \delta \frac{t+1}{2\alpha} + \rho \frac{T_0(t+1)}{1-\beta}\right]$$

$$\lambda(t) = \varkappa_0 t \left[1 - \delta \frac{t+1}{2\alpha} - \rho \frac{T_0(t+1)}{1-\beta}\right], \quad \beta = (1-\alpha)^{1/\alpha}.$$

By means of this solution it is possible to obtain a parametric representation of the norm of efficiency by coefficient δ which characterizes technological change and growth rate of labour resources ρ :

$$n(t) = \alpha \varkappa_0^{\alpha-1} t^{\alpha-1} \left[1 + \frac{\delta(1+\alpha)t + (1-\alpha)}{2\alpha} + \rho T_0 \frac{t+1}{1-\beta}\right].$$

In conclusion we shall note the possibility to introduce within the framework of the model the notion of variable transformability of stocks, that is incomplete transformability, whereby the degree of incompleteness is characterized by a variable coefficient.

Let us assume that the stocks removed from production (relative to its cost of reproduction) can be partly realized and the obtained capital directed into investments. Let us denote part of the realized value by π $(0 < \pi < 1)$. The equations of the model will take the following form:

$$\varphi(t) = \frac{dT(t)}{dt} + \rho[m(t)]\frac{dm(t)}{dt} \tag{8}$$

$$\varphi(t)\lambda(t) = \partial\ell(t) + \pi\,\partial\ell[m(t)]\frac{dm(t)}{dt} \tag{9}$$

$$\mathcal{U}[\lambda(t),1]\,\varphi(t) - \frac{\partial\mathcal{U}[\lambda(t),1]}{\partial x}\partial\ell(t) - $$

$$- e^{\delta[m(t)-t]}\varphi(t)\,\mathcal{U}[\lambda(m(t)),1] + \pi\frac{\partial\mathcal{U}[\lambda(t),1]}{\partial t}\partial\ell[m(t)]\frac{dm(t)}{dt} = 0. \tag{10}$$

The formula for the norm of efficiency of investments will be changed accordingly. Under the assumption of a small value of coefficient π we obtain for $n(t)$ the following formula

$$n(t) = n_0(t)\left[1 + \pi\frac{\partial\ell[m(t)]}{\partial\ell(t)}\frac{d^2T(t)}{dt^2}\right] \tag{11}$$

References

1. L.V. Kantorovich, Economic Calculation of Best Utilization of Resources M. USSR Akad. Nauk Publishing House (1959).
2. L.V. Kantorovich and A.L. Weinstein, On the Calculation of the Efficiency Norm on the Basis of a One-Product Model of Economic Development, Ekonomika i Mat. Metody 3 (1967), issue 5.
3. L.V. Kantorovich and V.I. Zhjanov, A One-Product Dynamic Model of Economics, Taking Account of the Change in Structure of stocks with Industrial Progress. Dokl. Akad. Nauk USSR, 211 (1973), No.6.
4. L.V. Kantorovich, Economic Problems of Technological Change, Ekonomika i Mat. Metody, 10 (1974), issue 3.

IDENTIFICATION AND CONTROL FOR LINEAR DYNAMIC SYSTEMS OF UNKNOWN ORDER

A.V.Medvedev

Institute of Automation, Academy of Sciences of Kirghiz SSR, Frunze, USSR

The linear dynamic systems theory appears to be the most developed part of the control theory. Comparative simplicity of this theory due to the superposition principle dominating there, makes it widely accessible when developing control systems for different kind of processes. Monographs [1, 2, 3] are devoted to numerous analysis and synthesis problems concerning these systems. In the following an approach is developed to linear dynamic plants control algorithms aimed at the estimation of the plant reverse operator for the case when the order of the equation describing the process is unknown.

I. Processes described by ordinary differential equations

Let a plant under study be described by the equation of the form

$$D^n q(t) = D^m U(t), \tag{1}$$

where $q(t)$ is plant output, $U(t)$ - control action, D - differential operator, n,m - order of operator D, $n>m$, the order of operators D^n, D^m being unknown. Considering that the operator connecting the output of plant $q(t)$ with the input of $U(t)$ and vice versa as well as $U(t)$ with $q(t)$ is described by convolution integral, it is necessary to estimate operator $U(t) = Aq(t)$. And now, by picking up $q^*(t)$ as forward action, we may calculate control $U^*(t)$ [2] which corresponds to it. Transient characteristics taken at the plant serve as a basis for operator A estimation.

The estimation of operator A is subdivided into two stages. Primarily, we solve the identification problem which consists in estimating operator B, $q(t) = BU(t)$ which is followed by estimating operator A.

It is known that at zero initial conditions B may be represented by the integral

$$q(t) = \int_0^t h(t-\tau)U(\tau)d\tau, \tag{2}$$

or in the discrete type

$$q_s = q(t_s) = \sum_{j=0}^{S-1} h(t_s-t_j)u(t_j)\Delta t_j \tag{3}$$

where $h(t-\tau)$, $h(t_s-t_j)$ is weight function of system (1) in the analogue and discrete variants correspondingly. Operator A, which is reverse to (2) alongside with B, is linear and bounded operating in continuous functions space and has the form [2]

$$U(t) = \int_0^t k(t-r)q(r)dr \tag{4}$$

or in the discrete form

$$U_s = U(t_s) = \sum_{j=0}^{S-1} k(t_s-t_j)q(t_j)\Delta t_j . \tag{5}$$

The method of estimation of operators (3), (5) (in connection with the all-round application of computers in control systems, we shall further apply a descrete registration only) is based on non-parametric estimation of unknown stochastic dependences.

Let the N-number of transient characteristics be taken on the plant described by linear differential equation(1). Consequantly, we shall have observations $q_{\nu k}$, mixed with disturbances at time moments t_i, $i = \nu k$, $k = 1,\ldots,\lambda$, $\nu = 1\ldots N$. For non-parametric estimate (3) we take the statistics class

$$q_\tau(t_s) = \sum_{j=0}^{S-1} u(t_j) \frac{\sum_{i,\rho=1}^{\tau} q_i\left[F\left(\frac{t_s-t_j-t_i}{c(\tau)}\right)\Phi\left(\frac{t_s-t_j-t_\rho}{c(\tau)}\right)-F\left(\frac{t_s-t_j-t_\rho}{c(\tau)}\right)\Phi\left(\frac{t_s-t_j-t_i}{c(\tau)}\right)\right]\Delta t_j}{\sum_{i,\rho=1}^{\tau}\Phi\left(\frac{t_s-t_j-t_i}{c(\tau)}\right)\Phi\left(\frac{t_s-t_j-t_\rho}{c(\tau)}\right)} \tag{6}$$

where $r = N\lambda$, $q_r(t_s)$ is the value of the system reaction to a disturbance $U(t_j)$, $j = 0,1,\ldots, S-1$.

Functions $F\left(\frac{t_s-t_j-t_i}{c(\tau)}\right)$, $\Phi\left(\frac{t_s-t_j-t_i}{c(\tau)}\right)$ satisfy conditions

$$c^{-1}(\tau)\int_{\Omega(t_s)}\Phi\left(\frac{t_s-t_j-t_i}{c(\tau)}\right)d\Omega=1 \quad , \quad \int_{\Omega(t_s)}(t_s-t_j-t_i)F\left(\frac{t_s-t_j-t_i}{c(\tau)}\right)d\Omega=-1 . \qquad (7)$$

If the type of disturbance U(t) is known and U(t) can be differentiated, then by estimating the known analogue of the integral convolution (2) we find (at zero initial conditions) that

$$q_\tau(t_s)=\sum_{j=0}^{S-1}U'(t_s-t_j)\sum_{i=1}^{\tau}q_i\Phi\left(\frac{t_j-t_i}{c(\tau)}\right)\left[\sum_{i=1}^{\tau}\Phi\left(\frac{t_j-t_i}{c(\tau)}\right)\right]^{-1}\Delta t_j . \qquad (8)$$

For nonparametric estimate (6) the following theorem is appropriate:

Theorem I. Let $h(t_s-t_j)$ be bounded for any $0 \leqslant t_j < t_s$ and have finite quantities of the second order, whereas random process

$$\xi_\tau=|q(t_s)-q_\tau(t_s)| \ , \ \xi_\tau \in E_1 \ , \ M\{\xi_\tau\}<\infty \ , \ \tau=1,2,\ldots$$

which is set at probability space $\{\Omega,\mathcal{U},P\}$ be measured relatively to the inclosed σ - algebras, U_r, r = 1,2... . In this case, with the probability one, the random process ξ_r, r = 1,2... converges to zero with c(r) = 0, rc(r) = ∞ and $\nu \to \infty$ N = ∞. Also l.i.m ξ_τ = 0 at $\nu = \infty$, N = ∞.

Thus, algorithm (6) enables us to calculate the plant reaction to arbitrary disturbance $U(t_j)$, j = 0,1,..., S-1 provided there are transient characteristics, taken at a particular plant.

As to the transient characteristics needed for the estimation of (5) they will be "taken" now from plant (6) simulator. Supposing $q_r(t_s)$ = 1, S=1,...,ℓ. Then U_s, S=1,2,..., ℓ will be determined from the matrix equation

$$U = H^{-1}Q ,$$

where

$$U=\begin{Vmatrix} u(t_1) \\ u(t_2) \\ \vdots \\ u(t_\ell) \end{Vmatrix} \qquad Q=\begin{Vmatrix} 1 \\ 1 \\ \vdots \\ 1 \end{Vmatrix} \qquad H=\begin{Vmatrix} h(t_1) & 0 & \ldots & 0 \\ h(t_2) & h(t_1) & \ldots & 0 \\ \vdots & \vdots & & \vdots \\ h(t_\ell) & h(t_{\ell-1}) & \ldots & h(t_1) \end{Vmatrix} .$$

And now, when we have a set of values $U(t_i)$, i=1,2,..., ℓ we find a non-parametric estimate (5). At zero initial conditions it will be of the form

$$u_\nu(t_s)=\sum_{j=1}^{S-1}q(t_j)\frac{\sum_{i,p=1}^{\nu}u_i\left[F\left(\frac{t_s-t_j-t_i}{c(\nu)}\right)\Phi\left(\frac{t_s-t_j-t_p}{c(\nu)}\right)-F\left(\frac{t_s-t_j-t_p}{c(\nu)}\right)\Phi\left(\frac{t_s-t_j-t_i}{c(\nu)}\right)\right]\Delta t_j}{\sum_{i,p=1}^{\nu}\Phi\left(\frac{t_s-t_j-t_i}{c(\nu)}\right)\Phi\left(\frac{t_s-t_j-t_p}{c(\nu)}\right)} , \qquad (9)$$

where $U=R\ell$,R is number of "taken" transient characteristics. The following proposal holds good for (9):

Theorem 2. Let theorem 1 take place and $k(t_s-t_j)$, $0\leqslant t_j<t_s$ have the second bounded derivatives. Then at $c(v)\to 0$, $vc(v)\to\infty$ with $R\to\infty$, $\ell\to\infty$

$$\underset{\substack{R\to\infty\\ \ell\to\infty}}{\text{l.i.m.}} |U_v(t_s)-U(t_s)| = 0, \quad \lim_{\substack{R\to\infty\\ \ell\to\infty}} P(\sup|U_v(t_s)-U(t_s)|<\varepsilon)=1,$$

where ε is a certain positive constant.

Formula (9) just represents the control algorithm for linear dynamic system.

Thus, the expounded approach to the dynamic plants control algorithms synthesis as apriori information requires only an assumption of its being linear and does not require knowledge of the equation order describing the plant. In this case in order to find the control algorithm (9), it is necessary to take several transient characteristics from the plant. Though convergence theorems are of asymptotic character, estimate (5) can be obtained in practice with a sufficient accuracy rate by having a limited number of transient characteristics. The advantage of the method proposed lies in the fact that the identification and control algorithms can be applied in the systems described by the linear differential equations of the general type (1).

2. Processes described by differential equations in partial derivatives

Let the condition of linear controllable system with distributed parameters be described by the function $q(x,t)$, where x is space variable belonging to interval $[0,1]$, while time $t\in[0,T]$, T is termination time for transient process. Disturbing action $U(t)$ is applied to the input of the system. Let function $U(t)\in L_2$ be measurable. At zero initial condition, i.e. $q(x,0)=0$, $0\leqslant x\leqslant 1$ the plant condition function $q(x,t)$ can be expressed by the formula [3]

$$q(x,t) = \int_0^t G(x,t,r)U(r)dr, \tag{10}$$

where $G(x,t,r)$ is pulse transient function of the system.

Let unique disturbing action $1(t)=1$, $t\geqslant 0$ and $1(t)=0$, $t<0$, $0\leqslant t\leqslant T$ be applied to the distributed system input. Let us carry on a series of observations $q_\nu(x,t_k)$ in the point $0\leqslant x\leqslant 1$ at moments t_i, $i=k\nu$, $K=1,\dots,\lambda$, $\nu=1,\dots,N$, where N is the number of taken transient characteristics. Based on these observations we build a statistic estimate (10) which being within the class of non-parametric type approximations, has the form [4,5]

$$q_\tau(x,t)=\int_0^t u(\tau)\sum_{i,p=1}^{\tau} q_i(x)\left[F(c^{-1}(\tau)(t-\tau-t_i))\Phi(c^{-1}(\tau)(t-\tau-t_p))-F(c^{-1}(\tau)(t-\tau-t_i))\times \right. \tag{11}$$
$$\left.\Phi(c^{-1}(\tau)(t-\tau-t_p))\right]\Phi(c^{-1}(\tau)(x-x_i))\left[\sum_{i,p=1}^{\tau}\Phi(c^{-1}(\tau)(t-\tau-t_i))\Phi(c^{-1}(\tau)(t-\tau-t_p))\Phi(c^{-1}(\tau)(x-x_i))\right]^{-1}d\tau,$$

where functions $F(c^{-1}(r)(t-r-t_i))$ and $\Phi(c^{-1}(r)(t-r-t_i))$ satisfy conditions (7).

For the statistics of the class under consideration holds good the following

Theorem 3. Let $G(x,t,\tau)$ be twice differentiable by $\tau\in[0,t]$ in point x, $0<x<1$. Then the random process $\zeta_\tau=|q(x,t)-q_r(x,t)|$, $\zeta_\tau\in E_1$, $r=1,2,\ldots$ given on the probability space $\{\Omega,\mathcal{U},P\}$ and measurable in relation to the inclosed σ - algebras, U_r, $r=1,2\ldots$ is convergeable in the mean and with the probability zero ζ_r, $r=1,2..$ strives to zero at $\lambda\to\infty$, $N\to\infty$.

Now considering the formula (11) as the distributed plant simulator, we take transient characteristics analogically to the above mentioned way, picking up for q(x,t) some given value of the plant output $q^*(x,t)$, $0\leqslant t\leqslant T$. And only now on the basis of the transient characteristics obtained on the simulator (11) we may build up a non-parametric algorithm for distributed plant control.

In conclusion it should be observed that here too the method of building up identification and control algorithms is just the same as that for the processes described by usual differential equations. Primarily transient characteristics are taken from the plant under study and its simulator is made and only then on the simulator "reverse" transient characteristics are taken and the plant reverse operator is estimated which just represents the plant control algorithm. The class of control and identification algorithms can be realized only on digital computers and therefore can be applied for control of complicated technological processes in computer controls and digital computers.

3. Digital computer simulation. Discussion. Prospects.

Above it has been pointed out that dynamic system non-parametric identification and control algorithms prove to be fit for unknown order plants. This fact we illustrate by simulation results. The plant is described by the equation of the 1,2 and 3rd order correspondingly. Three transient plant characteristics have been "taken" from a digital computer. When "taking" a transient characteristic we have

carried out 50 measurements of the plant output subjected to the action of random disturbances belonging to the interval $[-\alpha, \alpha]$ at different intervals of Δt discretization.
As the result we have a set of points $\{q_i, t_i\}$, $i=1,\ldots,150$, $\lambda=50$, $N=3$. Based on these observations the plant operator was estimated in accordance with the formula (8). Further, sinusoidal disturbing action was applied to the system input (in Fig.I - curve 1) and the plant response was calculated in accordance with the ordinary differential equations theory (curve-2) as well as with the non-parametric estimate of the system response (curve-3).

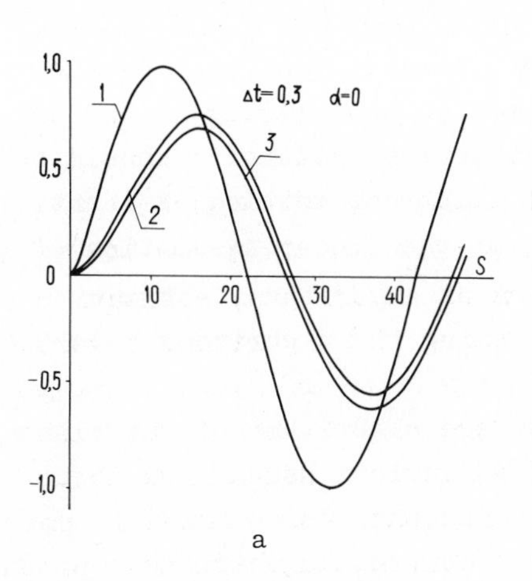

a

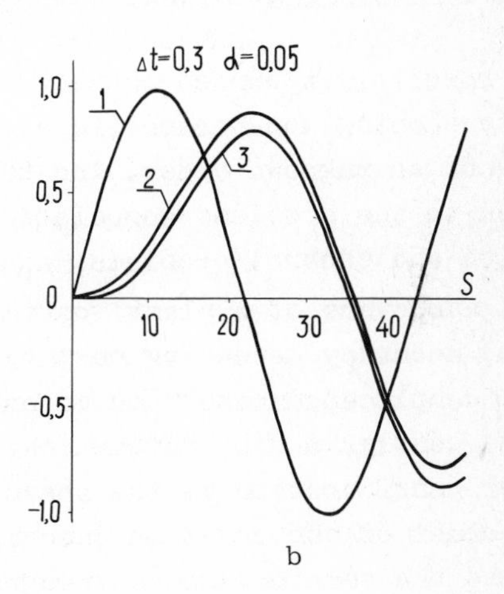

b

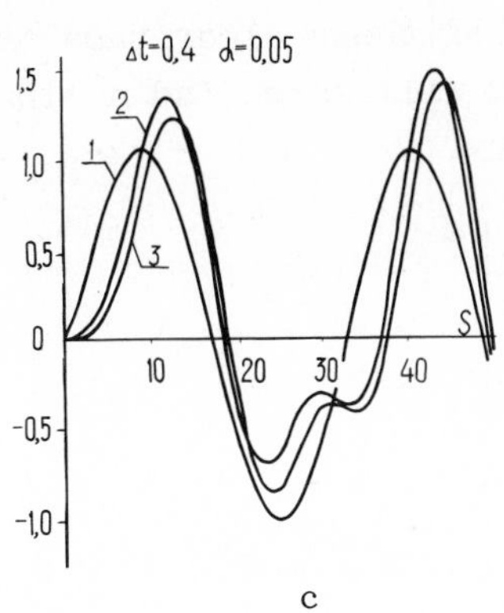

c

Fig. 1

Cases a, b, c in Fig.1 correspond to the plant simulators of 1, 2 and 3rd order, S is discrete time. Fig.2 shows the changes of quadratic mean error $\sigma = \nu_1^{-1}\sum_{i=1}^{\nu_1}(q(t_i)-q_2(t_i))^2$, $\nu_1=50$ depending on Δt.
The curve in Fig.2a corresponds to the case when linearly increasing disturbance was applied to the plant input, while 2b corresponds

to sinusoidal disturbance. Owing to the all-round application of computing machines in up-to-date information control and processing systems it seems impedient to use a non-parametric algorithm proceeding from the two reasons: first, owing to little apriori information (only the correctness of assumption about the plant linearity is required); and second, a certain universality and the unique type of these procedures makes it possible to develop a program for control computers aimed at controlling different technological plants of the class under consideration. In this case it is sufficient to keep in the control computer's memory a set of transient characteristics of each technological plant.

The results, expounded above, appear to be the first step on the way of developing non-parametric algorithms for controlling linear systems of an unknown order. And the most serious attention should be given to the problems associated with incorrect setting of identification and control problems, adequacy of the linear assumption of the real structure of a plant, optimization of algorithms, estimation of $q_r(t)$ accuracy depending on Δt; developing the algorithms covering more complicated cases and so on.

Consequently, a distinctive feature of the algorithms of the class under consideration is the absence of an apriori assumption about the order of the equation describing a process, which makes it possible to use the same algorithm (program) for controlling different production processes. The other important feature of these algorithms is their orientation to digital technique which will ever find a wider application in different branches of industry.

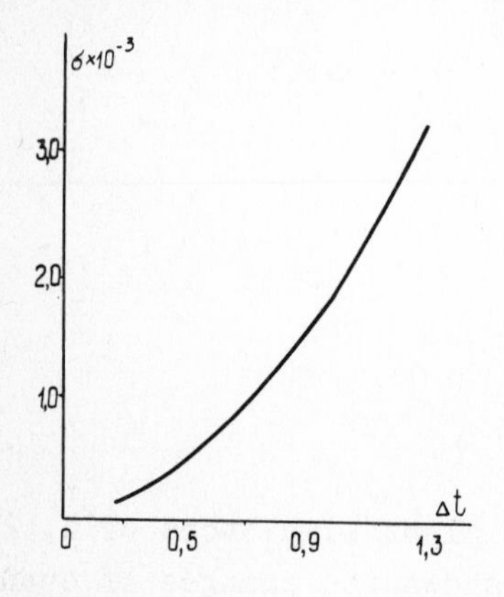

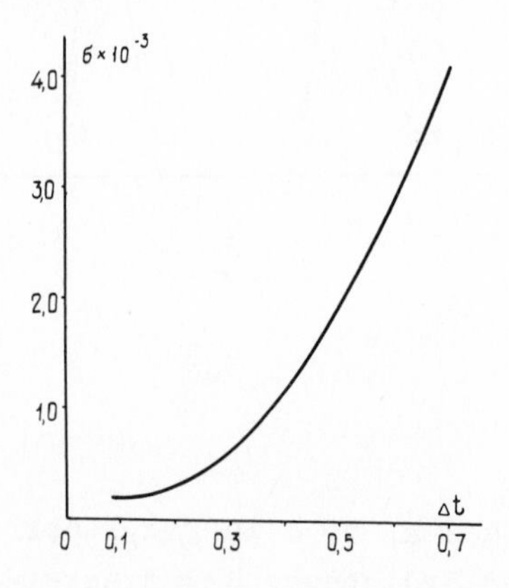

a Fig. 2 b

References

I. Заде Д., Дезоер Ч. Теория линейных систем. "Наука", М., I970.

2. Куликовский Р. Оптимальные и адаптивные процессы в системах автоматического регулирования. "Наука", М., I967.

3. Бутковский А.Г. Теория оптимального управления системами с распределенными параметрами. "Наука", М., I965.

4. Надарая Э.А. Непараметрические оценки кривой регрессии. Сб.: Некоторые вопросы теории вероятностных процессов. Тбилиси, "Мицниереба" I965.

5. Медведев А.В. Сходимость непараметрических алгоритмов адаптации. Сб.: Автоматизированные системы управления в цементном производстве. "Илим", Фрунзе, I973.

GROUP CHOICE AND EXTREMAL PROBLEMS IN THE ANALYSIS OF QUALITATIVE ATTRIBUTES

B.G. Mirkin
Institute of Economics & Industrial Engineering
Siberian Department of the USSR Academy of Sciences
Novosibirsk, USSR

A problem of aggregating incoherent "individual" data on a fixed set of objects for obtaining their common description has gained recently an increasing urgency. In one case they can be individual opinions of jury members about preference order concerning objects considered which are to be reduced to a common "group" decision. In another, they may be inconsistent indicators of objects quality which are to be reduced to a single compromise indicator; in still another, they are different classifications of objects chosen as initial ones in building a reconciled classification.

According to the established notions of the group choice theory [1]-[3] we shall refer to this problem as a group choice problem, and to the rule of transition to a "group" set of data, as a reconciling principle.

We shall assume that the data are given non-quantitatively, i.e. they are described by binary relations on a given set of objects A. If $R \subseteq A \times A$ describes a preference relation, then the fact of belonging $(x,y) \in R$ denotes that x is not less preferable than y. If R corresponds to some quality indicator, $(x,y) \in R$ denotes that x is not worse than y by this indicator. If R is defined by some classification of objects, then $(x,y) \in R$ denotes the fact of x and y belonging to the same class. To different types of data, of course, different types of binary relations correspond. Ordinarily, examined are nominal attributes, set by equivalence relations and rank ones, set by complete preorders (rankings).

In ref. [4] four requirements to reconciling principles are described (monotony, universality, independence of objects, sovereignty of group choice) similar to the appropriate axioms of Arrow, and it was shown that these requirements entirely guarantee the so called reconciling principles. The reconciling principle $F(R_1,\ldots,R_n)$ is called trivial if there exists such a subset V of the set of indices $\{1,2,\ldots,n\}$ that F coincides with F_V defined by the formula:

$$F_V(R_1,\ldots,R_n) = \bigcap_{K\in V} R_K .$$

Trivial principles are broadly used in reconciling classifications.

Now we shall consider a more complicated reconciling principle, i.e. a rule of majority which does not satisfy the axiom of universality [4].

The rule of majority allows different interpretations. We shall denote by a_{xy} the number of such relations in the set $R_1,\ldots$ $\ldots, R_n$ which contain the pair of objects (x,y). According to one interpretation, the group relation R is defined by the relation

$$(x,y)\in R \longleftrightarrow a_{xy} \geq a_{yx} \tag{1}$$

and according to another, by relation

$$(x,y)\in R \longleftrightarrow a_{xy} \geq \frac{n}{2} \tag{2}$$

The difference between the rules (1) and (2) can be explained by the following example. Suppose 100 persons have expressed their opinions about comparative preference concerning two objects x and y. For 99 persons the objects turned out to be similar by preference, and for one person x was better than y. This means that 99 individual relations contain both (x,y) and (y,x), and one relation contains only (x,y) so that a_{xy} =100, a_{yx} =99. Then, according to the rule (1), x is better than y, while x and y are similar by preference in the sense of (2). Roughly speaking, the rule (1) expresses the opinion of the "active minority", while the rule (2), the opinion of the "passive majority". If, however, only strict preferences are considered, it is easy to see that both rules yield the same results.

We shall examine the following natural modification of the rule (2). The relation R will be called majoritary if it satisfies the condition (2) for all pairs of objects (x,y) with possible exceptions of such pairs (x,y) for which $a_{xy} = \frac{n}{2}$.

The set of relations D will be called permissible if for any set of relations $R_1, \ldots, R_n$ from D (not necessarily different) there exists a transitive majoritary relation. The set of three strict rankings of type $(x,y,z),(y,z,x),(z,x,y)$ derived from each other through cyclic permutation of objects will be referred to as cyclic. Each of the cyclic set orderings generates the remaining ones through the permutation of the first object to the last place and of the last object to the first place; these two objects will be called cyclic. A set of three non-strict rankings $\{xQyQz, yQzQx, zQxQy\}$ (here symbol xQy denotes a relation: x is preferable or indifferent to y) will be called cyclic if it satisfies the following conditions:

a) if in one of these rankings all three objects x, y, z are indifferent between themselves, then for the cyclic objects of the remaining rankings a strict preference occurs;

b) all three rankings cannot be uniformly dichotomic, i.e. they must not partition the objects x,y and z into two classes so that a one-element class in each ranking precedes a two-element one (or in each ranking follows the two-element one).

<u>Theorem 1</u>. The set of rankings D is permissible if and only if it does not involve a cyclic set for any three objects.

Now we shall consider the natural extension of the majority rule (2) in terms of distance between the relations, applicable also to inadmissible sets of relations. For two relations R and P for distance $d(P,R)$ the number of pairs in their symmetric difference

$$d(P,R) = |P \Delta R| = |(P-R) \cup (R-P)|$$

will be taken.

Let $R_1, \ldots, R_n$ be given relations from D. The relation $R^* \in E$ minimizing the sum of distances $f(R) = \sum_{k=1}^{n} d(R, R_k)$ on all $R \in E$ will be referred to as a median of system $R_1, \ldots, R_n$ in class E. As a reconciling principle the procedure of taking the median in class E will be considered.

If rankings are only examined, then the distance between the appropriate relations is proportional to the distance axiomatically inserted in [5] (see [6]).

It is easy to show that the procedure of taking the median indeed generalizes the majority rule: if majoritary relation belongs to the class E, it is the median in the class E.

The procedure of taking the median can be stated in terms of variables a_{xy} characterizing the number of relations $R_1, \ldots, R_n$ containing the pair (x, y). Thus the following statement takes place.

Theorem 2. The relation R^* is the median of the system $R_1, \ldots, R_n$ in class E if and only if it maximizes the function

$$g(R) = \sum_{(x,y)\in R} (a_{xy} - \frac{n}{2}), \quad R \in E. \qquad (3)$$

In particular, if E consists only of relations corresponding to strict rankings, so that any $R \in E$ is marked by strict ranking of objects $(i_1, \ldots, i_n)$ (N the number of objects), i.e.

$R = \{(i_\kappa, i_l) \mid \kappa \leq l,\ \kappa, l = 1, \ldots, N\}$ the function g is

$$g(R) = g(i_1, \ldots, i_N) = \sum_{\kappa < l} a_{i_\kappa i_l} + \text{const}.$$

In this case, the maximization problem (3) is equivalent to the problem of such simultaneous permutation of columns and lines in the matrix $\| a_{ij} \|_1^N$ which would maximize the sum of overdiagonal elements $\sum_{\kappa<l} a_{i_\kappa i_l}$. The algorithms for this problem are given in [7], [8].

In search for the median in a class of equivalence relations corresponding to certain partitions of a set of objects (in solving the problem of reconciling different classifications or nominal attributes [9-10]) the problem (3) is concretized into a problem of constructing such a partition $R = \{R^1, R^2, \ldots, R^m\}$ of the set of objects into non-overlapping classes R^s which would maximize the function

$$g(R) = \sum_{s=1}^{m} \sum_{i,j \in R^s} (a_{ij} - \frac{n}{2}). \qquad (4)$$

The problem (4) is quite analogous to the known "intuitive" statements of the problem of automatic classification of a set of interconnected objects, and as indicators of interconnection, the variables $a_{ij} - \frac{n}{2}$ serve here.

The median in the context of data analysis is interpreted as "internal" factor representing "external" attributes $R_1, \ldots, R_n$.

Of interest is a search for internal factor in a class of relations characterizing the partitions $R = \{R^1, \ldots, R^m\}$ of the set of objects with a structure of "essential" connections between the classes set by the relation $\varkappa \subseteq \{1,2,\ldots,m\} \times \{1,\ldots,m\}$. The median in a class of these relations is a solution to a problem of maximization of the following function:

$$g(R, \varkappa) = \sum_{(s,t)\in\varkappa} \sum_{i\in R^s} \sum_{j\in R^t} (a_{ij} - \frac{n}{2}).$$

The optimal partition R with a "structure" $\varkappa$ characterizes the aggregated representation of "essential" connections in the matrix $\| a_{ij} \|_1^N$ which is, generally speaking, different from the unordered partition obtained within automatic classification.

In collaboration with V.L.Kupershtokh and V.A.Trofimov an algorithm has been developed for building "practically" optimal R and $\varkappa$ which was applied to the analysis of real association matrices in economics and biology.

We shall describe only two examples:

1. The analysis of the structure of intershop deliveries (in collaboration with G.V. Grenback).

Considered was a matrix a_{ij} where i, j are numbers of shops on a Novosibirsk plant, and a_{ij} the quantity of titles of parts going from the i-th shop to the j-th shop. In this example from a_{ij} we subtracted the arithmetic mean $\bar{a} = [1/N(N-1)] \sum_{i\neq j} a_{ij}$ instead of $\frac{n}{2}$. The obtained structure basically coincided with the existing structure of the plant, and existing deviations made it possible to invite the management of the plant to introduce certain changes into the organization structure which aimed at the relaxation of tensions for the management.

2. The analysis of genetic structures (in collaboration with S.N.Rodin).

As objects, there were considered genetic mutations $i, j = 1, \ldots, 36$ and $a_{ij} = 1$ if the interbreeding of mutants i and j does not lead to descendants, and $a_{ij} = 0$ otherwise.

The structure analysis has supported the genetic hypothesis that mutations affecting the same functional centers (sites) of albumen must in the same way interact with other mutations. Apart from this, it became possible to reveal the number of sites and the structure of their interaction.

References

1. Luce, R., H.Raiffa
   Games and Decisions.
   Wiley. N.Y. 1957.
2. Arrow, K.J.
   Social Choice and Individual Values.
   Wiley. N.Y. 1951.
3. Sen, A.K.
   Collective Choice and Social Welfare.
   Oliver and Boyd, 1970.
4. Mirkin, B.G.
   On Principles of Reconciling Relations. (Russian)
   Kibernetika. Kiev. 1973, no.2.
5. Kemeny, J., J. Snell.
   Mathematical Models in the Social Sciences.
   Elsevier. N.Y. 1963.
6. Mirkin, B.G., L.B.Chorny
   Some Properties of Space of Partitions.(Russian)
   Mat. Analiz Ekon. Model., part III, Institute of Economics of the Siberian Dept. of the USSR AS. Novosibirsk. 1972.
7. Burkov, V.N., V.O.Groppen
   Cuts in Heavily Associated Graphs and Permutation Potentials (Russian).
   Avt. i telem. 1972, no.6.
8. Litvak, B.G.
   On Ordering Objects by Preferences (Russian).
   Mat. met. upr. proizv., issue 5, the University of Moscow, Moscow, 1973.
9. Mirkin, B.G.
   Approximation Problems in the Space of Relations and the Analysis of Qualitative Attributes (Russian).
   Avt. i telem. 1974, no.9.
10. Mirkin, B.G.
    On an Approach to the Analysis of Primary Sociological Data.
    Paper submitted at the VII World Congress of Sociology, Institute of Economics of the Siberian Dept. of the USSR AS. Novosibirsk. 1970.

STUDIES IN MODELLING AND IDENTIFICATION OF DISTRIBUTED PARAMETER SYSTEMS+

L. LE LETTY

Département d'études et de recherches
en Automatique, C.E.R.T., 2 Avenue
Edouard Belin, 31000 TOULOUSE, France

INTRODUCTION

The paper presents a summary of some recent studies conducted in the department of automatic control in the field of modelling and identification of complex distributed parameter systems corresponding to engineering applications. These are :

- compact cross-flow heat exchangers in aircraft air-conditioning systems
- fixed catalyst bed chemical reactor
- identification in microwave guides (microstrip lines)

I. DYNAMIC MODELLING AND IDENTIFICATION OF A CROSS-FLOW COMPACT HEAT EXCHANGER

The general subject is a detailed analysis of the dynamic behaviour of an on-board air conditioning system in aircraft industry. An example

+ Part of this work has been supported by the "Direction des Recherches et Moyens d'Essais"; MM. J.P. Chretien and Le Pourhiet are the research engineers associated with these studies.

of such a system, a boot-strap system is given in Figure 1.

In this system, there is a regulation to be determined . The scaling of the air-conditioning system is made by guaranteeing the static performances in the flight domain. The regulation study is always built upon a simple transfer function between the cabin temperature and the position of the regulator valve. As for the dynamic behaviour of the system, the most interesting elements are the heat exchangers. These exchangers - precooling and main exchangers - are compact cross flow compact exchangers for which the performances are high for a given volume but for which the geometrical complexity is also high and unknown from the user. An example with two elements is given in Figure 2.

While the statics of such exchangers have been extensively studied for all kinds of configurations and internal geometry (see, for example, KAYS and LONDON, Ref. 2), the dynamics are not well documented, one of the reasons being that a cross flow set up leads to a characterization with three or four independant variables. The subject presented in this paper is then a detailed modelling in temperature of the exchanger.

I.1 - Characterization of an element and equations

Geometric and physical simplifying hypothesis are needed to obtain a tractable model of the object. The main assumptions are :

- the exchange surface is plane and rectangular
- the temperatures are uniform on the z-axis
- the physical parameters (velocities,of the fluids, specific heats,...) are constant.

Several models can be written according whether one considers outside walls or not, or whether one or two fluids are considered as mixed (uniformisation of the temperature on the direction perpendicular to the flow)

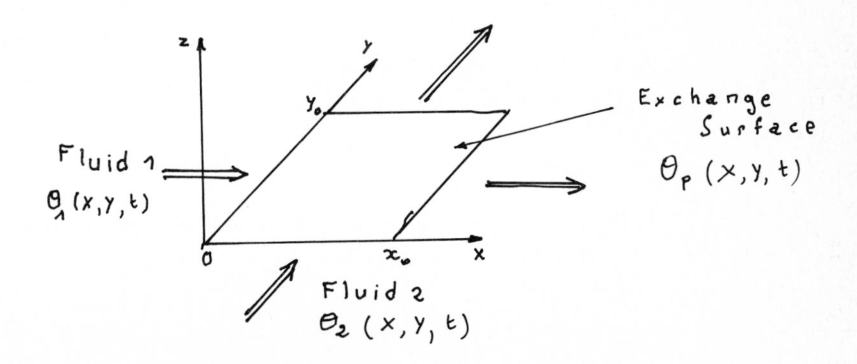

The practical case (no mixing, perfect insulation) is described by three temperatures (the two fluids and the separation wall) with the three independant variables x, y, t. Written in dimensional form, the equations are :

$$\begin{cases} \dfrac{\partial \theta_1}{\partial x} + \dfrac{\partial \theta_1}{\partial t} = a_1 \ (\theta_p - \theta_1) \\[2ex] \dfrac{\partial \theta_2}{\partial y} + r\dfrac{\partial \theta_2}{\partial t} = a_2 \ (\theta_p - \theta_2) \\[2ex] \dfrac{\partial \theta_p}{\partial t} = b \ (\theta_1 - \theta_p) + r'b \ (\theta_2 - \theta_p) \end{cases}$$

with the following definitions :

$$x = \frac{x}{x_0}, \ y = \frac{y}{y_0}, \ t = \frac{"t"}{\tau_1} \quad \tau_1 = \text{transit time} \quad 1 = \frac{x_0}{v_1}$$

$$\tau_2 = \frac{y_0}{v_2}, \ a_1 = \frac{\tau_1}{R_1 \ C_1}, \ a_2 = \frac{\tau_2}{R_2 \ C_2}, \ r = \frac{\tau_2}{\tau_1}, \ r' = \frac{R_2}{R_1}, \ b = \frac{\tau_1}{R_1 \ C_p}$$

("RC" = thermal time constants)

I.2 - Solution by Laplace transforms

The analysis of the dynamic behaviour leads to solve the above equations. This has been done after successive Laplace transformations giving the transfer functions of the system. The advantages of this approach are the possibility of obtaining independently the transforms of the temperatures in any point (x, y), to calculate the mean temperatures at the outputs x = 1 and y = 1 which are the only temperatures usually measurable, and also to avoid the difficulties due to discontinuities and time consuming of difference methods. The transfer functions are however quite complex.

With zero initial conditions and the boundary conditions :

$\theta_1 \ (0, y, t) = u \ (t)$ = step function

$\theta_2 \ (x, 0, t) = 0$

We have the output transfer functions :

- direct response :

$$H_1 (p, y) = \frac{\theta_1 (1, y, p)}{\theta_1 (0, p)}$$

$$= e^{-\lambda_1} \left[e^{-\lambda_2 y} I_0 (2 \sqrt{\alpha_1 \alpha_2 y}) + \lambda_2 \int_0^y e^{-\lambda_2 u} I_0 (2 \sqrt{\alpha_1 \alpha_2 u}) \, du \right]$$

- cross response

$$H_2 (p, x) = \frac{\theta_2 (x, 1, p)}{\theta_1 (0,p)}$$

$$= \frac{\alpha_2 \ell^{-\lambda_1 x}}{\lambda_2} \left[\ell^{\frac{\alpha_1 \alpha_2}{\lambda_2} x} - e^{-\lambda_2} \left\{ I_0 (2 \sqrt{\alpha_1 \alpha_2 x}) + \right.\right.$$

$$\left.\left. + \frac{\alpha_1 \alpha_2}{\lambda_2} e^{\frac{\alpha_1 \alpha_2 x}{\lambda_2}} \int_0^x I_0 (2 \sqrt{\alpha \alpha_2 u}) \, \ell^{\frac{-\alpha_1 \alpha_2 u}{\lambda_2}} du \right\} \right]$$

where :

$$\lambda_1 = p + a_1 - \frac{b a_1}{p + b (1 + r')}$$

$$\lambda_2 = rp + a_2 - \frac{r' b a_2}{p + b (1 + r')}$$

$$\alpha_1 = \frac{r' b a_1}{p + b (1 + r)}$$

$$\alpha_2 = \frac{b a_2}{p + b (1 = r')}$$

I_0 = Bessel function

Some interesting properties can be deducted from these transfer functions in the simplest cases, especially the separation of transport terms (time lag and discontinuity). However, the expressions are too complex to deal with analytically. Numerical inversion, through Dubner and Abate algorithm coupled with a Fast Fourrier transform, has been implemented to obtain the transient responses. Typical responses are represented in Figure 3.

I.3. - Approximate characterization by rational transfer functions through identification

The transfer functions are too complex to deal with and to be a practical representation when inserting the element in a complete system or for synthesis purpose. Furthermore the temperatures are not homogeneous at the outputs (in case of no mixing) and what is needed is a global representation as simple as possible; we will then consider the output mean temperatures which are in the other hand, the only measurable variables. After integrating and expanding the Bessel functions, the exact expressions of the corresponding transfer functions are :

$$H_{1m}(p) = e^{-\lambda_1} \left(1 - \frac{\alpha_1 \alpha_2}{\lambda_2} \sum_{n=0}^{\infty} \frac{(\alpha_1 \alpha_2)^n}{n!\,\lambda_2^n} + e^{-(\lambda_1+\lambda_2)} \sum_{n=1}^{\infty} \frac{(\alpha_1 \alpha_2)^n}{n!\,\lambda_2^{n+1}} \sum_{k=0}^{n-1} \sum_{j=0}^{k} \frac{\lambda_2^j}{j!}\right.$$

$$H_{2m}(p) = \alpha_1 \left[\sum_{n=0}^{\infty} \frac{(\alpha_1 \alpha_2)^n}{(\lambda_1 \lambda_2)^{n+1}} \left(1 - e^{-\lambda_2} \sum_{j=0}^{\infty} \frac{\lambda_2^j}{j!}\right) \left(1 - e^{-\lambda_1} \sum_{k=0}^{n} \frac{\lambda_1^k}{k!}\right) \right]$$

These expressions have been approximated by rational transfer functions after extracting the time lag and discontinuity due to transport phenomenom according to (other cases have been investigated) :

$$H_{1m}(p) = e^{-p}\ (e^{-a_1} + F(p)) \text{ with } F\ (p) = \frac{K_1\ (1+ \frac{p}{Z_0})}{1 + 2z\ \frac{p}{\omega_1} + \frac{p^2}{\omega_1{}^2}}$$

$$H_{2m}(p) = F_2(p) = \frac{K_2\left(1 + \frac{p}{z'_0}\right)}{(1 + 2z'\ \frac{p}{\omega_2} + \frac{p^2}{\omega_2{}^2})(1 + T_p)\ (1 + \frac{p}{p_0})}$$

These approximations are justified by the shape of the transient responses and the properties of the limiting case, - case without separation wall.

The identification has been made in the time domain by a quasi-Newton algorithm with a mean square error criterion. An example of results is given in figure 4. Comparison in the frequency domain has laso been made in order to observe the influence of high frequency modes not contained in the domain of interest and, in fact, over all frequencies, the high modes being severely attenuated.

I.4 - Further investigations and conclusion

Different kinds of setting-up of elementary units have been theoretically and numerically investigated -parallel, counter current, mixed, two or three elements - . Each element is then represented as a linear two-input, two-output multi-variable system with its exact and approximate matrix transfer function. Comparisons of the two representations are very good and then justify the method of analyzis taken for the elementary unit.

In order to reduce the order of characterization of the assembly which is probably too rich, we applied again the identification procedure. It is interesting to note then that all the transfer functions are satisfactorily represented by generalized second order functions (plus a lag-time for the direct response). It is this representation which has been used for the study of an experimental heat exchanger.

Then the important problem of establishing relations between physical parameters (introduced in the internal representation by partial differential equations) and the dynamic parameters obtained by identification and the problem of reducing then the number of parameters has been investigated. This type of investigation has been made by KAYS and LONDON, in the static field but is here more difficult and akward. A systematic approach by varying the physical parameters has given coherent results but the relations which are established are too approximate and concern too few data to conclude.

A simpler representation by gains and time constants has then also been performed. The validity of the representations by frequency domains has been established. Experimental work has been conducted at the Centre d'Essais Aéronautique de Toulouse on a heat exchanger whose diagram is represented below.

The study presents a method of approach for the detailed analyzis and identification of the dynamic behaviour of cross flow compact heat exchangers used in aircraft industry and also in other industries (gas turbine applications.....) The approach has proved to be consistant on a concrete heat exchanger but more data, especially higher frequency components in the available inputs and other applications on real exchangers are necessary to conclude.

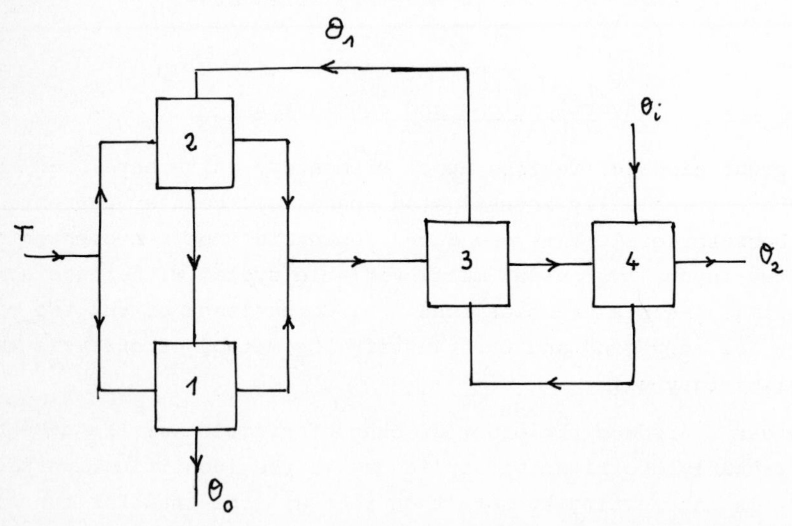

II. MODELLING - SIMULATION - IDENTIFICATION OF A FIXED BED CATALYST CHEMICAL REACTOR

II. 1 The bidimensional model

The model that we consider in this application concerns a cylindrical tubular reactor with diffusion terms in the stationary case. A real pilot reactor is available for the measurements.

The balance equations for mass and energy are :

1. fluid

$$\begin{cases} \varepsilon_R Da \dfrac{\partial^2 C}{\partial X^2} + \varepsilon_R D_r \left\{ \dfrac{\partial^2 C}{\partial R^2} + \dfrac{1}{R} \dfrac{\partial C}{\partial R} \right\} - \varepsilon_R u \dfrac{\partial C}{\partial X} = S_V K_g (C - C_s) \\ \varepsilon_R K_a \dfrac{\partial^2 T}{\partial X^2} + \varepsilon_R K_R \left[\dfrac{\partial^2 T}{\partial R^2} + \dfrac{1}{R} \dfrac{\partial R}{\partial R} \right] - \varepsilon_R \rho_f Z_f \dfrac{\partial T}{\partial X} = S_V h_g (T - T_S) \end{cases}$$

where C and T are the concentration and temperature on the fluid, C_s and T_s the concentration and temperature at the surface of catalyst particles

2. catalyst

$$\begin{cases} S_V K_g (C - C_s) = V_V f (C_s, T_s) \\ S_V h_g (T - T_s) = V_V \Delta H f (C_s, T_s) \end{cases}$$

where f represents the kinetics law.

The boundary conditions are :

1. $0 < X < L$

$$R = 0 : \left\{ \frac{\partial C}{\partial R} = \frac{\partial T}{\partial R} = 0 \right.$$

$$R = R_0 : \begin{cases} \dfrac{\partial C}{\partial R} = 0 \\ - K_r \dfrac{\partial T}{\partial R} = h_w (T - T_W) \end{cases} \qquad T_W : \text{wall temperature}$$

2. $O < R < R_0$

$$X = X_0 : \begin{cases} p_f u z_f (T_0 - T) = - K_a \dfrac{\partial T}{\partial X} \\ u (Co - C) = - D_a \dfrac{\partial C}{\partial X} \end{cases}$$

$$X = L : \left\{ \dfrac{\partial C}{\partial X} = \dfrac{\partial T}{\partial X} = 0 \right.$$

II. 2 Equations in normalized form

With the definitions :

$$\xi = \frac{C_0 - C}{C_0} , \quad \theta = \frac{T}{T_0} , X = \frac{X}{L} , \quad r = \frac{R}{R_0}$$

The equations are written :

$$\begin{cases} a_1 \dfrac{\partial^2 \xi}{\partial x^2} + a_2 \left[\dfrac{\partial^2 \xi}{\partial r^2} + \dfrac{1}{r} \dfrac{\partial \xi}{\partial r} \right] - \dfrac{\partial \xi}{\partial x} = a_3 (\xi - \xi_s) \\ a_3 (\xi - \xi_s) = \alpha_1 f (\theta_s, \xi_s) \\ b_1 \dfrac{\partial^2 \theta}{\partial x^2} + b_2 \left[\dfrac{\partial^2 \theta}{\partial r^2} + \dfrac{1}{r} \dfrac{\partial \theta}{\partial r} \right] - \dfrac{\partial \theta}{\partial x} = b_3 (\theta - \theta_s) \\ b_3 (\theta - \theta_s) = \gamma_1 f (\theta_s, \xi_s) \end{cases}$$

with the boundary conditions :

$$\left.\frac{\partial \xi}{\partial r}\right|_{r=0} = \left.\frac{\partial \theta}{\partial r}\right|_{r=0} = 0$$

$$\left.\frac{\partial \xi}{\partial r}\right|_{r=1} = 0, \quad \left.\frac{\partial \theta}{\partial r}\right|_{r=1} = - Bi\ (\theta - \theta_w)$$

$$\left.\xi - a_1 \frac{\partial \xi}{\partial x}\right|_{x=0} = 0, \quad \left.\theta - b_1 \frac{\partial \theta}{\partial x}\right|_{x=0} = 1$$

$$\left.\frac{\partial \xi}{\partial x}\right|_{x=1} = \left.\frac{\partial \theta}{\partial x}\right|_{x=1} = 0$$

Kinetics laws :

The model can be used for a reaction with a kinetics law which is an arbitrary function of the temperature and concentrations of chemical species occurring in the reaction. Two main laws have been used for the oxydation of SO_2

1. A first law which has been determined on a laboratory reactor (Ref. 4)

$$f(\xi_s, \theta_s) = g(\xi_s, \theta_s) \times F_{eq} \times \phi$$

with :

$$g(\xi_s, \theta_s) = 72500\ \ell^{-6.18.10^3/T_0\theta_s} . P_{SO_2}^{0,75} . P_{O_2}^{1.1} . P_{SO_3}^{-0.67}$$

$$F_{eq} = 1 - \frac{P_{SO_3}/K_p}{P_{SO_2} \, P_{O_2}^{\;1/2}} \quad , \quad k_p = 10^{(5005/T_0 \theta_s - 4.743)} \quad , \quad \phi = 1$$

and where :

$$P_{SO_2} = \frac{a\,(1 - \xi_s)}{1 - a\,\xi_s/2} \quad , \quad P_{SO_3} = \frac{a\,\xi_s + \varepsilon}{1 - a\,\xi_s/2} \quad , \quad P_{O_2} = \frac{1 - a - a\,\xi_s/2}{1 - a\,\xi_s/2}$$

2. Mars and Maesson law

$$f = g\,(\xi_s \;\; \theta_s) \times F_{eq} \times \phi$$

with :

$$g = 6.56.10^{9} \; e^{-\frac{13853}{T_0 \theta s}} \times \frac{P_{O_2} \cdot P_{SO_2}}{P_{SO_3}} \times \left(1 + \sqrt{2.3\ 10^{-8} \; e\left(\frac{13689}{T_0 \theta_s}\right) \frac{P_{SO_2}}{P_{SO_3}}} \right)^{-2}$$

$$\phi = 72.10^{-4} \left(11.72 \times 2.808\ 10^{17} \; e^{-\frac{27542}{T_0 \theta_s}} \times T_0 \; \frac{1 + k_p}{Kp} \right)^{1/2} \times \eta$$

$$\eta = \left(\frac{3}{Th3\phi} - \frac{1}{\phi} \right) / \; 3\,\phi$$

II.3 Problems on study

The problems which have been studied or are still on study are :

- simulation of the model
- sensitivity studies
- simulation of different kinetics laws
- comparison with experimental results
- identification of the Peclet and Stanton numbers (mass and heat) a_i, b_i, and the Biot number (B_i)

The measurements available are :

- the temperatures θ_{ij} at a limited number of discrete points

$$\theta_{ij} = \theta\,(x_i, r_j)$$

- the mean concentration of SO_2 at the output x = 1

$$\xi_m = 2\int_0^1 r\,\xi\,(1, r)\,dr$$

II. 4 Discretization - Simulation

After finite difference discretization, we obtain :

$$\begin{cases} L_\xi\,\xi = -\xi_S + Z_\xi \\ L_\theta\,\theta = -\theta_S + Z_\theta \\ \xi = g_\xi\,(\xi_S, \theta_S) \qquad g_\xi = \xi_S + k_1\,f\,(\xi_S, \theta_S) \\ \theta = g_\theta\,\;\xi_S, \theta_S) \qquad g_\theta = \text{-----------} \end{cases}$$

where L_ξ , L_θ are block tridiagonal matrix and Z represent the boundary conditions

The system can be written :

$A\,\underline{V} = \underline{B}\,(\underline{V}) + \underline{Z}$ with

$$A = \begin{bmatrix} L_\xi & O & O & O \\ I & -I & 0 & 0 \\ O & O & L_\theta & O \\ O & O & I & -I \end{bmatrix} \qquad B = \begin{bmatrix} O & -I & O & O \\ O & NL(\xi_S) & 0 & 0 \\ O & O & O & -I \\ O & O & O & NL(\theta_S) \end{bmatrix} \qquad \underline{V} = \begin{bmatrix} \underline{\xi} \\ \underline{\xi}_S \\ \underline{\theta} \\ \underline{\theta}_S \end{bmatrix}$$

where NL (ξ_S) and NL (θ_S) represent the non-linear kinetics law.

The method of resolution associates two imbricate iterative schemes, one for the linear equation (Gauss - Seidel - relaxation) one explicit type scheme for the non linearities but imbricate with the first scheme and with relaxation. We have, for example, for ξ

$$\underline{\xi}\,\Big|^{k} = L_{\xi}^{-1}\left(\underline{Z}_{\xi} - \xi_{s}\,\Big|^{k-1}\right)$$

$$\underline{\xi}\,\Big|_{i}^{k,l} = \underline{\xi}\,\Big|_{i}^{k,l-1} + \gamma\, D_{i}^{-1}\left[-E_{i}\underline{\xi}\,\Big|_{i-1}^{k,\,l-1} - F_{i}\underline{\xi}\,\Big|_{i+1}^{k,l-1} - D_{i}\underline{\xi}\,\Big|_{i}^{k,l-1}\right.$$

$$\left. + \underline{Z}_{\xi}\Big|_{i} - \xi_{s}\Big|_{i}^{k,\,l-1}\right]$$

and $\underline{\xi}_{s}$ by :

$$\begin{cases} \underline{\xi}_{s}\,\Big|^{k} = \underline{\xi}\,\Big|^{k-1} - k_{1}\underline{f}\left(\underline{\xi}_{s}\,,\,\theta_{s}\,\Big|^{k-1}\right) \\ \underline{\xi}_{s}\,\Big|^{k+1,l=1} = \underline{\xi}_{s}\Big|_{i}^{k,l=1} + \omega_{2}\left(\widetilde{\underline{\xi}_{s}}\,\Big|_{i}^{k;l=l_{max}} - \underline{\xi}_{s}\,\Big|_{i}^{k,l=1}\right) \end{cases}$$

II.5 Results and discussion

A typical result is presented in Fig. 5 (fluid temperature and SO_2 concentration). The temperature evolution in the fluid and at the catalyst surface characterizes a quasi-adiabatic reactor: fast increase of the temperature on the x-axis, up to a top corresponding to the end of the reaction, much slower increase near the reactor wall due to thermal losses. As for SO_2 concentration, it decreases in the first part of the reactor due to the temperature increase; but rapidly it approaches its equilibrium value and the reaction speed is negligible; the reaction then takes place in a domain near the wall where the equilibrium concentration is slower due to a much lower temperature. There is then a radial displacement of the maximum reaction speed in spite of the homogeneization effect of diffusion terms.

Comparison of the two first kinetics laws have been made for several experimental results. Figure 6 shows an example for the axial temperature (at r = 0). The second law is much better, the first law giving a too high reaction speed at the input. Other experiences are on study.

An identification algorithm has been emplemented; it is based on the minimization by a Gauss-Newton algorithm of the following index :

$$J(\underline{p}) = \frac{1}{2}\left(\xi_m(\underline{p}) - \xi_{m\ meas.}\right)^2 + \frac{\lambda}{2}\sum_{i=1}^{I}\sum_{j=1}^{J}\left(\theta(x_i,r_j) - \theta_{ij meas}\right)^2$$

where $\underline{p}$ represents the vector of parameters a_i, b_i, and B_i. This identification procedure is now under study but first variations of parameter and a sensitivity study has been made together with the kinetics study in order to approach the experimental results on the pilot reactor.

III. IDENTIFICATION IN MICROWAVE GUIDES (MICROSTRIP LINES)

III. 1 - INTRODUCTION

This research is motivated by the increasing use of microstrip lines at higher microwave frequencies. The classical analysis is based on the T E M approximation from which is derived the characteristic impedance and the propagation wave number . However, this analyzis is inadequate at higher frequencies where one has to consider hybrid mode configurations.

In this paper we present briefly the problem of identifying the propagation wave number in microstrip lines. A direct method (but iterative in resolution) by a non linear eigen value problem and an approach by identification from measurements of the electric field (or magnetic field when the measurements are available). A third method, Mittra's method, has also been implemented and has given excellent results.

III.2 - Propagation equations

The electromagnetic wave is defined by its components $\vec{E}_T$, $\vec{H}_T$ in Oxy and $\vec{E}_Z u$, $\vec{H}_Z u$ on Z axis. Maxwell equations give (in homogeneous field) :

$$\begin{cases} \vec{E}_T = \dfrac{1}{k^2} \left[\omega\mu \, (\vec{u} \wedge \vec{\text{grad}}\, H_Z) - \beta \;\; \vec{\text{grad}}\, E_Z \right] \\ \vec{H}_T = \dfrac{1}{k^2} \left[\omega\varepsilon \, (\vec{u} \wedge \vec{\text{grad}}\, E_Z) + \beta \;\; \vec{\text{grad}}\, H_Z \right] \end{cases}$$

$$\begin{cases} \Delta E_Z + k^2 E_Z = 0 \\ \Delta H_Z + k^2 H_Z = 0 \end{cases}$$

where $k^2 = \varepsilon\mu\omega^2 - \beta^2$

In the non-homogeneous case of microstrip lines k^2 depends (through ε) on the regions. A case with two regions is represented below.

The measurements will be :

E_y on OA

E on AB

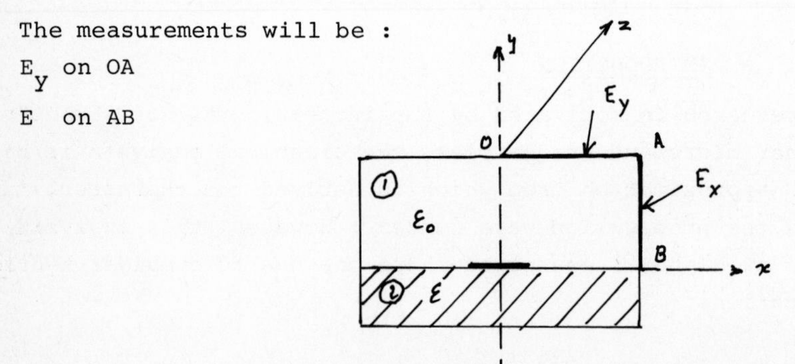

III.3 - Identification of β by an eigenvalue problem

In the general case of hybrid modes, the Helmotz equations

$$\Delta E_Z + k^2 E_Z = 0$$

$$\Delta H_Z + k^2 H_Z = 0$$

where k^2 is unknown have to be solved. Then k^2 gives β and E_Z, H_Z gives $\vec{E}_T$ which is compared to measurements.

A first difficulty is due to the fact that k^2 is not the same in the air (k^2_A) and the dielectric (k^2_D)

Writing the boundary conditions :

$$E_Z = \frac{\partial H_Z}{\partial n} = 0 \text{ on the conducting surfaces}$$

$H_Z = 0$, E_Z symetric on symetry y-axis

and the continuity relations at the separation surface (1)/(2)

$$(E_X)_{air} = (E_X)_{diel.} \; ; \; (\varepsilon_0 E_y)_{air} = (\varepsilon E_y)_{diel.}$$

$$H_{X\,air} = H_{X\,diel}$$

$$H_{Y\,air} = H_{Y\,diel}$$

We obtain the following eigenvalue problem, in terms of $\phi = H_Z$,

$$\eta = \frac{\omega \varepsilon_0}{\beta} E_Z$$

$$\begin{vmatrix} A & B(\tau) \\ C(\tau) & D \end{vmatrix} \begin{vmatrix} \phi \\ \eta \end{vmatrix} = k_A^2 \begin{vmatrix} \phi \\ \eta \end{vmatrix}$$

where $\tau = \dfrac{k^2_A}{k^2_D}$

This is a non linear eigenvalue problem. B and C depend on τ ,
The following procedure has been used :

1) τ is chosen
2) the linear eigenvalue problem is solved. The minimum modulus eigenvalue k_A^2 and ϕ , η are obtained.
3) ω and β are obtained by

$$\omega^2 = \frac{k^2_A}{\varepsilon_0 \mu} \quad \frac{1/\tau - 1}{\varepsilon_r - 1}, \beta^2 = - k_A^2 \frac{\varepsilon_r - 1/\tau}{\varepsilon_r - 1}$$

This gives ω, k_A^2, k_D^2 and β as functions of τ

The experimental conditions fixing ω, the procedure is repeated for several τ

The procedure has been implemented and has given some results but does not seem to be a valuable general approach due to difficulties coming from the apparition of spurious modes, the spectral radii constraints and the high dimensionality of the problem.

III.4 Identification from measurements

The identification procedure is as follows :

1. initialization of β
2. calculate E_Z and H_Z
3. from E_Z, H_Z, calculate $\vec{E}_T$ then E_y on OA and E_x on AB
4. compare to the experimental values. Define an error performance index.
5. minimization of the index with respect to β

A modal representation has been used here for the model

$$E_Z = \sum_1^N A_n f_n (x) g_n (y)$$

$$H_Z = \sum_1^N B_n p_n (x) q_n (y)$$

with the boundary conditions, we have :

$f_n = \cos \hat{k}_n x, \; g_n = \mathrm{Sh}\alpha_n y$

$p_n = \sin \hat{k}_n x, \; q_n = \mathrm{Ch}\alpha_n y$

where :

$$\hat{k}_n = (n - 1/2) \frac{\pi}{L} , \; \alpha_n = (\hat{k}_n^2 - k^2)^{1/2}$$

then :

$$E_x(y) = \sum_1^N X_n \mathrm{Sh}\, \alpha_n y$$

$$E_y(x) = \sum_1^N Y_n \cos \hat{k}_n x$$

where X_n and Y_n are known expressions of β

The identification of β is made from the measurements on AB :

$$\underline{E}_x = \begin{vmatrix} \mathrm{Sh}\alpha_1 y_1 & \cdots & \cdots & \mathrm{Sh}\,\alpha_N y_1 \\ & & & \\ \mathrm{Sh}\alpha_1 y_s & \cdots & \cdots & \mathrm{Sh}\,\alpha_N y_s \end{vmatrix} \underline{X} = M \underline{X}$$

where S is the number of measurements

Then a least square index gives :

$$\underline{X} = (M^T M)^{-1} M^T \underline{E}_x$$

and β is obtained by the minimization of the norm of the residue :

$$||R|| = || \underline{E}_x - \underline{M}_x ||$$

Then the coefficients A_n and B_n of the model are obtained from the measurements $\underline{E}_x$ and $\underline{E}_y$. The results are satisfactory. A special feature is that near T E M frequencies the minimum is flat (independency from β) while at higher frequencies we obtain a peak.

REFERENCES :

1. M. MASUBUCHI - "Dynamic response of cross flow heat exchangers" IVth IFAC Congress - Warsawa, June 1969
2. KAYS - LONDON : "Compact heat exchangers" - Mac Graw Hill Book Company - N.Y. 1964
3. J.P. CHRETIEN - LE LETTY - A. LE POURHIET - Rapport de Synthèse : "Commande et optimisation des systèmes à paramètres répartis" Convention D.R.M.E. - Mai 1974
4. HERNANDEZ LUNA : Thèse de doctorat d'Université - Toulouse, 1969
5. L. LE LETTY - A. LE POURHIET - J.B. GROS - M. ENJALBERT : "Modèle bidimensionnel d'un réacteur catalytique à lit "fixe" - 1er Colloque Franco-Soviétique "Simulation et Modélisation de processus et de réacteurs catalytiques - NANCY - Mai 1973
6. V.A. KOUZIN : 1er Colloque Franco-Soviétique "Simulation et Modélisation des processus et réacteurs catalytiques", Nancy - Mai 1973
7. R.S. VARGA "Matrix Iterative Analyzis" - Prentice Hall 1962
8. R. MITTRA and T. ITOH "A New Technique for the Analyzis of Dispersion Characteristics of Microstrip lines" - IEEE Trans Microwave Theory Techn - Vol MTT 19 - Jan. 1971
9. P. DALY " Hybrid mode analyzis of microstrip by finite element methods" - IEEE Trans Microwave Theory Tech Vol MTT-19, Jan.1971
10. Ph. CAMBON - L. LE LETTY : "Applications of decomposition and multi level techniques to the optimization of distributed parameter systems" - Symposium I.F.I.P., Rome, Mai 1973.

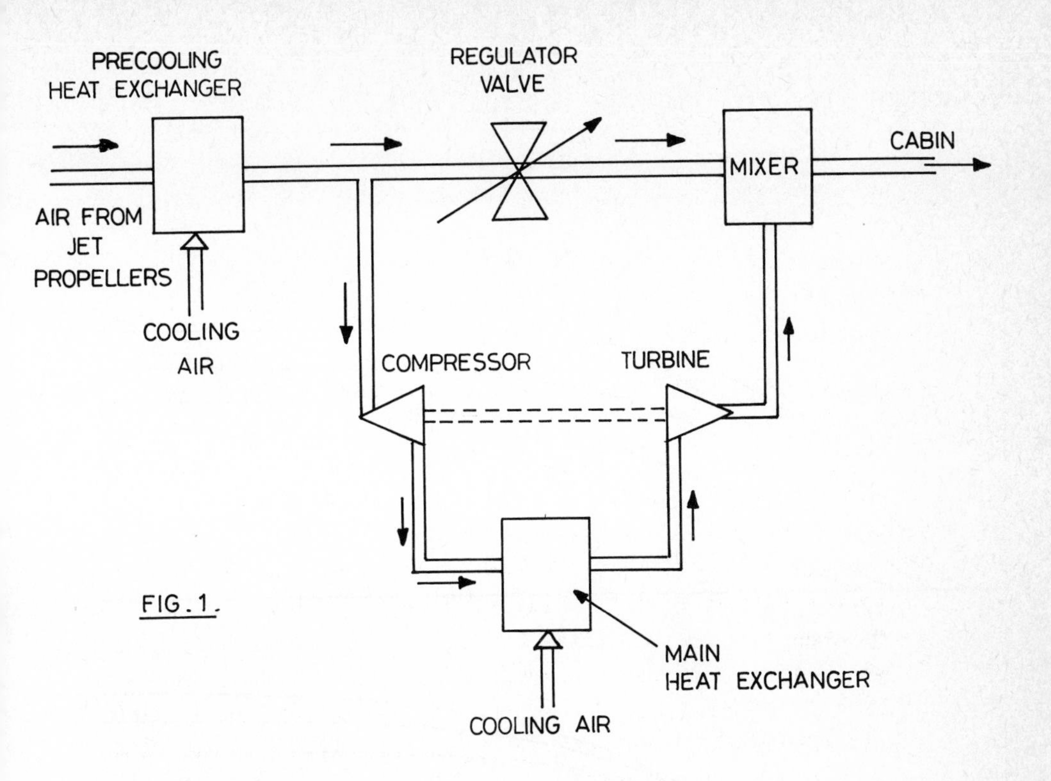
PRECOOLING
HEAT EXCHANGER
REGULATOR
VALVE
CABIN
MIXER
AIR FROM
JET
PROPELLERS
COOLING
AIR
COMPRESSOR
TURBINE
MAIN
HEAT EXCHANGER
COOLING AIR

FIG. 1.

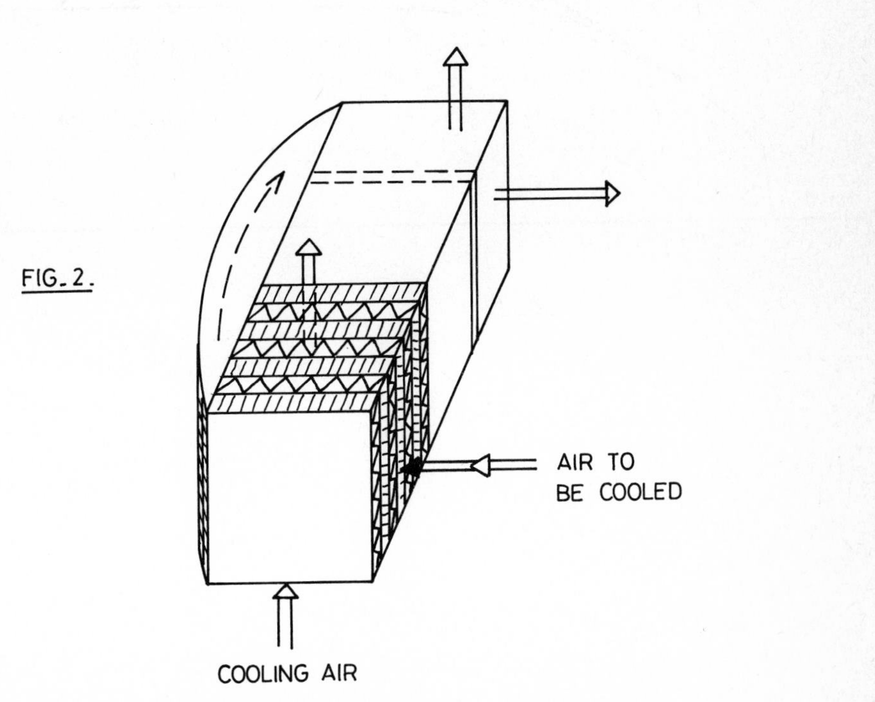
AIR TO
BE COOLED
COOLING AIR

FIG. 2.

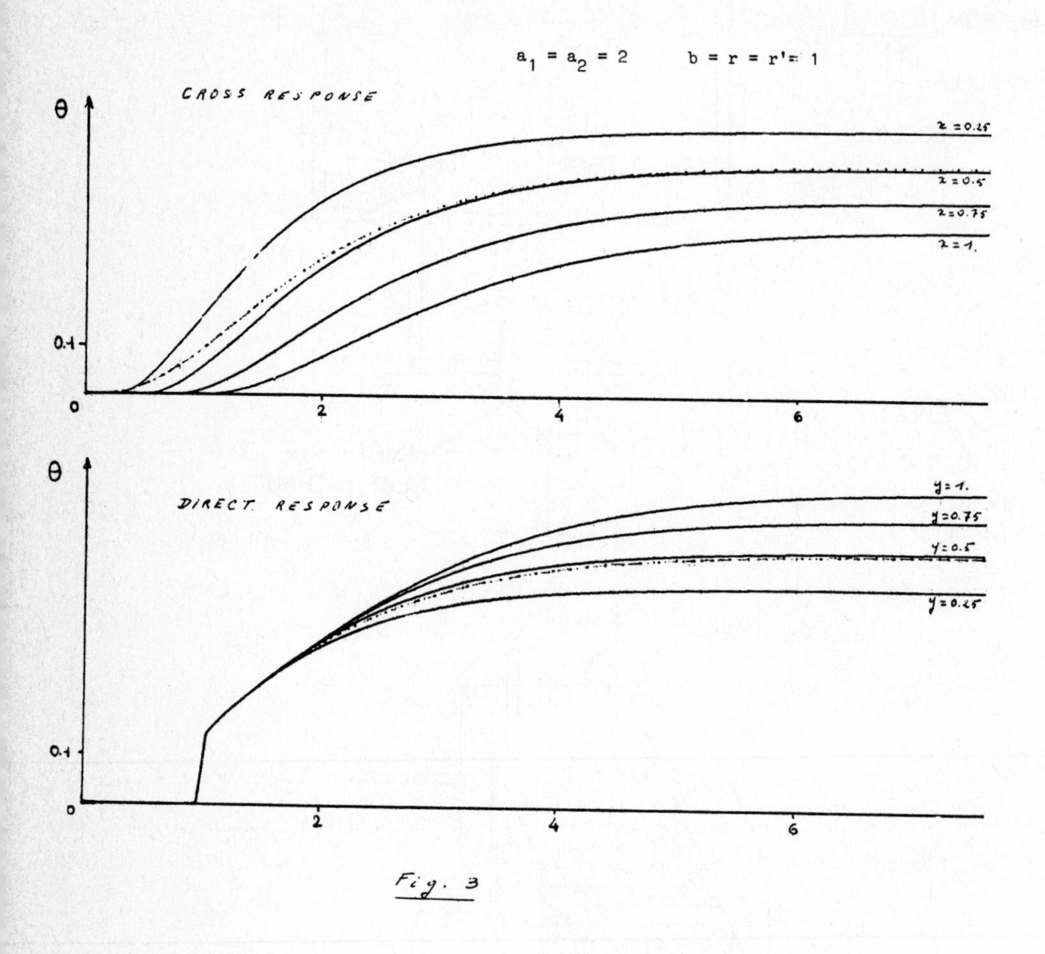
$a_1 = a_2 = 2$ $b = r = r' = 1$
CROSS RESPONSE
θ
0.1
0
2
4
6
x=0.25
x=0.5
x=0.75
x=1.
θ
DIRECT RESPONSE
y=1.
y=0.75
y=0.5
y=0.25
0.1
0
2
4
6

Fig. 3

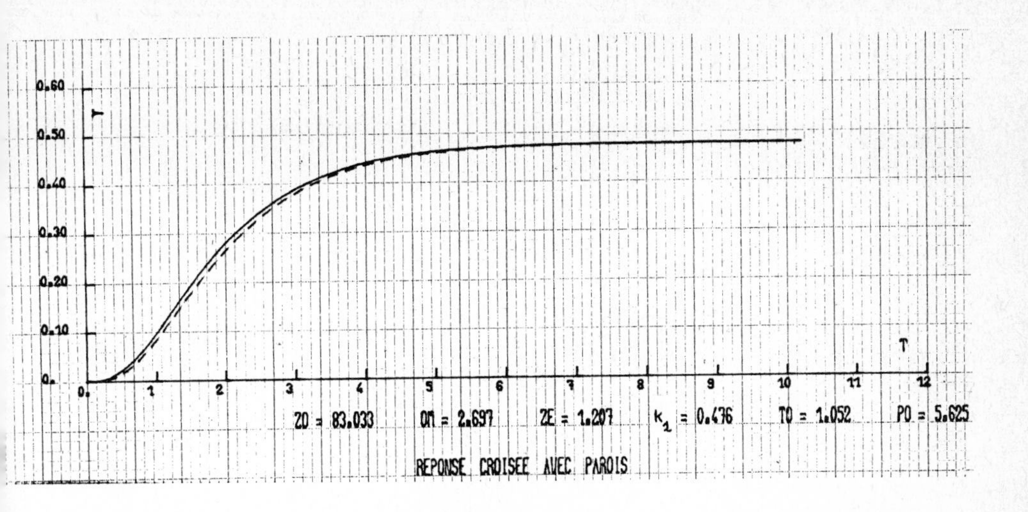

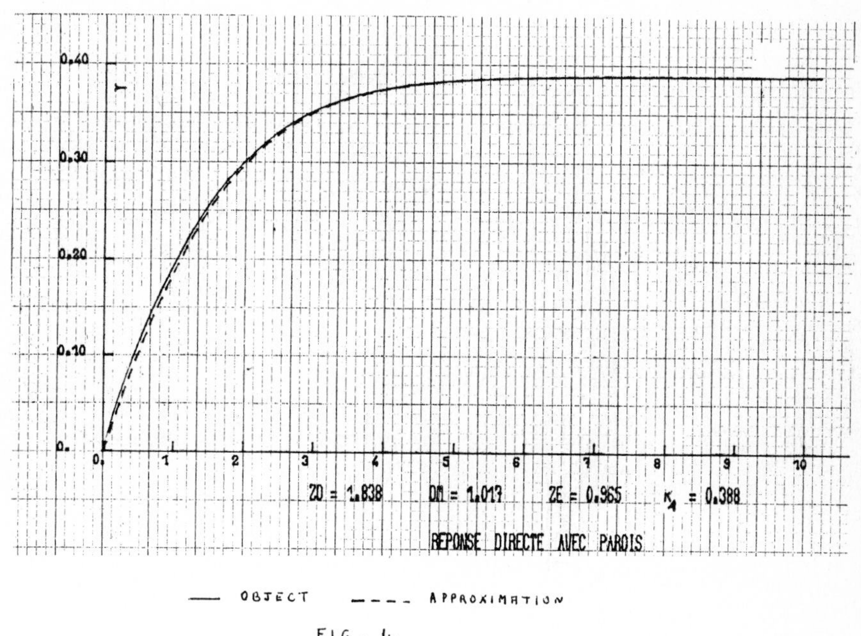

—— OBJECT ---- APPROXIMATION

FIG. 4

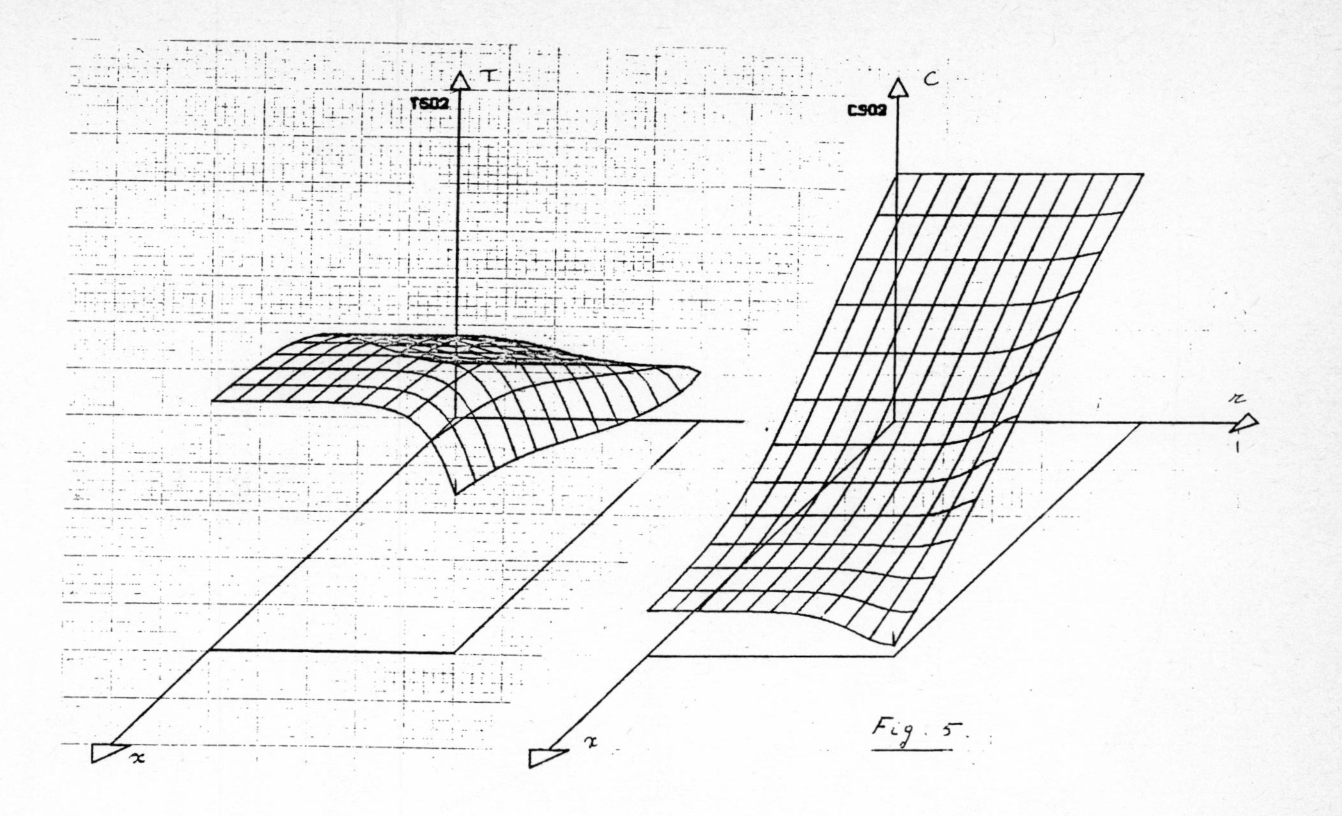

Fig. 5.

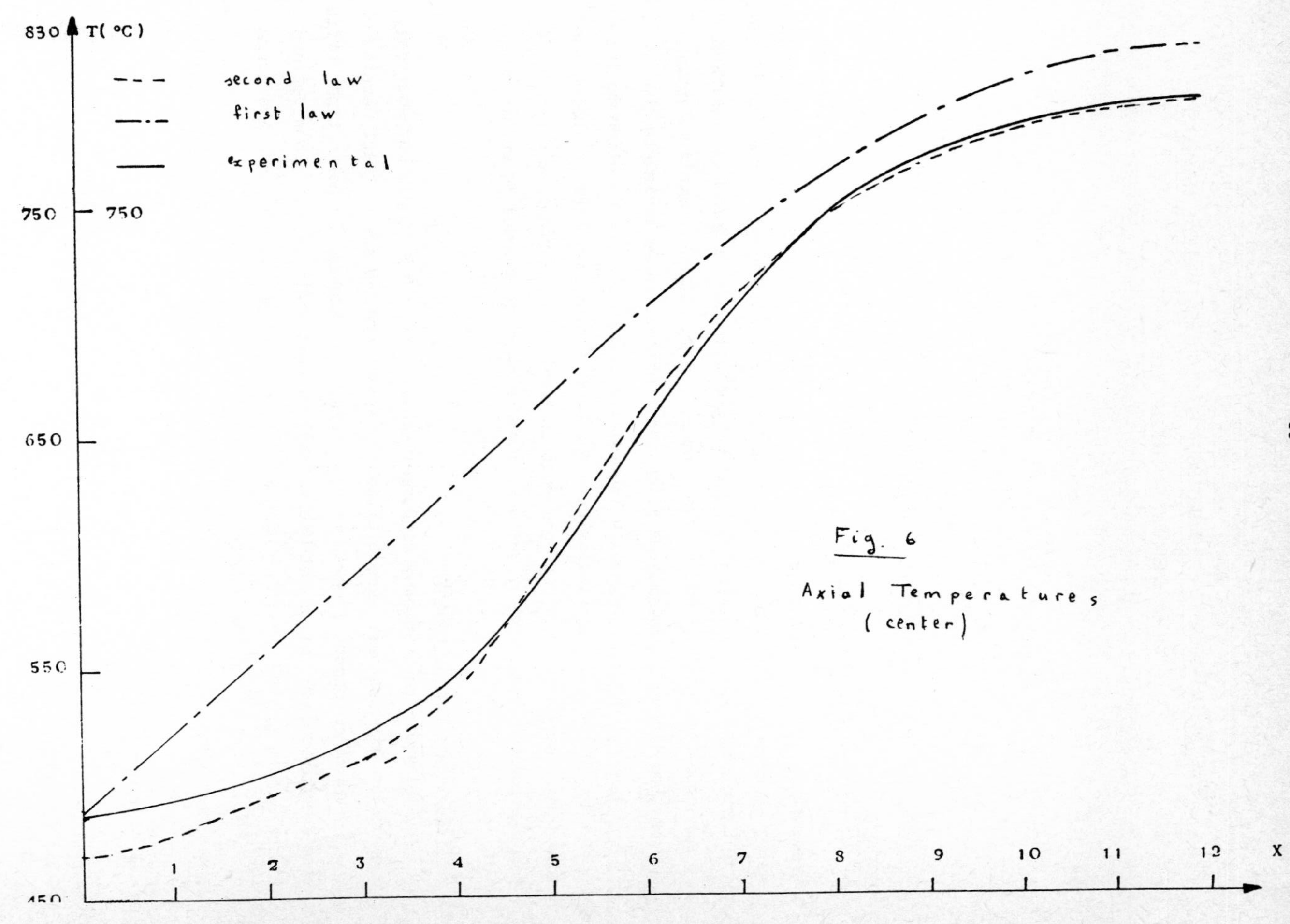

Fig. 6
Axial Temperatures
(center)

RECURSIVE SOLUTIONS TO INDIRECT SENSING MEASUREMENT PROBLEMS BY A GENERALIZED INNOVATIONS APPROACH

Edoardo Mosca
Facoltà di Ingegneria
Università di Firenze
50139 Firenze, Italy

ABSTRACT

For a wide class of applications referred to as indirect-sensing experiments, a systematic approach yielding solutions in recursive form is established. Indirect-sensing experiments include problems of estimation, filtering, system identification, and interpolation and smoothing by splines. Our approach is based on the novel notion of a discrete-time generalized (not necessarily stochastic) innovations process. The discrete-time linear least-squares filtering problem is used to relate the new concept to the familiar one of a stochastic innovations process. An application to the problem of identifying recursively impulse responses and system parameters by using pseudo-random binary sequences as probing inputs is considered. Further, the problem of interpolation and smoothing by splines is approached by the method developed.

1 - FORMULATION OF THE PROBLEM

In order to cast many different applications in a single mathematical framework and stress their essential features, we consider an abstract version of a problem that often occurs in experimental work, for istance, in estimation, filtering, system identification, etc.. Let H be a real Hilbert space of functions defined on a set Ω of points ω. The inner product of H is denoted by $\langle\cdot,\cdot\rangle$, and the corresponding norm by $\|\cdot\|$. Let H^P be the P-fold Cartesian product of H and $R^{P\times M}$ the space of all real-valued P×M matrices. We define an indirect-sensing linear measurement, or simply a measurement, on an element $w \in H^P$ to be the values $m \in R^{P\times M}$ taken on by an ordered set of M continuous linear functionals

$$m = \{\langle w , \rho\rangle\} \triangleq \begin{bmatrix} \langle w^1, \rho^1\rangle \cdots \langle w^1, \rho^M\rangle \\ \vdots \qquad\qquad \vdots \\ \langle w^P, \rho^1\rangle \cdots \langle w^P, \rho^M\rangle \end{bmatrix} \tag{1}$$

$$\rho \triangleq [\rho^1, \cdots, \rho^M]' \qquad w \triangleq [w^1, \cdots, w^P]'$$

where, by the Riesz representation theorem [1], $\rho \in H^M$ will be called the measurement representator. Notice that in (1) M stands for the number of distinct measurements executed on each of the P components of w.

It is assumed that a sequence of time-indexed measurements

$$m_t = \{ \langle w, \rho_t \rangle \} \quad , \qquad t \in I \triangleq \{ 1, 2, \cdots \} \tag{2}$$

with

$$\{ \rho_t^m \; , \; t \in I \; , \; m = 1, 2, \cdots, M \} \text{ linearly independent} \tag{3}$$

is available.

The set $\mathcal{E}_t$ made up of the first t representators and corresponding measurements defined by

$$\mathcal{E}_t \triangleq \{ \rho_\tau , m_\tau \; , \; \tau = 1, 2, \cdots, t \}$$

will be referred to as the experiment up to time t. Further,

$$\mathcal{E} \triangleq \{ \rho_t , m_t \; , \; t \in I \}$$

will simply be called the experiment. The problem is then to find a recursive formula for

$$\hat{w}_{|t} \triangleq [\hat{w}^1_{|t} , \cdots , \hat{w}^P_{|t}]'$$

where, for each p=1,2,...,P,

$\hat{w}^p_{|t} \triangleq$ the minimum norm element in H interpolating $\mathcal{E}_t$, or, in other words, the linear least-squares (l.l.s.) reconstruction of w^p based on the experiment up to time t.

Example 1 (l.l.s. estimation) - Let $H \triangleq L_2 (\Omega, \mathcal{A}, \mathbb{P})$, the Hilbert space of all second-order random variables (r.v.), viz. r.v'.s with finite second moments. Here the inner product of $u, v \in H$ is

$$\langle u, v \rangle = E [u v] \triangleq \int_\Omega u(\omega) \, v(\omega) \, \mathbb{P}(d\omega)$$

The experiment consists of acquiring the values of the covariance

$$m_t = \{ \langle w, \rho_t \rangle \} = E [w \rho_t']$$

and observing the realization of a second-order M-dimensional time-series ρ_t. For the sake of simplicity, the time series ρ_t and the P-dimensional r.v. w are assumed to have zero means. The problem is thus to obtain a recursive formula for $\hat{w}_{|t}$, the l.l.s. estimate of w based on the observations up to time t.

Example 2 (determination of system impulse-responses) - Consider a causal linear time-invariant system with Q inputs and P outputs. Let $\{ h_{pq}(\omega) \}$, $\omega \in [0, \infty)$, be its impulse-response matrix. Suppose that the given system is b.i.b.o. stable, then, for a sufficiently large $\omega_1 > 0$, $h_{pq}(\omega) = 0$, $\forall \omega > \omega_1$. Thus, if u_q denotes the system q-th input and m_t^p the system p-th output at time t,

$$m_t^p = \sum_{q=1}^{Q} \int_0^{\omega_1} h_{pq}(\omega) \, u_q(t-\omega) \, d\omega \, . \tag{4}$$

Setting

$$H \triangleq L_2(\Omega) \oplus L_2(\Omega) \oplus \dots \oplus L_2(\Omega) \quad \text{(Q times)},$$

the Hilbert space of all functions $v: \Omega \rightarrow R^Q$

$$v(\omega) \triangleq [v_1(\omega), \cdots, v_Q(\omega)]$$

such that

$$\|v\|^2 = \langle v, v \rangle = \sum_{q=1}^{Q} \int_{\Omega} [v_q(\omega)]^2 d\omega < \infty \tag{5}$$

we can write (4) as

$$m_t = \{\langle w, \rho_t \rangle\} \tag{6}$$

with

$$m_t \triangleq [m_t^1, \cdots, m_t^P]' \in R^P$$

$$w \triangleq [w^1, \cdots, w^P]' \in H^P$$

$$w^p(\omega) \triangleq [h_{p1}(\omega), \cdots, h_{pQ}(\omega)]', \qquad p = 1, \cdots, P \tag{7}$$

$$\rho_t(\omega) \triangleq [u_1(t-\omega), \cdots, u_Q(t-\omega)] \tag{8}$$

with t fixed in I and $\omega \in [0, \omega_1]$.

Here the experiment consists of sending into the system the "inputs" or representators $\{\rho_t\}$ and recording the values of the corresponding outputs $\{m_t\}$. The problem is thus to obtain a recursive formula for $\hat{w}_{|t}$, the l.l.s. reconstruction of the system impulse-response matrix from input-output data up to time t.

Let $\mathcal{R}_t$ be the linear manifold in H spanned by the measurement representators up to t

$$\mathcal{R}_t \triangleq \mathrm{Span}\{\rho_\tau, \forall \tau \leq t\} \triangleq \mathrm{Span}\{\rho_\tau^m, m = 1, \cdots, M, \forall \tau \leq t\}$$

It is well-known that $\hat{w}^P_{|t}$ coincides with the orthogonal projection of the unknown $w^P \in H$ onto $\mathcal{R}_t$

$$\hat{w}^p_{|t} = \Pi[w^p | \mathcal{R}_t]$$

Further, $\hat{w}_{|t}$ is uniquely specified by the two requirements:

$$\hat{w}^p_{|t} \in \mathcal{R}_t \qquad p = 1, \cdots, P \tag{9a}$$

$$\{\langle \tilde{w}_{|t}, \rho_t \rangle\} = 0, \quad \forall \tau \leq t \tag{9b}$$

where

$$\tilde{w}_{|t} \triangleq w - \hat{w}_{|t} \tag{10}$$

is the error of the l.l.s. reconstruction of w based on $\mathcal{E}_t$.

Requirements (9), together with the information supplied by the experiment $\mathcal{E}_t$, enable one to write down the so-called normal equations [2]. In general, this set of equations yields the desired $\hat{w}_{|t}$ in a nonrecursive form in that, if $\hat{w}_{|t+1}$ is needed, an augmented system of normal equations has to be solved by performing the same number of computations as if $\hat{w}_{|t}$ were unknown.

2 - INNOVATIONS AS GRAM-SCHMIDT PROCESSES

As a preliminary step to the development of a systematic approach to the problem that has been posed, viz. recursive linear least-squares solution to the indirect-sensing problem, it is convenient to introduce the notion of causally equivalent experiments. We say that two experiments $\{\rho_t, m_t\}$ and $\{r_t, \mu_t\}$ are causally equivalent if

$$\forall t \in I, \ \text{Span}\{\rho_\tau, \forall \tau \leq t\} = \text{Span}\{r_\tau, \forall \tau \leq t\}.$$

This is equivalent to requiring, perhaps in more suggestive terms, the existence of a causal and causally invertible linear transformation $\mathcal{L}: H^{P \times I} \longrightarrow H^{P \times I}$ that converts the representators of the first into the representators of the second experiment in a causal way,

$$\mathcal{L}[\{\rho_\tau, \tau \leq t\}] = \{r_\tau, \tau \leq t\}, \quad \forall t \in I.$$

An obvious consequence of the given definitions is

Proposition 1 - Let $\hat{w}_{|t}(\mathcal{E}_i)$ be the l.l.s. reconstruction of $w \in H^P$ based on an experiment $\mathcal{E}_i$, i = 1,2. Thus,

$$\left.\begin{array}{l} \hat{w}_{|t}(\mathcal{E}_1) = \hat{w}_{|t}(\mathcal{E}_2) \\ \forall t \in I, \ \forall w \in H^P \end{array}\right\} \Longleftrightarrow \left\{\begin{array}{l} \mathcal{E}_1 \ \& \ \mathcal{E}_2 \text{ are} \\ \text{causally equivalent.} \end{array}\right.$$

Let us now construct from the representators $\{\rho_t, t \in I\}$ of the given experiment (2) an orthonormal sequence $\{\nu_t, t \in I\}$ of the elements in H^M by the Gram-Schmidt procedure [1,2]. By orthonormality here we mean that

$$\{<\nu_t, \nu_\tau>\} \triangleq \begin{bmatrix} <\nu_t^1, \nu_\tau^1> & \cdots & <\nu_t^1, \nu_\tau^M> \\ \vdots & & \\ <\nu_t^M, \nu_\tau^1> & \cdots & <\nu_t^M, \nu_\tau^M> \end{bmatrix}$$

$$= I_M \, \delta_{t,\tau}, \qquad \forall t, \tau \in I.$$

We get

$$e_t \triangleq \rho_t - \sum_{\tau=1}^{t-1} \{<\rho_t, \nu_\tau>\} \nu_\tau, \quad t = 2, 3, \cdots, \tag{11a}$$

$$e_1 \triangleq \rho_1 \tag{11b}$$

$$\nu_t \triangleq G_t^{-1/2} e_t, \quad \forall t \in I. \tag{11c}$$

where $G_t^{-\frac{1}{2}}$ is the inverse of the positive square-root of the matrix

$$G_t \triangleq \{<e_t, e_t>\} .$$

The sequence $\{e_t, t \in I\}$ will be called the sequence of the <u>innovations</u> of the representators $\{\rho_t, t \in I\}$, and $\{\nu_t, t \in I\}$ that of the <u>normalized innovations</u>.
By the way the Gram-Schmidt procedure works, the initial experiment turns out to be causally equivalent to the corresponding <u>innovations experiment</u>

$$\mathcal{J} \triangleq \{\nu_t, \mu_t, t \in I\}$$

where the ν_t's are defined by (11), and

$$\mu_t \triangleq \{<w, \nu_t>\}$$
$$= [m_t - \sum_{\tau=1}^{t-1} \mu_\tau \{<\nu_\tau, \rho_t>\}] G_t^{-1/2} , \quad t = 2,3,\cdots, \tag{12a}$$
$$\mu_1 \triangleq m_1 G_1^{-1/2} . \tag{12b}$$

By transforming the initial experiment $\mathcal{E}$ into the corresponding innovations experiment $\mathcal{J}$ we find immediately the desired $\hat{w}_{|t}$ in a recursive form

$$\hat{w}_{|t} = \sum_{\tau=1}^{t} \mu_\tau \nu_\tau = \hat{w}_{|t-1} + \mu_t \nu_t$$
$$= \hat{w}_{|t-1} + \mu_t G_t^{-1/2} e_t , \quad t \in I \tag{13a}$$
$$\hat{w}_{|0} = 0 \tag{13b}$$

<u>Theorem 1</u> - Let $\mathcal{E} = \{\rho_t, m_t, t \in I\}$ be an indirect-sensing experiment, and $\mathcal{J} \triangleq \{\nu_t, \mu_t, t \in I\}$, with ν_t and μ_t respectively defined by (11) and (12), be the corresponding innovations experiment. Then, $\mathcal{E}$ and $\mathcal{J}$ are causally equivalent, and a recursive formula for the l.l.s. reconstruction of $w \in H^P$ based on $\mathcal{E}_t$ is given by (12) and (13).

Let us apply (13) to get

$\hat{\rho}_{t|t-1}$ = the l.l.s. reconstruction of the representator at time t based on the experiment defined by

$$\{ m_\tau = \{<\rho_t, \rho_\tau>\} , \quad \tau = 1,\cdots,t-1 \} \tag{14}$$

We get

$$\hat{\rho}_{t|t-1} = \sum_{\tau=1}^{t-1} \{<\rho_t, \nu_\tau>\} \nu_\tau .$$

Comparing this with (11a), we arrive at justifing the term "innovations".

<u>Corollary 1</u> - The sequence of the innovations of the representators of an experiment can be written in the form

$$
\begin{aligned}
e_t &= \varrho_t - \hat{\varrho}_{t|t-1} \qquad t = 2,3,\cdots, \\
e_1 &= \varrho_1
\end{aligned}
\tag{15}
$$

Every term e_t of the innovations sequence is therefore obtained by substracting from the representator ϱ_t its l.l.s. one-step prediction, i.e. its l.l.s. reconstruction based on the experiment (14) up to the immediate past.

Example 3 (Kalman-Bucy formulas) - Let the random vector w of Example 1 be a t-dependent random vector x_t. Eqs. (13) give at once

$$
\begin{aligned}
\hat{x}_{t+1|t} &= \hat{x}_{t+1|t-1} + E[x_{t+1} \nu_t'] \\
&= \hat{x}_{t+1|t-1} + E[x_{t+1} e_t'] G_t^{-1} e_t
\end{aligned}
\tag{16}
$$

Further, if x_t is the solution of the stochastic difference state-equation

$$
\begin{aligned}
x_{t+1} &= \phi_t x_t + \xi_t \\
E[x_1] &= 0 \qquad E[x_1 x_1'] = \Pi
\end{aligned}
\tag{17}
$$

and the observations ϱ_t are given by

$$\varrho_t \triangleq z_t = C_t x_t + \varsigma_t$$

with ξ_t and ς_t zero mean vectors for every $t \in I$ uncorrelated with x_1 and

$$E[\xi_t \xi_\tau'] = Q_t \delta_{t\tau} \qquad E[\varsigma_t \varsigma_\tau'] = R_t \delta_{t\tau} \qquad E[\xi_t \varsigma_\tau'] = \Gamma_t \delta_{t\tau}$$

the discrete-time Kalman-Bucy formulas are quickly obtained

$$\hat{x}_{t+1|t} = \phi_t \hat{x}_{t|t-1} + K_t e_t \tag{18}$$

$$
\begin{aligned}
K_t &\triangleq (\phi_t P_t C_t' + \Gamma_t)(C_t P_t C_t' + R_t)^{-1} \\
P_{t+1} &= \phi_t P_t \phi_t' - K_t G_t K_t' + Q_t \\
\hat{x}_{1|0} &= 0 \qquad P_1 = \Pi .
\end{aligned}
\tag{19}
$$

Example 4 (recursive system identification by PRBS's) - Hereafter, the problem of determining impulse responses and system parameters is considered. To this end the setting of Example 2 will be used throughout. Our first comment is that, though solution (13) is completely general and hence can immediately be applied to the problem posed in Example 2, the proposed algorithm becomes very complicated for large t unless some special input is used. This is so because: first, the number of computations required by (11) to get e_t increases linearly with t; and second, an ever expanding Span $\{ \varrho_\tau , \forall \tau \le t\}$ makes eventually the reconstructed impulse response extremely sensitive to measurement noise [3,5]. On the other hand, the given solution becomes particularly convenient if the system output is uniformly sampled every Δ sec. and a periodic input with period $L\Delta > \omega_1$ is used. In this way, if the measurements start at least $L\Delta$ sec. after the test input has been applied to the system, there are only L lineraly independent representators to consider, and ideally the experiment is completed in the next $L\Delta$ sec.

In the single-input single-output case, attractive input signals are the pseudorandom binary sequences (PRBS) [6] of length

$$L = 2^i - 1, \qquad i = 2,3,\ldots$$

and amplitude +V and -V. They look attractive essentially because of the following property of their autocorrelation function

$$\langle \varrho_t , \varrho_\tau \rangle = \begin{cases} \|\varrho\|^2 & t = \tau + mL\Delta \\ -\|\varrho\|^2/L & \text{elsewhere} \end{cases}$$

where, for a system with an input excited by a PRBS of period $L\Delta$, $\|\varrho\|^2 = V^2 L\Delta$. This feature greatly semplifies Eqs. (11) - (13). In fact, after some further manipulations, we get the recursive l.l.s. reconstruction of the system impulse response according to the following steps:

$$\begin{aligned} e_t(\omega) &= \varrho_t(\omega) - \varrho_{t-1}(\omega) + \alpha_t e_{t-1}(\omega) \\ \epsilon_t &= m_t - m_{t-1} + \alpha_t \epsilon_{t-1} \\ \hat{w}_{|t}(\omega) &= \hat{w}_{|t-1}(\omega) + L(L+1)^{-1} \|\varrho\|^{-2} \alpha_{t+1} \epsilon_t e_t(\omega) \end{aligned} \qquad (20)$$

where: $\alpha_t \triangleq (L-t+3)(L-t+2)^{-1}$; $t = 1,2,\ldots,L$; and the initial values are

$$\varrho_o(\omega) = 0 \qquad e_o(\omega) = 0 \qquad \hat{w}_{|0}(\omega) = 0$$

$$\epsilon_o = 0 \qquad m_o = 0 \; .$$

PRBS's have been used for a long time as probing inputs for identifying systems [7,8]. However, all previous algorithms used in connection with the identification experiment of this section essentially relied on the PRBS resemblance to white noise and were based on crosscorrelation-type arguments. Our success in getting in a neat way the recursions (20) has been due to the systematic procedure developed in this paper and based on the notion of a generalized innovation process.

4 - RECURSIVE INTERPOLATION AND SMOOTHING

Let $K(t,\tau)$ be a real-valued nonnegative definite kernel defined for t and on some interval T of the real line. Hereafter, the Hilbert space H of Sect. 2 will be identified with the reproducing kernel Hilbert space (RKHS) $H(K)$ with reproducing kernel (RK) $K(t,\tau)$. As for RKHS theory and applications, the reader is referred to [9] and [10]. The only property of $H(K)$ that will be repeatedly used in the sequel is the so-called reproducing property, viz.

$$y(t) = \langle y, K(\cdot,t)\rangle \qquad \forall y \in H(K).$$

The interpolation problem we intendo to pose can be formulated as follows. Given a sequence of numbers

$$y_i \triangleq y(t_i) = \langle y, K(\cdot,t_i)\rangle, \quad i \in I, \; t_i \in T,$$

find

$\hat{y}_{|n} \triangleq$ the minimum-norm element in $H(K)$ interpolating $y_1, y_2, \ldots, y_n$, in a recursive form. This problem is clearly a particular version of the indirect-sensing measurement problem formulated in Sect. 2.

Taking into account the reproducing property of $H(K)$, from (11) -(13) we get at once

$$\begin{aligned} e_n(\cdot) &= K(\cdot,t_n) - \sum_{i=1}^{n-1} e_i(t_n)\|e_i\|^{-2} e_i(\cdot) \\ \|e_n\|^2 &= K(t_n,t_n) - \sum_{i=1}^{n-1}[e_i(t_n)]^2\|e_i\|^{-2} \\ \mu_n &= \|e_n\|^{-1}\left[y_n - \sum_{i=1}^{n-1}\mu_i\|e_i\|^{-1}e_i(t_n)\right] \\ \hat{y}_{|n}(\cdot) &= \hat{y}_{|n-1}(\cdot) + \mu_n\|e_n\|^{-1}e_n(\cdot) \end{aligned} \qquad (21)$$

Example 5 (interpolation by splines)[1)] - Let y be the output of a one-input one-output finite-dimensional linear system

$$\mathcal{S}: \begin{cases} \dot{x}(t) = A(t)x(t) + b(t)u(t) \\ x(t_o) = 0 \\ y(t) = c(t)x(t) \end{cases}$$

Thus, the set of all outputs y on $T \triangleq [t_o, t_f]$ corresponding to all possible square-integrable inputs u on T, coincides [12] with the RKHS $H(K)$ with RK given by

$$K(t,\tau) = \int_{t_o}^{t\wedge\tau} H(t,\sigma)H(\tau,\sigma)\,d\sigma \qquad (22)$$

where $\wedge$ denotes minimum, $H(t,\sigma) \triangleq c(t)\phi(t,\sigma)b(\sigma)$ and $\phi(t,\sigma)$ is the state-transition matrix of $\mathcal{S}$. Moreover, the transformation $\mathcal{S}: u \to y$ from $L_2(T)$ onto

1) The results that follow can be generalized [11] to the case of unknown initial state $x(t_o)$

H(K) is a congruence (isometric isomorphism), i.e.

$$\mathcal{J} u = y \implies \| y \|^2 = \int_T u^2(t)\, dt \tag{23}$$

In particular, if

$$\mathcal{J} : \begin{cases} [Ly](t) = u(t) \\ x(t_o) \triangleq [y(t_o), y^{(1)}(t_o), \cdots, y^{(m-1)}(t_o)]' = 0 \end{cases}$$

with L a differential operator ($D \triangleq d/dt$)

$$L \triangleq D^m + a_{m-1} D^{m-1} + \cdots + a_1 D + a_o$$

(23) yields an explicit formula for the H(K)-norm of y, viz.

$$\| y \|^2 = \int_T [Ly(t)]^2 dt \tag{24}$$

and [2] $\hat{y}_{|n}$ is [11, 13] the L-spline interpolating $x(t_o), y_1, y_2, \ldots, y_n$. If $L \triangleq D^m$, $\hat{y}_{|n}$ is called the polynomial spline of order m interpolating $x(t_o), y_1, y_2, \ldots, y_n$.

Strictly related to the above interpolation problem, we now consider the following smoothing problem. Let $K(t, \tau)$ be again a nonnegative definite kernel, H(K) the associated RKHS and $\| \cdot \|$ the corresponding norm. Given a sequence of real numbers

z_i, $i \in I$,

find $\hat{y}_{|n} \triangleq$ the element in H(K) minimizing

$$\sum_{i=1}^{n} \sigma_i^{-2} (z_i - y_i)^2 + \| y \|^2 \tag{25}$$

$$y_i \triangleq y(t_i), \qquad t_i \in T,$$

in a recursive form. This is essentially a problem of smoothing by generalized splines. It has been shown [12] that (25) is equivalent to the following problem of statistical smoothing. Given the discrete-time observations

$$z_i = y_i + \zeta_i \quad , \quad i \in I, \tag{26}$$

where $y_i \triangleq y(t_i)$ are samples from a stochastic process y(t) with zero mean and covariance kernel

$$K(t, \tau) \triangleq E[y(t)\, y(\tau)]$$

and ζ_i r.v.'s uncorrelated with y(t) with zero mean and covariance

2) The L-spline interpolating $y_1, y_2, \ldots, y_n$, is the function passing through $y_1, y_2, \ldots, y_n$ and minimizing (26).

$$E[\zeta_i \zeta_j] = \sigma_i^2 \delta_{ij}$$

find the l.l.s. smoothed estimate $\hat{y}_{|n}(t)$ of $y(t)$, $t \in T$, based on $z_1, z_2, \ldots z_n$, in a recursive form. To solve this problem without resorting to a dynamic representation of the process y, we rephrase it in a suitable form. First, notice that by the reproducing property of $H(K)$ the unknown $y \in H(K)$ must be such that

$$y_i = \langle y, K(\cdot, t_i) \rangle, \qquad i \in I$$

From (21a) on the other hand we get

$$K(\cdot, t_i) = \sum_{j=1}^{i} \alpha_{ij} e_j(\cdot)$$

where

$$\alpha_{ij} \triangleq \begin{cases} \| e_j \|^{-2} e_j(t_i), & j < i \\ 1, & j = i \end{cases}$$

Therefore,

$$y_i = \sum_{j=1}^{i} \alpha_{ij} \langle y, e_j \rangle$$

Hence, setting

$$\theta_i = \theta \triangleq [\langle y, e_1 \rangle, \langle y, e_2 \rangle, \cdots]'$$

$$c_i \triangleq [\alpha_{i1}, \alpha_{i2}, \cdots, \alpha_{ii}, 0, 0, \cdots]'$$

we have

$$\begin{cases} \theta_{i+1} = \theta_i \\ z_i = c_i \theta_i + \zeta_i, \quad i \in I \end{cases} \tag{27}$$

from which the Kalman-Bucy formulas (18) and (19) give the l.l.s. estimate $\hat{\theta}_{|n}$ of θ based on $z_1, z_2, \ldots, z_n$, viz.

$$\begin{aligned} \hat{\theta}_{|n} &= \hat{\theta}_{|n-1} + F_n [z_n - c_n \hat{\theta}_{|n-1}] \\ F_n &= (c_n P_n c_n' + \sigma_n^2)^{-1} P_n c_n' \\ P_{n+1} &= P_n - (c_n P_n c_n' + \sigma_n^2) F_n F_n' \end{aligned} \tag{28}$$

with P_1 equal to a symmetric nonnegative definite matrix, e.g. $P_1 = \sigma_0^2 I$ with a sufficiently large σ_0^2. Finally, we obtain the desired recursive formula for $\hat{y}_{|n}$,

$$\hat{y}_{|n} = G \hat{\theta}_{|n} = \hat{y}_{|n-1} + G F_n [z_n - c_n \hat{\theta}_{|n-1}] \tag{29}$$

where

$$G \triangleq [\| e_1 \|^{-2} e_1(\cdot), \| e_2 \|^{-2} e_2(\cdot), \cdots].$$

5 - CONCLUSIONS

Indirect sensing experiments are defined and shown to encopass a large class of applications such as estimation, filtering, system identification, and interpolation and smoothing by splines. When a recursive solution to the indirect-sensing experiment problem is desired, the notion of a discrete-time generalized innovations process, or innovations experiment, appear to be a natural and effective one to use. The problem of estimating the state of a finite-dimensional linear system from discrete-time noisy measurements appears to be but one of the possible applications of the theory developed. The problem of determining the impulse response of a Q-input P-output system is approached by the use of the notion of an innovations experiment. When PRBS's are used as probing inputs, attractive formulas of recursive type are obtained by the proposed method easily and in a direct way. Finally, it is shown that problems of interpolation and smoothing by splines can be approached by the theory developed.

REFERENCES

1 A.W.Naylor and G.R.Sell, Linear Operator Theory in Engineering and Science, New York Holt, Rinehart and Winston, 1971, pp. 345.

2 D.G.Luenberger, Optimization by vector space methods, New York: Wiley, 1969, ch.6.

3 E.Mosca, "System identification by reproducing kernel Hilbert space methods", in Preprints 2nd Prague IFAC Symp. Identification and Process Parameter Estimation, Prague, Czechoslovakia: Academia, 1970, paper 1.

4 E.Mosca, "A deterministic approach to a class of nonparametric system identificatic problems", IEEE Trans. Inform. Theory, vol. IT-17, pp.686-696, Nov. 1971.

5 E.Mosca, "Determination of Volterra Kernels from input-output data", Int.J.Systems Sci., vol.3, pp.357-374, 1972.

6 D.Graupe, Identification of System, New York: Van nonstrand Reinhold, 1972, Ch.4.

7 K.R.Godfrey, "The application of pseudo-random sequences to industrial processes and nuclear power plant", in Preprints 2nd Prague IFAC Symp. Identification and Process Parameter Estimation, Prague, Czechoslovakia: Academia, 1970, paper 7.1.

8 P.N.Nikiforku, M.M. Gupta and R.Hoffman, "Identification of linear discrete systems in the presence of drift using pseudorandom sequences", in Preprints 2nd Prague IFAC Symp. Identification and Process Parameter Estimation, Prague, Czechoslovakia: Academia, 1970, paper 7.

9 N.Aronszajn, "Theory of reproducing kernels", Trans.Amer.Math.Soc., vol.68, pp.337-404, 1950.

10 E.Parzen, Times Series Analysis Papers, San Francisco: Holden-Day, 1971.

11 H.L.Weinert, "A reproducing kernel Hilbert space approach to spline problems with applications in estimation and control", Info.System Lab.Stanford

Univ., Tech.Rept.7001-6, May 1972.

12 E.Mosca, "Stochastic extension and functional restrictions of ill-posed estimation problems", Lecture Notes in Computer Scien.: 5th IFIP Conf.Optimiz. Techniques, vol.3, New York: Springer-verlag, 1973, pp.57-68.

13 T.E.N.Greville, Theory and applications of spline functions, New York: Academic Press, 1969.

A SYSTEM OF MODELS OF OUTPUT RENEWAL

N.B. Mironosetskii
Institute of Economics & Industrial Engineering
Siberian Department of the USSR Academy of Sciences
Novosibirsk, USSR

Regular renewal of output produced is a law in contemporary industrial production. The processes of output renewal are very diverse; peculiar to them are intricate hierarchical interrelationships between their components and uncertainty of parameters which characterize both the components and the processes as a whole. For these, inherent are certain specific features which, taken together, make possible their representation and analysis with discrete means, and, in particular, the effectiveness of applying models based on different types of graphs.

The diversity of output renewal processes and their specificity necessitate the use of a system of problem-directed models instead of a single universal model. Of the initial stages in output designing it is typical to have numerous alternative situations representing a set of various technical solutions. Of special interest are those competing variants whose preference is dependent either on random parameters, or on conditions needing additional research. According to the laws of combinatorics, the total number of variants on an item can be large, but the main point is that combinations of individual variants generate situations which are far from being obvious.

A model applied to the analysis of this kind of processes is based on a representation of a process as a multi-variant alternative stochastic directed graph $G(X, U)$ where X is a set of its vertices, U - a set of arcs. In the graph $G(X,U)$ the vertices correspond to different stages in the process of designing and are, for this reason, heterogeneous. In this model provided are 9 types of vertices representing all possible situations found in implementing complex projects. To represent different kinds of alternatives, at inputs and outputs of vertices logical conditions $\wedge$, $\vee$ and $\overline{\vee}$ (symbol $\overline{\vee}$ denoting the excluding $\vee$) are realized. To represent statistical and dynamic rules of preference and choice of alternatives, to the arcs (i,j) going from the vertices i with logical conditions $\vee$ or $\overline{\vee}$ at the output, the probabilities P_{ij} are confronted which may, in their turn, be connected by complicated Bayesian relations with the probabilities of the realization of inputs into the given vertex i :

$$\forall_j \in \Gamma i, \quad P_{ij} = f_j \,(P_{\kappa i} / \kappa \in \Gamma i^{-1}).$$

The development of these models is a multi-stage process in which the joint experience and intellect of all the participants in program developing is used. At the first stage, a structural scheme of technical complex under study is made, with a certain degree of detailing, as a graph $G_S(X_S, U_S)$. On the basis of expertise vertices $a \in A \subset X_S$ are discerned for which alternative solutions are permissible. An essential element of this stage is determining a type and logical conditions at input and output for $\forall a \in A$. For all the vertices $a \in A$ a set of permissible alternatives is determined, and each of them is represented by an arc (a,e), and $e \bar{\in} X_S$. For each alternative (a,e) a subgraph $G_e(X_e, U_e)$ of its realization is built. It is allowed in the model that any subgraph $G_e(X_e,U_e)$ in its turn can be stochastic alternative graph, and , hence, for its construction one needs analogous procedures.

The stochastic alternative graph $G(X, U)$ representing the process as a whole is obtained through adjoining, on the basis of the graph G_S, all the graphs of type G_e and through successively replacing the arcs $(a,j) \in U_S$ by a set of subgraphs G_e representing the alternatives provided for the vertex a.

The closing stage in building the stochastic alternative graph $G(X,U)$ is determining the parameters of all of its arcs. The parameters of arcs representing jobs are determined according to the normative approach of /1/, and for the probabilities P_{ae} of the alternatives (a,e) special procedures of group judges' evaluations treatment are used. Depending on the type of the event a there can be either vector $\{P_{ae}\}, e\in\Gamma a, \{a\}=\Gamma a$, or matrix $\{P_{ae}^{\kappa}\}$ $e\in\Gamma a, \kappa\in\Gamma a^{-1}, \{a\}=\Gamma a, \{\kappa\}=\Gamma a^{-1}$ in case of Bayesian relations. Apart from this, for conditions $\overline{V}$ at the vertex output there must be provided a rating $\sum_{e} P_{ae}=1$ or $\forall\kappa: \sum_{e} P_{ae}^{\kappa}=1$. In certain instances parameters P representing the probabilities of the choice of alternatives are, in their turn, random variables and, according to this, the expertise results will not be the numbers P_{ae} and P_{ae}^{κ} but the distribution functions of these variables set in one form or another.

The algorithm for the analysis of the stochastic alternative graph $G(X,U)$ is based on Monte Carlo ideas. To simulate the choice of alternatives, the procedures given in /5/ are used. The results of the algorithm are characteristics of graph $G(X,U): F(T), \Psi(S)$, i.e. distribution functions of designing duration T and of its cost S; $E(R_t), D(R_t)$ mathematical expectations and variances of demands for resources allocated by time intervals t where $0\leq t\leq T^{0}$ and T^{0} is a given variable; for $\forall_e$, $e\in X$, in particular, for $\forall_\beta$, $\Gamma_\beta=\emptyset$, P_e is determined, i.e. the probability of reaching the vertex e in the project implementation, and some other characteristics. There are some procedures for discerning the most probable structure of the graph $G(X,U)$.

In developing the software for this model a crucial requirement was its openness. This stemmed from the diversity of processes and problems studied with the help of this model; therefore, each user should have a possibility to extend the software of the model to new problems. The modular modification of the model software procedure realizes this principle.

The analysis and the obtaining of integral characteristics of a single alternative stochastic graph does not resolve the problem of the process implementation in that case where the plant performs a number of projects comparable in priority and complexity. One of the approaches to resources allocation for supplying several projects is based on the application of linear-programming models. The characteristics of each from L graphs L subject to implementation $G^{\nu}(X^{\nu},U^{\nu}), \nu=1(1)L: E(S^{\nu}), E(R_t^{\nu})$ obtained by the above described model are incorporated as information into a linear-programming model. The obtained solution gives, in the first approximation, information about the location of resources in accordance with a certain optimization criterion. Another approach in which the linear-programming problem serves as a procedure in a general scheme of simulation of implementing L projects is also used. The obtained characteristics S^{ν} and R_t^{ν} of the graph G^{ν}, $\nu=1(1)L$ in a single Θ step of modelling enable one to solve the linear programming problem and to estimate the value of the objective function K_{Θ}. These experiments generating $K_1, K_2, \dots, K_{\Theta}, \dots, K_N$ of the objective function K make it possible to obtain the function of the distribution of the values of the objective function K and to estimate on the whole the effectiveness of the implementation of the plant productive program under optimal allocation of resources at random demand.

At a stage in output renewal process where the main alternatives of the project have been basically resolved, a special stochastic graph of the project implementation is built up. In technology of producing complicated items an advanced system of stage control, matching other kinds of coordination is provided. The results of these stages which will be conventionally called control stages, are, generally speaking, fortuitous and, along with the planned ones and those allowing the continuation of developing, there may also occur results requiring backtracks to the stages in development already passed. The popular network model permits one to represent only the ideal course of a process. According to this, for adequate representation of a real process, the network graph $G_g(X_g, U_g)$ is first built which is then transformed into a stochastic graph with backtracks $G_e(X_c, U_c)$. The sets of vertices X_g and X_c may be the same, but the sets of arcs U_g and U_c are essentially different: $U_c = U_g \cup U_b$, here U_b is a set of backtrack arcs of the type (e_b, j), e_b the vertex generating a backtrack to the vertex j. With the help of multiple interview technique and processing judges' evaluations, the parameters of the arcs of the graph G_c are determined. Of interest are the coefficients of changes in parameters of the arc (i,j): $\mu_t(i,j)$, $\mu_s(i,j)$, $\mu_k(i,j)$ which determine respectively time, cost, resources needed in repeated executions of work (i,j) and parameters set on the backtrack arcs (e_b, j): $P_{e_b,j}$, the probability of the backtrack arc, $\mu_p(e_b, j)$, the coefficient of the change in probability of events e_b generating the backtracks. These parameters are recognized in the model and allow one to obtain the clarified characteristics of the process.

The algorithm for the analysis of a graph with fortuitous backtracks based on Monte Carlo technique and using the specific features of the built model, makes it possible to get on computers various information about the process, in particular, the characteristics of a number of parameters, among them: the density function $\Phi(T)$, $\Psi(S)$ of time and cost of the project execution, mathematical expectation and the variance of demands for resources of inaccumulative type of power $E(R_t)$, $D(R_t)$. These demands can be defined for each interval t of the planning period $[0,T]$. This information and structure of the graph $G(X,U)$ is starting for scheduling. Scheduling programs have been worked out on algorithms /7/ and conducted with the observance of modularity principle. Scheduling plans-graphs are given on computers as working documents, the form of which takes into consideration the requirements of the backfeed principle in the control over the processes described.

Building up the modular complexes of programs makes it possible to initiate the effective functioning of the system in different regimes, including the regime "man - computer". The use of a set of programs in a regime of a business game enables one to reveal on models extreme conditions of the process and to foresee necessary measures and their effectiveness beforehand, before the actual implementation of a certain process of output renewal under productive conditions.

References

1. Basic Principles in the Development and Application of Network Planning and Management Systems. (Russian). Ekonomika. Moscow. 1967.
2. Buslenko, N.P. Complicated Systems Modelling (Russian). Nauka. Moscow. 1968.
3. Pospelov, G.S. and Barišpolets, V.A. On Stochastic Network Planning (Russian). Tekhn. Kibern. (1966), no.6.
4. Golenko, D.I. Statistical Methods of Network Planning and Management (Russian). Nauka. Moscow. 1968.
5. Mikhailov, G.A. Some Problems in Monte Carlo Technique Theory (Russian). Nauka. Novosibirsk. 1974.
6. Mironosetskii, N.B. and Mogulskii, A.A. The Analysis of Stochastic Graphs of a Special Type (Russian). Matem. Anal. Ekon. Model. Novosibirsk. 1971.
7. Mironosetskii, N.B. Economic-mathematical Methods of Schedule Planning (Russian). Nauka. Novosibirsk. 1973.

BILINEAR SOCIAL AND BIOLOGICAL CONTROL SYSTEMS

by

R. R. Mohler
Department of Electrical & Computer Engineering
Oregon State University; Corvallis, Oregon 97331

OBJECT

The purpose of this paper is to:

1. Review briefly some relevant results of bilinear system theory and its socioeconomic application.
2. Develop a bilinear model for the realization of a biological process from experimental tracer data.

INTRODUCTION

Finite state bilinear systems are described by a system of differential equations which are linear in state, linear in control, but not jointly linear in both as are traditional linear systems. An introduction to the physical relevance of these systems, their controllability and optimal control is analyzed in references [1] and [2]. It is shown that linear system controllability and performance may be improved by the addition of a bilinear control mode. This is as it would be expected since the bilinear system does exhibit a form of adaptive control and variable structure. These are the reasons that bilinear systems, or at least their approximation, are so common in natural and societal processes as well as in engineering.

NEUTRON AND CELL FISSION

Theoretically, any atomic nucleus may undergo fission if it has a mass which is greater than the sum of masses of particles allowed to exist upon division. This situation yields an unstable nucleus, and its fission consequently liberates energy from the reduced mass in subdivision. An activation energy is normally required to deform the original nucleus sufficiently so that the electrostatic repulsion force overcomes the binding force of the nucleons. This is not an unusual phenomenon since chemical combustion processes and biological cell division take place in a similar manner. All of these processes are analogous to liquid-drop division which roughly proceeds through geometrical stages of a sphere, an ellipsoid, a dumbbell and finally two spheres by the application of an external force to overcome surface tension.

The point neutron dynamics from nuclear fission, with neutron population level n proportional to power level, is described by the following bilinear system of equations:

$$\frac{dn}{dt} = \frac{[u(1-\beta)-1]}{\ell} n - \sum_{i=1}^{6} \lambda_i c_i \tag{1}$$

and

$$\frac{dc_i}{dt} = \frac{u\beta_i}{\ell} n - \lambda_i c_i \ , \ i=1,\ldots,6, \tag{2}$$

where ℓ is mean neutron generation time, c_i is ith precursor level, λ_i is ith precursor decay constant, β_i is the portion of neutrons generated from the ith precursor with $\sum_{i=1}^{6} \beta_i = \beta$ and n is neutron multiplication [3]. The control u represents the net evolution of population from generation to generation and must be at least unity for the population to be self perpetuating, i.e., a chain reaction in nuclear fission.

Again, the dynamics of cell population is analogous with cells dying, cells generated from precursors and prompt new cells generated all in a controlled manner. While both processes involve variable structure control feedback terms (in some cases bilinear), the basic control mechanisms are considerably different. In a nuclear reactor, for example, inherent temperature feedback alters neutron multiplication or fission effectiveness, and external feedback control loops of a conventional or optimal design may be synthesized [3]. Cell processes, on the other hand, are controlled according to genetic code passed along successive cell generations, according to the cell's function in the synthesis or the maintenance of an organ and according to the cell's environmental conditions. Obviously, there is communication between cells in this controlled process. (It appears that the immune response mechanism utilizes T-cells to establish communication lines with alien cells and then breaks these lines leaving holes in the cellular membranes through which low molecular weight molecules pass out and water rushes in to burst the invader.)

In biology, cells, or combinations of cells, forming organs or functional units within organs may form natural compartmental structures to describe the dynamics of many important transfer processes.

COMPARTMENTAL MODELS IN ENGINEERING AND SOCIETY

It is found that a compartmental structure arises for processes in engineering and society as well as biology. In a nuclear reactor, for example, the dynamics are distributed and nearly bilinear, but ordinary bilinear differential equations (1) and (2) or difference equations are usually used to describe the behavior in different spatial compartments. These may include compartmentation according to such regions as fuel-loaded core, moderated core, homogeneous core, reflector, support structure, control and heat exchange. Neutrons and heat are generated and absorbed in certain compartments and transferred between compartments in the form of flux. Similar transfer mechanisms take place for mass transport, chemical reaction, migration of biological species and numerous generalizations of such processes.

Frequently, the dynamical equations are derived by a balance of energy, mass or force. Suppose a single quantity or substance X is distributed within and moves among n compartments according to conservation equations of the form

$$\frac{dx_i}{dt} = \sum_{j=1}^{n}{}' \phi_{ij} - \sum_{k=1}^{n}{}' \phi_{ki} + \phi_{ia} - \phi_{ai} + p_i - d_i \ , \quad i = 1,\dots,n, \tag{3}$$

where x_i a generalized state variable may be the amount of substance X in the ith compartment, ϕ_{ij} is the flux of X from the jth to the ith compartment, ϕ_{ia} (ϕ_{ai}) is the flux of X from (to) the environment to (from) the ith compartment, p_i is the rate of production, and d_i is the rate of destruction of X in the ith compartment. The primed summation denotes deletion of the ith term.

Assume that compartmental volumes or capacities are constant and that X moves between compartments and between the system and its environment by simple diffusion. Also, assume that X is at uniform concentration c_a in the environment. Then, the system fluxes are described by

$$\phi_{ij} = \rho_{ij} x_j \ ,$$

$$\phi_{ai} = \rho_{ai} x_i \ , \tag{4}$$

and

$$\phi_{ia} = \rho_{ia} c_a \ , \quad i,j = 1,\dots,n, \ i \neq j \ ,$$

where the coefficients ρ_{ij} and ρ_{ai} are exchange parameters which may be constant or may be multiplicative controls. Additive control may be present through the net production of substance $(p_i - d_i)$ or through $\rho_{ia} c_a$.

It is readily seen from Equations (3) and (4) that a collection of terms leads to a bilinear system of the following form:

$$\frac{dx}{dt} = Ax + \sum_{k-1}^{m} B_k u_k x + Cu + y , \tag{5}$$

where $x \varepsilon R^n$ is the state vector, $u \varepsilon R^m$ is the control vector, A, B_k, and C are nxn , nxn , and nxm matrices respectively, and $y \varepsilon R^n$ is a constant vector [1]. Again, x is composed of compartmental quantities of substance X; additive control may arise from net production or adjustable environmental influx, and multiplicative control from those intercompartmental and environmental efflux exchange parameters which are manipulated. Additive and multiplicative controls may or may not be independent variables for a particular bilinear system. When they are independent, zeros appear in the appropriate positions of B_k and C where necessary. Those exchange parameters which cannot be manipulated result in the constant coefficients of the A matrix and of vector y . While the process may include transport of more than one substance, no generality is lost in utilizing the above form of the equations when there is no coupling of substance fluxes.

Socioeconomic compartments may take the form of geographical regions partitioned according to administrative units or according to land topology. In particular, the former is quite convenient since demographic data, business transactions, etc. are readily available, and necessary surveys can be made within the organizational structure. Here, for example, a multiplicative control variable may be migration rate manipulated by means of attractiveness multipliers such as availability of jobs, housing, transportation, etc.

Compartments also may appear in a non-spatial manner. For demographic components, income levels or household sizes may designate compartments, with transfer between compartments by means of birth, death, or job change. In economics, compartments may separate producers from consumers with the appropriate interflux of capital, products, fuels, materials, services and labor. The mathematical model may take an adaptive form based on supply and demand forces for a market economy or based on government goals for a command economy.

TRACER METHODOLOGY

For many biological processes and certain socioeconomic processes it is impossible or at least very difficult to measure the necessary inputs and outputs to estimate compartmental system parameters. Sometimes, a tracer or state label may be observed, and the system parameters arrived at from the tracer dynamics [4].

If it is assumed that the tracer in every compartment is distributed uniformly through substance X , that labeled and unlabeled substances behave identically, and that the tracer does not appreciably affect the compartmental system behavior, then the tracer system is described by

$$\frac{d(x_i a_i)}{dt} = \sum_{j=1}^{n}{}' \phi_{ij}(t)a_j - \sum_{k=1}^{n}{}' \phi_{ki}(t)a_i + \phi_{ia}(t)a_{ia} - \phi_{ai}(t)a_i + p_i(t)a_i$$

$$- d_i(t)a_i + f_i(t) , \quad i=1,\ldots,n, \tag{6}$$

where $a_i(t)$ is the specific activity (fraction of X which is labeled) in compartment i , $a_{ia}(t)$ is the specific activity of substance in influx $\phi_{ia}(t)$, and $f_i(t)$ is the influx of tracer which is inserted directly into compartment i . Equation (6) includes, for example, the possibility both of direct tracer insertion by injection into compartments and of tracer uptake by natural flux routes from a labeled environment.

PHYSIOLOGICAL WATER BALANCE

A two-compartment, water balance model is presented here. Experimental data is used from the so-called wild house mouse in a controlled environment. The plasma compartment contains all the body water, $w_1(t)$, in equilibrium with the blood plasma. Similarly, the second compartment or "evaporate" compartment, contains the body water, $w_2(t)$, in equilibrium with insensible evaporation from the animal [$w_1(t) + w_2(t) = w_b(t)$ total water weight]. The plasma compartment receives water from the environment, $\phi_g(t)$, via the gastro-intestinal tract in the form of drinking water, as well as hygroscopic and metabolic water from the animal's food. Water is lost to the environment from the plasma compartment in the urine, $\phi_u(t)$. The evaporate compartment exchanges water with the environment via efflux $\phi_e(t)$ and influx $\phi_a(t)$. The two compartments, of course, exchange water inside the animal, with flux $\phi_{ij}(t)$ to compartment i from compartment j .

By inspection, the water conservation equations for this model are

$$\dot{w}_1 = -\phi_{21}(t) + \phi_{12}(t) - \phi_u(t) + \phi_g(t)$$
$$\dot{w}_2 = \phi_{21}(t) - \phi_{12}(t) - \phi_e(t) + \phi_a(t) , \tag{7}$$

and in water balance $(\dot{w}_1 = \dot{w}_2 = 0)$,

$$\phi_{21}(t) - \phi_{12}(t) = \phi_g(t) - \phi_u(t) = \phi_e(t) - \phi_a(t) . \tag{8}$$

Since the system is closed and conservative, the associated specific activity kinetics for an ideal tracer introduced into the system (6) by natural flux routes are given by

$$a = S(t)a \quad , \tag{9}$$

$$S(t) = \begin{bmatrix} -\dfrac{[\phi_{12}(t) + \phi_g(t)]}{w_1(t)} & \dfrac{\phi_{12}(t)}{w_1(t)} \\ \\ \dfrac{\phi_{21}(t)}{w_2(t)} & \dfrac{[\phi_{21}(t) + \phi_a(t)]}{w_2(t)} \end{bmatrix}$$

and $a \varepsilon R^2$. It can be shown that original system state can be obtained by an adjoint equation of (9).

Also, it can be shown from Gersgorin's first matrix theorem [5] that the eigenvalues are real and distinct in this case. By measurement of the eating and drinking water influx, the urinary and fecal efflux, total body water (estimated from body weight), and the tracer eigenvalues, it is possible to obtain the water in each compartment and the water exchanges with the environment and between compartments.

The measured and unknown quantities for the two-compartment equilibrium model are summarized in Table I. The unknown parameters, in turn, are related to the measured quantities by the equations which follow.

TABLE I

Parameters for the two-compartment water model.

Measured	ϕ_g ϕ_u w_b σ_1 σ_2 m_1
Unknown	ϕ_{21} ϕ_{12} ϕ_e ϕ_a w_1 w_2

$$\phi_{21} - \phi_{12} = \phi_g - \phi_u \tag{10}$$

$$\phi_e - \phi_a = \phi_g - \phi_u \tag{11}$$

$$w_b = w_1 + w_2 \tag{12}$$

$$w_1 w_2 \sigma_i^2 + [(\phi_{21} + \phi_a)w_1 + (\phi_{21} + \phi_u)w_2]\sigma_i + \phi_{21}\phi_a + \phi_{21}\phi_g$$

$$+ \phi_u\phi_a = 0 \ , \ i = 1,2 \tag{13}$$

$$m_1 = \frac{\phi_{21}}{\phi_{21} + \phi_a + w_2\sigma_1} \quad . \tag{14}$$

From the measured parameters shown in Table II for two equilibrium washout test conditions, a simulation was made of the two compartmental equations with the corresponding parameters shown in Table III. The eigenvalues and eigenvector slopes may be used to check the accuracy of the model with the measurements made.

TABLE II

Measured Parameters for Ad Libitum and 1/8 Ad Libitum Drinking Water Conditions.

	Ad Libitum	1/8 Ad Libitum
ϕ_g (g/day)	5.0	1.9
ϕ_u (g/day)	2.3	1.2
w_b (g)	14.5	11.6
σ_1 (·/day)	- 0.3	-0.2
σ_2 (·/day)	-13	---
m_1	0.9	0.7

TABLE III

Parameters for Simulations of Ad Libitum and 1/8 Ad Libitum Drinking Water Conditions.

	Ad Libitum	1/8 Ad Libitum
ϕ_{21} (g/day)	10.0	2.0
ϕ_{12} (g/day)	7.3	1.3
ϕ_e (g/day)	4.0	1.5
ϕ_a (g/day)	1.3	0.8
w_1 (g)	13.2	10.5
w_2 (g)	1.3	1.1
ϕ_1 (·/day)	-0.35	-0.19
ϕ_2 (·/day)	-9.2	-2.7
m_1	0.91	0.73

Mohler [1] shows that the water excretion in urine, ϕ_u, and body water, w_b may then be related to the control variable, u_k, by a bilinear equation of the form

$$\phi_u = u_k(c_g + b_g w_b) ,$$

where u_k is a combination of complicated control processes by means of manipulated permeability, arteriolar resistance and osmotic water pressure in the kidneys. Other measurements would have to be made to arrive at estimates of basic physiological parameters such as arteriolar resistance or membrane permeability, and more basic constants than c_g and b_g.

A preliminary analysis of identification in medical diagnosis for processes of this class is presented by Cohen [6].

ACKNOWLEDGEMENT

The author wishes to acknowledge the contributions of W. D. Smith and the support of the National Science Foundation for this work.

REFERENCES

1. Mohler, R. R., Bilinear Control Processes, Academic Press, New York and London, 1973.
2. Mohler, R. R. and Ruberti, A., Eds., Theory and Applications of Variable Structure Systems, Academic Press, New York and London, 1972.
3. Mohler, R. R. and Shen, C. N., Optimal Control of Nuclear Reactors, Academic Press, New York and London, 1970.
4. Sheppard, C. W. and Householder, A. S., "The Mathematical Basis of the Interpretation of Tracer Experiments in Closed Steady-State Systems", J. Applied Physics, 22 (1951), 510-520.
5. Lancaster, Peter, Theory of Matrices, Academic Press, New York and London, 1969.
6. Cohen, A., "Parameter Estimation for Medical Diagnosis", Proc. IFAC Identification Conference, The Hague, (1973), 239-241.

A PROBLEM FROM THE THEORY OF OBSERVABILITY

M.S.NIKOLSKIĬ

Steklov Institute of Mathematics,
Academy of Sciences of the USSR, Moscow

When studying control problems, the information aspect is important. What do we know about the motion and what is it possible to know from the available information for the purpose of control?
In classical models of the optimal control theory and of the theory of differential games, the presence of a large amount of information about the motion and the dynamics of objects is postulated. In practice, such information is not always accessible because of technical reasons. Perhaps it is for this reason that the theory of observability and of filtration arose inside the control theory.

One of the first problems studied in the theory of observability was Kalman's problems about observation of a linear object. The problem is discussed below. Later N.N.Krasovskiĭ studied the problem of the best observability of a linear system with disturbances in measurement. Many papers were published in recent years on problems of the optimal control theory and the theory of differential games with incomplete information. As a rule, these works did not do specially with the observable aspect, but the latter was taken into account under the exact formalization of process under study. We will mention here the works of N.N.Krasovskiĭ, A.B.Kuržanskii, Yu.S.Osipov, F.L.Chernous'ko, A.A. Melikjan, I.Ya.Kač, V.B.Kolmanovskii, G.S.Šelementjev and other authors. The problem of combination of observation and control was studied specially in some of these papers (see, for example, the works of A.B.Kuržanskii, A.A.Melikjan, V.B.Kolmanovskii, G.S.Šelementjev and others).

The aim of this paper is to study the problem of observability arising in pursuit differential games in the case when the phase vector of the evading object is not known exactly. I have begun studying such games on the advice of L.S.Pondrjagin.

Kalman's problem of observability

In the n-dimensional Euclidean space R^n, motion of the linear object takes place in accordance with the differential equation

$$\dot{x} = Ax + Bu, \tag{1}$$

where $x \in R^n$, $x(0) = x_0$, $u \in R^p$; A, B are constant matrices; $u = u(t)$ is an arbitrary measurable Lebegue's summable function.

The following signal is accessible for observer's measurement

$$y(t) = Gx(t), \quad 0 \leq t \leq \Delta, \tag{2}$$

where $y \in R^m$, G is the constant matrix, $\Delta > 0$.
It is assumed that the observer knows A, B, G and $y(t)$, $u(t)$ on $0 \leq t \leq \Delta$. His aim is to find $x(0) = x_0$.

The Problem of ideal observability.

It is assumed here unlike the Kalman's problem that the observer does not know $u(t)$ on $0 \leq t \leq \Delta$. His aim is to find $x(0) = x_0$.

The class of ideal observable systems belongs to the class of systems which are observable in Kalman's sense. The problem of ideal observability arises when the observer is outside of the system (1). In differential games of pursuit, one of the players is such an observer in regards to the other player.
It will be noted that in problem of ideal observability the observer knows too little, that is why the class of such systems is naturally rather narrow.

This problem and its generalizations were considered by the author (see Ref. [I]-[3]).

Now we shall go to the general problem of ideal observability.

The general problem of ideal observability

It is assumed that the observer knows the signal:

$$y(t) = C(t)x + \int_0^t \mathcal{D}(t-s)u(s)ds, \quad 0 \leq t \leq \Delta, \tag{3}$$

where $C(t)$ is the $m \times n$ - analytical entire matrix function, $\mathcal{D}(t)$ is the $m \times p$ - analytical entire matrix function, x is an arbitrary constant vector from R^n, $u \in R^p$, $u(t)$ is an arbitrary measurable function which is summable in Lebegue's sense.

It is assumed that the observer knows $C(t), \mathcal{D}(t)$ but does not know $u(\cdot)$. The aim of the observer is to find x.

Remark 1. The presence of convolution in (1) is essentially for the obtaining of effective criterion of ideal observability.

Remark 2. In case (1), (2)

$$y(t) = Ge^{tA}x_0 + \int_0^t Ge^{(t-s)A}Bu(s)ds. \tag{4}$$

Definition. The system (3) is ideal observable iff for every possible function $y(\cdot)$ vector x is defined in a unique manner.

The necessary (but in general not sufficient) condition for the ideal observability of system (3) consists in fulfilment of the condition $\det \int_0^\Delta C^*(s)C(s)ds > 0$.

Theorem 1. The system (3) is ideal observable iff the function $y(t) \equiv 0$, $0 \leq t \leq \Delta$, is generated only by $x = 0$.

Remark 3. This theorem has a simple form but it is rather difficult to use it directly in practice.

Let us consider the function $y(t) \equiv 0$ and a couple $(x, \tilde{u}(\cdot))$ which generates it. It is possible to prove that the vector x and an analytical function $u(\cdot)$ generate $y(t) \equiv 0$ also.

Differentiating $y(t) \equiv 0$ consistently and substituting $t=0$, we have

$$
\begin{aligned}
&C(0)z &&= 0\\
&C^{(1)}(0)z + \mathcal{D}(0)u(0) &&= 0\\
&C^{(2)}(0)z + \mathcal{D}^{(1)}(0)u(0) + \mathcal{D}(0)u^{(1)}(0) &&= 0\\
&\cdot \quad \cdot \quad \cdot \quad \cdot \quad \cdot \quad \cdot \quad \cdot \quad \cdot
\end{aligned}
$$

Let us put

$$
F_{\mathcal{N}} = \begin{pmatrix} C(0) \\ C^{(1)}(0) \\ \cdot\ \cdot\ \cdot\ \cdot \\ C^{(\mathcal{N})}(0) \end{pmatrix}, \quad G_{\mathcal{N}} = \begin{pmatrix} 0, & 0, & \cdots, & 0 \\ \mathcal{D}(0), & 0, & \cdots, & 0 \\ \cdot & \cdot & \cdot & \cdot \\ \mathcal{D}^{(\mathcal{N}-1)}(0), & \mathcal{D}^{(\mathcal{N}-2)}(0), & \cdots, & \mathcal{D}(0) \end{pmatrix}
$$

$$
\mathcal{N} = 1, 2, \cdots .
$$

Theorem 2 (main theorem). The system (3) is ideal observable iff there exists $\mathcal{N} \geq 1$ such that rank $F_{\mathcal{N}} = n$ $(n = \dim z)$ and rank $(F_{\mathcal{N}}, G_{\mathcal{N}})$ = rank $F_{\mathcal{N}}$ + rank $G_{\mathcal{N}}$.

Lemma. One of the following conditions is valid; either exists $\mathcal{N}_0 \geq 1$ such that the conditions of theorem 2 are fulfilled for all $\mathcal{N} \geq \mathcal{N}_0$, or one of the conditions of theorem 2 is not fulfilled for any $\mathcal{N} \geq 1$.

Theorem 3. In the case of system (4) and $n \geq 2$, it is sufficient to verify the conditions of theorem 2 only for

$$
\mathcal{N}_0 = (m+1)(n-2) + 1 .
$$

Remark 4. The Operational Calculus of Mikusinskiǐ and the theorem of R.V.Gamkrelidze, G.L.Kharatishvili about canonical factorization of entire operational matrix function (see Ref. [4]) are used for the proof of theorem 2.

Example 1. The system $\dot{x}_1 = x_2$, $\dot{x}_2 = u$, $y(t) = x_1(t)$ is ideal observable.

Example 2. The system $\dot{x}_1 = x_2 + u_1$, $\dot{x}_2 = u_2$, $y(t) = x_1(t)$ is not ideal observable.

Example 3. $$y(t) = C(t)z + \int_0^t \mathcal{D}(t-s)\,u(s)\,ds, \qquad (5)$$

where $y \in R^m$, $z \in R^n$, $C(t) \not\equiv 0$, $\mathcal{D}(t) \not\equiv 0$;

$$C(t) = \sum_{i=k}^{\infty} C_i \frac{t^i}{i!}, \quad k \geq 0; \quad \mathcal{D}(t) = \sum_{i=\ell}^{\infty} \mathcal{D}_i \frac{t^i}{i!}, \quad \ell \geq 0.$$

Let rank $\mathcal{D}_\ell = m$. It is possible to prove that under our assumptions the necessary and sufficient conditions for ideal observability of (5) are the following:

a) $k \leq \ell$; b) rank $(C_k^*, \cdots, C_\ell^*) = n$.

References

1 M.S.Nikol'skiĭ, Dokl. Akad.Nauk SSSR, v.191, N6,1970.

2 M.S.Nikol'skiĭ, J.Differencial'nye Uravnenija, v.YII,N4,1971

3 M.S.Nikol'skiĭ, Trudy Vicheslitelnogo Čentra Akad.Nauk SSSR, collection "Issledovanie Operaciĭ", issue N 3,1972.

4 R.V.Gamkrelidze, G.L.Kharatishvili, SIAM J.Control,v.12,N 2, 1974.

SOME QUESTIONS OF THE MODELLING OF COMPLEX CHEMICAL SYSTEMS

G.M.Ostrovskij, Yu.M.Volin
Karpov Institute of Physical Chemistry
ul. Obukha 10, 107120, Moscow, B-120/U.S.S.R.

The problem of complex chemical system (CCS) optimization is considered.

Mathematically the CCS optimization problem in many cases may be defined as follows:

$$y_i^{(k)} = f_i^{(k)}(x_j^{(k)}, u_r^{(k)}), \tag{1}$$

$$k = 1, \dots, N; \quad i = 1, \dots, n_k; \quad j = 1, \dots, m_k; \quad r = 1, \dots, r_k$$

$$x_i^{(k)} - y_p^{(q)} = 0, \tag{2}$$

$$F = \sum_{k=1}^{N} F^{(k)}(x_j^{(k)}, y_i^{(k)}, u_r^{(k)}), \tag{3}$$

$$\varphi_l^{(k)}(u_r^{(k)}) \leq 0 \quad l = 1, \dots, p_k, \tag{4}$$

$$y_i^{(k)} = b_i^{(k)} \quad i = 1, \dots, \bar{g}_k \leq g_k, \tag{5}$$

where $x_j^{(k)}$, $y_i^{(k)}$ are input and output state variables and $u_r^{(k)}$ are decision variables of the k-th block. Equations (1), (2) represent block models and a table of connections between input and output variables. It is assumed that in the k-th block the first s_k (g_k) input (output) variables are CCS input (output) variables: $\bar{x}_j^{(k)}$ ($\bar{y}_i^{(k)}$). The block decision variables $u_r^{(k)}$ and CCS input variables $\bar{x}_j^{(k)}$ are CCS decision variables. These variables must be determined to give F the maximum value.

In this paper we shall consider the direct optimization approach using first derivatives of the criterion F. The solution algorithm in this case consists of three parts: the computation of criterion, the computation of criterion derivatives and a searching strategy. The last part has been examined in many papers (see, for example, /1/), therefore the attention will be concentrated on the first two ones.

Optimization Criterion Computation

There exist two different criterion computation approaches. At first the fulfilment of conditions (5) is realized through the searching strategy algorithm and all CCS output variables are considered as free at the criterion computation stage. The search of the optimum solution is carried out in the CCS decision variable space:

$$u_r^{(k)}, \bar{x}_j^{(k)} \qquad r = 1, \ldots, r_k; \; j = 1, \ldots, s_k; \; k = 1, \ldots, N. \tag{6}$$

The criterion computation is the CCS steady-state regime calculation, the values of variables (6) being fixed /2/.

Let us introduce now the function $F^*(u_r^{(k)}, \bar{x}_j^{(k)})$ which is obtained from F with the help of substitution $x_j^{(k)}$, $y_i^{(k)}$ as functions of CCS decision variables, and consider the second approach. In this case the fulfilment of conditions (5) is realized at the criterion computation stage, so there exist $\bar{g} = \sum_{k=1}^{N} \bar{g}_k$ additional equations. On such a value the number of variables must be increased to fulfil CCS steady-state regime equations. Let us assume that in each block the first $\bar{s}_k$ input variables are used to satisfy conditions (6) and designate $\bar{s} = \sum_{k=1}^{N} \bar{s}_k$. Then

$$\bar{s} = \bar{g} . \tag{7}$$

The search of the optimal solution is carried out in the space of variables

$$u_r^{(k)}, \bar{x}_j^{(k)} \qquad k = 1, \ldots, N; \; r = 1, \ldots, r_k; \; j = \bar{s}_k + 1, \ldots, s_k . \tag{8}$$

The criterion computation is the calculation of the CCS steady-state regime with additional conditions (5), the values of variables (8) being fixed.

Criterion Derivatives Computation

The usual method of criterion derivatives computation (with

the help of differences) has two defects: it is inaccurate and requires lengthy calculations, the number of decision variables is large First let us consider the case when fulfilment of conditions (5) is realized through the searching strategy algorithm.

The criterion derivatives with respect to decision variables may be expressed in the following form /2/:

$$\frac{\partial F^*}{\partial u_r^{(k)}} = \sum_{i=1}^{n_k} \lambda_i^{(k)} \frac{\partial f_i^{(k)}}{\partial u_r^{(k)}} + \frac{\partial F}{\partial u_r^{(k)}}, \tag{9}$$

$$k = 1,\dots,N; \quad r = 1,\dots,r_k; \quad j = 1,\dots,s_k$$

$$\frac{\partial F^*}{\partial \bar{x}_j^{(k)}} = \bar{\mu}_j^{(k)} \tag{10}$$

where $\lambda_i^{(k)}$, $\mu_j^{(k)}$ satisfy the adjoint process equations:

$$\mu_j^{(k)} = \sum_{i=1}^{n_k} \lambda_i^{(k)} \frac{\partial f_i^{(k)}}{\partial x_j^{(k)}} + \frac{\partial F^{(k)}}{\partial x_j^{(k)}} \quad j = 1,\dots,m_k, \tag{11}$$

$$\mu_i^{(k)} = \lambda_p^{(q)} - \frac{\partial F^{(q)}}{\partial y_p^{(q)}}, \tag{12}$$

$$\lambda_i^{(k)} = \frac{\partial F^{(k)}}{\partial \bar{y}_i^{(k)}} \qquad i = 1,\dots,g_k \quad \text{(boundary conditions)}. \tag{13}$$

The number of equations (11)-(13) is equal to that of the unknown variables $\lambda_i^{(k)}$, $\mu_j^{(k)}$.

Let us consider now the case when the fulfilment of conditions (5) is realized at the criterion computation stage. It may be shown that the expression of the criterion derivatives has now the form of equations (9), (10) with $j = \bar{s}_k + 1,\dots,s_k$ (in (10)), and the adjoint process is represented with the equations (11), (12),

$$\bar{\mu}_j^{(k)} = 0 \qquad j = 1,\dots,\bar{s}_k, \tag{14}$$

$$\bar{\lambda}_i^{(k)} = \frac{\partial F^{(k)}}{\partial \bar{y}_i^{(k)}} \qquad i = \bar{g}_k + 1,\dots,g_k. \tag{15}$$

It may be easily shown that the number of boundary conditions (14),(15) is equal to that of (13) of the first case. So there exists again the equality between the number of unknown variables and that of equations. But the solution of the adjoint process equations in this case is a more difficult problem as the boundary conditions are connected both with the input and output adjoint process variables.

In order to use the adjoint process method it is necessary to have formulae for matrices of partial derivatives

$$\left\| \frac{\partial f_i^{(k)}}{\partial x_j^{(k)}} \qquad \frac{\partial f_i^{(k)}}{\partial u_r^{(k)}} \right\|. \tag{16}$$

The programmer must obtain these formulae and the corresponding programs beforehand. In the case of complex block models this may require a lot of preparatory work. Of course the matrix (16) may be computed with the help of differences. The analysis has shown that this modification of the adjoint process method has a definite advantage compared to the method of decision variables differences. But some defects of the last method still remain.

In this connection the following algorithm has been proposed /3/. Let us assume that a sufficiently large set of simple computation operations is available: addition, subtraction, multiplication, sin(x), exp(x) etc (so-called conditionally elementary operations- CEO). Then an arbitrary nonlinear system of equations may be considered as a complex computational system with CEO as blocks of such a system:

$$y^{(k\ell)} = f^{(k\ell)}(x^{(k\ell)}, u^{(k\ell)}) \tag{17}$$

$$x_i^{(k\ell)} = y_p^{(kq)} \tag{18}$$

where k is the number of CCS block and ℓ is that of subblock corresponding to some CEO. The computation of the matrix (16) is equivalent to that of partial derivatives output variables with respect to input variables for the system (17), (18). So the adjoint process method may be used:

$$\mu_j^{(k\ell)} = \sum_{\ell=1}^{n_{k\ell}} \lambda_i^{(k\ell)} \frac{\partial f_i^{(k\ell)}}{\partial x_j^{(k\ell)}}, \tag{19}$$

$$j = 1, \dots, m_{k\ell}; \quad \ell = 1, \dots, L_k$$

$$\mu_i^{(k\ell)} = \lambda_p^{(k\ell)}.$$

In order to compute the matrix (16) it is necessary n times to compute the adjoint process (19), elements of the i-th column of the unit matrix being taken as input adjoint variables at the i-th computation. So in the proposed algorithm the system of equations (1) is replaced by the system of blocks described with simple equations. The CEO and partial derivatives $\partial f_i^{(k\ell)}/\partial x_j^{(k\ell)}$ programs and the organization program (which make it possible on the basis of the structure of the equations (17), (18) to create the program both for the computation of complex CEO system (17), (18) and for that of corresponding adjoint process system (19) for each block of the original CCS) may be made beforehand. In this case the programmer must only describe the structure of the equations (17), (18). Some defect of the method consists in necessity to compute n times the adjoint process system (19) for calculation of partial derivatives (16). It is possible however to avoid this defect. Let the computation system (17), (18) be written for each block of the system (1), (2). Each system (1) being replaced by the corresponding system (17), (18), one gets a new two-level system, CEO being the elements of the first level, and blocks of the original system-those of the second. The adjoint process for the new system includes N adjoint systems (19). Mathematically the new two-level system and its adjoint process are equivalent to the original ones. So in this case it is necessary to compute only once (as before) the two-level adjoint process. That means that the adjoint process corresponding to each block of the CCS must be computed only once.

REFERENCES

1. Fletcher R., Powell M.I.D., Computer Journal, v. 6, N°2, 163(1963).
2. Островский Г.М., Волин Ю.М. Методы оптимизации сложных химико-технологических схем. М., "Химия", 1970.
3. Островский Г.М., Борисов В.В., Волин Ю.М. Автоматика и вычислительная техника, 1973, № 2, 43.

EVALUATION MODEL OF HUMANISTIC SYSTEMS BY FUZZY MULTIPLE INTEGRAL AND FUZZY CORRELATION

Eiichiro Tazaki, Associate Professor
and Michio Amagasa, Research Associate
Faculty of Science and Technology
Science University of Tokyo
Noda City, Chiba 278, Japan

1. Introduction

The concept of fuzzy sets defined by Zadeh [1] gives an important mathematical clue for an approach to studies of systems with no sharp boundaries. Lately, the fuzzy integral [2] based on the fuzzy set theory has been proposed and applied to the measurement of fuzzy objects, especially, pattern recognitions.

We consider the fuzzy integral is also available to be applied to the evaluation of fuzzy systems, because the fuzzy measure in the fuzzy integral will be regarded as a preference measure of evaluation and the fuzzy integral will represent a conflicting process between the preference measure and the evaluated object. When we consider the evaluation problem of complex systems composed of several subsystems, especially humanistic type ones, it is very difficult to set up the utility model of overall system. Because, in general, the human preference has a hierarchical, strongly nonlinear and qualitative property. For such a case, we propose an utility model of evaluation by applying a fuzzy multiple integral. In this model the preference measure for the overall system is given by a composition of the subsystems' measure.[3]

Applying the proposed method to a practical cases, the preference measure has to be determined by experiment. To realize such a process, we introduce a fuzzy distribution function and give an algorithm to calculate the measure. Further, we introduce a fuzzy correlation among subsystems in order that the fuzzy measure will be identified effectively by experimental data.

The proposed method is successfully applied to the subjective evaluation of a class of figures' largeness.

2. Evaluation of complex systems by fuzzy multiple integral

At first, we shall consider an evaluation of simple system composed of one subsystem as a preliminary for the evaluation of complex system.

The proposed model of evaluation is based on the fuzzy integral, where the fuzzy measure g is considered as a measuring scale of preference, The fuzzy measure g is given on a family of subsets 2^X with respect to X as follows:

$$0 \leq g(X') \leq 1 \ , \ X' \varepsilon 2^X \qquad (1)$$

$$X' \subset X'' \subset X \ \rightarrow g(X') \leq g(X'') \qquad (2)$$

where the system X is a finite set.

Now, let $h(x)$ be a membership function with respect to the object of evaluation. Then the fuzzy integral is defined by

$$\int h(x)\circ g(\cdot) = \vee_{i=1}^{N}[\, h(x_i)\wedge g(X_i)\,] \tag{3}$$

where $X=\{x_1,x_2,\text{----},x_N\}$, $X_i=\{x_1,x_2,\text{----},x_i\}$.

The fuzzy integral will represent a conflicting process between the preference measure and the evaluated object. The fuzzy measure g is called, hereafter, the preference measure of evaluation. Therefore, if the fuzzy measure $g(X_i)$ is given for any subset of 2^X, the fuzzy integral (3) can be calculated. In order to simplify the successive discussion, we assume that the rule for the generation of subsets with respect to the measure is defined as follows:

$$g_\lambda(X'\cup X'')=g_\lambda(X')+g_\lambda(X'')+\lambda g_\lambda(X')\cdot g_\lambda(X'') \qquad -1<\lambda<\infty \tag{4}$$

On the basis of the above mentioned results, we consider the evaluation of complex system Z composed of two subsystems X and Y. When the fuzzy measure for the subsystems X and Y are expressed by g_X and g_Y, respectively, we will consider a product measure g_λ in the product space X×Y. Let H be a family of subsets in Z, then H is expressed the following form,

$$H=\cup_{i=1}^{N}[X_i\times Y_i] \tag{5}$$

where $\{X_i\}$,$\{Y_i\}$ are monotone increasing sequences formed by $X_i = \{x_1,x_2,\text{----},x_i\}$, $Y_i = \{y_1,y_2,\text{----},y_i\}$, respectively.

The product measure g_λ with respect to H is described as follows:

$$g_\lambda(H) = \vee_{i=1}^{N}[g_X(X_i)\wedge g_Y(Y_i)] \tag{6}$$

Hence, the double integral with respect to the two variables function $h(z)=h(x,y)$ is defined as follows:

$$\vee_{j=1}^{N}[\vee_{i=1}^{N}\{h(x_i,y_j)\wedge g_X(X_i)\}\wedge g_Y(Y_j)] \tag{7}$$

$$(=\vee[\vee\{h(x_i,y_j)\wedge g_Y(Y_j)\}\wedge g_X(X_i)]) \qquad \text{, where } 0 \le h \le 1.$$

The equality in (7) is held when h(z)is fuzzy measurable with respect to the monotone family $H\subset Z$. Thus, if the evaluation of overall system is given by the fuzzy double integral and if we apply such a method to the practical problems, the preference measure has to be identified. In order to realize the process, it is necessary to obtain the fuzzy distribution function and the fuzzy frequency function of each subsystem, respectively, given by the experiment.

Here we call the function F defined by the following equation the fuzzy distribution function.

$$F(Z_i) = \vee_{i=1}^{N}[F_X(X_i)\wedge F_Y(Y_i)] \tag{8}$$

The function F_X and F_Y are, respectively, defined on the monotone sequence sets $\{X_i\}$ and $\{Y_i\}$, and these functions have the following several properties.

$$0 \le F_X(X_i) \le 1 \text{ , } i=1,2,\text{----},N \tag{9}$$

$$F(X_i) \le F_X(X_\kappa) \text{ , } i \le \kappa \tag{10}$$

$$F_X(X_N) = 1 \tag{11}$$

where $X_i = \{x_1,x_2,\text{----},x_i\}$, and $Y_i = \{y_1,y_2,\text{----},y_{N+1-i}\}$

F_Y has also the same properties as F_X, but the inequality sign is converse of (10), and then $F_Y(Y_1)=1$. Hence, if the fuzzy distribution function is determined in each subsystem, the distribution function of overall system can be determined.

On the other hand, when the rule (4) is applied to the functions g_1 and g_2, respectively, the fuzzy frequency functions can be expressed by

$$g_1^1 = F_X(X_1),\quad g_1^2 = (F_X(X_2)-F_X(X_1))/(1+\lambda_X F_X(X_1)),$$
$$\text{-------------},\ g_1^i = (F_X(X_i)-F_X(X_{i-1}))/(1+\lambda_X F_X(X_{i-1})) \qquad (12)$$
$$g_2^1 = F_Y(Y_1),\quad g_2^2 = (F_Y(Y_1)-F_Y(Y_2))/(1+\lambda_Y F_Y(Y_2)),$$
$$\text{-------------},\ g_2^i = (F_Y(Y_{i-1})-F_Y(Y_i))/(1+\lambda_Y F_Y(Y_i)) \qquad (13)$$

where $-1 < \lambda_X, \lambda_Y < \infty$.

In consequence, we define $g=g_1 \times g_2$ on the set $S(Z)=\cup_{i=1}^{N}(X_i \times Y_i)$ composed of the monotone sequences of subsets in the system Z, and we define (14) as the fuzzy product frequency function.

$$g(S(Z)) = \vee_{i=1}^{N} g_1^i \wedge g_2^i \qquad (14)$$

Therefore, if the fuzzy distribution function of each subsystem can be given by using the experimental data, the fuzzy distribution function and the fuzzy frequency function of overall system can be, respectively, determined by equations (8), (12), (13) and (14).

3. Fuzzy correlation

We introduce a fuzzy correlation among subsystems in order that the fuzzy measure will be identified effectively by experimental data. We define the fuzzy correlation η_{XY} between the subsystems X and Y as well as it in the sense of statistics.

$$\eta_{XY} = \zeta_{XY}/\zeta_X \zeta_Y \qquad (15)$$

where ζ_{XY}, ζ_X^2 and ζ_Y^2 are the fuzzy covariance and the fuzzy variances, respectively, and are represented as follows:

$$\zeta_X^2 = \vee_{i=1}^{N}(X_i-\mu_X)^2 g_1^i \wedge g_X(X_i) \qquad (16)$$
$$\zeta_Y^2 = \vee_{i=1}^{N}(Y_i-\mu_Y)^2 g_2^i \wedge g_Y(Y_i) \qquad (17)$$
$$\zeta_{XY} = \vee_{j=1}^{N}\{\vee_{i=1}^{N}((X_i-\mu_X)(Y_j-\mu_Y)g(X_i,Y_j)\wedge g_X(X_i))\wedge g_Y(Y_j)\} \qquad (18)$$
$$\mu_X = \vee_{i=1}^{N} X_i g_1^i \wedge g_X(X_i) \qquad (19)$$
$$\mu_Y = \vee_{i=1}^{N} Y_i g_2^i \wedge g_Y(Y_i) \qquad (20)$$

The fuzzy correlation η_{XY} shows the degree of connection among subsystems X and Y in fuzzifical sense. The range of its value is $|\eta_{XY}| \leq 1$. In particular, if η_{XY} equals to zero, the subsystem X is sufficiently independent from Y. Therefore, both of the fuzzy distribution function and the fuzzy frequency function of overall system can be, respectively, formulated by the product of isolated ones with respect to the subsystems X and Y. Then, the evaluation of overall system is given by the product of the respective fuzzy integral for the subsystems X and Y.

4. Numerical example

We will consider the length and extent of figures as the object of evaluation. We synthesize a mathematical model of the subjective evaluation as is shown in Fig.1. The proposed method will be inspected by experiments in the following section.

4.1 Construction of subjective evaluation

The model of subjective evaluation is shown in Fig.1. The p and $\tilde{p}$ are stimulus variables from the object of evaluation and are input variables to the external and internal mechanisms, respectively. The Π and $\tilde{\Pi}$ convert p and $\tilde{p}$ to the inner evaluation x^i and $\tilde{x}$, respectively. That is

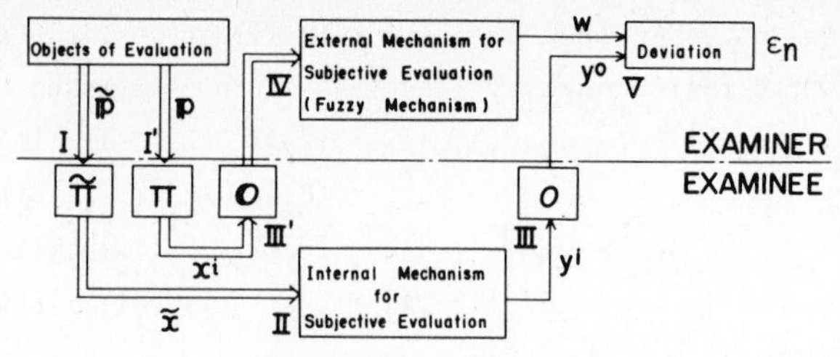

FIG. I Construction of Subjective Evaluation

$$\Pi : p \rightarrow x^i \qquad \tilde{\Pi} : \tilde{p} \rightarrow \tilde{x} \tag{21}$$

The y^i is an integrated inner evaluation which can be determined by means of the internal mechanism. The 0 and **0** are the interfaces so as to replace the inner evaluation with numerical value. And sense and ability of examinee with respect to number are also involved in it. The w is an outer evaluation computed by the external mechanism. The y^0 is the outer evaluation obtained through the interface 0.

The path I → II → III → V shows the process of internal evaluation and the path I → III′→ IV → V shows the process of external evaluation. The ε_n is a deviation of w and y^0. In the external mechanism, we explain the fuzzy mechanism for the external evaluation in order to obtain the value of evaluation w.
The fuzzy mechanism is depicted in Fig.II.

4.2 Computational algorithm

In this section, we attempt to show a computational algorithm to calculate the fuzzy measure from experimental data. The algorithm will be easily obtained with reference to Fig.II as follows:

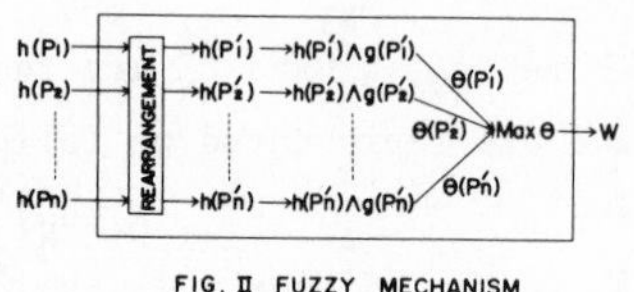

FIG. II FUZZY MECHANISM

STEP 1 Formulate the membership function $h(p_i)$ for the objects of evaluation and rearrange the elements $h(p_i)$:i=1,2,----,n according to the ordering of largeness as shown in Fig.II.

STEP 2 Calculate the fuzzy distribution and frequency functions according to (8),(12) ,(13) and (14), respectively, from experimental data.

STEP 3 According to (4), determine the fuzzy preference measure using the fuzzy frequency function.

STEP 4 Compute the value of evaluation w by means of the fuzzy mechanism.

STEP 5 Determine an optimal λ so that the deviation ε_n is minimized.

4.3 Formulation of the membership function with respect to shape and color

We consider shapes and colors as attributes standing for length and extent of figures. The attribute of shape consists of the ratio of length and breadth, the ratio of the shortest and the longest length and the length of circuit etc., and the attribute of color consists of brightness, chroma and hue. We define the composed objective membership functions of length and extent, respectively, subject to these attributes.

Shape : $X=(X_1,X_2,X_3,X_4)$=(the longest length, ratio of length and breadth, length of a circuit, ratio of the shortest length and the longest length)

Color : $Y=(Y_1,Y_2,Y_3)$=(brightness, chroma, hue)

$Z = (X,Y) = (Z_1,Z_2,Z_3,Z_4,Z_5,Z_6,Z_7)$

A) Membership function of length or extent with respect to shape and color

$$h_1 : Z_1 \times Z_2 \times \text{-----} \times Z_7 \rightarrow [0,1] \tag{22}$$

The h_1 is the membership function for the length or extent of figure.

B) Membership function of length with respect to shape

$$h_2 : X_1 \times X_2 \rightarrow [0,1] \tag{23}$$

C) Membership function of length with respect to color

$$h_3 : Y_1 \times Y_2 \times Y_3 \rightarrow [0,1] \tag{24}$$

We know that the color does not almost give an effect to the membership of length in general.

D) Membership function of extent with respect to shape

$$h_4 : X_3 \times X_4 \rightarrow [0,1] \tag{25}$$

E) Membership function of extent with respect to color

$$h_5 : Y_1 \times Y_2 \times Y_3 \rightarrow [0,1] \tag{26}$$

F) Membership function of length and extent with respect to shape and color

$$h_6 : h_2 \times h_3 \rightarrow [0,1] \qquad h_7 : h_4 \times h_5 \rightarrow [0,1] \tag{27}$$

In practice, the proposed formulations of membership functions are, respectively, given by the followings:

A') $h_1=\Sigma_{i=1}^{K} M_i/K$ The M_i is a membership function corresponding to the fractionized part of shape.

B') $h_2=1/(1+\alpha)\cdot(h_i+h_j)-0.05$,where $0.1<\alpha<1.0$ and h_i and h_j are the membership functions for the longest length of figure and the ratio of length and breadth, respectively, and $h_i=\log x$, $h_j=\log y$.

C') The membership function of length with respect to color is constant, that is $h_3=0.5$

D') $h_4=\alpha h_i+\beta h_j$, $(\alpha+\beta=1)$, where h_i is membership function of length of circuit and h_j is the membership function of the ratio of the shortest length and the longest length. Further, α and β are constant scalars, respectively.

E') $h_5=\alpha h_i+\beta h_j+\gamma h_k+c$, $(\alpha+\beta+\gamma=1)$, where α,β and γ are constant scalars, respectively and h_i, h_j and h_k are the membership functions for brightness,chroma and hue, respectively.

Hence, the objective membership functions of length and extent can be composed by the above described membership functions. According to the above formulation, the objective memberships are obtained as shown in Tables 1,2,3 and 4. Further, on the basis of these results, the composed objective memberships with respect to shape and color are given by Tables 5 and 6.

NO.	Shape	M.F
1		0.66
2		0.64
3		0.81
4		0.94
5		0.96
6		0.92
7		0.63
8		0.95
9		0.50
10		0.61

Table 1. Membership function of length with respect to shape. Basic Shape : NO. 9

NO.	Color
1	Black
2	Brown
3	Red
4	Purple
5	Yellow
6	Orange
7	Blue
8	Green

Table 2. Membership function of length with respect to color. Basic Color : NO. 8

NO.	Shape	M.F.
1		0.28
2		0.44
3		0.78
4		0.49
5		0.56
6		0.64
7		0.58
8		0.51
9		0.78
10		0.70
11		0.47

Table 3. Membership function of extent with respect to shape. Basic Shape : NO. 11

NO.	Color	M.F.
1	Red	0.64
2	Orange	0.69
3	Yellow Orange	0.71
4	Yellow	0.90
5	Yellow Green	0.75
6	Green	0.54
7	Green Blue	0.52
8	Blue Green	0.43
9	Blue	0.46
10	Blue Purple	0.27
11	Purple	0.30
12	Red Purple	0.31

Table 4. Membership function of extent with respect to color Basic Color : NO. 7

NO.	M.F.
1	0.58
2	0.57
3	0.65
4	0.72
5	0.73
6	0.71
7	0.56
8	0.72
9	0.50
10	0.55

Table 5. Composed membership function of length with respect to shape and color.

Shape \ Color	1	2	3	4	5	6	7	8	9	10	11	12
1	0.46	0.48	0.49	0.59	0.51	0.41	0.40	0.35	0.37	0.27	0.29	0.29
2	0.54	0.56	0.57	0.67	0.59	0.49	0.48	0.43	0.45	0.35	0.37	0.37
3	0.71	0.73	0.74	0.84	0.76	0.66	0.65	0.60	0.62	0.52	0.54	0.54
4	0.56	0.59	0.60	0.69	0.62	0.51	0.50	0.45	0.47	0.37	0.39	0.39
5	0.60	0.62	0.63	0.73	0.65	0.55	0.54	0.49	0.50	0.41	0.43	0.43
6	0.64	0.66	0.67	0.77	0.69	0.59	0.58	0.53	0.55	0.45	0.47	0.47
7	0.67	0.63	0.64	0.74	0.66	0.56	0.55	0.50	0.51	0.42	0.43	0.44
8	0.57	0.60	0.61	0.70	0.63	0.52	0.51	0.47	0.48	0.38	0.40	0.41
9	0.71	0.73	0.74	0.84	0.76	0.66	0.65	0.60	0.62	0.52	0.54	0.54
10	0.67	0.69	0.70	0.80	0.72	0.62	0.61	0.50	0.58	0.48	0.50	0.50
11	0.55	0.58	0.59	0.68	0.61	0.50	0.49	0.45	0.46	0.37	0.38	0.39

Table 6. Composed membership function of extent.

NO.	Shape	Color
1	0.50	0.50
2	0.60	0.50
3	0.30	0.50
4	0.30	0.50
5	0.70	0.50
6	0.60	0.50
7	0.40	0.50
8	0.70	0.50
9	0.50	
10	0.60	

Table 7. Results of an experiment for length with respect to shape and color.

NO.	Shape	Color
1	0.40	0.50
2	0.60	0.40
3	0.70	0.50
4	0.50	0.40
5	0.60	0.40
6	0.50	0.50
7	0.60	0.50
8	0.50	0.60
9	0.60	0.70
10	0.70	0.60
11	0.50	0.70
12		0.70

Table 8. Results of an experiment for extent with respect to shape and color.

On the other hand, in order to identify the preference measure of evaluation, we had an experiment such that examinees evaluate the length and extent of figures. As the results of experiment, we have obtained the values of evaluation with respect to shape and color as shown in Tables 7 and 8.

According to the computational algorithm described in section 4.2, an optimal λ for the preference measure is given by Table 9.

In order to inspect an advantage of the proposed method, the evaluation of figures is executed by using the calculated measure. The comparison of computed results and experimental results is shown by Table 10. It shows that a good consistency is held within the limit of this problem.

	W	y^{o}	λ
Length (Shape)	0.61	0.60	-0.98
Length(Shape,Color)	0.57	0.60	-0.20
Extent(Shape,Color)	0.45	0.40	1.00

Table 9. Example of optimal λ

$\lambda = 1$

Extent	W	y^{o}	ϵ_n
Figure 1	0.565	0.500	0.065
Figure 2	0.505	0.600	0.095

Table 10. Comparison of a computed result and an experimental result.

5. Conclusion

We have proposed a mathematical model of subjective evaluation on the basis of fuzzy integral. In particular, we have developed the evaluating method of complex systems composed of several subsystems by virtue of fuzzy multiple integral.

Furthermore, for the sake of identifying the preference measure, that is fuzzy measure , effectively, we have introduced the fuzzy correlation.

In order to show how the proposed method works, an example of the subjective evaluation for a class of figures has been illustrated and the consistency of experimental values and computed values has been successfully obtained as shown in Table 10.

Applications to the marginal evaluation problem to meet specification will be a future extension of this approach.

References

1) L.A.Zadeh : Fuzzy Sets, Information and Control, 8, 338/353 (1965)

2) M.Sugeno : Fuzzy Measure and Fuzzy Integral, Trans. SICE(Japan), 8, 218/226 (1972)

3) E.Tazaki and M.Amagasa : Evaluation of Complex Systems by Fuzzy Multiple Integral and Fuzzy Correlation, Paper presented to the 16-th Joint Annual Meeting of Automatic Control, Tokyo, Japan (1973)

PENALTY FUNCTION METHOD AND NECESSARY OPTIMUM CONDITIONS IN OPTIMAL CONTROL PROBLEMS WITH BOUNDED STATE VARIABLES

Yu. M. Volin
Karpov Institute of Physical Chemistry
ul. Obukha 10, 107120, Moscow, B-120/U.S.S.R.

An approach to determine necessary optimum conditions (n.o.c.) for generalized solutions (g.s.) of mathematical programming problem based on penalty function method is considered. The results are used to derive generalized maximum principle (g.m.p.) for optimal control problem with ordinary differential equations and bounded state variables. For linear state variables problems g.m.p. has also turned out to be sufficient optimal condition and in this case generalized duality theorem (g.d.th.) takes place.

1. Mathematical Programming Problem

Let us consider the following problem

$$I(u) \rightarrow \sup, \tag{1.1}$$

$$u \in U \subset B_u, \tag{1.2}$$

$$F(u) \in K \subset B_F \tag{1.3}$$

(B_u, B_F are Banach spaces, K is a convex closed cone) and assume that $I(u) < C < \infty$ when $u \in U$.

Let $\mathcal{H}$ be a set of sequences $\{u^{(k)}\}$ for which: $\{u^{(k)}\} \in \mathcal{H}$, $\|\tilde{u}^{(k)} - u^{(k)}\| \rightarrow 0$, $k \rightarrow \infty \Rightarrow$ $\{\tilde{u}^{(k)}\} \in \mathcal{H}$. Designate with $\mathcal{V}$ the subset of all $\{u^{(k)}\} \in \mathcal{H}$ satisfying the condition: $R(F(u^{(k)})) \rightarrow 0$, $k \rightarrow \infty$, where $R(x) = \inf \|x-y\|^2$ $(y \in K)$ ($\mathcal{V}$ may be called the g.s. set of the system of conditions (1.2), (1.3)). Let $\underline{I}(\{u^{(k)}\}) = \lim\limits_{k\rightarrow\infty} \underline{I}(u^{(k)})$, $\bar{I} = \sup \underline{I}(\{u^{(k)}\})$ $(\{u^{(k)}\} \in \mathcal{V})$. A sequence $\{u^{(k)}\} \in \mathcal{V}$ will be called g.s. of the problem (1.1)-(1.3) if

$$\lim I(u^{(k)}) = \bar{I}, \quad k \rightarrow \infty.$$

Let us assume further n.o.c. of g.s. to be known for every $\mathcal{I}(u)$ from a sufficiently wide class of functionals in the following asymptotical form

$$P_{\mathcal{F}}\ (\mathcal{U}^{(k)}) \to 0,\ k \to \infty, \tag{1.4}$$

where $P_{\mathcal{F}}(\mathcal{U})$ is a functional. In papers /1-5/ an approach based on penalty function method and diagonal transference procedure has been developed. This approach makes it possible for a wide class of optimization problems to pass from condition (1.4) to n.o.c. of problems with additional restrictions. Generalizing the method of these papers it is possible to obtain the following results.

Let $\{\mathcal{U}^{(k)}\}$ be g.s. of the problem (1.1)-(1.3) and

$$I_{k\alpha\beta}\ (\mathcal{U}) = I(\mathcal{U}) - \alpha R(F(\mathcal{U})) - \beta \|\mathcal{U} - \mathcal{U}^{(k)}\|^2 \qquad (k = 1,2,\dots;\ \alpha,\beta \geq 0). \tag{1.5}$$

Let us consider the set of problems (1.5), (1.2). Let $I_{k\alpha\beta} \in \{\mathcal{F}\}$ Then for a g.s. $\{\mathcal{U}^{(m)}_{k\alpha\beta}\}$ of the problem $I_{k\alpha\beta}(\mathcal{U}) \to \sup$, (1.2) the n.o.c. is realized:

$$P_{k\beta}\,(\mathcal{U}^{(m)}_{k\alpha\beta},\alpha) = P_{I_{k\alpha\beta}}\,(\mathcal{U}^{(m)}_{k\alpha\beta}) \to 0,\ m \to \infty. \tag{1.6}$$

Let us presume further that the following inequality takes place

$$|\,P_{k\beta}\,(\mathcal{U},\alpha) - P\ (\mathcal{U},\alpha)| \leq A\beta \quad (\forall k,\alpha) \quad (P = P_{k0}). \tag{1.7}$$

<u>Theorem 1</u>. If $\{\mathcal{U}^{(k)}\}$ is a g.s. of (1.1)-(1.3) then there exist sequences $\{\tilde{u}^{(k)}\},\ \{\alpha_k\}$ $(\alpha_k \to \infty,\ k \to \infty)$ for which

$$P(\tilde{u}^{(k)},\alpha_k),\ \|\tilde{u}^{(k)} - \mathcal{U}^{(k)}\|,\ \alpha_k R(F(\tilde{u}^{(k)})) \to 0,_{k\to\infty.} \tag{1.8}$$

<u>Proof (sketch)</u>. Let $\bar{I}_{k\alpha\beta} = \sup I_{k\alpha\beta}\,(\mathcal{U})\ (\mathcal{U} \in \mathbf{U})$ and $\varepsilon_k \to +0$. For $\beta > 0$ there are $\{\alpha_k\},\ \{m_{k\beta}\}$ for which $\alpha_k \to \infty$, $\alpha_k\ R(F(\mathcal{U}^{(k)})) \to 0,\ k \to \infty,\ |\,I_{k\alpha_k\beta}\,(\mathcal{U}^{(k)}_\beta) - \bar{I}_{k\alpha_k\beta}| + |P_{k\beta}\,(\mathcal{U}^{(k)}_\beta,\alpha_k)| < \varepsilon_k$, $\mathcal{U}^{(k)}_\beta = \mathcal{U}^{(m_{k\beta})}_{k\alpha_k\beta}$. As $\bar{I}_{k\alpha_k\beta} > C_1$ (C_1 is some constant) it follows: $\alpha_k R\ (F(\mathcal{U}^{(k)}_\beta)) + \beta\|\mathcal{U}^{(k)}_\beta - \mathcal{U}^{(k)}\|^2 \to 0.$ Now with the help of diagonal transference procedure with respect to β it is possible to show the validity of (1.8).

2. Optimal Control Problem

Let us consider now the following optimal control problem

$$\left.\begin{aligned} &I(u(\cdot)) = \varphi_0\,(x^1) \to \sup\ \ (x^1 = x(t_1)),\\ &\frac{dx}{dt} = f(x,u,t)\ \ (t \in T = [t_0,t_1]),\ x(t_0) = x^0, \end{aligned}\right\} \tag{2.1}$$

$$u(\cdot) \in \mathbf{U} \tag{2.2}$$

$$\varphi(x^1)=0, \quad g(x(t),t)\le 0. \tag{2.3}$$

x, u, f, φ, g are vectors with dimensions n, m, n, q, s. Functions $f, \frac{\partial f}{\partial x}, g, \varphi_i$ are supposed to be continuous on $E^n \times U$; $\frac{\partial^2 f}{\partial x^2}$ is assumedto be restricted. With respect to t these functions are assumed to be measurable and uniformly restricted.

Let a set of $u(\cdot)$ with values in bounded set V and dense with respect to measure in the set of all measurable $u(\cdot)$ $(u(t)\in V)$ be the set of admissible controls. Suppose also that Lipschitz condition for the function f with respect to x is fulfilled uniformly with respect to $u\in V$, $t\in T$.

The problem (2.1)-(2.3) may be considered as a special case of (1.1)-(1.3) if, for example, we put $\mathcal{U}=u(\cdot)$, $B_{\mathcal{U}}=L_2(R^m, T)$,

$$B_F=R^q\times L_2(R^s, T),\ K=\{y\in R^q, z(\cdot)\in L_2(R^s, T): y=0, z(t)\le 0, t\in T\}.$$

In this case $R(y, z(\cdot))=\|y\|^2+\int_T\|\tilde{\tilde{z}}(t)\|^2dt$, where $\tilde{\tilde{z}}(t)=\begin{cases} z(t), & z(t)>0\\ 0, & z(t)\le 0.\end{cases}$ Let conditions (2.3) correspond to the restriction (1.3) and $\mathcal{U}$ be the set of such sequences $\{u^{(k)}(\cdot)\}$ ($u^{(k)}(\cdot)$ is measurable; $u^{(k)}(t)\in V, t\in T$) that the corresponding $x^{(k)}(\cdot)$ in $C(R^n, T)$ converges to some $x(\cdot)$: $x^{(k)}(\cdot)\rightrightarrows x(\cdot)$. $\{\ ^{(k)}(\cdot)\}\in\mathcal{U}\Rightarrow(2.3)$ for $x(\cdot)$.

Let $\{u^{(k)}(\cdot)\}$ be g.s. of the problem (2.1)-(2.3). Designate

$$I_{k\alpha\beta}(u(\cdot))=\varphi_0(x^1)-\alpha(\|\varphi(x^1)\|^2+ \int_T\|\tilde{\tilde{g}}(x(t),t)\|^2dt)-\beta\int_T\|u(t)-u^{(k)}(t)\|^2dt. \tag{2.4}$$

The n.o.c. in the problem (2.4), (2.1-II), (2.2) has the form (1.6) if

$$P_{k\beta}(\mathcal{U},\alpha)=\int_T(\bar{H}_{k\beta}(x,\psi,t)-H_{k\beta}(x,\psi,u,t))\,dt \tag{2.5}$$

where $H_{k\beta}(x,\psi,u,t)=\psi'f(x,u,t)-\beta\|u-u^{(k)}(t)\|^2$, $\bar{H}_{k\beta}(x,\psi,t)=\sup H_{k\beta}(x,\psi,u,t)$ $(u\in V$ and $\psi(\cdot)$ satisfies the equations

$$\frac{d\psi'}{dt}=-\psi'\frac{\partial f}{\partial x}+2\alpha\tilde{\tilde{g}}'\frac{\partial g}{\partial x},\quad \psi'(t_1)=\frac{\partial\varphi_0}{\partial x}-2\alpha\varphi'\frac{\partial\varphi}{\partial x}. \tag{2.6}$$

„ ′ " designates transposition. This n.o.c. will be called g.m.p. For the problem (2.4), (2.1-II) g.m.p. has been proved in /2/.

The fulfilment of (1.7) follows from the boundedness of V . Now the following theorem is a simple consequence of theorem 1.

Theorem 2(g.m.p. of the problem (2.1)-(2.3)). Let $\{u^{(k)}(\cdot)\}$ be g.s. of (2.1)-(2.3). Then there exist sequences $\{\tilde{u}^{(k)}(\cdot)\}$ $\{\lambda^{(k)}\}, \{\mu^{(k)}(\cdot)\}$ $(\tilde{u}^{(k)}(\cdot)\in U, \lambda^{(k)}\in R^q, \mu^{(k)}(\cdot)\in C(R^s, T))$ for which

$$\int_T (\bar{H}(\tilde{x}^{(k)}, \tilde{\psi}^{(k)}, t) - H(\tilde{x}^{(k)}, \tilde{\psi}^{(k)}, \tilde{u}^{(k)}, t))dt \to 0, \quad (2.7)$$

$$\int_T \|\tilde{u}^{(k)}(t) - u^{(k)}(t)\| dt \to 0, \quad (2.8)$$

$$|(\lambda^{(k)})'\varphi(\tilde{x}^{(k)}(t_1))| + \int_T (\mu^{(k)})' g(\tilde{x}^{(k)}(t), t)dt \to 0 \quad (2.9)$$

if $k\to\infty$ and $\tilde{\psi}^{(k)}(\cdot), \mu^{(k)}(\cdot)$ satisfy the equations

$$\frac{d\psi'}{dt} = -\psi'\frac{\partial f}{\partial x} + \mu'\frac{\partial g}{\partial x}, \quad \psi'(t_1) = \frac{\partial \varphi_0}{\partial x} + \lambda'\frac{\partial \varphi}{\partial x} \quad (2.10)$$

(with transpositions $\mu^{(k)}(\cdot), \lambda^{(k)}, \tilde{\psi}^{(k)}(\cdot), \tilde{u}^{(k)}(\cdot), \tilde{x}^{(k)}(\cdot)$),

$$\mu_i^{(k)}(t) \geq 0, \ \mu_i^{(k)}(t) = 0 \ \text{ if } \ g_i(\tilde{x}^{(k)}(t), t) < 0. \quad (2.11)$$

In (2.7) $H(x, \psi, u, t) = \psi' f(x, u, t).$ (2.12)

$\lambda^{(k)}, \mu^{(k)}(\cdot)$ being normed, the following theorem immediately follows from theorem 2.

Theorem 3. Let $\{u^{(k)}(\cdot)\}$ be g.s. of the problem (2.1)-(2.3) and $x^{(k)}(\cdot) \rightrightarrows x(\cdot)$. Then there exist $\lambda_0, \lambda, \mu(\cdot)$ $(\lambda_0 \geq 0, \lambda\in R^q, \mu_i\ (i=\overline{1,s})$ is a nonnegative measure concentrated on the set $\{t : t\in T, g_i(x(t), t) = 0\}$) for which

$$\int_T (\bar{H}(x, \psi^{(k)}, t) - H(x, \psi^{(k)}, u^{(k)}, t))\, dt \to 0, \quad (2.13)$$

$$\lambda_0 + \|\lambda\| + \int_T \sum_{i=1}^{s} d\mu_i(t) > 0 \quad (2.14)$$

where $\psi^{(k)}(\cdot)$ satisfies (2.10-I) (with transposition $u^{(k)}(\cdot), x(\cdot)$) and

$$\psi'(t_1) = \lambda_0\frac{\partial \varphi_0}{\partial x} + \lambda'\frac{\partial \varphi}{\partial x}. \quad (2.15)$$

The theorem similar to theorem 3 has been proved in /6/ (analogous theorem with the replacement of asymptotical equality (2.13) by the precise one is valid for classical solutions of the problem (2.1)-(2.3) /7/ and for stationary g.s., the last assertion being a simple corollary of theorem 3).

Let us emphasize that n.o.c. of theorem 3 are weaker than those of theorem 2 (see the example given below).

3. Linear state variable problem

Let

$$f(x,u,t)=a(t)x+b(u,t),\quad g(x,t)=c(t)x+d(t),$$
$$\varphi_i(x)=p_i'x+q_i,\quad i=\overline{0,q}. \tag{3.1}$$

In this case g.m.p. gives also sufficient optimal condition.

<u>Theorem 4</u>. Suppose $\{\tilde{u}^{(K)}(\cdot)\}, \{\lambda^{(K)}\}, \{\mu^{(K)}(\cdot)\}$ satisfying (2.7)-(2.12) exist for a given $\{u^{(K)}(\cdot)\}$. Then $\{u^{(K)}(\cdot)\}$ is the g.s. of the problem (2.1)-(2.3).

<u>Proof</u> (sketch). Let $\{\bar{u}^{(K)}(\cdot)\}$ be a g.s. of (2.1-II), (2.2), (2.3). Without decreasing the generality of the results it is possible to assume that

$$\overline{\lim_{K\to\infty}}\left(|(\lambda^{(K)})'\varphi(\bar{x}^{(K)})|+\int_T(\mu^{(K)})'g(x^{(K)},t)dt\right)\le 0. \tag{3.2}$$

As f, g and φ_i are linear with respect to x it is valid (see /2,3/) that

$$I_{\lambda^{(K)},\mu^{(K)}}(\bar{u}^{(K)}(\cdot))-I_{\lambda^{(K)},\mu^{(K)}}(\tilde{u}^{(K)}(\cdot))= \tag{3.3}$$
$$\int_T\left(H(\tilde{x}^{(K)},\tilde{\psi}^{(K)},\bar{u}^{(K)},t)-H(\tilde{x}^{(K)},\tilde{\psi}^{(K)},\tilde{u}^{(K)},t)\right)dt$$

where

$$I_{\lambda,\mu}(u(\cdot))=\varphi_0(x^1)+\lambda'\varphi(x^1)-\int_T\mu'g(x,t)\,dt. \tag{3.4}$$

From (2.9), (3.2)-(3.4) follows the assertion of the theorem.

<u>Example</u>.

$$x(2)\to\sup,\quad \frac{dx}{dt}=u\ (t\in[0,2]),\ x(0)=-0{,}5;$$
$$|u|\le 1,\quad g(x,t)=(1-t)x\le 0. \tag{3.5}$$

Here $\frac{d\psi}{dt}=(1-t)\mu$, $H(x,\psi,u,t)=\psi u$ and (2.13),(2.14) are valid for every $u(\cdot)$ satisfying the condition $x(1)=0$ (with $\lambda_0 = \lambda = 0$, $\mu(t)=\delta(t-1)$). But g.m.p. is valid only for the optimal $u(\cdot)$.

4. Generalized Duality Theorem

Different results concerning g.d.th. for mathematical programming problem have been obtained by Golstein /8/. However it is interesting to show that for the problem (2.1)-(2.3), (3.1) g.d.th. is a very simple corollary of g.m.p.

Theorem 4. Let f, g, φ_i satisfy (3.1) and (see (3.4))

$$J(\lambda, \mu(\cdot)) = \sup_{u(\cdot)\in U} I_{\lambda,\mu}(u(\cdot)). \tag{4.1}$$

Then

$$\bar{I} = \underline{I}(\{u^{(k)}(\cdot)\}) = \inf J(\lambda, \mu(\cdot)) = \bar{J} \tag{4.2}$$

where $\lambda \in R^q$, $\mu(\cdot) \in C(R^s, T)$, $\mu_i(t) \geq 0 \; (i = \overline{1,s})$.

Proof. $I_{\lambda,\mu}(\{u^{(k)}(\cdot)\}) \geq \underline{I}(\{u^{(k)}(\cdot)\})$. So $\bar{J} \geq \bar{I}$. Let $\{\lambda^{(k)}\}$, $\{\mu^{(k)}(\cdot)\}$, $\{\tilde{u}^{(k)}(\cdot)\}$ be those given in theorem 2. From (3.3) (for an arbitrary $\tilde{u}^{(k)}(\cdot) \in U$), (2.7) and (2.9) $J(\lambda^{(k)}, \mu^{(k)}(\cdot)) - \underline{I}_{\lambda^{(k)},\mu^{(k)}}(\tilde{u}^{(k)}(\cdot)) \to 0$,

$I_{\lambda^{(k)},\mu^{(k)}}(\tilde{u}^{(k)}(\cdot)) - \underline{I}(\tilde{u}^{(k)}(\cdot)) \to 0$, $k \to \infty$. So

$J(\lambda^{(k)}, \mu^{(k)}(\cdot)) \to \bar{I}$, $k \to \infty$ and $\bar{J} = \bar{I}$.

Theorem 4 makes it possible to get solution algorithms for the problem (2.1)-(2.3), (3.1).

The author is grateful to A.V.Finkelstein for some useful remarks concerning parts 2 and 3 of this work.

REFERENCES

1. Волин Ю.М., Островский Г.М. Управляемые системы, вып.9, 43(1971).
2. Волин Ю.М. там же, вып. II, 83 (1973).
3. Волин Ю.М., Островский Г.М., Финкельштейн А.В. там же, 88 (1973)
4. Волин Ю.М., Островский Г.М., Финкельштейн А.В. Дифференциальные уравнения, т. 9, № 3, 423 (1973).
5. Волин Ю.М. В сб. "Математические методы в химии", Новосибирск, ч. 2, 142 (1973).
6. Schwarzkopf A.B. SIAM Journal Control, v.10, N°3, 487 (1972).
7. Дубовицкий А.Я., Милютин А.А. Журн. вычис. мат. и матем. физики, т. 5, № 3, 395 (1965).
8. Гольдштейн Е.Г. Теория двойственности в математическом программировании и ее приложения. Москва, Физматгиз, 1971.

MULTILEVEL OPTIMAL CONTROL OF INTERCONNECTED DISTRIBUTED PARAMETER SYSTEMS

M. Amouroux - J.P. Babary - B. Pradin - A. Titli
Laboratoire d'Automatique et d'Analyse des Systèmes du C.N.R.S.
7, Avenue du Colonel Roche
B.P. 4036
31055 TOULOUSE CEDEX
France

I - INTRODUCTION

The control of distributed parameter systems presents a great many theoretical problems and has already given rise, over the last few years, to some fundamental research. However, the results obtained so far do not, in general, enable one to get round the difficulties which arise when the proposed control laws are applied. For this reason, a number of research teams have turned their attention to the use of new concepts such as that of classical decomposition-coordination in hierarchical control. This idea has been introduced either from a somewhat mathematical point of view [1, 2] or from more of a "control" angle [3 to 5] . In the latter case, techniques of dynamic and static hierarchical control (two level) are used to control a collection of interconnected subsystems. This collection is obtained by discretising the initial partial differential problem in space, or in time and space. It would seem at present that research in this direction should be limited to the field of applications. It is concerned at any rate with the resolution of an overall problem, whilst at the present time there is good reason for considering a more important problem, numerous cases of which are to be found in the economic sector, that of the optimal control of a group of interconnected sub-systems the behaviour of each of which is defined by partial differential equations.

The use of hierarchical control techniques to solve such a problem is being examined at the present time at the Laboratoire d'Automatique et d'Analyse des Systèmes in Toulouse as a joint project, undertaken by two groups : "Hierarchical Control" and "Distributed Parameter Processes".

The aim of this paper is to show how the problem has been tackled, and to present some of the results which have already been obtained. In the first part, we show how, for a certain class of distributed parameter systems, it is possible to decompose the overall problem into sub-problems, whilst at the same time retaining their "distributed parameter" nature ; in this part, the coordination task necessary to reach the overall solution is defined. The second part is devoted to the study of the sub-system and its controllability particularly as regards the coordination problems ; the implementation of the actuators is in fact the second essential aspect of this study.

II - OPTIMISATION OF INTERCONNECTED SYSTEMS

II-1 - Brief survey of the principles of decomposition-coordination [4]

As a result of the problems presented by the control of complex systems composed of interconnected sub-systems, (when the overall approach is too costly) or because of theoretical considerations (difficulties of convergence of the algorithms for large scale problems) to introduce new methods, particularly the decomposition-coordination methods which are found in hierarchical control. These techniques use multi-level and multi-objective pyramid shaped control structures, and are able to employ different types of decomposition (or division) of work :

- horizontal division based on the complexity of the process and its "interconnected sub-systems" aspect ;
- vertical division based on the complexity of the control with the levels : regulation, optimisation, self-adaptation, self-organisation ;
- functional division according to the situations which the process will come up against.

The principle of hierarchical control therefore consists of decomposing an overall problem into a certain number of sub-problems $P_i(\mu)$ with parameters related to μ ("intervention vector" or "coordination parameter") in such a way as to satisfy :

$$\text{Sol } P_1(\mu) \dots P_i(\mu) \dots\dots P_N(\mu) \implies \text{Sol } P \text{ (overall problem)}$$

The entire coordination problem consists, on the higher level, of forcing μ to approach a value μ^* which will lead to the solution of the overall problem. For the optimisation problem with which we are concerned, only the horizontal division can be used in, for example, a two level control structure. Each sub-problem is then defined by a group of two functions, essentially : the mathematical model of the process and the criterion which is associated with it. There are three possible types of coordination :

- manipulation of the criterion function ($\mu = \beta$, Lagrangian parameters associated with the interconnexion constraint ; modification of the criterion function by the coupling terms) ;
- action on the model ($\mu = X$, coupling variables between sub-systems ;
- simultaneous action on the two functions ($\mu = \begin{bmatrix} \beta \\ X \end{bmatrix}$)

It can be shown that the choice of the coordination variables leads to a separable form, the additive $H = \sum_i H_i$ for the Hamiltonian associated with the optimisation problem (H_i = Hamiltonian i, bringing in only index i variables, except for the coordination variable). The examination of H_i enables the sub-problems to be formulated in optimisation terms.

II-2 - Formulation of the overall problem

Consider an overall system composed of N linear invariant, interconnected sub-systems defined by the following equations :

$$\left\{ \begin{array}{l} \dfrac{\partial Y_i(x,t)}{\partial t} = M_i \left[Y_i(x,t) \right] + X_i(x,t) + B_i(x)\, U_i(x,t) \\ \text{initial conditions : } Y_i(x,0) = Y_{i0}(x) \qquad i = 1, \ldots, N \\ \text{boundary conditions : } L_i \left[Y_i(x',t) \right] = 0 \end{array} \right. \tag{1}$$

where $Y_i(x, t)$ is the state variable of the ith sub-system

$x \in \Omega \subset E^m$; $x' \in \partial\Omega$; $t \in [0, T]$

M_i, L_i are matrix differential operators bringing in only derivatives related to the space variables x

$B_i(x)$ are matrices which are functions of x only.

The interconnexion between these N sub-systems is represented by a linear coupling :

$$X_i = \sum_{j=1}^{N} C_{ij} Y_j \qquad i = 1 \ldots N \tag{2}$$

where C_{ij} are interconnexion matrices.

Several types of control are possible ; this study will be limited to the following types :

- distributed controls on the domain Ω : $U_i(x, t)$, $x \in \Omega$

- pointwise control defined on a finite number of points of the space domain :

$$U_i\ (x_k, t),\ k = 1, 2, \ldots p \text{ or } U_i\ (x, t) = \sum_{k=1}^{p} U_k\ (t) \, . \, \delta\ (x - x_k)$$

The objective function of the overall system is assumed to be given in a separable additive quadratic form :

$$J = \sum_{i=1}^{N} J_i = \sum_{i=1}^{N} \int_0^T \int_{\Omega} F_i \left[Y_i\ (x, t),\ U_i\ (x, t)\right] dx\, dt \tag{3}$$

The overall problem is to minimise J subject to the constraints (1) and (2).

II-3 - Decomposition of the problem

Define the Hamiltonian H of the optimisation problem as

$$H = \sum_{i=1}^{N} F_i\left[Y_i\ (x, t),\ U_i\ (x, t)\right] + \sum_{i=1}^{N} \psi_i^T (x, t) \left\{ M_i\left[Y_i\ (x,t)\right] + X_i\ (x,t) + B_i\ (x)\ U_i\ (x, t)\right\} + \sum_{i=1}^{N} \beta_i\ (x,t) \left[X_i\ (x, t) - \sum_{j=1}^{N} C_{ij}\ Y_j\ (x, t)\right] \tag{4}$$

The conditions of optimality are obtained using the Maximum Principle applied to systems governed by partial differential equations.

Since the coupling between the sub-systems is a state variable one, the decomposition method chosen -in fact, the only one possible- is the infeasible method of coordination using the criterion function [4] . In this method, the coupling equations are treated on the coordination level which fixes the $\beta_i\ (x, t)$ for the first level sub-problems. In order to avoid the singular problems which could arise (X_i appearing linearly in H_i) $Y_i\ (x, t)$ is replaced in F_i by its expression in terms of the $X_i\ (x, t)$. The sub-problems which must be solved can be written as follows :

minimise the criterion

$$J'_i = \int_0^T \int_{\Omega} \left\{ F_i\left[X_i\ (x, t),\ U_i\ (x,t)\right] + \beta_i^T\ (x,t)\ X_i\ (x, t) - \sum_{j=1}^{N} \beta_j^T\ (x, t)\ C_{ji}\ Y_i\ (x, t)\right\} dx\, dt \tag{5}$$

subject to the constraints and the initial and boundary conditions

$$\frac{\partial Y_i}{\partial t} = M_i\ \left[Y_i\right] + X_i + B_i\ U_i$$

On the coordination level there is the possibility of using different algorithms to determine the $\beta_i\ (x, t)$. The gradient algorithm has been chosen here because of its ease of application

$$\beta_i^{n+1}(x,\ t) = \beta_i^{n}(x,\ t) - K_c \left[\frac{\partial H}{\partial \beta_i(x,t)}\right]^n \tag{6}$$

where n is the iteration index on the coordination level and K_c is the iteration constant.

II-4 - Resolution of the sub-problems

Discretisation of the sub-system equations with respect to the space variables (or with respect to the space and time variables) enables one to arrive at dynamic (or static) interconnected sub-systems, and to obtain optimum control of these ; or rather, it enables one to define interconnected dynamic (or static) sub-problems which are solved by classical hierarchical control techniques [3, 5, 6].

Here, the sub-problem will be solved by the application of the Maximum Principle [7]. The conditions of optimality are written as [8] :

$$\begin{cases} \dfrac{\partial Y_i}{\partial t} = \dfrac{\partial H}{\partial \psi_i} = M_i\left[Y_i\right] + X_i + B_i U_i \\[2ex] \dfrac{\partial \psi_i}{\partial t} = -\dfrac{\partial H}{\partial Y_i} - (-1)^{\ell} \dfrac{\partial^{\ell}}{\partial x^{\ell}} \left(\dfrac{\partial H}{\partial \left[\dfrac{\partial^{\ell} Y_i}{\partial x^{\ell}}\right]}\right) \\[2ex] \dfrac{\partial H}{\partial U_i} = 0 \\[2ex] \dfrac{\partial H}{\partial X_i} = 0 \end{cases} \tag{7}$$

with the initial and boundary conditions on Y_i as defined earlier those relative to $\psi_i(x,\ t)$:

$$\psi_i(x,\ T) = 0 \quad ; \quad \left.\frac{\partial^{\ell-1}}{\partial x^{\ell-1}}\left(\frac{\partial H}{\partial \frac{\partial^{\ell} Y_i}{\partial x^{\ell}}}\right)\right|_{\partial\Omega} = 0 \tag{8}$$

These distributed parameter are solved by the eigenfunctions method by seeking a solution in the form :

$$\begin{aligned} U_i(x,\ t) &= \sum_n u_{in}(t)\, \varphi_{in}(x) \\ Y_i(x,\ t) &= \sum_n y_{in}(t)\, \varphi_{in}(x) \end{aligned} \tag{9}$$

It will be supposed that such Eigenfunctions exist. Thus the problem amounts to the resolution of ordinary differential systems in $u_{in}(t)$ and $y_{in}(t)$. Several types of coordination could then be considered :

- the decomposition of the $\beta_i(x, t)$ in terms of the $\varphi_{in}(x)$:

$$\beta_i(x, t) = \sum_n \beta_{in}(t)\,\varphi_{in}(x) \qquad (10)$$

In this case, the coordination is carried out by the $\beta_{in}(t)$ which considerably reduces the transfer of information between levels.

- coordination by the functions $\tilde{\beta}_i(x_k, t)$, defined as a certain number of points in the space domain. In this case, the $\tilde{\beta}_i$ can be considered as "pseudo-controls" for sub-system i. This type of coordination presents the problem of finding optimal actuating points x_k and this problem is tackled in the second part of the paper.

II-5 - Example

In this section, the results obtained on an example made up of two sub-systems defined on the domain $\Omega =]0, 1[$ with :

$$M_i = \left[k_i \frac{\partial^2}{\partial x^2} + 1\right] \quad ; \quad B_i = 1 \quad ; \quad C_{ij} = \begin{cases} 0 & \text{if } i = j \\ -1 & \text{if } i \neq j \end{cases}$$

$$L_i = \frac{\partial}{\partial x} \quad ; \quad J_i = \int_0^T \int_\Omega \sum_{i=1}^{2} \left[Y_i^2 + k_3 U_i^2\right] dx\, dt$$

The conditions of optimality are written as :

$$\begin{cases} \dfrac{\partial Y_i}{\partial t} = k_i \dfrac{\partial^2 Y_i}{\partial x^2} + Y_i - X_i + U_i \\ \dfrac{\partial \psi_i}{\partial t} = -k_i \dfrac{\partial^2 \psi_i}{\partial x^2} - \psi_i - \sum_{j=1}^{2} C_{ji}\beta_j \end{cases} \qquad \begin{cases} 2k_3 U_i + \psi_i = 0 \\ -\psi_i + \beta_i + 2X_i = 0 \end{cases} \qquad i = 1, 2$$

with the boundary conditions :

$$L_i[Y_i] = 0 \quad ; \quad L_i[\psi_i] = 0$$

$$Y_i(x, 0) = Y_{i0}(x) \quad ; \quad \psi_i(x, T) = 0$$

The following criterion is to be minimised :

$$J'_i = \int_0^T \int_\Omega \left[Y_i^2 + k_3 U_i^2 + \beta_i X_i - \sum_{j=1}^{2} \beta_j^T C_{ji} Y_i\right] dx\, dt$$

The resolution of the sub-problems by decomposition into Eingenfunctions and the use of the coordinator [10] gave the results shown in figure 1.

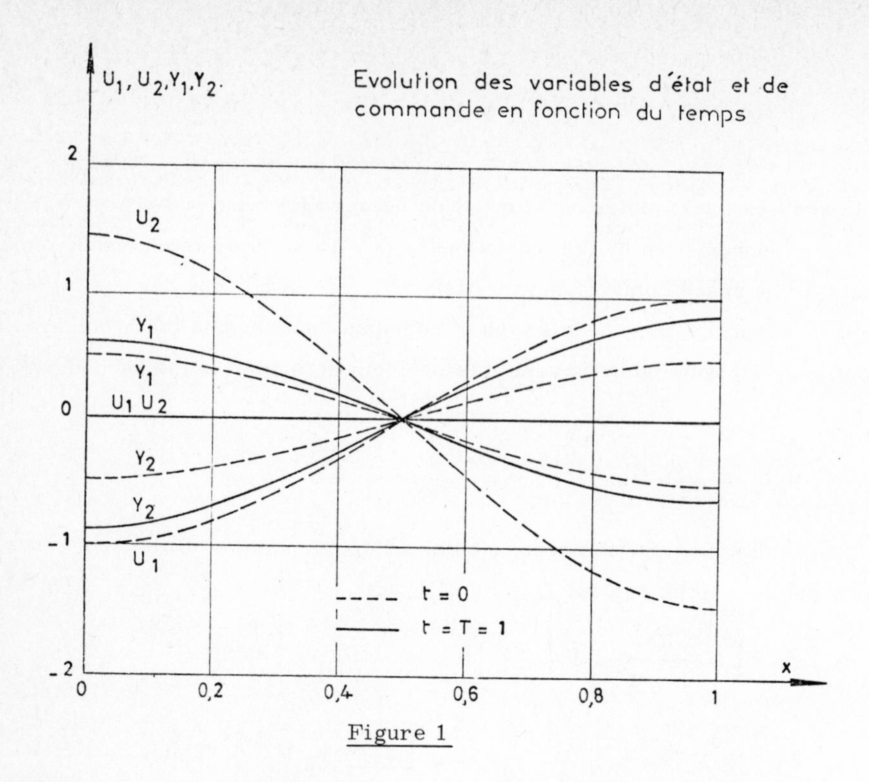

Figure 1

III - STUDY OF THE SUB-SYSTEM

In order to apply control to a distributed parameter system it is necessary to study a certain number of problems such as the observability of the system in relation to the choice of the type and number of sensors, and its controllability, depending on the control action chosen.

Here, the case in which the sub-systems (1) are controlled at a certain number of points situated in the interior of the space domain is considered :

$$\frac{\partial Y}{\partial t}(x,t) = M\left[Y(x,t)\right] + X(x,t) + \sum_{j=1}^{p} U_j(t)\,.\,\delta(x - x_j) \tag{11}$$

$$L\left[Y(x',t)\right] = 0 \quad ; \quad Y(x,0) = Y_0(x)$$

Using the hypothesis of the separability of the criterion formulated in the previous paragraph, the criterion :

$$J(U) = \int_0^T \int_\Omega F(Y(x,t),\ U(t))\, dx\, dt \tag{12}$$

with $\quad U = \left[U_1, \ldots, U_p\right]^T$

is associated with equations (11).

The problem is one of minimising J (u) subject to the constraint (11), with respect to U and to x_j (j = 1 ... p). In what follows, the problem of obtaining the control law is decoupled from that of determining the optimal positions of the actuating points. The former can be tackled in different ways [9], for example :

(i) by minimising the criterion $J(u_{opt})$ with respect to x_j (j=1...p) ; in this case, in general, the solution depends on the state of the system (cf. III-1).

(ii) by extremalising a characteristic function which depends on the system only and sometimes on the type of criterion considered (cf. III-2).

The state mode is obtained from the transformation defined in the preceding paragraph ; it is in the form :

$$\dot{y}(t) = A\, y(t) + B\, u(t) + C\, \xi(t)$$

$$A = \begin{bmatrix} \lambda_1 & & 0 \\ & \ddots & \\ 0 & & \lambda_n \\ & & & \ddots \end{bmatrix}_{\infty \times \infty} \qquad B = \begin{bmatrix} \varphi_1^*(x_1) & \cdots & \varphi_1^*(x_p) \\ \vdots & & \vdots \\ \varphi_n^*(x_1) & \cdots & \varphi_n^*(x_p) \\ \vdots & & \vdots \end{bmatrix}_{\infty \times p} \tag{13}$$

where φ^* are the Eigenfunctions of the adjoint operator M^*.

Certain intrinsic properties of the system can be studied from this model. Indeed, whatever the approach adopted, it is necessary to study the controllability of the system so as to define the collection of admissible positions for the actuating points ; more precisely, the hypothesis of the controllability of the system is linked to the property [10] :

$$\text{row}\left[\varphi_i^*(x_1) \ldots \varphi_i^*(x_p)\right] = 1 \qquad i = 1, 2 \ldots \tag{14}$$

In practice, the application of such a property gives results which are unusable. Because of this, a modal reduction of the system is considered and in particular the influence of this reduction on the optimal position of the actuating points is studied.

III-1 - Minimisation of the criterion $J(u_{opt})$

Two types of criteria are considered :

III-1-a - Quadratic criterion

$$J(U) = \int_0^{\infty} \left[\int_{\Omega} Y^T(x, t)\, Q\, Y(x, t)\, dx + U^T R\, U(t) \right] dt \tag{15}$$

The use of the transformation (9) leads to the equivalent criterion :

$$J(u) = \int_0^{\infty} \left[y^T(t)\, Q'\, y(t) + u^T(t)\, R\, u(t) \right] dt \tag{16}$$

with $Q' = \int_{\Omega} \varphi^T(x)\, Q\, \varphi(x)\, dx$

The optimal control, minimising the criterion J (U) with respect to U is given by the formula [11] :

$$U_{opt} = -R^{-1} B^T(x_1, x_2, \ldots, x_p)\, K(x_1, x_2, \ldots, x_p)\, y(t) \tag{17}$$

where K is the matrix solution of the algebraic Riccati equation.

The optimal distribution, if it exists, is the solution of

$$\min_{(x_1, \ldots, x_p)} J_{opt}(x_1, \ldots, x_p) = \min_{x_1, \ldots, x_p} \left[\frac{1}{2} y^T(0)\, K(x_1, \ldots, x_p)\, y(0) \right] \tag{18}$$

III-1-b - Energy criterion

$$J(U) = \int_0^T U^T(t)\, U(t)\, dt \tag{19}$$

The minimisation of the criterion with respect to U leads to the formulation of the control law [12] :

$$U_{opt}(x_1, \ldots, x_p, t) = B^*(x_1, \ldots, x_p)\, \Phi^*(T-t) \left[\int_0^T \Phi(T-\tau)\, BB^*\, \Phi^*(T-\tau)\, d\tau \right]^{-1} yd \tag{20}$$

where B^* represents the adjoint matrix of B, Φ^*, the adjoint matrix of the state transition matrix Φ and yd the desired state.

Assume that :

$$U_{opt}(x_1, \ldots, x_p, t) \triangleq M(x_1, \ldots, x_p, t)\, yd \tag{21}$$

The optimal distribution is the solution of :

$$\min_{(x_1, \ldots, x_p)} \left\| U_{opt}(x_1, \ldots, x_p, t) \right\| \tag{22}$$

III-2 - Extremalisation of a characteristic function

To each type of criterion defined above it is possible to associate a characteristic function, and this makes it possible to obtain an optimal distribution of the actuating points independently of the state of the system.

III-2-a - Quadratic criterion

Since the matrix K is positive definite and symmetric :

$$0 < \frac{J(U_{opt})}{y^T(0)\,y(0)} \leqslant \frac{1}{2} \| K \| \tag{23}$$

where $\| K \| = (\sum_i \sum_j k_{ij}^2)^{1/2}$

The problem of the minimisation of the criterion $J(u_{opt})$ with respect to $(x_1, \ldots, x_p)$ is therefore transformed into that of minimising its upper limit :

$$\min_{x_1, \ldots, x_p} \| K(x_1, \ldots, x_p) \| \tag{24}$$

III-2-b - Energy criterion

In the same way the upper limit on the norm of the control U on (O, T) is minimised :

$$\min_{x_1, \ldots, x_p} \| M(x_1, \ldots, x_p, t) \| \tag{25}$$

III-2-c - Criterion of controllability [10]

If W represents the upper limit of $\| U(t) \|$, the collection of states obtainable from this control is defined by :

$$y^T(t)\, P^{-1}(x_1, \ldots, x_p)\, y(t) \leqslant W \tag{26}$$

where P is a square matrix the general term of which is given by :

$$P_{ij} = \sum_{k=1}^{P} \varphi_i^*(x_k)\, \varphi_j^*(x_k) \int_0^T e^{(\lambda_i + \lambda_j)(T-\tau)}\, d\tau \tag{27}$$

This collection forms a hyper-ellipsoid the square of whose volume is proportional to the determinant of the matrix P. The optimal distribution of the actuating points can be linked to the maximisation with respect to $(x_1, \ldots, x_p)$ of the volume of the attainable domain, therefore to that of the determinant P :

$$\max_{x_1, \ldots, x_p} \left[\det P (x_1, \ldots, x_p) \right] \tag{28}$$

III-3 - Example

From a practical point of view, these methods can be applied to distributed systems whose model is reduced to the order ν . In particular, the influence of this order on the distribution of the actuating points can be studied. Consider the example (with only one actuating point) :

$$\begin{cases} \dfrac{\partial Y}{\partial t} = \beta \dfrac{\partial^2 Y}{\partial x^2} + \gamma Y + U(t)\, \delta (x - x_1) \\ Y(0, t) = Y(1, t) = 0 \quad ; \quad x, x_1 \in]0, 1[\quad ; \quad t \in [0\ T] \end{cases}$$

The optimal distribution of the actuating points in relation to the different criteria considered above is shown in figures 2 and 3.

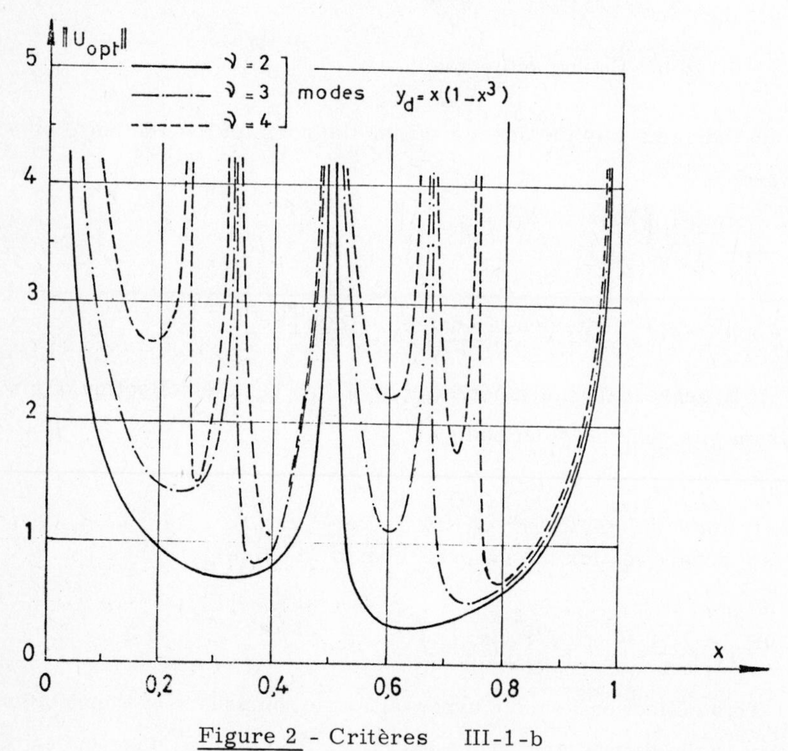

Figure 2 - Critères III-1-b

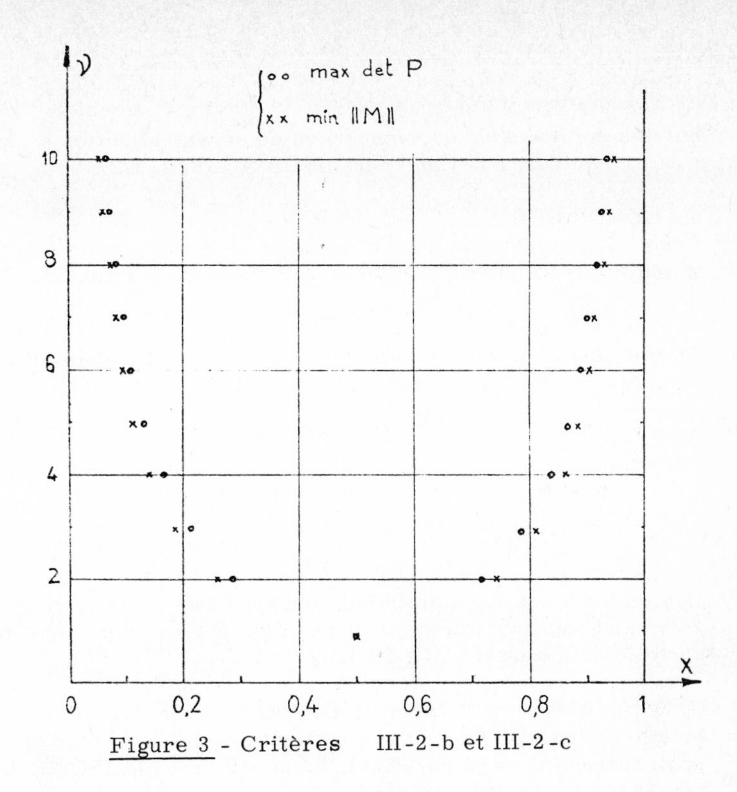

Figure 3 - Critères III-2-b et III-2-c

IV - CONCLUSION

In this paper, we have shown how the problem of the optimal control of interconnected distributed parameter systems using hierarchical control techniques can be tackled. The techniques given here enable one to obtain a collection of control sub-problems which retain their "distributed parameter" nature.

In this decomposition, it is necessary to choose a set of coordination variables which lead to an additive separable form for the Hamiltonian. Each of the sub-problems can be solved using the Maximum Principle. On the level of each sub-system, different types of criteria have been defined and these enable an optimal distribution of a collection of actuating points to be determined. In addition, it was considered necessary to include the study of a sub-system (control, controllability, application of actuators, and the dual problem of observation, observability and the implementation of sensors) taking into account the exchange of information between the different levels of the hierarchical structure .

REFERENCES

[1] A. BENSOUSSAN - J.L. LIONS - R. TEMAM
Sur les méthodes de décomposition,de décentralisation et de coordination et applications. Cahier n° 11, IRIA, Juin 1972.

[2] P. LEMONNIER
Résolution numérique d'équations aux dérivées partielles par décomposition et coordination. Cahier n° 11, IRIA, Juin 1972.

[3] J.D.A. WISMER
Optimal control of distributed parameter systems using multi-level techniques. Ph.D Los Angeles, 1966.

[4] A. TITLI
Contribution à l'étude des structures de commande hiérarchisée en vue de l'optimisation des processus complexes. Thèse d'Etat, Université Paul Sabatier, Toulouse, Juin 1972.

[5] P. CAMBON - L. LE LETTY
Applications des techniques de décomposition et de hiérarchisation à la commande optimale des systèmes régis par des équations aux dérivées partielles. Revue RAIRO, Décembre 1972.

[6] G. GRATELOUP - A. TITLI - B. PRADIN
Solution of partial derivatives optimal control problems by multi-level control techniques (à paraître). Symposium IFAC/IFORS, 8 au 11 Octobre 1974, Varna (Bulgarie).

[7] A.G. BUTKOVSKY
Distributed control systems. n° 11, Elsevier, Pub. Comp. 1969.

[8] A.P. SAGE
Optimum systems control. Prentice Hall, Londres 1968.

[9] A.M. FOSTER - P.A. ORNER
A design procedure for a class of distributed parameter control systems. Transactions of the ASME, Juin 1971.

[10] M. AMOUROUX - J.P. BABARY
Sur la commandabilité de systèmes linéaires à paramètres répartis et son utilisation pour la détermination de points d'action optimaux. Revue RAIRO, Novembre 1973, J3, pp. 120-132.

[11] M. ATHANS - P.L. FALB
Optimal control. Mc Graw Hill Book Company, 1966.

[12] M. FAHMY
A solution technique for a class of optimal control problems in distributive systems. Ph D Univ. of Michigan, 1966.

TIME OPTIMAL CONTROL PROBLEM FOR DIFFERENTIAL INCLUSIONS

V.I.Blagodatskih

Mathematical Institute of USSR Academy of Sciences, Moscow, USSR

I. Introduction

Let E^n be Euclidean space of state-vectors $x=(x_1,\dots,x_n)$ with the norm $\|x\|=\sqrt{\sum_{i=1}^n x_i^2}$ and $\Omega(E^n)$ be metric space of all nonempty compact subsets of E^n with Hausdorff metric

$$h(F,G)=\min\{d: F\subset S_d(G),\ G\subset S_d(F)\}$$

where $S_d(M)$ denotes a d - neighborhood of a set M in the space E^n.

Let us consider an object with behaviour described by the differential inclusion

$$\dot{x}\in F(t,x) \tag{I}$$

where $F: E^1\times E^n \to \Omega(E^n)$ is a given mapping. The absolutely continuous function $x(t)$ is the solution of the inclusion (I) on the interval $[t_0,t_1]$, if the condition $\dot{x}(t)\in F(t,x(t))$ is valid almost everywhere on this interval.

On the one hand the differential inclusion is the extension of ordinary differential equations

$$\dot{x}=f(t,x) \tag{2}$$

when function $f(t,x)$ is multivalued. On the other hand this extension is not formal, for many different problems may be transformed to differential inclusions and the development of differential inclusion permits us to solve these problems. For example, A.F.Filippov with the help of differential inclusions investigated [I] the solutions of differential equation (2) on the sets where function $f(t,x)$ had discontinuities. N.N.Krasovski used the differential inclusion [2] for constructing a strategy in differential games. Let us consider the connection of differential inclusions with some other problems.

Optimal control problem was considered first [3] by L.S.Pontryagin and others for systems described by the equation

$$\dot{x}=f(t,x,u),\quad u\in V. \tag{3}$$

This problem may be transformed to determination of optimal solution $x(t)$ of the differential inclusion

$$\dot{x}\in f(t,x,V)=\{f(t,x,u): u\in V\}.$$

Knowing the optimal solution $x(t)$ it is possible with help of Filippov's implicit functions lemma [4] to construct for system (3) a control $u(t)$ which produces this optimal solution. Note that control system (3) may be transformed to differential inclusion form even in the case when the set V depends on time and state, i.e. is of the form $V(t,x)$. On the other hand inclusion (I) may be considered as a control system with changing control domain

$$\dot{x}=v,\quad v\in F(t,x).$$

It will be noted that optimal control problems stimulated very much the development of the theory of differential inclusions.
The implicit differential equation

$$f(t, x, \dot{x}) = 0$$

may be transformed to a differential inclusion form, too,

$$\dot{x} \in F(t,x) = \{v: f(t,x,v) = 0\}.$$

On the other hand inclusion (I) may be considered as the implicit differential equation

$$\rho(\dot{x}, F(t,x)) = 0,$$

where $\rho(p, A)$ denotes the distance from a point p to a set A:

$$\rho(p, A) = \min_{a \in A} \| p - a \|.$$

A system of differential inequalities

$$f_i(t, x, \dot{x}) \leq 0, \quad i = 1, \dots, k,$$

may be transformed to the differential inclusion

$$\dot{x} \in F(t,x) = \{v: f_i(t,x,v) \leq 0, \ i = 1, \dots, k\}.$$

On the other hand inclusion (I) may be considered in the case of $F(t,x)$ convex as an infinite system of differential inequalities

$$(\dot{x}, \psi) \leq c(F(t,x), \psi), \quad \psi \in S_1(0),$$

where $c(F, \psi)$ is the support function of the set F:

$$c(F, \psi) = \max_{f \in F} (f, \psi).$$

2. Time optimal control problem

Let M_0, M_1 be nonempty closed subsets of E^n. The solution $x(t)$ given on the interval $[t_0, t_1]$ transfers M_0 to M_1 in time $t_1 - t_0$ if the conditions $x(t_0) \in M_0$, $x(t_1) \in M_1$ are satisfied. The time optimal control problem is to determine the solution of the inclusion (I) transferring the set M_0 to the set M_1 in a minimum time.

Maximum principle. Let the support function $c(F(t,x), \psi)$ of the inclusion (I) be continuously differentiable in x and the solution $x(t)$, $t_0 \leq t \leq t_1$, transfer the set M_0 to the set M_1. We shall say that the solution $x(t)$ satisfies the maximum principle on interval $[t_0, t_1]$ if there exists such nontrivial solution $\psi(t)$ of the adjoint system

$$\dot{\psi} = - \frac{\partial c(F(t, x(t)), \psi)}{\partial x} \tag{4}$$

that the following conditions are satisfied:

A) the maximum condition

$$(\dot{x}(t), \psi(t)) = c(F(t, x(t)), \psi(t))$$

is satisfied almost everywhere on the interval $[t_0, t_1]$;

B) the transversality condition on the set M_0: vector $\psi(t_0)$ is the support vector for the set M_0 at the point $x(t_0)$, that is

$$c(M_0, \psi(t_0)) = (x(t_0), \psi(t_0));$$

C) the transversality condition on the set M_1: vector $-\psi(t_1)$ is the support vector for the set M_1 at the point $x(t_1)$ that is

$$c(M_1, -\psi(t_1)) = (x(t_1), -\psi(t_1)).$$

3. Necessary conditions of optimality

Multivalued function $F: E^1 \times E^n \to \Omega(E^n)$ is called measurable if for any closed set $P \subset E^n$ the set $\{x: F(x) \cap P \neq \emptyset\}$ is Lebesgue measurable. The continuity and lipschitzability of multivalued function $F(x)$ is defined in the usual way. For example, the function $F(x)$ satisfies Lipschitz's condition with constant L if for any points $x, x' \in E^n$ the inequality

$$h(F(x), F(x')) \leq L \|x - x'\|$$

is valid. The number $|F| = h(\{0\}, F\}$ is called modulus of set F.

Theorem I. Let the multivalued function $F(t,x)$ of inclusion (I) be measurable in t and satisfy Lipschitz's condition in x with a summable constant $L(t)$ and $|F(t,x)| \leq g(t)$ where $g(t)$ is a summable function. Assume that the support function $c(F(t,x), \psi)$ is continuously differentiable in x, the sets M_0, M_1 are convex and solution $x(t)$, $t_0 \leq t \leq t_1$, transferring the set M_0 to the set M_1 is optimal. Then this solution satisfies maximum principle on the interval $[t_0, t_1]$. Moreover the condition

$$c(F(t_1, x(t_1)), \psi(t_1)) \geq 0$$

is valid.

The proof of the theorem I follows the plan suggested in the book [3] for systems of the type (3). The main difficulty is to define the variation of the solution. Here instead of the classical theorem on differentiability of solution with respect to initial condition (see, for example[5]) it's necessary to use theorem 2 stated below.

Let function $f: E^n \to E^1$ satisfy Lipshitz'a condition. The set of all partial limits of the gradient of this function at the point $x+h$ when $h \to 0$, that is

$$\partial f(x) = \overline{\lim_{h \to 0}} \nabla f(x+h)$$

is called the subdifferential of function $f(x)$ at the point x.

Theorem 2. Let $x(t)$, $t_0 \leq t \leq t_1$ be a solution of inclusion (I) with the initial condition x_0, $\psi(t)$ be a solution of the adjoint system (4) corresponding to $x(t)$ and $\delta x(t)$ be a solution of differential inclusion

$$\delta \dot{x} \in \partial_\psi \left[\frac{\partial c(F(t, x(t)), \psi(t))}{\partial x} \delta x \right]$$

with initial condition $\delta x(t_0)=h$, $\varepsilon>0$. Then there exists such $y_\varepsilon(t)$ - solution of inclusion (I) with initial condition $y_\varepsilon(t_0)=x_0+\varepsilon h$ defined on interval $[t_0,t_1]$ that the following condition is valid

$$y_\varepsilon(t)=x(t)+\varepsilon\,\delta x(t)+o(\varepsilon).$$

The idea of the proof is contained in paper [6].

Remark. The sets M_0, M_1 in theorem I may be nonconvex. It is sufficient that there exist the approximating cones to the sets M_0, M_1 at the points $x(t_0)$, $x(t_1)$, respectively. In this case the conditions B), C) in the maximum principle have to be replaced by conditions that the vectors $\psi(t_0), -\psi(t_1)$ are supports to respective approximating cones at the points $x(t_0), x(t_1)$.

4. Convexity of the set of solutions

Naturally the question arises: when is the maximum principle not only necessary but also sufficient condition of optimality? This seems to be very important to know when the set of solutions $\Sigma_{[t_0,t_1]}(P)$ of inclusion (I) with initial condition $x(t_0)\in P$ is convex in the space $C[t_0,t_1]$ of all continuous functions on the interval $[t_0,t_1]$. Denote by Z_τ the intersection of the set $\Sigma_{[t_0,t_1]}(P)$ by the plane $t=\tau$. The set $Z(\tau)$ is the set of all points at which it's possible to arrive along the solutions of the inclusion (I) from initial set P at a moment τ.

Multivalued function $F(t,x)$ is concave in x on the set M if the condition

$$\alpha F(t,x)+\beta F(t,x')\subset F(t,\alpha x+\beta x')$$

is valid for any points $x, x'\in M$ and for any numbers $\alpha,\beta\geqslant 0, \alpha+\beta=$

Theorem 3. Assume the initial set P be convex. Then the set of solutions $\Sigma_{[t_0,t_1]}(P)$ is convex if multivalued function $F(t,x)$ is concave in x on the sets $Z(\tau)$ for any $\tau\in[t_0,t_1]$.

The idea of the proof is contained in paper [7].

5. Sufficient conditions of optimality

For the maximum principle to be also a sufficient condition of optimality of the solution $x(t)$ one needs to put two additional conditions on the solution $x(t)$.

We shall say that support function $c(F(t,x),\psi)$ is concave in x at the point x_0 in the direction ψ_0 if the condition

$$\left(\frac{\partial c(F(t,x_0),\psi_0)}{\partial x}, x-x_0\right)\geqslant c(F(t,x),\psi_0)-c(F(t,x_0),\psi_0)$$

is valid for any point $x\in E^n$.

Note that the concavity of a multivalued function $F(t,x)$ yields the concavity of the support function $c(F(t,x),\psi)$ at any point and in any direction.

Let $x(t)$, $t_0\leqslant t\leqslant t_1$ be a solution of inclusion (I) and $\psi(t)$ be a respective solution of the adjoint system (4). We shall say that the solution $x(t)$ satisfies strong transversality condition on the set M_1 with the adjoint function $\psi(t)$ if the condition

$$c(M_1,-\psi(t))<(x(t),-\psi(t))$$

is valid for any moment $t: t_0\leqslant t<t_1$.

Theorem 4. Assume that M_0, M_1 are nonempty closed subsets of E^n, the solution $x(t)$ of inclusion (I) transfers the set M_0 to the set M_1 on the interval $[t_0, t_1]$ and satisfies the maximum principle on this interval and $\psi(t)$ is the respective solution of adjoint system (4). Assume that the support function $c(F(t,x),\psi)$ is concave in x at point $x(t)$ in direction $\psi(t)$ for any $t \in [t_0, t_1]$ and the solution $x(t)$ satisfies the strong transversality condition on the set M_1 with adjoint function $\psi(t)$. Then the solution is optimal.

The proof is contained in paper[8].

6. Uniqueness of optimal solution

In the case of continuous differentiability of the support function $c(F(t,x),\psi)$ in ψ the maximum condition A) and the adjoint system (4) may be written as the system of differential equations

$$\dot{x}(t) = \frac{\partial c(F(t,x(t)), \psi(t))}{\partial \psi}$$

$$\dot{\psi}(t) = -\frac{\partial c(F(t,x(t)), \psi(t))}{\partial x} .$$

Initial conditions for solution $(x(t), \psi(t))$ of this system may be determined from conditions B), C) of the maximum principle and from inclusions $x(t_0) \in M_0$, $x(t_1) \in M_1$. The question arises naturally in this case: when is for given initial conditions the unique solution $(x(t), \psi(t))$ determined from the maximum principle? The following theorem is an answer to this question.

Theorem 5. Assume that the support function $c(F(t,x),\psi)$ is measurable in t, continuously differentiable in (x,ψ) except for the points $\psi = 0$ and derivatives $c'_x(F(t,x),\psi)$, $c'_\psi(F(t,x),\psi)$ satisfy Lipschitz's condition in (x,ψ). Then for any given initial conditions for functions $x(t)$, $\psi(t)$ the optimal solution is unique.

7. Concluding remarks

Pontryagin's maximum principle was proved [3] for control system (3) for function $f(t,x,u)$ continuously differentiable in x. These systems may be transformed in differential inclusion form and support function in this case is

$$c(F(t,x),\psi) = \max_{u \in U} (f(t,x,u), \psi).$$

Theorem I is applied when this support function is continuously differentiable in x. It is not very difficult to show that there exists the function $f(t,x,u)$ continuously differentiable in x but for which the corresponding support function $c(F(t,x),\psi)$ does not have this property and vice versa. That is Pontryagin's maximum principle and theorem I, are intersected over some class of control systems of type (3). It is possible to formulate as a hypothesis the following theorem which includes the theorem I and Pontryagin's maximum principle.

Theorem 6. Suppose that the multivalued function $F(t,x)$ for inclusion (I) is measurable in t and satisfies Lipschitz's condition in x, sets M_0, M_1 are convex and solution $x(t)$, $t_0 \leq t \leq t_1$, is optimal. Then there exists such nontrivial absolutely continuous function $\psi(t)$ that the following conditions are valid.

A) $$\frac{d}{dt}\begin{pmatrix} x(t) \\ \psi(t) \end{pmatrix} \in \partial c_{(x,\psi)}\left(F(t, x(t)), \psi(t)\right)$$

almost everywhere on the interval $[t_0, t_1]$;

B) $$c(M_0, \psi(t_0)) = (x(t_0), \psi(t_0));$$

C) $$c(M_1, -\psi(t_1)) = (x(t_1), -\psi(t_1)).$$

References

I. A.F.Filippov, Differential equations with discontinuous right-hand side (Russian) , Matem. sbornik, 5I, N I, I960.

2. N.N.Krasovski, Game problems on encounter of motions. (Russian), Nauka, 1970.

3. L.S.Pontryagin et al., Mathematical theory of optimal processes (Russian) , Nauka, I96I.

4. M.Q.Jacobs, Remarks on some recent extensions of Filippov's implicit functions lemma, SIAM, Control, 5,N 4, I967.

5. L.S.Pontryagin, Ordinary differential equations (Russian) , Nauka, I970.

6. V.I.Blagodatskih, On differentiability of solutions with respect to initial conditions (Russian), Different. Uravn., 9, N I2, I973.

7. V.I.Blagodatskih, On convexity of domains of reachability (Russian), Different. Uravn., 8, N I2, I972.

8. V.I.Blagodatskih, Sufficient conditions of optimality for differential inclusions (Russian), Izv. AN SSSR, ser. Mathemat., N 3, I974.

APPLICATION OF MAXIMUM PRINCIPLE FOR OPTIMIZATION OF PSEUDO-STATIONARY CATALYTIC PROCESSES WITH CHANGING ACTIVITY

V.I. Bykov, G.S. Yablonskii, M.G. Slin'ko

Institute of Catalysis

Siberian Branch of the USSR Academy of Sciences

630090, Novosibirsk 90, USSR

The experience of using the necessary conditions of optimality in the form of L.S. Pontryagin's maximum principle for qualitative investigation and numerical calculations of optimal control for the distributed-parameter systems is discussed in the paper. Such sets of equations describe the widespread in industry class of pseudo-stationary processes with changing catalyst activity. A conception of non-local optimal control is introduced and also numerical algorithm is proposed. This algorithm allows the conditions of maximum principle to be realized on a computer.

Statement of a problem, optimality conditions. Initial equations are:

$$\frac{\partial C}{\partial \tau} = f(C,\theta,U), \qquad 0 \leq \tau \leq \tau_k,$$

$$\frac{\partial \theta}{\partial t} = g(C,\theta,U), \qquad 0 \leq t \leq t_k, \qquad (I)$$

with boundary conditions:

$$C(0,t) = C^{\circ}(t), \qquad 0 \leq t \leq t_k,$$

$$\theta(\tau,0) = \theta^{\circ}(\tau), \qquad 0 \leq \tau \leq \tau_k, \qquad (2)$$

where C is n-vector-function, which charactirizes the state of the process (concentrations, temperature, pressure in a reactor), m-vector-function Θ describes change of catalytic activity, t is astronomic time, τ is contact time, piecewise-continuous vector-function U characterizes control influence, its separate components being able to depend on τ or t or both τ and t; f,g are supposed to have sufficiently smooth arguments. Denote as D the field of changing independent variables τ , t, given by inequality (I).

The field of permissible controls is given in the form

$$\Omega = \left\{ U_* \leqslant U \leqslant U^* \right\}, \tag{3}$$

where U_*, U^* are constant vectors. Inequality (3) is by component one. Criterion of optimality can be presented in the most cases as follows

$$\max_{U\in\Omega} \left\{ J = \int_0^{t_k} \int_0^{\tau_k} G(C,\Theta,U)\, d\tau dt \right\}. \tag{4}$$

Pontryagin's maximum principle gives necessary conditions of optimality which are formulated for our case as follows / 1-3/ : it is necessary for optimal control $U(\tau,t)$ in the sense of problem (I-4) that there exist such non-zero vector-functions Ψ and $\mathcal{X}$ which satisfy the set of equations in D:

$$\frac{\partial \Psi_i}{\partial \tau} = -\frac{\partial G}{\partial C_i} - (\Psi, \frac{\partial f}{\partial C_i}) - (\mathcal{X}, \frac{\partial g}{\partial C_i}), \quad i=1,2,\ldots,n,$$

$$\frac{\partial \mathcal{X}_j}{\partial t} = -\frac{\partial G}{\partial \Theta_j} - (\Psi, \frac{\partial f}{\partial \Theta_j}) - (\mathcal{X}, \frac{\partial g}{\partial \Theta_j}), \quad j=1,2,\ldots,m, \tag{5}$$

with boundary conditions:

$$\Psi_i(\tau_k,t)=\mathcal{X}_j(\tau,t_k)=0, \quad i,j=1,2,\ldots,n;m, \tag{6}$$

that on $U_k(\tau,t)$ $\forall (\tau,t)\in D$ function

$$H = G + (f,\Psi) + (g,\mathcal{X}) \tag{7}$$

reaches its maximum as function of variable $U_k \in \Omega$.

<u>Qualitative investigations.</u> On the basis of maximum principle it is possible for some cases to carry out an investigation of optimal solutions, determine their properties of common character, obtain *a priori* estimations /4a-5/ . In addition the knowledge of qualitative picture of optimal strategy facilitates the choice of initial approximation already close to optimal one /4b,6/ when realizing the maximum principle as a numerical algorithm.

Let $U_{t_*}(\tau,t)$ and $U_{t_k}(\tau,t)$ be optimal controls on the segments $[0, t_*]$ and $[0, t_k]$ respectivly at $\tau \in [0, \tau_k]$, then the optimal regime is called to be local, if

$$\forall\ t_* \in (0, t_k) \qquad U_{t_*}(\tau,t) \equiv U_{t_k}(\tau,t).$$

Non-locality of optimal control means that the control depends on a strategy during the whole cycle at each moment and varies with changing t_k. In chemical technology such regimes are conditioned ones /3/. For optimization problem of catalytic processes with changing activity it was shown /4a/ that optimal temperature regimes were non-local ones in time, excluding only limit permissible isothermal regimes. From this fact it is clear that from a given class of problems, application of algorithm of optimal control by determination of optimal regime for stationary conditions is invalid . In addition, for non-local control it is necessary to take strictly into account the limitations on control parameters. Imposition of limitations after calculation of the controls can lead to considerable mistakes in given case /3/.

An example of a problem of singular control that arises for considered class of optimization problems (I-4) can be the problem of cooler optimal temperature determination T_x in tube reactor with cooling and decaying catalyst activity:

$$\frac{\partial C}{\partial \tau} = f_1(C,\theta,T), \qquad \frac{\partial \theta}{\partial t} = g(C,\theta,T), \tag{8}$$

$$\frac{\partial T}{\partial \tau} = f_2(C,\theta,T) - B(T - T_x), \tag{9}$$

where

$$T_{x*} \leq T_x(\tau, t) \leq T_x^*$$

and it is required to obtain

$$\max \int_0^{t_k} C(\tau_k, t)\,dt \,. \tag{10}$$

Control T_x goes to (9) in a linear form and that is why it may have singular sections. The difficulty of singular control determination here can be overcome in the following way. Optimal temperature profile $T(\tau, t)$ is determined by carriying out the stage of theoretical optimization, that is by solution of the problem (8), (10) with control T. Substituting $T(\tau, t)$ in thermal balance equation (9) we obtain T_x :

$$T_x = T + \frac{1}{B}\left[\frac{\partial T}{\partial \tau} - f_2(\tau, t)\right] \quad ,$$

which is optimal singular control after satisfying given limitations.

Numerical algorithm. Necessary conditions of optimality in the form of maximum principle make up the boundary problem (1,2,5, where the controls at each point of D are determined from H function maximum condition, H being known from (7). Boundary problem as stated here in operator form can be written as follows

$$\mathcal{D}X + F(X, U_{opt.}) = 0 \,, \tag{11}$$

where vector of optimal controls $U_{opt.}$ is determined from condition

$$H(X, U_{opt.}) = \max_{U \in \Omega} H(X, U) \,.$$

Side by side with (11) consider a set of equations

$$\frac{\partial Y}{\partial \xi} = \mathcal{D}Y + F(Y, U_{opt.}) \,, \quad Y\big|_{\xi=0} = Y^0, \tag{12}$$

with boundary conditions (2,6). If $Y(\xi, t, \tau) \to X(\tau, t)$ at $\xi \to \infty$, (that is solution of the problem (12) converges to solution of (11) at the same boundary conditions irrespective of initial data choice) then sought-for optimal control can be found by realization on computer of difference scheme approximating equations (12).

The algorithm proposed was used for optimal temperature determination for different schemes of reactions in plug flow and incomplete mixing reactors. Convergence was observed in all considered cases after50-60 iterations. The required computing time decreased by a factor of 5-10 as compared to the trial and error method /7/. Algorithm can easily be generalized also for the problems of terminal control.

The solution of a number of optimization problems of important industrial processes with decaying catalyst activity /5,6/ can state that apparatus of maximum principle gives not only possibility to carry out qualitative investigation of optimal control, but also on its basis to build effective calculative algorithms for seeking optimal conditions to conduct a class of catalytic processes.

References

1. Yu.M.Volin, G.M.Ostrovskii. Avtomatika i telemekhanika, v.25, No10, 1964.
2. A.F.Ogunye, W.H.Ray. AIChE Journal, v.16, No6, 1970.
3. A.V.Fedotov, Yu.M.Volin, G.M.Ostrovskii, M.G.Slin'ko. TOKhT, v.2, No1, 1968.
4. V.I.Bykov, G.S.Yablonskii, M.G.Slin'ko. TOKhT: a) v.8, No1, 1974, b) v.7, No5, 1973.
5. M.G.Slin'ko, V.I.Bykov, G.S.Yablonskii et al. Problemy kibernetiki, No27, 1973.
6. V.I.Bykov, A.V.Fedotov, Yu.Sh.Matros and all. TOKhT, v.8, No3, 1974.
7. V.I.Bykov, V.A.Kuzin, A.V.Fedotov. Sb."Upravlyaemye sistemy", No1, 1968.

APPROXIMATE SOLUTION OF OPTIMAL CONTROL PROBLEMS USING THIRD ORDER HERMITE POLYNOMIAL FUNCTIONS

by

ERNST D. DICKMANNS* and KLAUS H. WELL†

Abstract

An algorithm for the approximate solution of two point boundary value problems of Class C^2 is given. A simple version having one check point at the center of each polynomial segment results in an algorithm which is easy to program and very efficient. Computer test runs with a Newton-Raphson iterator and numerical differentiation to generate the partial derivatives required show a fast convergence compared to extremal field methods and gradient methods in function space.

Introduction

Optimal control problems with smooth continuous solutions will be treated. They are transformed into mathematical programming problems in two steps. First applying the calculus of variations or the maximum principle a two-point boundary value problem results. This is then solved approximately by parameterization using piecewise polynomial approximations.

Assuming that the frequency content of the solution can be estimated the range of the independent variable is subdivided into sections within which the solution may be well approximated by third order polynomials. For each segment and each variable the four coefficients of the polynomial are determined from the function values at each end - which are the unknown parameters that have to be estimated initially - and from the derivative obtained by evaluating the right hand side of the differential equations with these values. By this the approximating function is continuous and has continuous first derivatives. At one or more check points within each segment the interpolated function values are computed. The derivatives evaluated with these values from the right hand side of the differential equations are then compared to the slope of the interpolating polynomial at this point. The sum of the squares of all these errors plus the errors in the prescribed boundary conditions is chosen as the payoff quantity to be minimized.

---

*Head, †Scientist, Trajectory Section, Institut für Dynamik der Flugsysteme, Deutsche Forschungs- und Versuchsanstalt für Luft- und Raumfahrt e.V., 8031 Oberpfaffenhofen, FRG.

In this paper an algorithm taking one checkpoint in the middle of each segment is developed using a modified Newton-Raphson scheme for iterative parameter adjustment. In connection with the third order polynomial which can be determined from the function values at adjacent gridpoints only (parameters) this leads to especially simple relations. Higher order approximations over more than one segment and more than one check point are of course feasable but not investigated here.

Statement of the Problem

The extremal value of the functional

$$J = \phi(x_f, t_f) \tag{1}$$

under the differential equation constraints

$$\dot{x} = t_f \cdot f(x,u) \qquad \text{(n-vector)} \tag{2}$$

with the control vector u having m components, the initial values

$$x(o) = x_0 \qquad \text{(n-vector)} \tag{3}$$

and the final constraints

$$\psi(x_f, t_f) = 0 \qquad \text{(q-vector)} \tag{4}$$

has to be found.

The solution is assumed to be continuous with continuous first derivatives; t_f is a final time parameter allowing to treat open final time problems in a formulation with the independent variable normalized to the range $0 \le t \le 1$.

Reduction to a Boundary Value Problem

Applying the calculus of variations or the maximum principle [1] the determination of the optimal control is transformed into solving a two-point boundary value problem. The differential equation constraints (2) lead to an additional set of time varying multipliers which are given by

$$\dot{\lambda} = -t_f \cdot \left(\frac{\partial f}{\partial x}\right)^T \lambda \qquad \text{(n-vector)} . \tag{5}$$

The final constraints (4) invoke constant multipliers μ (q-vector) and transversality conditions have to be satisfied

$$T = [\lambda + \phi_x + \mu^T \psi_x]_{t=1} = 0 . \qquad \text{(n-vector)} \tag{6}$$

Here the subscript x means partial differentiation with respect to x. For open final value of the independent variable the Hamiltonian function

$$H = t_f \cdot \lambda^T f \tag{7}$$

has to satisfy the condition

$$R = \left[-\frac{\partial H}{\partial t_f} + \frac{\partial \phi}{\partial t_f} + \mu^T \frac{\partial \psi}{\partial t_f} \right]_{t=1} = 0 \quad \text{(scalar)} \quad . \tag{8}$$

Eqs. (2) and (5) may be written in the form

$$\dot{z} = g(z)$$

where $z^T = (x^T, \lambda^T)$ is a (2n)-vector which has to satisfy the boundary conditions (3), (4), (6) and (8). This is a nonlinear boundary value problem. The functions z(t), the multipliers μ and the parameter t_f have to be determined.

Parameterization and Iteration Scheme

In figure 1 the basic idea of the algorithm is displayed. Three segments have been chosen (NS = 3) resulting in a total of 4 gridpoints per variable (full dots and empty or full squares). The slope evaluated by introducing the estimated function values into the right hand side of the differential equations (9) is given by solid straight lines at the gridpoints j. The resulting interpolating cubic polynomials are shown as the wavelike solid curves. The function values at the check points are marked by empty circles and the resulting slopes from the differential equations by dashed straight lines. Both the initial estimate and the converged curves are given.

Changing one function value at a gridpoint (z_{22} in fig. 1, empty triangle) affects only the two bordering check points (full triangles), however, for all variable z. For each segment a new time variable $0 \leq t' \leq (T_{j+1} - T) = \tau$ is introduced which yields as interpolated function value at the (central) check point

$$C_j = \frac{1}{2} \left[z_j + z_{j+1} + t_f \frac{\tau}{4} (f(z_j) - f(z_{j+1})) \right] \tag{10}$$

and

$$\dot{C}_j = \frac{3}{2\tau} (z_{j+1} - z_j) - \frac{t_f}{4} (f(z_j) + f(z_{j+1})) . \tag{11}$$

The difference in the slopes at the check point then is

$$\Delta_j = \frac{3}{2\tau}(z_{j+1} - z_j) - \frac{t_f}{4}(f(z_j) + f(z_{j+1})) - t_f \cdot f(c_j) \tag{12}$$

(2n-vector).

With this the contribution of the segment j to the convergence measure is

$$S_j = \frac{1}{2} \Delta_j^T \Delta_j \quad . \tag{13}$$

As total convergence measure the sum

$$M = \sum_{j=1}^{NS} S_j + \frac{1}{2} [\psi^T \psi + T^T T + R^2] \tag{14}$$

is chosen, where ψ, T and R are given by eqs. (4), (6), and (8). Convergence is considered to be achieved for $M \leq \varepsilon$, where ε is a predetermined small number.

Starting from estimated values z_{ji} for all gridpoints j and variables i=1...2n, for the multipliers μ and for the parameter t_f improved values of the total parameter vector

$$p^T = [\lambda_1^T, z_2^t \ldots z_{NGP}^T, \mu^T, t_f] \tag{15}$$

have to be found to drive M towards 0. Using a modified Newton-Raphson iteration scheme the linearized iteration equations may be written

$$\frac{\partial M}{\partial p} \delta p = -\alpha \cdot M \quad ,$$

$$\left(\frac{\partial \psi}{\partial z}\right)_{NGP} \delta z_{NGP} + \frac{\partial \psi}{\partial t_f} \delta t_f = -\alpha \psi \quad ,$$

$$\left(\frac{\partial T}{\partial z}\right)_{NGP} \Delta z_{NGP} + \frac{\partial T}{\partial \mu} \delta \mu + \frac{\partial T}{\partial t_f} \delta t_f = -\alpha T \quad , \tag{16}$$

$$\left(\frac{\partial R}{\partial z}\right)_{NGP} \delta z_{NGP} + \frac{\partial R}{\partial \mu} \delta \mu + \frac{\partial R}{\partial t_f} \delta t_f = -\alpha R \quad ,$$

where α is a factor to improve convergence. Taking advantage of the fact that each element S_j of M depends only on the values z_j and z_{j+1} adjacent to it and on the parameter t_f, the total variation of S is

$$dS = \Delta_j^T \left(\frac{\partial \Delta_j}{\partial z_j} \delta z_j + \frac{\partial \Delta_j}{\partial z_{j+1}} \delta z_{j+1} + \frac{\partial \Delta_j}{\partial t_f} \delta t_f \right) \quad . \tag{17}$$

Introducing this into (16) yields for each segment

$$\frac{\partial \Delta_j}{\partial z_j}\,\delta z_j + \frac{\partial \Delta_j}{\partial z_{j+1}}\,\delta z_{j+1} + \frac{\partial \Delta_j}{\partial t_f}\,\delta t_f = -\alpha \cdot \Delta_j \tag{18}$$

and as set of iteration equations there follows (with $\frac{\partial \Delta_j}{\partial z_k} = A_{j,k}$, and $\frac{\partial \Delta_j}{\partial t_f} = N_j$; for indices see fig. 1).

$$\begin{array}{c|l}
\text{Dim.} & \quad n \qquad 2n \;\cdots\cdots\; 2n \qquad q \quad 1 \\
\begin{array}{c} 2n \\ \vdots \\ 2n \\ q \\ n \\ 1 \end{array} &
\begin{bmatrix}
A_{1,1} & A_{1,2} & 0 & 0 & & & & N_1 \\
0 & A_{2,2} & A_{2,3} & 0 & & \mathbf{0} & & N_2 \\
0 & 0 & A_{3,3} & A_{3,4} & & & & \vdots \\
 & & & \ddots & \ddots & & & \vdots \\
 & & & & A_{NS,NS} & A_{NS,NGP} & 0 & N_{NS} \\
 & & & & & \psi_z & 0 & \psi_{tf} \\
 & \mathbf{0} & & & & T_z & T_\mu & T_{tf} \\
 & & & & & R_z & R_\mu & R_{tf}
\end{bmatrix}
\begin{bmatrix} \\ \\ \\ \\ \delta p \\ \\ \\ \\ \end{bmatrix} = -\alpha
\begin{bmatrix} \Delta_1 \\ \Delta_2 \\ \vdots \\ \vdots \\ \Delta_{NS} \\ \psi \\ T \\ R \end{bmatrix}
\end{array} \tag{19}$$

The submatrices have the following dimensions:

A_{11}	$2n \times n$	$T_\mu = \frac{\partial T}{\partial \mu}$	$n \times q$
$A_{i,j}$	$2n \times 2n$	$T_{tf} = \frac{\partial T}{\partial t_f}$	$n \times 1$
$N_j = \frac{\partial N_j}{\partial t_f}$	$2n \times 1$	$R_z = \frac{\partial R}{\partial z}$	$1 \times 2n$
$\psi_z = \frac{\partial \psi}{\partial z}$	$q \times 2n$	$R_\mu = \frac{\partial R}{\partial \mu}$	$1 \times q$
$\psi_{tf} = \frac{\partial \psi}{\partial t_f}$	$q \times 1$	$R_{tf} = \frac{\partial R}{\partial t_f}$	1×1
$T_z = \frac{\partial T}{\partial z}$	$n \times 2n$		

The partial derivative matrices $A_{j,k}$ are computed by numerical differentiation. The dimension of the linear system of equations (19) is $2n \times NS + n + q + 1$. Because of the bidiagonal form in the upper left part it is conveniently reduced for solution to a $(3n + q + 1)$-system independent of the number of segments NS chosen.

Numerical Examples

The algorithm has been tested on a variety of problems such as time minimal accelerated turns of a Hovercraft (n = 2, one control, figure 1), optimal landing approach trajectory of an aircraft (n = 4, one control), maximum lateral range of gliding entry vehicles (n = 5, two controls) and threedimensional skips with prescribed heading change and minimum energy loss at the exit of the atmosphere for the same class of vehicles.

The last problem will be given here. The differential equations are [2, 3]

$$\begin{aligned}
\dot{x}_1 &= [-a\cdot b\cdot \exp(-\beta x_4)\cdot(C_{D0}+ku_1^n)x_1^2-G/(R+x_4)^2\cdot \sin x_3]\cdot t_f \\
\dot{x}_2 &= [a\cdot b\cdot x_1\cdot \exp(-\beta x_4)u_1 \sin u_2/\cos x_3-x_1/(R+x_4)\cdot \tan x_5\cdot \cos x_2\cdot \cos x_3]\cdot t_f \\
\dot{x}_3 &= [a\cdot b\cdot x_1 \exp(-\beta x_4)u_1 \cos u_2-(G/((R+x_4)^2 x_1)-x_1/(R+x_4))\cdot \cos x_3]\cdot t_f \qquad (20) \\
\dot{x}_4 &= [x_1/(R+x_4)\cdot \cos x_3\cdot \sin x_2]\cdot t_f \\
\dot{x}_5 &= [x_1 \sin x_3]\cdot t_f
\end{aligned}$$

where $a = 1/550\ m^2/kg$, $C_{D0} = .04$, $k = 1.$, $n = 1.86$ are vehicle parameters and $b = 1.54\ kg/m^3$, $G = 3.9865\cdot 10^5\ km^3/s^2$, $R = 6371$ km, $\beta = .0145\ km^{-1}$ are parameters of the planet and its atmosphere. For the

initial values	and	the final conditions
$x_1(o) = 8.18$ km/s		$\psi_1 = x_2(1) - 2.5 = 0$
$x_2(o) = 0$ deg		$\psi_2 = x_3(1) - 1.25 = 0$
$x_3(o) = -1.25$ deg		$\psi_3 = x_4(1) - 80.0 = 0$
$x_4(o) = 80$ km		$x_5(1)$ open
$x_5(o) = 0$ deg		$x_1(1)$ to be maximized

the control time history of the lift coefficient (u_1) and the aerodynamic bank angle (u_2) are to be found which yield the final value of x_1 to be maximal.

Initial estimates were found by linear interpolation between given boundary values or by physical reasoning in the other cases. The convergence behaviour is shown in fig. 2. Table 1 gives the intentionally bad initial estimates and the converged values of the parameters. The estimated initial controls are seen to be very poor. The result achieved with NS = 5 segments is in very good agreement with results obtained by multiple shooting [4] and a refined gradient algorithm based on [5, 6]. Computer time needed was only a fraction (about 1/5 or less) of that of the other methods. Systematic investigations of the radius of convergence are being performed.

Conclusion

An algorithm for the approximate solution of two point boundary value problems of class C^2 has been given. It is based on third order Hermite polynomial approximation. With one check point in the center of each segment it results in an algorithm which is simple to program and very efficient. Numerical test runs with a Newton-Raphson iterator and numerical differentiation to generate the partial derivatives required showed fast convergence compared to extremal field methods and gradient methods in function space.

Literature

[1] Pontryagin,L.S., Boltyansky,V.G., Gamkrelidse,R.V., Mishchenko, E.F.: Mathematische Theorie optimaler Prozesse. Oldenbourg Verlag (1964).

[2] Dickmanns,E.D.: Gesteuerte Drehung von Satellitenbahnen durch Eintauchen in die dichtere Atmosphäre. Dissertation, R.W. T.H. Aachen (1969) also as DFVLR-Sonderdruck No. 42.

[3] Dickmanns,E.D.: Maximum Range Threedimensional Lifting Planetary Entry. NASA TR-382 (1972).

[4] Stoer,J., Bulirsch,R.: Einführung in die Numerische Mathematik II. Springer Verlag, Heidelberger Taschenbücher, Band 114 (1973), pp. 170-191.

[5] Bryson,A.E., Denham,W.F.: A Steepest Ascent Method for Solving Optimum Programming Problems. J. Applied Mech., Vol. 84, No. 2 (1962).

[6] Gottlieb,R.G.: Rapid Convergence to Optimum Solutions Using a Min-H Strategy. AIAA J., Vol. 5 (Feb. 1967), No. 2, pp. 322-329.

[7] Ciarlet,P.G., Schultz,M.H., Varga,R.S.: Numerical Methods of High Order Accuracy for Nonlinear Boundary Value Problems. Numerische Mathematik 9, pp. 394-430 (1967).

j	t/t_f		1 $X_1=V$	2 $X_2=\chi$	3 $X_3=\gamma$	4 $X_4=\Lambda$	5 $X_5=h$	6 $Z_{j,6}=\lambda_v$	7 $Z_{j,7}=\lambda_\chi$	8 $Z_{j,8}=\lambda_\gamma$	9 $Z_{j,9}=\lambda_\Lambda$	10 $Z_{j,10}=\lambda_h$	controls $U_1=C_L$	$U_2=\mu'$/deg
1		e	8.18	0.	-.021817	0.	80.	1.	7.2578	-4.1313	0	0	.5	120
		c	8.18	0.	-.021817	0.	80.	1.12447	3.1684	1.4157	-1.2005	1.955'-3	.156	65.9
		a	8.18	0.	-.021817	0.	80.	1.12689	3.1635	1.4401	-1.2015	1.965'-3	.156	65.5
2	0.2	e	8.144	.008727	-.013090	.002	80.	1.	7.2270	-4.1729	0	0	.5	120
		c	8.1752	.003715	-.014742	.000107	71.322	1.10666	3.2476	1.1900	-.96595	1.535'-3	.158	69.9
		a	8.1753	.003695	-.014785	.000171	71.301	1.10863	3.2429	1.2125	-.96679	1.542'-3	.158	69.5
3	0.4	e	8.108	.017453	-.004363	.004	80	1.	7.1956	-4.1544	0	0	.5	120
		c	8.1413	.013712	-.005840	.000700	66.388	1.09288	3.3094	1.2778	-.72649	6.153'-4	.167	68.9
		a	8.1414	.013680	-.005879	.000698	66.340	1.09445	3.3049	1.2996	-.72713	6.154'-4	.166	68.5
4	0.6	e	8.072	.026180	.004363	.006	80	1.	7.1637	-4.1360	0	0	.5	120
		c	8.0812	.028585	.005309	.002229	66.200	1.07557	3.3533	1.7994	-.48456	-1.500'-4	.186	61.8
		a	8.0810	.028580	.005333	.002229	66.142	1.07671	3.3490	1.8258	-.48498	-1.507'-4	.186	61.4
5	0.8	e	8.036	.034906	.013090	.008	80.	1.	7.1312	-4.1175	0	0	.5	120
		c	8.0271	.039594	.015655	.004728	71.194	1.04480	3.3795	2.4409	-.24208	4.008'-5	.216	54.1
		a	8.0266	.039609	.015710	.004734	71.168	1.04538	3.3753	2.4677	-.24229	5.327'-5	.216	53.8
6	1.0	e	8.0	.043633	.021817	.01	80	1.	7.0981	-4.0991	0	0	.5	120
		c	7.9958	.043633	.021815	.007738	80.000	1.00000	3.3881	2.8848	0.0000	4.525'-4	.246	49.6
		a	7.9952	.043633	.021817	.007751	80.000	1.00000	3.3839	2.9043	0.0000	4.736'-4	.247	49.4

t_f = 300.0 (e), 288.38 (c), 288.69 sec (a); M = 1.02 · 10^4 (e), 7.69 · 10^{-8} (c)

Table 1: Threedimensional atmospheric skip, numerical example: initial estimate (e); converged solution with presented method (c) and accurate numerical solution by multiple shooting (a)

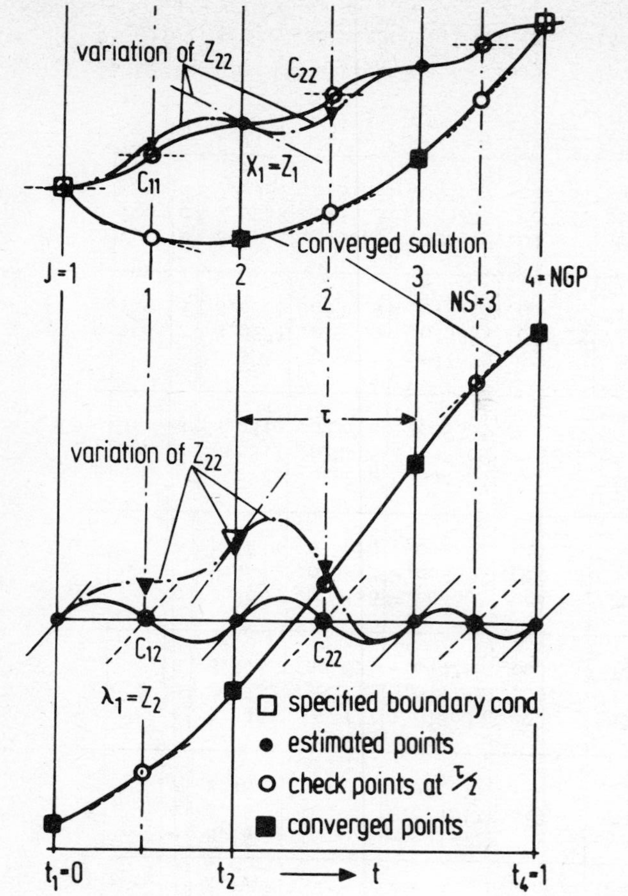

Figure 1: Basic Idea

Figure 2: convergence behaviour for three-dimensional atmospheric skip problem

OPTIMAL STABILIZATION OF THE DISTRIBUTED PARAMETER SYSTEMS

A.I.Egorov
Dnepropetrovsk Railway Transport Institute
Dnepropetrovsk, USSR.

The methods of investigating the optimal stabilization problems occupied an important place in the optimal control theory. The fundamentals of the methods for the finite-dimensional systems have been described by N.N.Krasovsky in the supplement to the monograph [1]. Some particular results for the distributed parameter systems have been obtained in [2,3]. They can be generalized and concretized with the help of Bellman's equations with the functional derivatives and on the basis of the functional derivatives the second Ljapunov's method can be employed.

In this article the method is used for controlling heat process. Other problems of controlling distributed parameter systems can be considered in the same way.

1. THE PROBLEM STATEMENT. Let the controlled process be described by the boundary-value problem

$$\frac{\partial u}{\partial t}=\sum_{i,j=1}^{n}\frac{\partial}{\partial x_i}\left[a_{ij}(t,x)\frac{\partial u}{\partial x_j}\right]+f\left(t,x,u,\frac{\partial u}{\partial x_1},\dots,\frac{\partial u}{\partial x_n},\alpha\right),\ t>t_0,\ x\in\Omega, \tag{1}$$

$$a\frac{du}{d\nu}+cu=\varphi(t,x,u,\beta),\ x\in S,\ t>t_0 \tag{2}$$

$$u(t_0,x)=\psi(x),\ x\in\Omega+S, \tag{3}$$

where a_{ij}, f, c, φ and ψ are the given functions,

$$\sum_{i,j=1}^{n}a_{ij}\xi_i\xi_j\geq\gamma^2\sum_{i=1}^{n}\xi_i^2,\ a_{ij}=a_{ji},\ i,j=1,2,\dots n$$

for $x\in\Omega+S$, $t\geq t_0$ and

$$a=\left[\sum_{i=1}^{n}\left(\sum_{j=1}^{n}a_{ij}\cos(n,x_j)\right)^2\right]^{1/2}.$$

Ω is a limited area in E^n with partly-smooth boundary S, n is an external normal to S, and ν is a conormal. α and β are the scalar control parameters, which can take any real values.

The functions $\alpha(t,x)$ and $\beta(t,x)$ will be considered as admissible controls measured and bounded in respect to t if they satisfy the conditions

$$0 \leq \mathcal{Y}_1(t) \equiv \int_{\Omega} r(t,x)\alpha^2(t,x)d\Omega < \infty, \quad 0 \leq \mathcal{Y}_2(t) \equiv \int_{S} q(t,x)\beta^2(t,x)dS < \infty,$$

where r and q are given functions, such that $\mathcal{Y}_1(t)$ and $\mathcal{Y}_2(t)$ are locally integrable in respect to t for the admissible controls.

According to [5] the only function $u(t,x) \in W_2^1(Q)$, $Q = \Omega \times [t_0, T]$ corresponds to every pair of the admissible controls α and β with some restrictions for the data of the problem. This function $u(t,x)$ is called a generalized solution and it meets both the integral identity

$$\int_{\Omega} [u\Phi]_{t_1}^{t_2} d\Omega = \int_{Q_1^2} \{u\Phi_t' - \sum_{i,j=1}^{n} a_{ij} \frac{\partial u}{\partial x_i} \frac{\partial \Phi}{\partial x_j} + f\Phi\} dQ - \int_{S_1^2} [cu - \varphi]\Phi dS \tag{4}$$

for almost all t and t_2 from $[t_0, T]$, and the condition

$$u(t,x) \xrightarrow{a.} \Psi(x) \in L_2(\Omega), \text{ when } t \to t_0. \tag{5}$$

Here $Q_1^2 = \Omega \times [t_1, t_2]$, and S_1^2 is the boundary of this cylinder; Φ is any function from $W_2^1(Q)$; T is an arbitary number, but always fixed $/t_0 \leq T/$.

Let the state of the object be measurable at any point $p \in Q$. The control functions $\alpha[t,x] = \alpha(t,x,u(t,x))$ and $\beta(t,x) = \beta(t,x,u(t,x))$ are formed on the basis of the information obtained. It is priori clear that for ensuring the optimal controls it is necessary to take α and β as the functionals determined on the $u \in L_2$. In this case the arguments t and x in the functions α and β are put into the square brackets.

Then let the functions $M(t,x)$, $M_{ij}(t,x)$ and $N(t,x)$ be such that

$$0 \leq \mathcal{Y}_3(t) \equiv \int_{\Omega} [Mu^2 + \sum_{i,j=1}^{n} M_{ij} \frac{\partial u}{\partial x_i} \frac{\partial u}{\partial x_j}] d\Omega + \int_{S} Nu^2 dS < \infty$$

on the solution of the problems (1)-(3), corresponding to any pair of the admissible controls, and the function $\mathcal{Y}_3$ be locally integrable.

It is necessary to find such values $\alpha[t,x]$ and $\beta[t,x]$, that the functional

$$\mathcal{Y}[t_0, u(t_0,x)] = \int_{t_0}^{\infty} [\mathcal{Y}_1(t) + \mathcal{Y}_2(t) + \mathcal{Y}_3(t)] dt$$

have the least possible value.

2. BELLMAN's EQUATION. The designation

$$S[t,u(t,x)] = \min_{\alpha,\beta} \mathcal{I}[t,u(t,x)]$$

is introduced in accordance with the general Bellman's method. Let's suppose that the function $S(t,u)$ is differentiable with respect to t and as a functional of u it is differentiable by Freshe on $L_2(\Omega)$. Then

$$S[t+\Delta t, u(t+\Delta t,x)] = S[t,u(t,x)] + \frac{\partial S[t,u]}{\partial t}\Delta t + \Phi(t,\Delta u) + O(\Delta t,\Delta u), \quad (6)$$

where Φ is the linear functional with respect to Δu in L_2, calculated at the point (t,u), $O(\Delta t,\Delta u)$ is small by Δt and $\|\Delta u\|$ is a magnitude of the order which is higher than the first one. In the case when $\Delta u \in L_2(\Omega)$ almost to all t there is the function $v(t,x) \in L_2(\Omega)$ and

$$\Phi(t,u) = \int_\Omega v(t,x)\Delta u(t,x)\, d\Omega = \int_\Omega v(t,x)(u)_t^{t+\Delta t}\, d\Omega. \quad (7)$$

So according to the Bellman's optimal principle we have

$$S[t,u(t,x)] = \min_{\alpha,\beta}\left\{\mathcal{I}(t)\Delta t + S[t,u] + \frac{\partial S[t,u]}{\partial t} + \int_\Omega (u)_t^{t+\Delta t} v\, d\Omega + O\right\} \quad (8)$$

As

$$(u)_t^{t+\Delta t} v = (uv)_t^{t+\Delta t} - (v)_t^{t+\Delta t} u(t+\Delta t,x), \quad (9)$$

we suppose that $v(t,x) \in W_2^1(\Omega)$, and the integral in (8) can be substituted by its value from the formula (4) where $t = t_1$ and $t_2 = t+\Delta t$

From (8) at $\Delta t \to 0$ we obtain the Bellman's equation in the functional derivatives

$$-\frac{\partial S}{\partial t} = \min_{\alpha,\beta}\left\{\mathcal{I}(t) - \int_\Omega \left[\sum_{i,j=1}^n a_{ij}\frac{\partial u}{\partial x_i}\frac{\partial v}{\partial x_j} - f v\right] d\Omega - \int [cu-\varphi] v\, dS\right\}. \quad (10)$$

This equation makes it possible to show the different procedure of the approximate solution of the problem. The solution of the problem for the linear objects with quadratic criterion of the optimality has been done by the author together with G.Bachoi and M.Rakhimov. The authors have received boundary value problems, which are the analogues of the Rikkarti's equation. The methods of the approximate solution are given too.

3. THE APPLICATION OF THE SECOND LJAPUNOVŚ METHOD. Let $f(t,x,0,...,0)=0$ $\varphi(t,x,0,0)\equiv 0$ and, hence, $u\equiv 0$ be the solution of the boundary-value problem (1)-(3) for $\alpha=\beta=0$. This solution will be called stable according to Ljapunov in the metric $W_2^1(Q)$ for fixed admissible controls $\alpha(t,x)$ and $\beta(t,x)$, if there is such $\delta>0$ for any small $\varepsilon>0$, that from the inequality $\|u(t_0,x)\|_{L_2}<\delta$ for $t>t_0$ will follow, that $\|u(t,x)\|_{W_2^1}<\varepsilon$ for $t>t_0$, where $u(t,x)$ is the solution of the problem (1)-(3), defined from the integral identity (4) and condition (5). The solution is called asymptotically stable, if, besides, $\|u(t,x)\|_{W_2^1}\to 0$ for $t\to 0$. Therefore in accordance with the theory of stability the considered problem should be formulated as follows:

Both the admissible controls $\alpha^0[t,x]$ and $\beta^0[t,x]$ and the solution $u^0(t,x)$ of the problem (1)-(3), corresponding to them, must be determined in such a way, that the trivial solution $u\equiv 0$ would be asymptotically stable for $\alpha=\alpha^0$ and $\beta=\beta^0$, and the functional would have the least possible value.

The functional $V(t,x)$, determined on the elements $u\in L_2(\Omega)$, in which t is a numerical parameter, will be called the Ljapunovś functional (4), if it is differentiable in t and if we can show such a value $C_1>0$, that $|v(t,u)|<C_2$ for all $t\geq t_0$ and $\|u\|_{L_2}<C_1$.

Then letś determine the concept of full derivative with respect to t, made according to the boundary-value problem (1)-(3) by the rule

$$\frac{dV}{dt}=\lim_{\Delta t\to 0}\frac{V[t+\Delta t,u(t+\Delta t,x)]-V[t,u(t,x)]}{\Delta t},$$

where $u(t,x)$ meets the identity (4) and condition (5).

Taking into consideration the properties of V we have:

$$V[t+\Delta t,u(t+\Delta t,x)]-V[t,u(t,x)]=\frac{\partial V[t,u]}{\partial t}\Delta t+\Phi(t,\Delta u)+o(\Delta t,\Delta u),$$

where Φ is uniquely determined by the formula (7). As u meets the identity (4), then, according to the identity (4) and (9), we obtain

$$\Phi(t,\Delta u)=\int_{Q_1^2}\Big\{u v_t-\sum_{i,j=1}^{n}a_{ij}\frac{\partial u}{\partial x_i}\frac{\partial v}{\partial x_j}+f v\Big\}dQ-\int_{S_1^t}(cu-\varphi)v\,dS+ +\int_{\Omega}(v)_t^{t+\Delta t}u(t+\Delta t,x)\,d\Omega .$$

Taking into account, that

$$\frac{1}{\Delta t}\Big(\int_{Q_1^2}u v_t'\,dQ-\int_{\Omega}(v)_t^{t+\Delta t}u(t+\Delta t,x)\,d\Omega\Big)\to 0,\ \text{when } \Delta t\to 0$$

and proceeding from the last three formulas we obtain the formula of the full derivative

$$\frac{dV}{dt} = \frac{\partial V}{\partial t} - \int_{\Omega} \left(\sum_{i,j=1}^{n} a_{ij} \frac{\partial u}{\partial x_i} \frac{\partial v}{\partial x_i} - f v \right) d\Omega - \int_{S} (cu - \varphi)\, v\, dS. \qquad (11)$$

Then we determine the Bellman's functional

$$B(V, t, u, \alpha, \beta) = \frac{\partial V}{\partial t} - \int_{\Omega} \left(\sum_{i,j=1}^{n} a_{ij} \frac{\partial u}{\partial x_i} \frac{\partial v}{\partial x_j} - f v \right) d\Omega - \int_{S} (cu - \varphi)\, v\, dS + \mathcal{I},$$

where $\mathcal{I}$ is the functional introduced into the determination of the criterion of optimality, and $v \in W_2^1(Q)$.

We shall mark $u(t,x)$ as the solution of the boundary-value problems (1)-(3), according to Krasovsky [1]. This solution corresponds to the control $\alpha[t,x]$, $\beta[t,x]$.

THEOREM. Let the positive definite Ljapunov's functional V° for the boundary-value problems (1)-(3) be found in such a manner that the function $V(t,x)$ uniquely determined by its Freshe's differential, refers to $W_2^1(Q_T), Q_T = \Omega \times [t_0, T]$, where T is a positive arbitrary number, which is greater, than t .

Let such functionals $\alpha^{\circ}[t,x,u] \in L_2(Q_T)$ and $\beta^{\circ}[t,x,u] \in L_2(S_T)$ be for any $u \in W_2^1(Q_T)$, that

1) $B(V^{\circ}, t, u, \alpha^{\circ}[t,x,u]), \beta^{\circ}[t,x,u]) = 0$ for $u \in W_2^1$ and almost for all $t \in [t_0, T]$

2) $B(V^{\circ}, t, u, \alpha(x), \beta(x)) \geq 0$ for any $\alpha(x) \in L_2(\Omega)$, $\beta(x) \in L_2(S)$,

then $\alpha^{\circ}[t,x,u]$ and $\beta^{\circ}[t,x,u]$ solve the problem of the optimum stability. In this case

$$\int_{t_0}^{\infty} \left\{ \mathcal{I}_1(\alpha^{\circ}[t,x]) + \mathcal{I}_2(\beta^{\circ}[t,x]) + \mathcal{I}_3(u^{\circ}[t,x]) \right\} dt = \min_{\alpha,\beta} \int_{t_0}^{\infty} \left\{ \mathcal{I}_1(\alpha[t,x]) + \mathcal{I}_2(\beta[t,x]) + \mathcal{I}_3(u[t,x]) \right\} dt = V^{\circ}(t_0, u^{\circ}[t_0,x]).$$

The proof of this theorem can easily be obtained by the method, accounted in [1], p.p. 486-487, based on the properties of the functional V° and the theorem of the asymptotic stability [4].

REFERENCE. The formula (11) of the full derivative of the functional V gives concrete recommendation for investigating the problems of the stability of distributed parameter systems with the help of the second Ljapunov's method. From that, in particular , sufficient conditions can be easily obtained for the stability of the first approximation systems.

R e f e r e n c e s

1. Malkin I. Theory of Motion Stability. "Nauka" Press 1966.
2. Sirazetdinov T. Stability of The Distributed Parameter Systems. Kazan Aircraft Institute Press, 1970.
3. Zubov V. Motion Stability. "Visshaya Shkola" Press, 1973.
4. Plotnikov V. Energetic Inequality and Property of Super-definiteness of System of Eigenvalue Functions. "Izvestia AS USSR, S, Math., V32, issue 4, 1968, pp 743-755.

ABOUT ONE PROBLEM OF SYNTHESIS OF OPTIMUM CONTROL BY THERMAL CONDUCTION PROCESS

A.I.Egorov, G.S.Bachoi
Dnepropetrovsk Railway Transport Institute,
Dnepropetrovsk, USSR

The formal procedure of obtaining Bellman equation in the problem of heat conductivity control is stated in [1]. Here we show how the problem of synthesis of optimum control with quadratic criterion of optimum is solved with the help of this equation. For simplification of formulas we used the simplest example which can be easily generalized. It should be noted, that during the solution the nonlinear boundary-value problem for integral-differential equation is derived, which is infinite dimensional analog of well-known Rikkati equation for finite dimensional systems.

1. STATEMENT OF PROBLEM. BELLMAN EQUATION. Let us assume that the control process is described by the function $u(t,x)$ which inside the region of $Q=\{0\leq t\leq T,\ 0\leq x\leq 1\}$ satisfies the thermal conduction equation

$$u_t = u_{xx} + q(x)p(t) + f(t,x), \tag{1}$$

and on the boundary Q the homogeneous conditions

$$u(0,x) = 0, \tag{2}$$

$$u_x(t,0) = u_x(t,1) + \alpha u(t,1) = 0, \quad \alpha = const > 0, \tag{3}$$

where f and q are given functions, and control $P(t)$ belongs to $L_2(0,T)$ with open or closed region of values, which we shall term by means of P.

The problem under consideration lies in determining control of such $P[t,u(t,x)]$, that the functional

$$J[P] = \int_0^1 [u(T,x) - \psi(x)]^2 dx + \beta \int_0^T p^2(t)\,dt, \quad \beta = const > 0$$

should have the least possible value. Here T is fixed moment of time, and ψ is given function from $L_2(0,1)$.

With the help of the method, stated in [1], we can show, that

Bellman functional

$$S[t,u] = \min_{p\in P}\left\{\int_0^1 [u(T,x)-\psi(x)]^2 dx + \beta\int_t^T p^2(t)\,dt\right\}$$

satisfies the equation

$$-\frac{\partial s}{\partial t} = \min_{p(t)\in P}\left\{\beta p^2(t) + p(t)\int_0^1 v(t,x)\,q(x)\,dx - \alpha u(t,1)\,v(t,1) + \int_0^1 [f(t,x)\,v(t,x) - u_x(t,x)\,v_x(t,x)]\,dx\right\} \tag{4}$$

In this case it is supposed, that function $v(t,x)$, which is uniquely determined by Freshe differential $\varphi(u,h)$ of functional S :

$$\varphi(u,h) = \int_0^1 v(t,x)\,h(t,x)\,dx \tag{5}$$

is absolutely continuous on x .

In particular case, when the set of P coincides with the axle $(-\infty,+\infty)$, we can have

$$p(t) = \frac{1}{2\beta}\int_0^1 v(t,x)\,q(x)\,dx \tag{6}$$

$$\frac{\partial s}{\partial t} = \int_0^1 [u_x v_x - f v]\,dx + \frac{1}{4\beta}\left[\int_0^1 v(t,x)\,q(x)\,dx\right]^2 + \alpha u(t,1)\,v(t,1) \tag{7}$$

from the equation (4).

From the definition of S :

$$S[T,u] = \int_0^1 [u(T,x) - \psi(x)]^2\,dx \tag{8}$$

2. THE CONSTRUCTION OF OPTIMAL CONTROL. We shall find the solution of problem (7)-(8) in the form

$$S[t,u] = \int_0^1\int_0^1 K(t,x,s)[u(t,x)-\psi(x)][u(t,s)-\psi(s)]\,dx\,ds + \int_0^1 \varphi(t,x)[u(t,x)-\psi(x)]\,dx + \eta(t). \tag{9}$$

Calculating Freshe differential of this functional we find, that

$$\varphi(u,h) = \int_0^1\int_0^1 [K(t,x,s) + K(t,s,x)][u(t,s)-\psi(s)]\,h(t,x)\,dx\,ds + \int_0^1 \varphi(t,x)\,h(t,x)\,dx$$

and according to formula (5) we have

$$V(t,x)=\int_0^1[K(t,x,s)+K(t,s,x)][u(t,s)-\psi(s)]\,ds+\varphi(t,x). \tag{10}$$

Substituting (9) and (10) in the equation (7) we have

$$\int_0^1\int_0^1[-K_t(t,x,s)-K_{xx}(t,x,s)-K_{xx}(t,s,x)+\frac{1}{4\beta}K_1(t,x,s)][u(t,x)-\psi(x)]\times$$

$$\times[u(t,s)-\psi(s)]dx\,ds+\int_0^1\{-\varphi_t(t,x)-\varphi_{xx}(t,x)-K_2(t,x)-K_3(t,x)+$$

$$+\frac{1}{2\beta}\int_0^1\int_0^1[K(t,x,s)+K(t,s,x)]\varphi(t,y)q(y)q(s)\,dy\,ds\}[u(t,x)-\psi(x)]\,dx+$$

$$+\{-\eta_t(t)-\int_0^1\varphi_{xx}(t,x)\psi(x)dx-\int_0^1 f(t,x)\varphi(t,x)dx+\frac{1}{4\beta}\left(\int_0^1\varphi(t,x)q(x)dx\right)^2\}+$$

$$+\{[\varphi_x(t,1)+\alpha\varphi(t,1)]+\int_0^1[K_x(t,1,s)+\alpha K(t,1,s)][u(t,s)-\psi(s)]ds+$$

$$+\int_0^1[K_x(t,s,1)+\alpha K(t,s,1)][u(t,s)-\psi(s)]ds\}u(t,1)-$$

$$-\{\int_0^1[K_x(t,0,s)+K_x(t,s,0)][u(t,s)-\psi(s)]ds+\varphi_x(t,0)\}u(t,0)=0$$

where

$$K_1(t,x,s)=\int_0^1\int_0^1[K(t,y,s)+K(t,s,y)][K(t,z,x)+K(t,x,z)]q(y)q(z)\,dy\,dz$$

$$K_2(t,x)=\int_0^1[K(t,x,s)+K(t,s,x)]f(t,s)\,ds \tag{11}$$

$$K_3(t,x)=\int_0^1[K_{ss}(t,x,s)+K_{ss}(t,s,x)]\psi(s)\,ds$$

The functional (9) will satisfy the equation (7)for any function $u(t,x)-\psi(x)$, if we demand that $K(t,x,s)$, $\varphi(t,x)$ and $\eta(t)$ satisfy following equations:

$$K_t(t,x,s)+K_{xx}(t,x,s)+K_{xx}(t,s,x)-\frac{1}{4\beta}K_1(t,x,s)=0 \tag{12}$$

$$\left.\begin{aligned}K_x(t,0,s)&=K_x(t,1,s)+\alpha K(t,1,s)=0\\ K_x(t,s,0)&=K_x(t,s,1)+\alpha K(t,s,1)=0\end{aligned}\right\} \tag{13}$$

$$\varphi_t(t,x)+\varphi_{xx}(t,x)+K_2(t,x)+K_3(t,x)=$$

$$=\frac{1}{2\beta}\int_0^1\int_0^1[K(t,x,s)+K(t,s,x)]\varphi(t,y)q(y)q(s)\,dy\,ds \tag{14}$$

$$\varphi_x(t,0) = \varphi_x(t,1) + \alpha\varphi(t,1) = 0 \tag{15}$$

$$\eta_t(t) + \int_0^1 \varphi_{xx}(t,x)\,\psi(x)\,dx + \int_0^1 f(t,x)\,\varphi(t,x)\,dx = \frac{1}{4\beta}\left(\int_0^1 \varphi(t,x)\,q(x)\,dx\right)^2 \tag{16}$$

Besides this it follows from (8) and (9), that

$$K(T,x,s) = \delta(s-x),\ \varphi(T,x) = 0,\ \eta(T) = 0. \tag{17}$$

The boundary-value problem (12)-(13) is the infinite dimensional analog of Rikkati equation, which appears during the solution of the problem about the optimum stabilization of system with finite number of degree of freedom.

We find function K in the form

$$K(t,x,s) = \sum_{i,j=1}^{\infty} C_{ij}(t)\,X_i(x)\,X_j(s),\quad X_i(x) = \frac{\cos\lambda_i x}{\sqrt{\omega_i}} \tag{18}$$

where λ_i are the eigenvalues of boundary-value problem $X'' + \lambda^2 X = 0$, $X'(0) = 0$, $X'(1) + \alpha X(1) = 0$, which are the positive roots of equation $\lambda\,tg\,\lambda = \alpha$, ω_i^{-1} - the normalization factors.

Substituting the function (18) in equations (12) and (13) we obtain the system of equations with reference to coefficients C_{ij}:

$$\frac{dC_{ij}}{dt} = \lambda_i^2\left[C_{ij} + C_{ji}\right] + \frac{1}{4\beta}\sum_{k,\ell=1}^{\infty}\left[C_{kj} + C_{jk}\right]\left[C_{\ell i} + C_{i\ell}\right] q_k q_\ell \tag{19}$$

and thanks to the conditions (17) we shall have

$$C_{ij}(T) = \delta_{ij},\quad i,j = 1,2,\ldots \tag{20}$$

Here q_i are Fourier coefficients of function $q(x)$.

By analogy the solution of problem (14)-(15) is found in the form of

$$\varphi(t,x) = \sum_{i=1}^{\infty} a_i(t)\,X_i(x). \tag{21}$$

Then coefficients a_i are determined from equations:

$$\left.\begin{aligned} \frac{da_i}{dt} &= \lambda_i^2 a_i + \frac{1}{2\beta}\sum_{k,j=1}^{\infty} q_k q_j\left[C_{ij} + C_{ji}\right] a_k - \sum_{j=1}^{\infty}\left[f_j - \lambda_j^2\psi_j\right]\left[C_{ij} + C_{ji}\right] \\ a_i(T) &= 0,\ i = 1,2,\ldots \end{aligned}\right\} \tag{22}$$

where $f_i(t)$ and ψ_i are Fourier coefficients of functions $f(t,x)$ and $\psi(x)$ correspondingly.

Taking into consideration (21), the equation (16) with the last condition (17) can be rewritten as

$$\eta_t(t)=\sum_{i=1}^{\infty}[\lambda_i^2\psi_i - f_i]a_i + \frac{1}{4\beta}\Big(\sum_{i=1}^{\infty}a_i q_i\Big)^2, \quad \eta(T)=0 \qquad (23)$$

So, we obtain a total system of equations for definition of all undetermined values.

From the equation (19) and conditions (20) we can find, that $C_{ij}(t)\equiv 0$ by $i\neq j$ and

$$C_{ii}(t)=\frac{2\beta\lambda_i^2 \exp[2\lambda_i^2(t-T)]}{2\beta\lambda_i^2+q_i^2(1-\exp[2\lambda_i^2(t-T)])} \qquad (24)$$

That is why we can have

$$a_i(t)=-\int_t^T C_{ii}(\tau)\Big\{\frac{q_i}{\beta}\sum_{k=1}^{\infty}q_k a_k(\tau)-2\gamma_i(\tau)\Big]e^{\lambda_i^2(t-\tau)}d\tau, \quad i=1,2,\ldots \qquad (25)$$

from (22), where $\gamma_i = f_i-\lambda_i^2\psi_i$.

Multiplying both parts of the i-th equation by q_i and summing them up, we find

$$r(t)=\int_t^T B(t,\tau)r(\tau)\,d\tau+\Gamma(t)$$

where

$$r(t)=\sum_{k=1}^{\infty}q_k a_k(t), \quad B(t,\tau)=-\sum_{i=1}^{\infty}\frac{2q_i^2\lambda_i^2\exp[\lambda_i^2(t+\tau-2T)]}{2\beta\lambda_i^2+q_i^2(1-\exp[2\lambda_i^2(\tau-T)])}$$

$$\Gamma(t)=4\beta\sum_{i=1}^{\infty}\int_t^T\frac{q_i\lambda_i^2[f_i(T)-\lambda_i^2\psi_i]\exp[\lambda_i^2(t+\tau-2T)]}{2\beta\lambda_i^2+q_i^2(1-\exp[2\lambda_i^2(\tau-T)])}d\tau .$$

Having determined $r(t)$ and substituted it into (25), we shall have $a_i(t)$, $i=1,2,\ldots$. After that we can easily find $\eta(t)$ from (24). Therefore the functions $K(t,x,\xi)$, $\varphi(t,x)$ and $\eta(t)$ (see $S[t,U]$) are determined uniquely, and by the formula (6) we can obtain optimum control as a functional U . It is not necessary to use the function $\eta(t)$ for the construction of P and therefore we need not solve the problem (23).

Since in every concrete case the problem under study is formally solved completely, then on the basis of obtained formulas we can easily ground the procedure mentioned above, imposing corresponding restrictions on functions, being in its enunciation. But the substantiation of the method of dynamic programming (the obtaining of Bellman equation) demands more profound analysis, similar to that which is proposed in [2] for finite dimensional systems.

In conclusion we must mention, that Bellman's method was used in [3 - 4] for solving analogous control problems. However the authors of these works could not receive optimum control and they had to limit themselves by giving the account of this method.

References

1. Egorov A.I. Optimal stabilisation of systems with distributed parameters. In this volume.
2. Boltyanski V.G. Sufficient conditions of optimum and the basis of dynamic programming method. News of the AS USSR, math., vol.28, N3,1964,481-514.
3. Sirazetdinov I.K. On analytical designing of regulators for magnetohydrodynamical procedures. 1-2. Automatics and Telemechanics, NN10,12,1967.
4. Erzbeger H. and Kim M. Optimum boundary control of distributed parameter systems. Information and Control, vol.9, N3,1966.

ABOUT THE PROBLEM OF SYNTHESIS OF OPTIMUM CONTROL BY ELASTIC OSCILLATIONS.

A.I.Egorov, M.Rakhimov
Dnepropetrovsk Railway Transport Institute
Dnepropetrovsk, USSR

The formal procedure of obtaining Bellman equation in the problem of heat conductivity control is stated in [1].

It is shown here, that in the same way one can obtain analogous equation for problems of control processes in other systems with distributed parameters.

Here, analizing a simple problem of elastic oscillations control, the application of this method is stated.

As a result the problem reduces to solving the non-linear boundary-value problems for matrix integro-differential equations, generalizing the well-known Rikkati equations.

Let us assume that the control process is described by the function $u(t,x)$, which inside the rigion $Q=\{0\le x\le 1, t_0\le t\le T\}$ satisfies the equation

$$u_{tt}-u_{xx}=p(t)q(x)+f(t,x) \tag{1}$$

and on the boundary Q - the conditions

$$u(t_0,x)=\varphi_0(x),\ u_t(t_0,x)=\psi_0(x) \tag{2}$$

$$u(t,0)=u(t,1)=0, \tag{3}$$

where $f(t,x)$, $q(x)$ and $\varphi_0(x)$ are assigned functions from L_2 , and $\psi_0(x)$ is absolutely continuous function, and control $p(t)$ belongs to $L_2(0,T)$ with the meanings at the interval (open or closed), which in future will be designated by P .

At these assumptions, each control $p(t)$ determines unique solution of the problem (1)-(3), as a function $u(t,x)\in W_2^1(Q)$, which satisfies the integral identity

$$\int_0^1 [u_t(t,x)\varphi(t,x)]_{t_1}^{t_2} dx = \int\int_{Q^2} [u_t\varphi_t - u_x\varphi_x + (f+q(x)p(t))\varphi]\, dQ \tag{4}$$

at any function $\varphi\in W_2^1(Q)$, which turns into zero in the neighbourhood of points $x=0$ and $x=1$.

Here t_1 and t_2 are arbitrary points from $[t_0, T]$, and $Q_1^2 = \{0 \le x \le 1, t_1 \le t \le t_2\}$.

The problem of optimal control under consideration lies in determining $p(t)$ and corresponding to it $u(t,x)$ such that the functional

$$\mathcal{J}[t_0 w(t_0,x)] = \int_0^1 [\alpha u^2(T,x) + \beta u_t^2(T,x)] dx + \gamma \int_{t_0}^{T} p^2(t) dt$$

takes the least possible meaning. Here w is the vector with components u and u_t; α, β, and γ are positive constants, and T is fixed moment of time, exceeding t_0.

Supposing

$$S[t, w(t,x)] = \min_{p \in P} \mathcal{J}[t, w(t,x)],$$

we find, by an ordinary method, that

$$-\frac{\partial S}{\partial t} \Delta t = \min \left\{ \gamma p^2(t) \Delta t + \Phi(t, w(t,x); \Delta w(t,x)) + O \right\}, \quad (5)$$

where Φ is linear on Δw functional, obtained in point (t, w), and O is small on Δt, $\|\Delta w\|$ is value of higher order than one. Since $\Delta u(t,x)$ and $\Delta u_t(t,x)$ belong to $L_2(0,1)$ almost at all t, the vector-function $v(t,x) = \{v_1, v_2\}$ is such that

$$\Phi = \int_0^1 v^*(t,x) \Delta w(t,x) dx. \quad (6)$$

We have

$$\Delta u = u(t+\Delta t, x) - u(t,x) = \frac{\partial u(t,x)}{\partial t} \Delta t + O$$

$$v_2 \Delta u_t = \Delta(v_2 u_t) - u_t(t+\Delta t, x) \Delta v_2$$

and according to formula (4) we obtain

$$\int_0^1 v_2 \Delta u_t dx = \int_{Q_1^2} [u_t v_{2t} - u_x v_{2x} + (f + q(x)p(t)) v_2] dQ - \int_0^1 u_t(t_2, x) \Delta v_2 dx,$$

where we take $t_1 = t$ and $t_2 = t + \Delta t$, and it is supposed, that v_2 has the properties of function Φ in (4).

That is why

$$\Phi(t, w(t,x), \Delta w(t,x)) = \int_0^1 v_1(t,x) \frac{\partial u(t,x)}{\partial t} dx \Delta t + \int_{Q_1^2} [u_t v_{2t} - u_x v_{2x} +$$
$$+ (f + q(x)p(t)) v_2] dQ - \int_0^1 u_t(t_2, x) \Delta v_2 dx + O.$$

Substituting the obtained expression Φ in equation (5), we **have** the Bellman equation

$$-\frac{\partial S}{\partial t}=\min_{p\in P}\left\{\gamma p^2(t)+\int_0^1[v_1u_t-u_x v_{2x}+(f+q(x)p(t))v_2]dx\right\}.$$

Let us assume now, that $P=(-\infty,+\infty)$ **and there exists, integrable** with square, derivative v_{2xx}. Then from Bellman equation we have

$$p(t)=-\frac{1}{2\gamma}\int_0^1 q(x)v_2(t,x)dx \tag{7}$$

$$-\frac{\partial S}{\partial t}=\int_0^1[v_1u_t+uv_{2xx}+fv_2]dx-\frac{1}{4\gamma}\left(\int_0^1 q(x)v_2(t,x)dx\right)^2 \tag{8}$$

We find the solution of equation (8) in the form

$$S[t,w(t,x)]=\int_0^1\int_0^1 w^*(t,x)K(t,x,s)w(t,s)dsdx+\int_0^1\varphi^*(t,x)w(t,x)dx+\eta(t), \tag{9}$$

where matrix K, vector φ and scalar function $\eta(t)$ must be determined.

From (9) we find, that

$$\Phi(t,w,h)=\int_0^1\int_0^1 w^*(t,s)N(t,s,x)h(t,x)dsdx+\int_0^1\varphi^*(t,x)h(t,x)dx,$$

where

$$N(t,x,s)=K(t,x,s)+K^*(t,s,x) \tag{10}$$

and vector $v=\{v_1,v_2\}$, a component part of formula (6), is determined as follows

$$v(t,x)=\int_0^1 N^*(t,s,x)w(t,s)ds+\varphi(t,x). \tag{11}$$

Suppose N_{ij} are elements of matrix N, and φ_i is the i- component of vector φ. Then, substituting the meanings of S and v from (9) and (11) into equation (8), we obtain

$$-K_t(t,x,s)=L(t,x,s)-\frac{1}{4\gamma}M(t,x,s), \tag{12}$$

$$-\varphi_{1t}(t,x)=\varphi_{2xx}(t,x)+\int_0^1 N_{12}(t,x,s)f(t,s)ds-\frac{1}{2\gamma}\int_0^1\int_0^1 q(y)q(s)N_{12}(t,x,y)\varphi_2(t,s)dyds \tag{13}$$

$$-\varphi_{2t}(t,x)=\varphi_1(t,x)+\int_0^1 N_{22}(t,x,s)f(t,s)ds-\frac{1}{2\gamma}\int_0^1\int_0^1 q(s)q(y)N_{22}(t,x,y)\varphi_2(t,s)dyds \tag{14}$$

$$\eta'(t) = \frac{1}{4\gamma}\left[\int_0^1 q(x)\varphi_2(t,x)dx\right]^2 \tag{15}$$

where

$$L(t,x,s) = \begin{pmatrix} N_{21xx}(t,x,s) & N_{22xx}(t,x,s) \\ N_{11}(t,x,s) & N_{12}(t,x,s) \end{pmatrix},$$

and $M(t,x,s)$ is matrix with components

$$M_{ij}(t,x,s) = \int_0^1\int_0^1 q(y)q(z)N_{2i}(t,y,s)N_{2j}(t,z,x)dydz$$

$$M_{12}(t,x,s) = M_{21}(t,s,x).$$

Since we assumed that $v_2(t,0) = v_2(t,1) = 0$ at any vector $w(t,x)$, then from formula (11) we obtain

$$N_{22}(t,0,s) = N_{22}(t,1,s) = N_{21}(t,0,s) = N_{21}(t,1,s) = 0 \tag{16}$$

$$\varphi_2(t,0) = \varphi_2(t,1) = 0. \tag{17}$$

Besides, directly from the definition of functional S it follows that

$$S[T, w(T,x)] = \int_0^1 [\alpha u^2(T,x) + \beta u_t^2(T,x)]dx$$

and from formula (9) we obtain

$$\left.\begin{array}{l} K_{ij}(T,x,s) = 0, \quad i \neq j \\ K_{11}(T,x,s) = \alpha\,\delta(s-x), \; K_{22}(T,x,s) = \beta\,\delta(s-x) \end{array}\right\} \tag{18}$$

$$\varphi_1(T,x) = \varphi_2(T,x) = \eta(T) = 0, \tag{19}$$

where $\delta(x)$ is Dirac function.

The boundary-value problem (12), (16) is the generalization of the analized case of the known Rikkati equation from the theory of optimal stabilization of systems with the finite number of the degrees of freedom. Taking into consideration that matrices K and N are connected in ratio (10), we find that (12) represents the system of non-linear integro-differential equations relative to K_{ij} . Here we shall not solve this boundary-value problem. However, we shall mention, that two last conditions in (18) show the necessity to analyze it in the space of distributions (the generalized functions). After the solution of Rikkati boundary-value problem (12), (16), (18) it is necessary to determine vector φ from (13), (14), (17) and (18). This will give the possibility to find the control according to formula (7). However the problem of belonging $P(t)$ to space $L_2(0,T)$ requires special investigations.

References

1. Egorov A.I. Optimal stabilization of systems with distributed parameters. In this volume.

ON THE PARTITIONING PROBLEM IN THE SYNTHESIS OF MULTILEVEL OPTIMIZATION STRUCTURES

G. GRATELOUP *
Professeur
I.N.S.A. Toulouse

M. RICHETIN *
Attaché de Recherche
C.N.R.S.

INTRODUCTION

Multilevel control structures are especially important when a large-scale system is to be controlled. The system is divided into interconnected sub-systems (Fig. 1) controlled by decision-making units hierarchically arranged (Fig. 2) (Mesarovic et al., (1). Multilevel structures are interesting since control is divided and thus simplified, and since the control system is much more reliable (Plander, (2)).

Up to now, most of the works have dealt with two-level optimization (Lasdon et al., (3), Brosilow et al., (4), Titli (8)). The different sub-systems are coordinated by coordination variables, calculated on the second level, and whose nature depends on the chosen coordination method. But in any case, these variables are linked with coupling equations of sub-systems. For large-scale systems, the number of coordination variables may be high. Consequently a multilevel structure is necessary if constraints exist on the size of coordinators or on the transmission capacity of channels. Several solutions are then available. The synthesis of these structures has been partly realized by Strasjak (6), Kulikowski (7).

At first, the decomposition of linear coupling equations associated with a multilevel structure is defined. The extension of classical coordination methods is done and it is shown that there is no feasible method for a n-level optimization $(n > 2)$.

In a second part, the effects of couplings on the convergence of coordination algorithms are studied and are illustrated by an example.

DECOMPOSITION OF THE COUPLING EQUATIONS

In a multilevel structure, each coordination deals with the couplings between groups of sub-systems. So if $n_v = 2$ (Fig. 2), then the vth level coordinator deals with the couplings between two groups of sub-systems.

The coupling equations are now decomposed according to the vertical division of the coordination and we suppose that they are linear.

* Laboratoire d'Automatique et d'Analyse des Systèmes du C.N.R.S.
B.P. 4036 - 31055 TOULOUSE CEDEX.

Let I^1 be the set of numbers of the N sub-systems. Suppose v = 2. The 2nd level-coordinator deals with n_2 groups of sub-systems and let I^2 be the n_2-partition of I^1 :

$$I^2 = \{P_1^2 \dots P_{n_2}^2\} ; \; P_i^2 \cap P_j^2 = \emptyset \; (i \neq j) ; \; \sum_{i=1}^{n_2} |P_i^2| = N$$

(| | = order of a set)

The coupling equations become :

$$X_i = \sum_j C_{ij} Z_j + \sum_k V_{ik}^2 \qquad \text{with} \quad \forall i \in I^1 \quad \forall j,\, j \in P_r^2 \text{ if } i \in P_r^2 \quad \forall k\, (k \neq r \text{ if } i \in P_r^2) \in \{1, \dots, n_2\} \tag{1}$$

$$V_{ik}^2 = \sum_{j \in P_k^2} C_{ij} Z_j \qquad \text{with} \quad \forall i \in I^1 \quad \forall k\, (k \neq r \text{ if } i \in P_r^2) \in \{1, \dots, n_2\} \tag{2}$$

For any v, this decomposition is then achieved from the n_w-partitions of I^1 (w = 1, ..., v).

<u>Example 1</u> : $N = 8 ; \; v = 3 ; \; n_2 = 4 ; \; n_3 = 2 ; \; i = 1$

$$I^2 = \{(1,2), (3,4), (5,6), (7,8)\} ; \; I^3 = \{(1,2,3,4), (5,6,7,8)\}$$

$$X_1 = \sum_{j=1}^{2} C_{ij} Z_j + V_{12}^2 + V_{12}^3$$

$$V_{12}^2 = C_{13} Z_3 + C_{14} Z_4$$

$$V_{12}^3 = C_{15} Z_5 + C_{16} Z_6 + C_{17} Z_7 + C_{18} Z_8$$

On Fig. 3, this decomposition procedure means successive partitions of the coupling matrix C.

In general, V_{ab}^c is the input coupling vector for the sub-system "a", made up from the outputs of a group of sub-systems "b" which is obtained after the cth partition of the coupling matrix.

<u>Coordination methods</u>

A static system is considered for which each sub-system has the following model and criteria : $Z_i = T_i(X_i, M_i) \; ; \; \min f_i(X_i, M_i)$

The overall optimization criteria is separable : $\min F = \min \sum_i f_i$. This enables to realize a multilevel optimization (Fig. 2). With constraints being of type (1) and (2), the Lagrangian of the optimization problem is :

$$L = \sum_{i=1}^{N} [f_i(X_i, M_i) + \mu_i^T (Z_i - T_i(X_i, M_i)) + \rho_i^T (X_i - \sum_j C_{ij} Z_j - \sum_k V_{ik}^2 - \sum_l V_{il}^3 - \dots) + \sum_k \lambda_{ik}^2 (V_{ik}^2 - \sum_j C_{ij} Z_j) + \sum_l \lambda_{il}^3 (V_{il}^3 - \sum_j C_{ij} Z_j) + \dots] \tag{3}$$

On the analogy of the feasible method, let us suppose that Z, V_{ik}^2, V_{il}^3, ... are respectively calculated by 1st, 2nd, 3rd ... level coordinators.

To give the Lagrangian a separable form, it can be shown that λ^2_{ik}, λ^3_{il} must be fixed for the local optimizations and consequently calculated on coordination levels. If λ^c_{ab} is obtained by a gradient algorithm :

$$\lambda^c_{ab}(k+1) = \lambda^c_{ab}(k) + M\frac{\partial L}{\partial \lambda^c_{ab}} \quad (M>0) \; ; \; \frac{\partial L}{\partial \lambda^c_{ab}} = V^c_{ab} - \sum_j C_{ij} Z_j$$

the variables Z_j which occur in $\frac{\partial L}{\partial \lambda_{ab}}$ must be known. If the information flows vertically in the structure and according to the hierarchy, the lowest coordination levels where λ^c_{ab} can be calculated is then determined (see Fig. 4 for example 1).

Consequently, all the coupling equations are satisfied only when all the coordinators have converged. Therefore, there is no feasible coordination method for a n-level optimization ($n > 2$).

Moreover, V^c_{ab} and λ^c_{ab} can be simultaneously calculated by the same coordinator This leads to the combined method (Grateloup et al., (5)). The stationnarity equations of the Lagrangian for V^c_{ab} and λ^c_{ab} are :

$$\frac{\partial L}{\partial V^c_{ab}} = \lambda^c_{ab} - \rho_a = 0 \; ; \; \frac{\partial L}{\partial \lambda^c_{ab}} = V^c_{ab} - \sum_j C_{ij} Z_j \tag{4}$$

As V^c_{ab} and λ^c_{ab} have been defined as vectors of same dimension, from (4), they can be directly calculated for given ρ_a, Z_j. So, at any v^{th} coordination-level ($v > 1$), a direct iteration algorithm can be implemented (Grateloup et al., (11)).

COORDINATION CONVERGENCE

The effects of couplings on the coordination convergence in multilevel structures are here studied and were first pointed out by Sprague (9).

Let us suppose $v = 2$, $n_2 = 2$. V and λ are the coordination variables on the 2nd coordination-level, and Z, X, M, ρ, μ, are the other variables of the optimization problem. The optimal value of a variable is marked $*$.

To determine V and λ, a combined method is used with a direct-iteration algorithm whose convergence is studied.

From (4) $\quad V = V(Z)$

$\qquad\qquad \lambda = \lambda(\rho)$ $\quad$ with $\quad V(Z) = \begin{pmatrix} 0 & C^{12} \\ C^{21} & 0 \end{pmatrix} Z \; ; \; C = \begin{pmatrix} C^{11} & C^{12} \\ C^{21} & C^{22} \end{pmatrix}$, $\quad \lambda(\rho) = \rho$

C^{11}, C^{22}, C^{12}, C^{21} = submatrices of C corresponding with the partition of the N subsystems into two groups ($n_2 = 2$).

At the ith iteration of the 2nd level-coordinator, we get :

$$\begin{aligned} (V^i - V^*) &= \frac{dV}{dZ}(Z^i - Z^*) \\ (\lambda^i - \lambda^*) &= \frac{d\lambda}{d\rho}(\rho^i - \rho^*) \end{aligned} \Rightarrow \begin{aligned} \varepsilon^i_V &= \frac{dV}{dZ}\varepsilon^i_Z \\ \varepsilon^i_\lambda &= \frac{d\lambda}{d\rho}\varepsilon^i_\rho \end{aligned} \tag{5}$$

For given V^i and λ^i, when local optimizations and 1st level-coordinators have converged, the following equations are satisfied :

$$L_X(X^{i+1}, M^{i+1}, \mu^{i+1}, \rho^{i+1}) = 0$$
$$L_M(X^{i+1}, M^{i+1}, \mu^{i+1}) = 0$$
$$L_Z(\mu^{i+1}, \rho^{i+1}, \lambda^{i}) = 0 \qquad \text{(where } L_r = \frac{\partial L}{\partial T}\text{)} \qquad (6)$$
$$L_\mu(X^{i+1}, M^{i+1}, Z^{i+1}) = 0$$
$$L_\rho(X^{i+1}, Z^{i+1}, V^{i}) = 0$$

A first-order development of the preceeding functions give :

$$\frac{\partial L_X}{\partial X}\Big|_* \varepsilon_X^{i+1} + \frac{\partial L_X}{\partial M}\Big|_* \varepsilon_M^{i+1} + \frac{\partial L_X}{\partial \mu}\Big|_* \varepsilon_\mu^{i+1} + \frac{\partial L_X}{\partial \rho}\Big|_* \varepsilon_\rho^{i+1} = 0$$
$$\frac{\partial L_M}{\partial X}\Big|_* \varepsilon_X^{i+1} + \frac{\partial L_M}{\partial M}\Big|_* \varepsilon_M^{i+1} + \frac{\partial L_M}{\partial \mu}\Big|_* \varepsilon_\mu^{i+1} = 0$$
$$\frac{\partial L_Z}{\partial \mu}\Big|_* \varepsilon_\mu^{i+1} + \frac{\partial L_Z}{\partial \rho}\Big|_* \varepsilon_\rho^{i+1} + \frac{\partial L_Z}{\partial \lambda}\Big|_* \varepsilon_\lambda^{i} = 0 \qquad \text{(where } f|_* = \text{optimal value of f).} \qquad (7)$$
$$\frac{\partial L_\mu}{\partial X}\Big|_* \varepsilon_X^{i+1} + \frac{\partial L_\mu}{\partial M}\Big|_* \varepsilon_M^{i+1} + \frac{\partial L_\mu}{\partial Z}\Big|_* \varepsilon_Z^{i+1} = 0$$
$$\frac{\partial L_\rho}{\partial X}\Big|_* \varepsilon_X^{i+1} + \frac{\partial L_\rho}{\partial Z}\Big|_* \varepsilon_Z^{i+1} + \frac{\partial L_\rho}{\partial V}\Big|_* \varepsilon_V^{i} = 0$$

From (5) and (7), it can be pointed out :

$$(\varepsilon_X^{i+1}, \varepsilon_M^{i+1}, \varepsilon_Z^{i+1}, \varepsilon_\mu^{i+1}, \varepsilon_\rho^{i+1})^T = E_* (\varepsilon_X^{i}, \varepsilon_M^{i}, \varepsilon_Z^{i}, \varepsilon_\mu^{i}, \varepsilon_\rho^{i})^T \qquad (8)$$
$$E_* = B_*^{-1} D_* A_*$$

$$B_* = \begin{pmatrix} \frac{\partial L_X}{\partial X} & \frac{\partial L_X}{\partial M} & 0 & \frac{\partial L_X}{\partial \mu} & \frac{\partial L_X}{\partial \rho} \\ \frac{\partial L_M}{\partial X} & \frac{\partial L_M}{\partial M} & 0 & \frac{\partial L_M}{\partial \mu} & 0 \\ 0 & 0 & 0 & \frac{\partial L_Z}{\partial \mu} & \frac{\partial L_Z}{\partial \rho} \\ \frac{\partial L_\mu}{\partial X} & \frac{\partial L_\mu}{\partial M} & \frac{\partial L_\mu}{\partial Z} & 0 & 0 \\ \frac{\partial L_\rho}{\partial X} & 0 & \frac{\partial L_\rho}{\partial Z} & 0 & 0 \end{pmatrix}_* ; \quad D_* = \begin{pmatrix} 0 & 0 \\ 0 & 0 \\ 0 & -\frac{\partial L_Z}{\partial \lambda} \\ 0 & 0 \\ -\frac{\partial L_\rho}{\partial V} & 0 \end{pmatrix}_* ; \quad A_* = \begin{pmatrix} 0 & 0 & \frac{dV}{dZ} & 0 & 0 \\ 0 & 0 & 0 & 0 & \frac{d\lambda}{d\rho} \end{pmatrix}_*$$

It can be shown that E_* has the following structure :

$$E_* = \begin{pmatrix} 0 & 0 & X & 0 & X \\ 0 & 0 & X & 0 & X \\ 0 & 0 & X & 0 & X \\ 0 & 0 & X & 0 & X \\ 0 & 0 & X & 0 & X \end{pmatrix}$$

where X = non-zero submatrix

Then : $\begin{pmatrix} \varepsilon_z^{i+1} \\ \varepsilon_\rho^{i+1} \end{pmatrix} = F_* \begin{pmatrix} \varepsilon_z^{i} \\ \varepsilon_\rho^{i} \end{pmatrix}$; $F_* = \begin{pmatrix} -b_{zz} \cdot \frac{\partial L_\rho}{\partial V} \cdot \frac{dV}{dz} & -b_{z\rho} \cdot \frac{\partial L_z}{\partial \lambda} \cdot \frac{d\lambda}{d\rho} \\ -b_{\rho z} \cdot \frac{\partial L_\rho}{\partial V} \cdot \frac{dV}{dz} & -b_{\rho\rho} \cdot \frac{\partial L_z}{\partial \lambda} \cdot \frac{d\lambda}{d\rho} \end{pmatrix}_*$ (9)

where, if $x = \sum_{i=1}^{N} x_i$ and $z = \sum_{i=1}^{N} z_i$

$b_{zz}(z \times x)$; $b_{z\rho}(z \times z)$; $b_{\rho z}(x \times x)$; $b_{\rho\rho}(x \times z)$: submatrices of B_*^{-1}

Since the coupling equations are linear :

$\frac{\partial L_\rho}{\partial V} = -I$; $\frac{dV}{dz} = \begin{pmatrix} 0 & C^{12} \\ C^{21} & 0 \end{pmatrix}$; $\frac{\partial L_z}{\partial \lambda} = -\begin{pmatrix} 0 & C^{12} \\ C^{21} & 0 \end{pmatrix}^T$; $\frac{d\lambda}{d\rho} = I$ (I = identity matrix)

$$F_* = \begin{pmatrix} b_{zz}\begin{pmatrix} 0 & C^{12} \\ C^{21} & 0 \end{pmatrix} & b_{z\rho}\begin{pmatrix} 0 & C^{12} \\ C^{21} & 0 \end{pmatrix}^T \\ b_{\rho z}\begin{pmatrix} 0 & C^{12} \\ C^{21} & 0 \end{pmatrix} & b_{\rho\rho}\begin{pmatrix} 0 & C^{12} \\ C^{21} & 0 \end{pmatrix}^T \end{pmatrix}_* \qquad (10)$$

To study the coordination convergence, from equations (5) and (8), is is sufficient to study the discrete dynamic system (9). If the modulus of all the eigenvalues of F_* are less than 1, then the coordination is stable.

This stability depends on the partition of the coupling matrix, as is shown in (10), explicitly by $\frac{dV}{dz}$, $\frac{\partial L_z}{\partial \lambda}$ and implicitly in the submatrices b_{zz}, $b_{z\rho}$, $b_{\rho z}$, $b_{\rho\rho}$. Indeed, the matrix B_* contains the submatrices $\frac{\partial L_z}{\partial \rho}$ and $\frac{\partial L_\rho}{\partial z}$.

$$\frac{\partial L_z}{\partial \rho} = -\begin{pmatrix} C^{11} & 0 \\ 0 & C^{22} \end{pmatrix}^T$$

For the best coordination convergence, an optimal partition is then difficult to be found. Nevertheless the nature of F_* suggests a suboptimal partitioning of matrix C.

Let $\begin{cases} A = [a_{ij}] \\ |A| = \sum_i \sum_j |a_{ij}| \end{cases}$

Let $\tau[A]$ be the spectral radius of A

It is known that (Varga, (12)):

$$\tau[A] \leq \min\left[\max_j \sum_i |a_{ij}|, \max_i \sum_j |a_{ij}|\right]$$

and also (Coviello, (13)) :

$$\tau[A]^2 \geq \max\left[\max_i A_i^T A_i, \max_j A^{jT} A^j\right]$$

(A_i = i^{th} line of A)
A_j = j^{th} column of A)

Let us consider the dynamic linear system (9). In the expression of elements f_{ij} of F_*, all the elements of C^{12}, C^{21}, are multiplicative terms.

By looking for a partition of C which minimizes $|C^{12}| + |C^{21}|$ we tend to decrease the modulus of f_{ij} and thus we tend to decrease the upper and lower bounds of $\tau[F_*]$ whose value mainly determine the dynamic of the system. Consequently, we tend to have a faster coordination convergence.

It can be noted that the effects of the partition arise in the submatrices b_{zz}, $b_{z\rho}$, $b_{\rho z}$, $b_{\rho\rho}$, but are difficult to estimate.

Example 2 : Consider N identical sub-systems (linear models, quadratic criteria)

$N = 6$; $v = 2$; $n_2 = 2$; $|P_1^2| = |P_2^2| = 3$

The coupling matrix C is given by Fig. 5.

There are 10 partitions of the N sub-systems into 2 equal groups. A direct iteration algorithm is implemented at the 2nd coordination-level. It is unstable for all the cases. So we look for k such as for $V = kV(Z)$ and $\lambda = k\lambda(\rho)$ the algorithm will be stable.

Let $K = |C^{12}| + |C^{21}|$. Results appear in table 1.

	K	k
(123)(456)	9	0,94
(124)(356)	16	0,48
(125)(346)	23	0,24
(126)(345)	16	0,36
(134)(256)	15	0,33
(135)(246)	24	0,15
(136)(245)	19	0,18
(145)(236)	15	0,44
(146)(235)	22	0,22
(156)(234)	21	0,14

Table 1 - Effects of the partition on coordination convergence

There is a good correlation between k and K (Fig. 6). For slightly different values of K, this correlation is not so strong. This comes from the suboptimality of the partitioning criteria. But these results show that less interactive groups of sub-systems give a faster coordination convergence.

CONCLUSION

In multilevel optimization, all the decomposition-coordination methods deal with the coupling equations of the sub-systems. The partition of the system is then an important factor in the synthesis of multilevel structures. In the study of the effects of couplings on coordination convergence, a suboptimal partitioning criteria has been proposed for linear coupling equations. For large scale systems, one will be faced with the partition of large coupling matrices and graphical methods seems to be useful.

• °
•

REFERENCES

1. Mesarovic, M.D., Macko, D. and Takahara, Y., 1970, "Theory of Hierarchical, Multilevel Systems", Academic Press, New-York, U.S.
2. Plander, I., 1972, "The reliability of a hierarchic multi-computer system for real time direct industrial process control", IFIP 1971, North-Holland Publishing Co, Amsterdam.
3. Lasdon, L.S., Shoeffler, J.D., 1965, "A multilevel technique for optimization", JACC Proceedings, Rensselaer Polytech. Inst., Troy, New-York.
4. Brosilow, C.B., Lasdon, L.S., Pearson, J.D., 1965, "Feasible optimization methods for interconnected systems", JACC Proceedings, Rensselaer Polytech. Inst.,New-York
5. Grateloup, G., Titli, A., 1973, Int. J. Systems Sci., 4, 577.
6. Strasjak, A., 1969, "On the synthesis of multi-level large-scale control systems", IFAC Fourth Congress Proceedings, Warsaw, Poland.
7. Kulikowski, R., 1970, Automatica, 6, 315.
8. Titli, A., 1972, "Contribution à l'étude des structures de commande hiérarchisée en vue de l'optimisation des processus complexes", Thèse de Doctorat d'Etat, Université Paul Sabatier, Toulouse, France.
9. Sprague, C.F., 1964, "On the reticulation problem in multivariable control systems JACC Proceedings, Stanford Univ., California.
10. Fossard, A.J., Clique, M., Imbert, N., 1972, RAIRO, J-3, 3.
11. Grateloup, G., Titli, A., Lefèvre, T., 1973, "Les algorithmes de coordination dans la méthode mixte d'optimisation à deux niveaux", Proceedings of 5th Conf. on Optimization Techniques, Springer-Verlag, Berlin, Germany.
12. Varga, R.S., 1962, "Matrix Iterative analysis", Prentice-Hall, U.S.
13. Coviello, G.J., "An organizational approach to the optimization of multivariable systems", Ph D. Dissertation, n° 64-12907, Case Institute of Technology, Cleveland U.S.

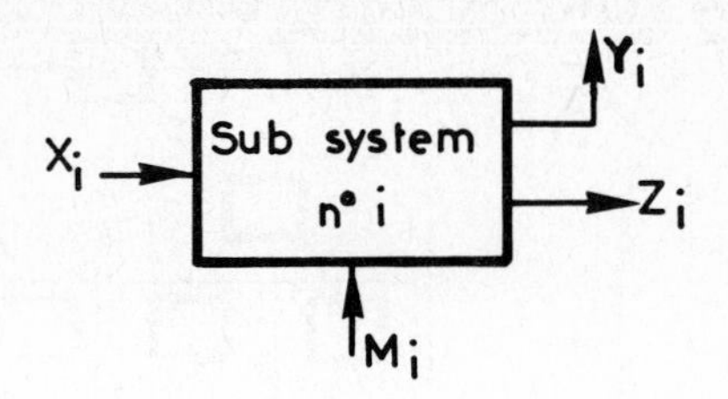

X_i = input coupling vector (R^{x_i})
Z_i = output coupling vector (R^{z_i})
Y_i = outpout vector (R^{y_i})
M_i = control vector (R^{m_i})

Fig. 1 : Sub-system n° i

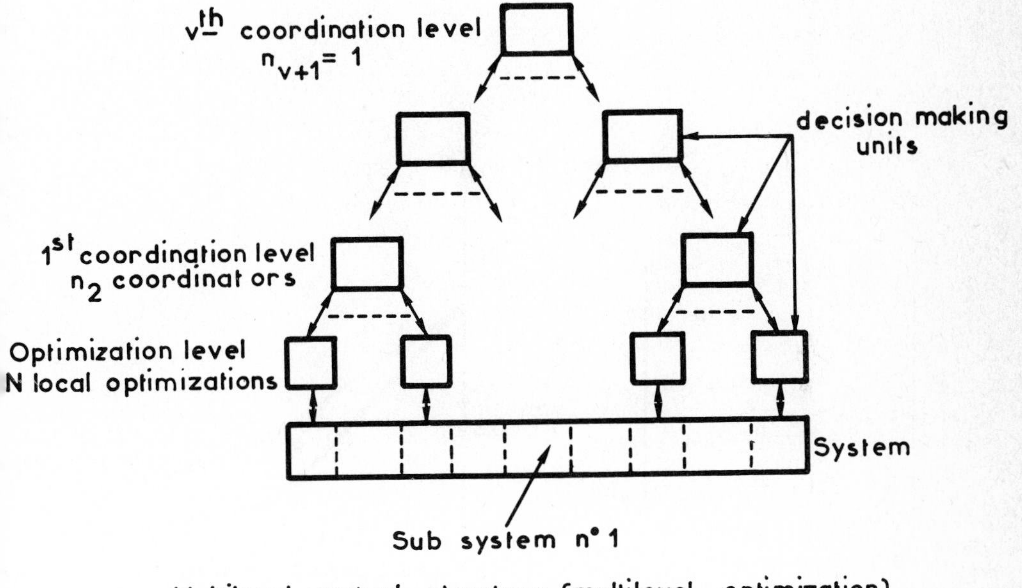

Multilevel control structure (multilevel optimization)

Fig. 2 : A coordination method

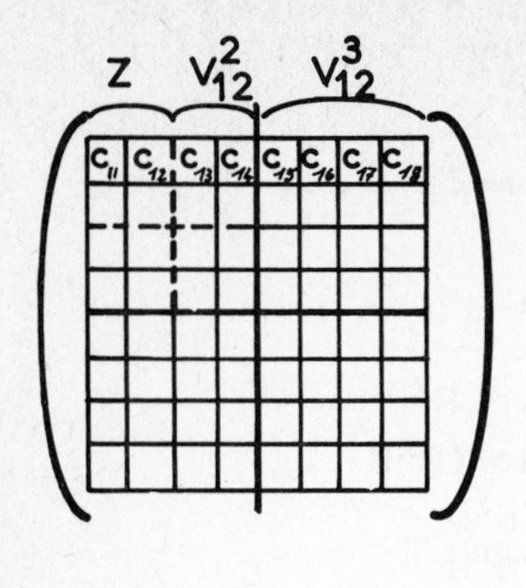

Partition of coupling matrix C

Fig. 3 :

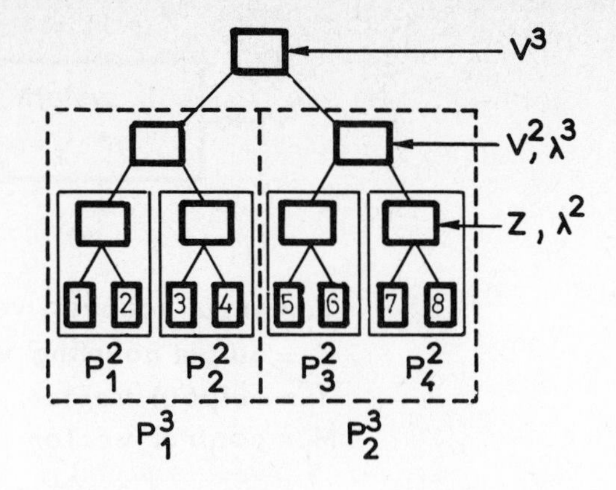

A coordination method

Fig. 4 :

$$
\begin{vmatrix}
0 & 3 & 4 & 2 & 0 & 0 \\
2 & 0 & 0 & 0 & 0 & 2 \\
0 & 3 & 0 & 1 & 0 & 0 \\
2 & 0 & 0 & 0 & 2 & 0 \\
0 & 0 & 0 & 3 & 0 & 2 \\
1 & 0 & 1 & 0 & 2 & 0
\end{vmatrix}
$$

Fig. 5 : Coupling matrix C.

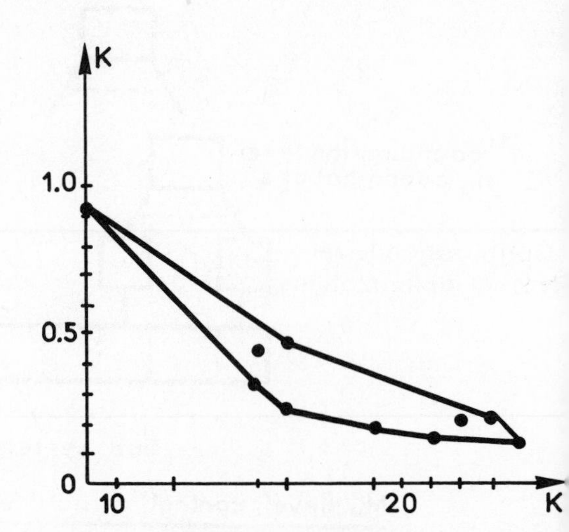

Fig. 6 : Coupling effects on coordination convergence

ON THE PROBLEM OF AN OPTIMAL THREE-DIMENSIONAL AIRCRAFT MOTION

V.K. Isaev, V.V. Sonin, L.I. Shustova
Moscow Physical-Technical Institute,
Moscow, USSR

The optimization problem of aircraft control during three-dimensional maneuver is considered. The equations for three-dimensional motion of a vehicle in a spherical speed system have the following form

$$\frac{dV}{dt}=\frac{C_p\cos(\alpha+\varepsilon)-C_x}{m}\cdot qS-g(H)\sin\Theta, \tag{1.1}$$

$$\frac{d\lambda}{dt}=\frac{V}{r}\cos\Theta\,\frac{\cos\eta}{\cos\varphi}, \tag{1.2}$$

$$\frac{d\varphi}{dt}=\frac{V}{r}\cos\Theta\sin\eta, \tag{1.3}$$

$$\frac{d\eta}{dt}=\frac{C_p\sin(\alpha+\varepsilon)+C_y}{mV\cos\Theta}\cdot qS\sin\gamma-\frac{V}{r}\cos\Theta\cos\eta\,\mathrm{tg}\,\varphi, \tag{1.4}$$

$$\frac{d\mu}{dt}=-\bar{C}_eC_p\frac{qS}{m},\quad \bar{C}_e=\frac{C_e}{3600\,g_0}. \tag{1.5}$$

$$\frac{d\Theta}{dt}=\frac{C_p\sin(\alpha+\varepsilon)+C_y}{mV}\cdot qS\cos\gamma-\frac{g(H)}{V}\cos\Theta\cdot\varkappa, \tag{1.6}$$

$$\frac{dH}{dt}=V\sin\Theta, \tag{1.7}$$

where V, Θ, H, m are instant values for speed, trajectory inclination angle, altitude, mass; λ - longitude, φ - latitude, C_p, C_x, C_y, C_e - the coefficients of thrust, aerodynamic drag, lift and fuel consumption; α, S - angle of attack, vehicle characteristic area; $\mu = \ln G$, $G = mg_0$ - aircraft weight, $\varkappa = 1-\frac{V^2}{g(H)r}$. The control functions are assumed to be: $\bar{R}$ - the thrust throttling coefficient, $C_p = C_R\bar{R}$, C_y - lift coefficient, γ-roll angle, measured from the trajectory plane. Now we shall consider the case, when the control $\bar{u} = (\bar{R}, C_y, \gamma)$ lies within of the closed fixed parallelepiped

$$0 \leq \bar{R}_{MIN} \leq \bar{R} \leq \bar{R}_{MAX}, \tag{2.1}$$

$$0 < C_{Y_{MIN}} \leq C_Y \leq C_{Y_{MAX}}, \tag{2.2}$$

$$-\frac{\pi}{2} < \gamma_{MIN} \leq \gamma \leq \gamma_{MAX} \tag{2.3}$$

(By linear transformation of control functions

$$U = U_{MIN}(x) + [U_{MAX}(x) - U_{MIN}(x)]\, u$$

it is possible to turn from convex constraints on control, depending on phase coordinates to unit cube $0 \leq u \leq 1$).

It is necessary to find an optimal control $\bar{u}_* = (\bar{R}_*, C_{Y*}, \gamma_*)$ which transfers the system (1) from the initial position, $x_0 \in G^0$, $x_0 = x(0) = (V_0, \lambda_0, \varphi_0, \eta_0, \mu_0, H_0, \Theta_0)$ into the end state

$$x_1 = x(T) \in G^1 \tag{3}$$

/ G^0 and G^1 are initial and end varieties/, some functional I reaching the maximum value under the constraints (2). Below the optimization problem for the end mass value $I = m(T)$ is considered as an example. Let us find an optimal control for the formulated problem, using the maximum principle [1] under the following assumptions:
1. The sum of all forces projections to the trajectory normal is equal to zero

$$\frac{d\Theta}{dt} = 0 \,. \tag{4.1}$$

Hence

$$\frac{qS}{m} = \frac{g(H)\,æ}{C_y \cos\gamma} \,.$$

2. The angles of attack α, engine incidence ε and trajectory inclination Θ are considered small values of the second order:

$$\sin(\alpha + \varepsilon) \simeq 0, \quad \cos(\alpha + \varepsilon) \simeq 1, \qquad \sin\Theta \simeq 0, \quad \cos\Theta \simeq 1. \tag{4.2}$$

The system of optimal motion equations has the first integrals [2]

$$p_\eta \frac{\cos\lambda}{\cos\varphi} + p_\varphi \sin\lambda - p_\lambda \operatorname{tg}\varphi \cos\lambda = C_1, \tag{5.1}$$

$$p_\eta \frac{\sin\lambda}{\cos\varphi} - p_\varphi \cos\lambda - p_\lambda \operatorname{tg}\varphi \sin\lambda = C_2, \tag{5.2}$$

$$p_\lambda = C_3. \tag{5.3}$$

The assumptions (4), relations (5) make it possible to exclude the variables Θ, r, p_Θ, p_r out of consideration and substitute known functions of coordinates and constants C_1 , C_2 , C_3 for $p_\eta, p_\varphi, p_\lambda$.

As the Hamiltonian function

$$\mathcal{H} = p_V \frac{C_p \cos(\alpha+\varepsilon) - C_x}{C_y \cos\gamma} \cdot g\varkappa + p_\eta \left[\frac{C_p \sin(\alpha+\varepsilon) + C_y}{V} \cdot \frac{g\varkappa \operatorname{tg}\gamma}{C_y} - \right.$$

$$\left. - \frac{V}{r} \cos\eta \operatorname{tg}\varphi \right] + p_\lambda \frac{V\cos\eta}{r\cos\varphi} + p_\varphi \frac{V}{r} \sin\eta - p_\mu \frac{\bar{C}_e C_p g\varkappa}{C_y \cos\gamma} .$$

does not contain μ explicitly, then

$$p_\mu(t) = p_{\mu_0} = \text{const} \tag{6}$$

and the conjugate variable p_V is still unknown.

Let C_x and C_y be connected by a quadratic polar $C_x = C_{x_0} + AC_y^2$. Taking (4.2) and (6.1) into consideration, transform the Hamiltonian into the following from

$$\mathcal{H} = \frac{a}{\cos\gamma} + b \operatorname{tg}\gamma + c , \tag{7.1}$$

where

$$a(p_V, C_y, \bar{R}) = g\varkappa \left[\frac{\tilde{\varphi}(p_V, \bar{R})}{C_y} - p_V A C_y \right] \tag{7.2}$$

$$\tilde{\varphi}(p_V, \bar{R}) = p_V (C_R \bar{R} - C_{x_0}) - p_\mu \bar{C}_e C_R \bar{R} . \tag{7.3}$$

The coefficients b and c do not depend on the control

$$b = p_\eta \frac{g\varkappa}{V} , \tag{7.4}$$

$$c = \frac{V}{r} \left(p_\lambda \frac{\cos\eta}{\cos\varphi} + p_\varphi \sin\eta - p_\eta \cos\eta \operatorname{tg}\varphi \right) . \tag{7.5}$$

For the case of stationarity under consideration the first integral takes place

$$\mathcal{H} = \frac{a}{\cos\gamma} + b \operatorname{tg}\gamma + c = h , \tag{8}$$

and the constant $h = 0$, when motion time T is not fixed. It is not difficult to prove, that the problem of defining the control values, minimizing the Hamiltonian function $\min_{\bar{R}, C_y, \gamma} \mathcal{H}$, comes to successive solution of three onedimensional problems for minimum point calculation

$$\min_{\gamma_{MIN} \le \gamma \le \gamma_{MAX}} \left(\frac{a}{\cos\gamma} + b\,tg\gamma + c\right), \tag{9}$$

$$\min_{\bar{R}_{MIN} \le \bar{R} \le R_{MAX}} \left[p_v(C_R\bar{R} - C_{x_0}) - p_\mu C_R \bar{C}_e(\bar{R})\bar{R}\right], \tag{10}$$

$$\min_{C_{y_{MIN}} \le C_y \le C_{y_{MAX}}} \left(\frac{\tilde{\varphi}}{C_y} - p_v A C_y\right). \tag{11}$$

Using (8) while defining γ it is possible to write corresponding optimality conditions in terms of values of b and c, depending on constants C_1, C_2, C_3 and phase coordinates

$$\gamma_* = \begin{cases} \arcsin \dfrac{-b}{\sqrt{b^2+c^2}} = \gamma_{OPT}, & \gamma_{MIN} \le \gamma_{OPT} \le \gamma_{MAX}, \\ \gamma_{MAX}, & \mathcal{H}(\gamma_{MAX}) < \mathcal{H}(\gamma_{MIN}), \\ \gamma_{MIN}, & \mathcal{H}(\gamma_{MIN}) < \mathcal{H}(\gamma_{MAX}). \end{cases} \tag{12}$$

During calculation of $\bar{R}_*$ it is necessary to find all the roots $\bar{R}_i$ of the equation (p_v is considered known):

$$\frac{\partial}{\partial \bar{R}}\left[p_v(C_R\bar{R} - C_{x_0}) - p_\mu C_R \bar{C}_e(\bar{R})\bar{R}\right] = 0, \tag{13}$$

which satisfy the condition $\bar{R}_{MIN} \le \bar{R}_i \le \bar{R}_{MAX}$. Add $\bar{R}_{MIN}$ and $\bar{R}_{MAX}$ to the obtained roots and select the meaning, to which the minor value of the function $\tilde{\varphi}$ corresponds. From the equality $\frac{\partial}{\partial C_y}\left(\frac{\tilde{\varphi}}{C_y} - p_v A C_y\right) = 0$ we derive

$$C_y = C_{y_{OPT}} = \sqrt{-\frac{\tilde{\varphi}}{A p_v}}. \tag{14}$$

Let us investigate the function $a(p_v, C_y, \bar{R})$, (7.2), versus C_y.

a) If $p_v < 0$ and $\tilde{\varphi} > 0$, then among three values of lift $C_{y_{OPT}}$, $C_{y_{MIN}}$, $C_{y_{MAX}}$ choose that one, to which minimum a corresponds;

b) if $p_v < 0$, $\tilde{\varphi} < 0$ then $C_{y_*} = C_{y_{MIN}}$;

c) if $p_v > 0$, $\tilde{\varphi} > 0$ then $C_{y*} = C_{y\,MAX}$;

d) if $p_v > 0$ and $\tilde{\varphi} < 0$, choose out of boundary values of $C_{y\,MIN}$ or $C_{y\,MAX}$ the meaning, to which minimum a corresponds. It remains to find p_v ; to this end we use the equation

$$a(p_v, C_y, \bar{R}) = -(b \sin\gamma_* + c\cos\gamma_*). \tag{15}$$

As a result we obtain the following algorithm for optimal control determination. Let constants C_i , defined by the solution of a boundary value problem, be known. During every integration step of differential equations system (I) the conjugate variables p_η , p_φ , p_λ are calculated from solution of algebraic equations (5.1) - (5.3); then from the formulae (7.4), (7.5) coefficients b and c are defined; according to conditions of optimal roll control the value of γ_* and right-hand side of (15) are found. After this the value of a conjugate value p_v is determined as a root of the equation (15), the values of thrust $\bar{R}_*$ and lift coefficient C_{y*} , obtained by the method described above, are substituted into its left-hand side. The values, corresponding to the obtained root, are assumed to be optimal control, that is they are used as instant controls. The procedure just described makes it possible to obtain an extermal in a sense of maximum principle [I]. In addition, to solve a given problem it is necessary to write boundary conditions and construct algorithm, which permits defining constants C_1 , C_2 , C_3, corresponding to these boundary conditions, by using the features of varieties G^0 and G^1 .

REFERENCES

I. Понтрягин Л.С., Болтянский В.Г., Гамкрелидзе Р.В., Мищенко Е.Ф. "Математическая теория оптимальных процессов", Физматгиз, М., I96I г.

2. Шкадов Л.М.,Плохих В.П.,Илларионов В.Ф.,Буханова Р.С. "Механика оптимального пространственного движения летательных аппаратов в атмосфере", Машиностроение, М.,I972 г.

A GENERAL STOCHASTIC EQUATION FOR THE NON-LINEAR FILTERING PROBLEM

GOPINATH KALLIANPUR

Department of Mathematics, University of Minnesota

1. Introduction.

A great deal of attention has been devoted in recent years to the theory of non-linear filtering, in particular, to the problem of deriving a stochastic differential equation for the filter. (see the bibliography in [2]). Perhaps the most general form of such an equation when the noise in the observation process model is the Wiener process is the one obtained in the paper of Fujisaki, Kallianpur and Kunita [2].

The work in [2] was motivated by applications in which the signal and observation processes are governed by an Ito stochastic differential equation or by a more general stochastic equation studied by Ito and Nisio (see [2]). However, the aim of the present paper is to show that the approach to filtering theory adopted in [2] is not limited to this kind of application and to give a generalization of the main result (Theorem 4.1) of the Fujisaki-Kallianpur-Kunita paper.

For reasons of brevity we shall consider real-valued observation processes but there is no difficulty whatever in making the appropriate changes to cover the vector-valued case.

2. Observation process model and the innovation process.

The system or signal process $x_t(\omega)$ taking values in a complete metric space S and the observation process $z_t(\omega)$ $(t \in [0,T])$ are assumed given on some complete probability space $(\Omega, \underline{\underline{A}}, P)$ and further related as follows.

$$(2.1) \qquad z_t(\omega) = \int_0^t h_u(\omega)\, du + w_t(\omega) ,$$

where

This work was supported in part by NSF Grant GP 30694X.

(2.2) $\quad w_t(\omega)$ is a real-valued standard Wiener process

(2.3) $\quad h_t(\omega)$ is a (t,ω) measurable real-valued process such that $\int_0^T E(h_t^2)dt$ is finite.

Let us introduce the following family of σ-fields.

(2.4) $$\underline{\underline{G}}_t = \sigma\{x_s, w_s, s \leq t\} , \underline{\underline{N}}_t^T = \sigma\{w_v - w_u, t \leq u \leq v \leq T\}$$

and

$$\underline{\underline{F}} = \sigma\{z_s, s \leq t\} .$$

It will be assumed that the σ-fields $\underline{\underline{F}}_t$ and $\underline{\underline{G}}_t$ are augmented by adding to $\underline{\underline{F}}_o$ and $\underline{\underline{G}}_o$ all P-null sets. In the model (2.1) the information about the signal process is carried by (h_t) by means of the measurability assumption

(2.5) $\quad$ For each t , h_t is $\underline{\underline{G}}_t$ measurable, i.e. (h_t) is $(\underline{\underline{G}}_t)$ - adapted

In order to take into account applications involving stochastic control we make the further assumption that for every t the σ-fields

(2.6) $\quad \underline{\underline{G}}_t$ and $\underline{\underline{N}}_t^T$ are stochastically independent.

Clearly (2.6) includes the case when the signal (x_t) and noise (w_t) are completely independent.

The derivation of the desired stochastic equation rests on two important results proved in Fujisaki-Kallianpur-Kunita [2]. We state them below without proof. From the assumptions made above on (h_t) it can be shown that one can work with a modification of the conditional expectation $E(h_t \mid \underline{\underline{F}}_t)$ which is jointly measurable and $(\underline{\underline{F}}_t)$ - adapted. This particular modification will be henceforth denoted by $\hat{h}_t$.

Let us now define the process (ν_t) by

$$(2.7) \qquad \nu_t = z_t - \int_0^t \hat{h}_s \, ds .$$

Proposition 1. $(\nu_t, \underline{\underline{F}}_t, P)$ is a Wiener martingale. Furthermore $\underline{\underline{F}}_t$ and $\sigma\{\nu_v - \nu_u ;\ t < u < v \leq T\}$ are independent.

(ν_t) is called the innovation process.

Proposition 2. (A martingale representation theorem). Under conditions (2.1), (2.2), (2.3), (2.5) and (2.6) every separable square integrable martingale $(Y_t, \underline{\underline{F}}_t, P)$ is sample continuous and has the Ito stochastic integral representation

$$(2.8) \qquad Y_t - E(Y_o) = \int_0^t \Phi_s \, d\nu_s$$

where

$$(2.9) \qquad \int_0^T E(\Phi_s^2) \, ds < \infty$$

and (Φ_s) a jointly measurable and adapted to $(\underline{\underline{F}}_s)$.

3. A stochastic differential equation for the general non-linear filtering problem. First we replace the conditions defining the class $D(\tilde{A})$ of [2] by a wider set of assumptions which make the theory applicable to more general types of signal processes (x_t). Let f be a real measurable function on S such that

$$(3.1) \qquad E[f(x_t)]^2 < \infty \quad \text{for all } t \text{ in } [0,T] .$$

The function f is said to belong to the class D if there exists a jointly measurable, real function $B_t[f](\omega)$ adapted to $(\underline{\underline{F}}_t)$ and having the following properties.

(3.2) Almost all trajectories of the process $(B_t[f])$ are right - continuous and of bounded variation over the interval $[0,T]$ with $B_o[f] = 0$.

(3.3) $\qquad E(\mathrm{Var}\ B[f])^2 < \infty$

where

$\mathrm{Var}\ B[f](\omega)$ is the total variation of the trajectory $B_t[f](\omega)$ $(0 \leq t \leq T)$.

(3.4) The process $M_t(f) \equiv f(x_t) - E[f(x_o) \mid \underline{\underline{F}}_o] - B_t[f]$ is a $(\underline{\underline{G}}_t, P)$ martingale.

Note that from conditions (3.1) and (3.3) it follows that $(M_t(f))$ is a square integrable martingale.

Theorem 1. Let the conditions of Section 2 and (3.1)-(3.4) hold. Then for every f in D there exists a jointly measurable process $(\overline{B}_t[f])$ adapted to the family $(\underline{\underline{F}}_t)$ such that almost all its trajectories are right-continuous and of bounded variation over the interval $[0,\tau]$ with $\overline{B}_o[f] = 0$. Furthermore,

(3.5) $\qquad E(\mathrm{Var}\ \overline{B}[f])^2 < \infty$,

and the process

(3.6) $\qquad \overline{M}_t(f) \equiv E[f(x_t) \mid \underline{\underline{F}}_t] - E[f(x_o) \mid \underline{\underline{F}}_o] - \overline{B}_t[f]$ is a square-integrable $(\underline{\underline{F}}_t, P)$ martingale.

Proof. Since f is fixed we shall suppress it for the time being and write B_t for $B_t[f]$, etc. The existence of $(\overline{B}_t)$ follows from the ideas of C. Dellacherie [1] and P.A. Meyer [3] concerning the "dual projection" or "compensator" associated with an increasing integrable process. Write $B_t = U_t - V_t$ where (U_t) , (V_t) are increasing processes with right-continuous trajectories and such that $U_o = V_o = 0$. From (3.3) we also have $E(U_T^2) < \infty$ and $E(V_T^2) < \infty$. The process $\xi_t = E(U_T - U_t \mid \underline{\underline{F}}_{t+})$ is a positive supermartingale of class (D) . Hence it has a Doob decomposition $Y_t - \overline{U}_t$ where (Y_t) is a martingale and $\bar{U}_t$ is a (uniquely determined) predictable, integrable increasing process adapted to $(\underline{\underline{F}}_{t+})$ with $\bar{U}_o = 0$. Hence for $s < t$, we have $E(U_t - U_s \mid \underline{\underline{F}}_{s+}) = E(\bar{U}_t - \bar{U}_s \mid \underline{\underline{F}}_{s+})$ which gives

$$(3.7) \qquad E(U_t - U_s \mid \underline{\underline{F}}_s) = E(\bar{U}_t - \bar{U}_s \mid \underline{\underline{F}}_s) .$$

The predictability of $(\bar{U}_t)$ also implies that $\bar{U}_t$ is actually $\underline{\underline{F}}_t$-measurable. Define the process $(\bar{V}_t)$ in a similar fashion and write $\bar{B}_t = \bar{U}_t - \bar{V}_t$. It is clear that $\bar{B}_o = 0$, almost all trajectories of $(\bar{B}_t)$ are right-continuous and of bounded variation over $[0, T]$ and further that $E(\mathrm{Var}\,\bar{B}) = E(\bar{U}_T) + E(\bar{V}_T) < \infty$. However, the deduction of the square integrability of $\mathrm{Var}\,\bar{B}$ from (3.3) is a bit more complicated and will not be given here. It, of course, implies the square integrability of $\bar{M}_t(f)$. The fact that $(\bar{M}_t(f), \underline{\underline{F}}_t, P)$ is a martingale follows easily. For $s < t$,

$$\begin{aligned}
&E\{\bar{M}_t(f) - \bar{M}_s(f) \mid \underline{\underline{F}}_s\} \\
&= E\{f(x_t) - f(x_s) \mid \underline{\underline{F}}_s\} - E(\bar{B}_t - \bar{B}_s \mid \underline{\underline{F}}_s) \\
&= E\{M_t(f) - M_s(f) \mid \underline{\underline{F}}_s\} + E(B_t - B_s \mid \underline{\underline{F}}_s) - E(\bar{B}_t - \bar{B}_s \mid \underline{\underline{F}}_s) \\
&= 0 \quad \text{from (3.4) and (3.7)} .
\end{aligned}$$

Since $(M_t(f), \underline{\underline{G}}_t, P)$ is a square integrable martingale there exists a unique sample continuous process $\langle M(f), w\rangle$ adapted to $(\underline{\underline{G}}_t)$ such that almost all of its trajectories are absolutely continuous with respect to Lebesgue measure in $[0, T]$. Furthermore there exists a modification of the Radon-Nikodym derivative which is (t,ω)-measurable and adapted to $(\underline{\underline{G}}_t)$ and which we shall denote by $\tilde{D}_t f(\omega)$. Then it follows that [2] $\langle M(f), w\rangle_t = \int_o^t \tilde{D}_s f \, ds$ a.s.
where

$$\int_o^T E(\tilde{D}_s f)^2 \, ds < \infty .$$

We now state the principal result which yields the stochastic equation of non-linear filtering. Conditions (3.3) and (3.5) are crucially used in the proof which will be presented in detail elsewhere. It will be understood that only separable versions of the martingales $M_t(f)$ and $\bar{M}_t(f)$ are considered. We shall also use the shorter notation $E^t(\cdot)$ for $E(\cdot \mid \underline{\underline{F}}_t)$.

Theorem 2. Assume conditions (2.1)-(2.3), (2.5), (2.6) and (3.1)-(3.4). Suppose f belongs to the class D and satisfies

$$(3.8) \qquad \int_0^T E[f(x_t)\, h_t]^2\, dt < \infty .$$

Then $E^t[f(x_t)]$ satisfies the following stochastic differential equation.

$$(3.9) \qquad E^t[f(x_t)] = E^0[f(x_0)] + \bar{B}_t[f] + \int_0^t [E^s(f(x_s)h_s) - E^s(f(x_s))E^s(h_s) + E^s(\tilde{D}_s f)]d\nu_s .$$

Theorem 4.1 of Fujisaki-Kallianpur-Kunita [2] is a particular case of the above result. According to the assumptions in Section 4 of [2] if $f \in D$ (which is denoted by $D(\tilde{A})$ in that paper) there exists a (t,ω)-measurable real function $\tilde{A}_t f(\omega)$ adapted to $(\underline{\underline{G}}_t)$ such that $\int_0^T E(\tilde{A}_t f)^2 dt$ is finite and $f(x_t) - E[f(x_0) \mid \underline{\underline{F}}_0] - \int_0^t \tilde{A}_s f\, ds$ is a (necessarily square integrable) $(\underline{\underline{G}}_t, P)$-martingale. Hence conditions (3.2),(3.3) and (3.4) are satisfied with

$$(3.10) \qquad B_t[f] = \int_0^t \tilde{A}_s f\, ds .$$

It is then easily verified that $\bar{B}_t[f] = \int_0^t E^s[\tilde{A}_s f]ds$ and that the stochastic equation (3.9) reduces to equation (4.12) of [2]. As explained in the Introduction the assumption (3.10) above made in [2] was suggested by applications of which the following is a typical example. The signal and observation processes form a Markov process (x_t, y_t) satisfying the stochastic differential equation

$$(3.11) \qquad \begin{aligned} dx_t &= a_1(t,x_t,y_t)dt + b_1(t,x_t,y_t)dw_t \\ dy_t &= a_2(t,x_t,y_t)dt + b_2(t,y_t)dw_t \end{aligned}$$

where (w_t) is a (vector) Wiener process, $y_0 = 0$ a.s., x_0 is an arbitrary random variable independent of $\sigma\{w_s, 0 \le s \le T\}$ and the coefficients satisfy suitable conditions ensuring the existence and uniqueness of the solution of

(3.11). Let (A_t) , $t \in [0,T)$ be the extended generator of the Markov process $\eta_t = (x_t, y_t)$ as defined in [2] and let $D(A)$ be the set of all real functions f depending only on the first variable x, belonging to the domain of (A_t) and satisfying the conditions $E[f(\eta_t)]^2 < \infty$ for each t and $\int_0^T E[A_t f(\eta_t)]^2 dt < \infty$. It can easily be shown that in Theorem 2 we may take $D(A)$ for D and $B_t[f] = \int_0^t A_s f(\eta_s) ds$. For details see [2].

It is hoped that the equation obtained in Theorem 2 will prove useful in the study of new types of stochastic filtering and control problems.

References

[1]. C. Dellacherie, Capacites et Processus Stochastiques, Ergebnisse der Mathematik und ihrer Grenzgebiete, Band 67 (1972).

[2]. M. Fujisaki, G. Kallianpur and H. Kunita, Stochastic differential equations for the nonlinear filtering problem, Osaka Journal of Mathematics, 9, (1972).

[3]. P.A. Meyer, Sur un problème de filtration, Séminaire de Probabilités VII, Lecture Notes in Mathematics, Springer-Verlag (1973).

ON SUFFICIENCY OF THE NECESSARY OPTIMALITY OF L.S.PONTRYAGIN'S MAXIMUM PRINCIPLE ANALOGUES TYPE

V.V.Leonov

Institute of Mathematics
Siberian Branch
U.S.S.R. Academy of Sciences
Novosibirsk

The creation of Pontryagin's maximum principle and R.Bellmann's dynamic programming method has promoted the intensive development of the optimal process theory in the recent time.The use of maximum principle analogues for the processes of quite different nature has proved to be most successful.But it has been established that the method has a limitation,for the maximum principle analogues give only necessary conditions of the control optimality.

We have found that for processes of a special form but of a different nature (such as descrete processes,processes with delay, distributed - parameter processes and those which obey ordinary differential equations) conditions of the corresponding maximum principle analogues are not only necessary but sufficient conditions of optimality.It turns out, that if the process describes a conflict situation the optimal behaviour of both parts is realized in pure strategies.

Because of restriction of the paper size we present two particular results only.

Let us use the notation

$$extr^{\{z\}} = \begin{cases} max \ldots, & \text{if } z > 0, \\ min \ldots, & \text{if } z < 0 . \end{cases}$$

We shall be concerned with the process

$$\frac{dx}{dt}=Q(t)x+\sum_{j=1}^{m_1-1} a_j(u_j,t)g_j(x,t) +$$

$$+\sum_{\ell=1}^{m_2-1} b_\ell(v_\ell,t)f_\ell(x,t)+R(t)u_{m_1}+S(t)v_{m_2}+w(t), \qquad (1)$$

where the control parameters u_j ($j=1,\dots,m_1$) and v_ℓ ($\ell=1,\dots,m_2$) belong accordingly to the limited closed sets U and V_ℓ of Euclidean spaces E^{p_j} and E^{q_ℓ}, the x are elements of the n-space, $Q(t)$ are continuous for $t\in[0,T]$ $(n\times n)$- matrices, a_j (u_j , t) and b_ℓ (v_ℓ , t) are n -space vectors continuously dependent on their arguments on sets $[0,T]\times U_j$ and $[0,T]\times V_\ell$, $R(t)$ and $S(t)$ are continous for $t\in[0,T]$ $n\times p_{m_1}$ and $n\times q_{m_2}$ - dimension matrices respectively, while $w(t)$ is apriori known n - space vector function, continuous for $t\in[0,T]$. The sclar functions $g_j(x,t)$ and $f_\ell(x,t)$ will be considered to be linear in regard to x :

$$g_j(x,t)=\sum_{\nu=1}^{n} g_{j\nu}(t)x_\nu+g_{j0}(t) \qquad (j=1,\dots,m_1-1), \quad (2)$$

$$f_\ell(x,t)=\sum_{\mu=1}^{n} f_{\ell\mu}(t)x_\mu+f_{\ell0}(t) \qquad (\ell=1,\dots,m_2-1), \quad (3)$$

where $g_{j\nu}(t)$ and $f_{\ell\mu}(t)$ are continuous functions of $t\in[0,T]$.

Let the initial state of system (1) be set:

$$x(0)=x_0 . \qquad (4)$$

Let us assume that the first part can choose $u_j\in U_j$ $(j=1,\dots,m_1)$ and the second - $v_\ell\in V_\ell$ ($\ell=1,\dots,m_2$) arbitrarily, if only controls are measurable and the 1-st part means to maximize the criterion

$$R(x(T)) = c^* x(T), \qquad (5)$$

the 2-nd part, on the contrary, - to minimize (5).

It appears that if the quantities (2) and (3) maintain their fixed signs along the trajectory for $t<T$ at any permissible system (1) control with the set initial state (4), then any measurable permissible control, which for almost every $t<T$ obeys the equalities

$$\operatorname*{extr}_{u_j\in U_j}^{\{g_j\,x(t),t\}} \sum_{\nu=1}^{n} \psi_\nu(t)a_{j\nu}(u_j,t) =$$

$$=\sum_{\nu=1}^{n} \psi_\nu(t)a_{j\nu}(u_j(t),t) \qquad (j=1,\dots,m_1-1), \qquad (6)$$

$$\max_{u_{m_1} \in U_{m_1}} \psi^*(t)R(t)u_{m_1} = \psi^*(t)R(t)u_{m_1}(t), \tag{7}$$

$$\operatorname{extr}^{\{-f_\ell(x(t),t)\}} \sum_{\mu=1}^{n} \psi_\mu(t) b_{\ell\mu}(v_\ell, t) =$$

$$= \sum_{\mu=1}^{n} \psi_\mu(t) b_{\ell\mu}(v_\ell(t), t) \qquad (\ell = 1, \ldots, m_2 - 1), \tag{8}$$

$$\min_{v_{m_2} \in V_{m_2}} \psi^*(t)s(t)v_{m_2} = \psi^*(t)s(t)v_{m_2}(t), \tag{9}$$

where $\psi^*(t) = (\psi_1(t), \ldots, \psi_n(t))$, is optimal for the one and the other parts, i.e. it realizes the saddle strategy of both parts.In a general case the similar controls are not defined uniquely, but all of them obey the "saddle" condition. Then either player may choose his strategy in advance in virtue of equalities (6)-(9), having no information about the other player's behaviour. Quantities $\psi_\nu(t)$ (ν = 1,..., n) are defined uniquely by the system

$$\frac{d\psi_\nu}{dt} = -\sum_{s=1}^{n} q_{s\nu}(t)\psi_\nu - \sum_{j=1}^{m_1-1} q_{j\nu}(t) \operatorname{extr}_{u_j \in U_j}^{\{g_j\}} \sum_{s=1}^{n} \psi_s a_{js}(u_j, t) - \tag{10}$$

$$- \sum_{\ell=1}^{m_1-1} f_{\ell\nu}(t) \operatorname{extr}^{\{-f_\ell\}} \sum_{s=1}^{n} \psi_s b_{\ell s}(v_\ell, t) \quad (\nu = 1, \ldots, n),$$

under the right-end conditions

$$\psi(t) = c. \tag{11}$$

An algorithm for determining the sign definiteness of quantities (2), (3) has been worked out.

The establishment of optimality conditions in the multi-stage processes is analogous.

Let us consider the multi-stage process

$$x(n) = Q_n x(n-1) - \sum_{j=1}^{\ell_n - 1} a_j^n(u_j(n)) g_j^n(x(n-1) +$$

$$+ \sum_{j=1}^{t_n - 1} b_j^n(v_j(n)) f_j^n(x(n-1)) + R_n u_{\ell_m}(n) + s_n v_{t_n}(n) + w_n, \tag{1'}$$

where $x(n) \in E^{m_n}$, $Q_n - m_n \times m_{n-1}$, is matrix, $a^n(u_j(n))$ and $b_j^n(v_j(n))$ depend continuously on their arguments on $u_j(n) \in U_j^n \subset E^{p_j^n}$, $v_j(n) \in V_j^n \subset E^{q_j^n}$, R_n and S_n are $m_n \times q_{e_n}^n$ -, $m_n \times p_{t_n}^n$ -dimension matrices respectively w_n is m_n - dimension vector $g_j^n(x(n-1))$ and $f_j^n(x(n-1))$ are assumed to be of the form (2) and (3), hence

$$g_j^n(x(n-1)) = \sum_{\nu=1}^{m_n} g_{j\nu}^n x_\nu(n-1) + g_{j0}^n \quad (j = 1, \dots, l_n - 1), \qquad (2')$$

$$f_j(x(n-1)) = \sum_{\mu=1}^{m_n} f_{j\mu} x_\mu(n-1) + f_{j0} \quad (j = 1, \dots, t_n - 1). \qquad (3')$$

The first part means to maximize the criterion

$$R(x(N)) = c^* x(N), \qquad (5')$$

by a permissible control u, the second - to minimize (5').

If the quantities (2') and (3') maintain the fixed sign on the process (1') at every n - stage, i.e. $sgn\, g_j^n(x(n-1)) \equiv (-1)^{\pi_j^n}$, $sgn\, f_j^n(x(n-1)) \equiv (-1)^{\delta_j^n}$ for any permissible control, then every optimum (saddle) control of both parts $\{\bar{u}_j(n)\}$, $\{\bar{v}_j(n)\}$ obeys the equalities

$$\underset{u_j(n) \in U_j^n}{extr}{}^{\{(-1)^{\pi_j^n}\}} \sum_{\nu=1}^{m_n} \varphi_\nu(n) a_{jn}^n(u_j(n)) = \sum_{\nu=1}^{m_n} \varphi_j(n) a_{j\nu}^n(\bar{u}_j(n)), \qquad (6')$$

$$\max_{u_{l_n}(n) \in U_{l_n}^n} \varphi^*(n) R_n u_{m_n}(n) = \varphi^*(n) R_n \bar{u}_{m_n}(n), \qquad (7')$$

$$\underset{v_j(n) \in V_j^n}{extr}{}^{\{(-1)^{\delta_j^n}+1\}} \sum_{\nu=1}^{m_n} \varphi_\nu(n) b_{j\nu}^n(b_j(n)) = \sum_{\nu=1}^{m_n} \varphi_\nu(n) b_{j\nu}^n(\bar{v}_j(n)) \qquad (8')$$

$$(j = 1, \dots, t_n - 1),$$

$$\min_{v_{t_n}(n) \in V_{t_n}^n} \varphi^*(n) s_n v_{t_n}(n) = \varphi^*(n) s_n \bar{v}_{t_n}(n), \qquad (9')$$

where $\varphi^*(n) = (\varphi_1(n), \dots, \varphi_{m_k}(n))$ and quantities $\varphi_\nu(n)$ $(\nu = 1, \dots, m_n)$ are defined by the system

$$\varphi_0(n-1) = \sum_{s=1}^{m_n} q_{s0}^n \varphi_s(n) + \sum_{j=1}^{b_n-1} q_{j\nu}^n \, extr^{\{(-1)^{\pi_j^n}\}} \sum_{s=1}^{l_n-1} a_{js}^n(u_j(n)) \varphi_s(n) +$$

$$+ \sum_{j=1}^{t_n - 1} b_{j\nu}\, extr \left\{(-1)^{\delta_j^n} + 1\right\} \sum_{s=1}^{m_n} \psi_s(n)\, b_{js}(v_j(n)) \quad (\nu = 1, \ldots, m_{n-1}), \tag{10'}$$

to be solved under the right-end conditions

$$\psi(N) = c \tag{11'}$$

It should be noticed in conclusion, that if (1), (1') have no members with variables $v - 1$ the problem becomes an ordinary optimal control one, and the above cited results are usually true.

ON FINAL STOPPING TIME PROBLEMS

(Summary) [1]

J.L. MENALDI and E. ROFMAN
Instituto de Matemática "Beppo Levi"
Universidad Nacional de Rosario
ARGENTINA

§1. Let us consider the state equation of a dynamical system

$$(1) \quad \begin{cases} \dot{y}_{xt}(s) = f(s, y_{xt}(s)) & 0 \leqq t \leqq s \leqq T < +\infty \\ y_{xt}(t) = x & x \in \mathbb{R}^n \end{cases}$$

with the cost function

$$(2) \quad J_{xt}(\tau) = g(\tau, y_{xt}(\tau)) + \int_t^\tau l(s, y_{xt}(s))ds \qquad 0 \leqq t \leqq \tau \leqq T, \quad x \in \mathbb{R}^n .$$

Our purpose is to find the function

$$(3) \quad \rho(t,x) = \min\left\{J_{xt}(\tau) \; / \; t \leqq \tau \leqq T\right\} \quad , \quad \forall \, (t,x) \in [0,T] \times \mathbb{R}^n$$

and the smallest time $\tau_{xt} \in [t,T]$ for which we get

$$(4) \quad \rho(t,x) = J_{xt}\left(\tau_{xt}\right) \quad , \qquad \forall \, (t,x) \in [0,T] \times \mathbb{R}^n .$$

(1) – (4) will be called Pb I .

In §2 we show (Theorem 1) properties of regularity of the function $\rho(t,x)$ and we get some necessary conditions (Theorem 2) that $\rho(t,x)$ must satisfy if (3) holds. As we intend to study Pb I with the tools of the theory of variational inequalities we introduce:

.) the new unknown

$$u(t,x) = [\exp(-b(t))][\rho(T-t, x) - g(T-t, x)]$$

---

[1] This paper was prepared for the IFIP–TC7 Technical Conference on Optimization Techniques, Novosibirsk, USSR, July 1974. It will be published in spanish as a "Rapport de Recherche" at IRIA, Rocquencourt, FRANCE.

with $\quad b(t) \in C^{\circ}[0,T] \quad$ and $\quad \frac{db}{dt} \in L^2(0,T)$;

.) the functions

$$f^*(t,x) = - f(T-t, x)$$

$$1^*(t,x) = [\exp(- b(t)]\left[\frac{\partial g}{\partial s}(T-t, x) + \frac{\partial g}{\partial y}(T-t, x) \quad . \quad f(T-t, x) + 1(T-t, x)\right] \quad ;$$

.) the weight functions

$$w_i(x) = (1 + |x|^2)^{-\frac{s_i}{2}} \quad , \quad i = 1,2 \quad ,$$

$$s_1 - \frac{n}{2} > \gamma + 1 \quad , \quad s_2 + 1 \leqq s_1 \quad , \quad \gamma \text{ given in Theorem 1} \quad ;$$

.) the spaces

$$H = \left\{v \,/\, v.w_1 \in L^2(\mathbb{R}^n)\right\} \quad ; \quad V = \left\{v \in H \,/\, \frac{\partial v}{\partial x_i} \,.\, w_2 \in L^2(\mathbb{R}^n) \; ; \; i = 1,2,\ldots,n\right\} \quad ;$$

.) the operator

$$A(t) : V \longrightarrow H \quad , \quad (A(t)v)_{(x)} = \frac{db}{dt}(t).v(x) + \sum_{i=1}^{n} f_i^*(t,x) \frac{\partial v}{\partial x_i}(x) \quad ;$$

.) the functional

$$\varphi(t) : H \longrightarrow \mathbb{R} \quad , \quad (\varphi(t), v)_H = \int_{\mathbb{R}^n} 1^*(t,x)\, v(x)\, w_1^2(x)\, dx \quad ;$$

.) and the cone

$$K_o = \{v \in H \,/\, v(x) \leqq 0 \quad \forall\, x \in \mathbb{R}^n \quad , \quad (a.e)\} \quad .$$

Finally, we put the

Pb II
$$\begin{cases} \text{to find } u(.) \in H^1(0,T\,;H) \cap L^2(0,T;V) \; ; \; u(0) = 0 \; ; \; u(t) \in K_o \;\; \forall t \in [0,T] \\ \text{such that a.e. in }]0,T[\text{ the following inequality holds:} \\ \left(\frac{du}{dt}(t)\, , v-u(t)\right)_H + \left(A(t)\, u(t)\, , v-u(t)\right)_H \geqq \left(\varphi(t)\, , v-u(t)\right)_H \quad , \quad \forall\, v \in K_o \; . \end{cases}$$

In Theorem 3 we show the equivalence between Pb I and Pb II , and the existence and uniqueness of the solution of these problems.

In §3 we study the analogous of Pb I for a bounded set $\Omega \subset \mathbb{R}^n$.

In this case we have to do with boundary conditions. Following the same ideas of §2 we obtain the existence and uniqueness of the solution of this new problem.

In §4 we present a numerical approach to solve the problems which have just been introduced in §2 and §3 (the proof of the convergence is included). It consist in a combination of the penalty method with an internal approximation technique.

ihm.-

EQUILIBRIUM AND PERTURBATIONS IN PLASMA-VACUUM SYSTEMS

C. MERCIER

ASSOCIATION EURATOM-CEA SUR LA FUSION

Département de Physique du Plasma et de la Fusion Controlée
Centre d'Etudes Nucléaires
92260 FONTENAY-AUX-ROSES (FRANCE)

INTRODUCTION.

High temperature plasmas of toroidal forms necessary to obtain fusion reactions between light elements (thermonuclear fusion) must be maintained far from all material contact. In order to realise this condition one uses, in the vacuum surrounding the plasma, a metal wall supposed perfectly conducting and currents I_i whose positions and intensities have to be suitably chosen. These currents are particularly important in the case of non-circular cross-section plasmas and in the case of discharges of long duration where the role of the wall diminishes with the passage of time since in reality the wall conductivity is not infinite.

EQUILIBRIUM WITH EXTERNAL CONDUCTORS. /1/ /2/ /3/

The problem of equilibrium consists of finding a toroidal solution of the system of equations

$$J \times B = \text{grad } P$$
$$\text{div } B = 0$$
$$J = \text{rot } B \; .$$

B, J and P are respectively the magnetic field, the current intensity and the plasma pressure. The pressure must be constant over nested tori and zero on the outermost torus.

This problem can be solved in symmetry of revolution. No toroidal-type solution is known for other geometries. In the following we will consider therefore symmetry of revolution and we will use in general the cylindrical coordinates $(X,\emptyset,Y)$. In this case the configuration is defined by the flux functions $F(r)$ and $f(F)$ which satisfy the following equations

$$
(I)\quad \begin{cases} \mathcal{L}F_P = -\frac{1}{2}\frac{df^2}{dF} - x^2\frac{dP}{dF} & \text{in the plasma} \\ \mathcal{L}F_V = X\sum_i I_i\ \delta(X-X_i)\ \delta(Y-Y_i) & \text{in vacuo .} \end{cases}
$$

At the plasma-vacuum interface Γ_P we have

$$
\begin{cases} F_P = 0 \\ F_V = \gamma = \text{const.} \\ \frac{1}{x}\frac{\partial F_V}{\partial n} = \frac{1}{x}\frac{\partial F_P}{\partial n} = B_{m\Gamma_P} \end{cases}
$$

and at the wall Γ_V

$$
F_V = 0
$$

with the operator $\mathcal{L} \equiv \frac{\partial^2}{\partial X^2} + \frac{\partial^2}{\partial Y^2} - \frac{1}{X}\frac{\partial}{\partial X}$ and $\frac{\partial}{\partial n}$ represents the normal derivative. The currents I_i are placed $P_i(X_i, Y_i)$. The functions $f(F)$ and $P(F)$ characterise the plasma and are supposed known.

First of all one solves the internal problem by specifying the external cross-section Γ_P of the plasma. To resolve the problem in the vacuum, only a knowledge of the meridian magnetic field $B_{m\Gamma_P}(\ell)$ is necessary which depends on the curvilinear abscissa ℓ of Γ_P .

Schematically, the method used to solve the problem consists of definning a complete set of functions $W_{n\Gamma_P}(\ell)$ which are orthogonal and periodic over the cross-section of the plasma. The given field $B_{m\Gamma_P}(\ell)$ is expanded over the cross-section of the plasma in terms of these functions.

$$
B_{m\Gamma_P}(\ell) = \frac{1}{X}\frac{\partial F}{\partial n}\bigg|_{\Gamma_P} = \sum_n a_n\, W_{n\Gamma_P}(\ell)
$$

where

$$
a_n = \oint \frac{W_{n\Gamma_P}}{X}\,\frac{\partial F}{\partial n}\, d\ell
$$

with the norm

$$
\oint W^2_{n\Gamma_P}\, d\ell = 1 .
$$

Let us now define $W_n(X,Y)$ by

$$
\mathcal{L}.W_n = 0
$$

with

$$W_n = 0 \quad \text{on} \quad \Gamma_V$$

$$W_n = W_{n_{\Gamma_P}} (\ell) \quad \text{on} \quad \Gamma_P \; .$$

Using the relation of Green's type for the operator $\mathcal{L}$

$$\iiint_V (W \mathcal{L} U - U \mathcal{L} W) \frac{dV}{X^2} = \iint_{\Gamma_P + \Gamma_V} (W \frac{\partial U}{\partial n} - U \frac{\partial W}{\partial n}) \frac{d\sigma}{X^2} \; ,$$

with $U = F$ and $W = W_n$, we obtain :

$$\sum_i I_i W_n (X_i Y_i) + \gamma \int_{\Gamma_P} \frac{d\ell}{X} \frac{\partial W_n}{\partial n} = a_n \; .$$

One therefore attempts to resolve optimally this infinite system of equations with γ and I_i as unknowns, the number of currents being limited and their position chosen to have at the same time a sum $\Sigma |I_i|$ a minimum (or some other similar condition) and, if possible, the wall currents very weak.

Such a program has been perfected by SOUBBARAMAYER (from the C.E.A.), BOUJOT and MORERA (from C.I.S.I.) and is now operational. The technique employed is the method of finite elements.

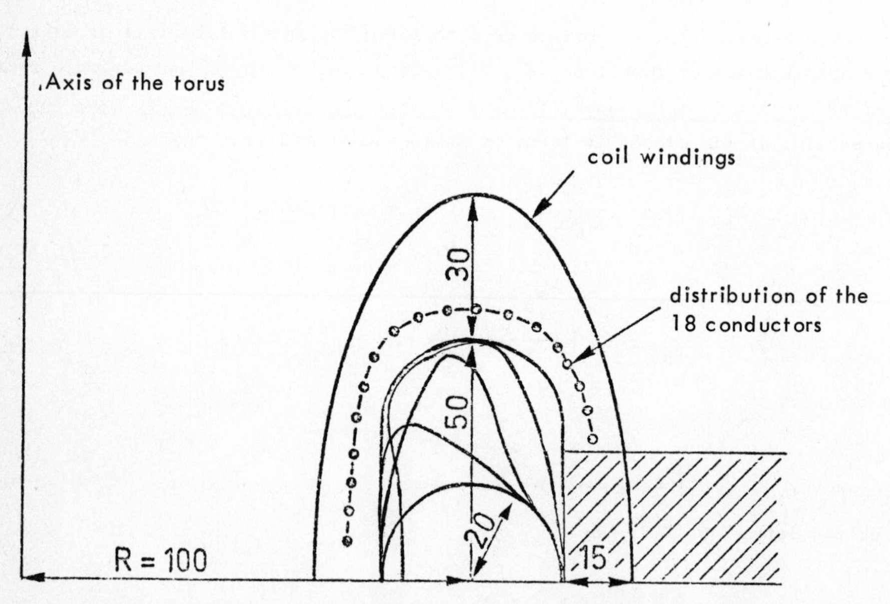

Fig.1- Numerical values and plasma cross-section forms for the example treated.

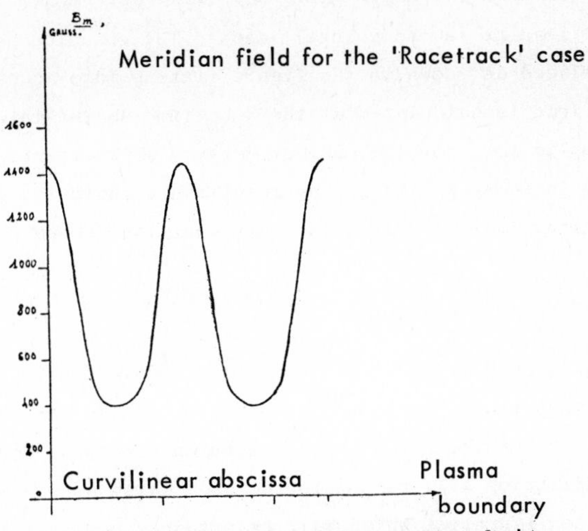

Fig.2- An example of the meridan field for the internal model.

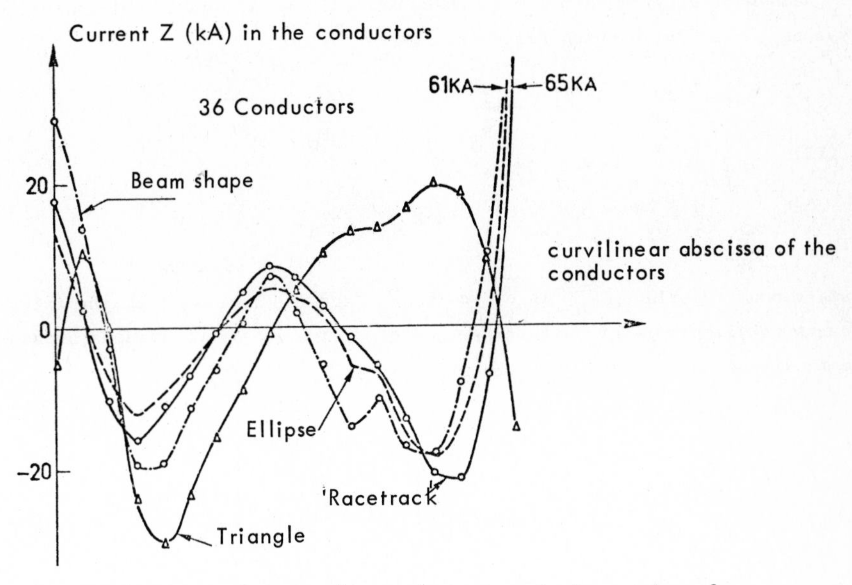

Fig.3- The currents in the conductors producing a given from of plasma cross-section.

As an illustration of the method, we have treated some cases whose results are presented in Figs.1-3. Fig.1 shows the different forms Γ_P of the plasma studied. The wall remains the same in all cases. The solution has been sought with 36 conductors placed as shown in the figure (taking into account the symmetry). The meridian field is calculated at the same time as the internal problem by supposing a parabolic form for the current density which is zero at the plasma boundary. This field is shown in Fig.2. The results are indicated in Fig.3 which gives the current intensities according to their position around the plasma.

PERTURBED EQUILIBRIA.

The preceding problem is solved numerically only to within a certain precision. Further, the arrangement and the intensity of the currents found will not be exactly realised due to, for example, technical reasons. Consequently this first problem of equilibrium will be considered as a first approximation. One therefore seeks the configuration which will be obtained under imposed real conditions. One can in addition suppose that the functions $f(F)$ and $P(F)$ characterising the plasma are only known approximatively.

Mathematically the problem is posed as follows. An approximate solution of the system (I) is known which satisfies

$$(II)\quad \begin{cases} \mathcal{L}.\bar{F}_{OP} = -\frac{1}{2}\dfrac{d\,\bar{f}^2}{d\,\bar{F}_O} - x^2\dfrac{d\,\bar{P}}{d\,\bar{F}_O} \\ \\ \mathcal{L}.\bar{F}_{OV} = X\sum_i I_i\,\delta(X-X_i)\,\delta(Y-Y_i)\ . \end{cases}$$

The plasma-vacuum interface $\bar{\Gamma}_P$ is close to Γ_P and the intensity of the external currents is given together with their position. The wall Γ_V remains unchanged. The condition on $\bar{\Gamma}_P$ and Γ_V are given by

$$\begin{cases} \bar{F}_{OP} = \bar{F}_{OV} = 0 \\ \dfrac{1}{X}\dfrac{\partial\bar{F}_{OV}}{\partial n} - \dfrac{1}{X}\dfrac{\partial\bar{F}_{OP}}{\partial n} = \delta\, B_{m.ie} \end{cases} \quad \text{on } \bar{\Gamma}_P$$

$$\bar{F}_{OV} = 0 \quad \text{on } \Gamma_V\ .$$

It is supposed that the quantities δB_{mie} , $\delta f^2 = f^2 - \bar{f}^2$, $\delta P = P - \bar{P}$ are small. The solution of the neighbouring system (I) will be obtained by introducing a vector field $\vec{\xi}$ which transforms every point of the plasma-vacuum configuration sought for (defined by system (I)) into points of the approximated plasma-vacuum configuration (defined by system (II)). Points on the boundaries correspond as well as points of each medium.

In order to specify the transformations, we call X, Y the coordinates of a point P in the space E defined by the solution of (I) , and $\overline{X}, \overline{Y}$ the corresponding point in the space $\overline{E}$ defined by the solution of (II). Let S denote this transformation. It is an infinitesimally small transformation.

$$S = \mathbb{1} + s = \mathbb{1} - \vec{\xi}.\nabla \qquad S^{-1} = \overline{\mathbb{1}} - \overline{s} \; .$$

The system (I) can thus be transformed in the space $\overline{E}$. For example

$$\mathcal{L} \Longrightarrow S \, . \, \mathcal{L} \, . \, S^{-1}$$

$$F \Longrightarrow S.F\,(XY) = \overline{F}(\overline{X}\,\overline{Y})$$

$$\Gamma \Longrightarrow \overline{\Gamma} \; .$$

Consequently in the known space $\overline{E}$, $\overline{F}(\overline{X},\overline{Y})$ is approximated by the solution $F_0(X,Y)$ of system (II) :

$$F(\overline{X}\,\overline{Y}) = \overline{F}_0(\overline{X}\,\overline{Y}) + \overline{F}_1(\overline{X}\,\overline{Y}) \; .$$

We now introduce two functions G and g whose knowledge will easily permit the solution of the problem posed.

$$G(\overline{X}\,\overline{Y}) = \overline{F}_{P_1} - s\,\overline{F}_{P_o} \qquad g(\overline{X}\,\overline{Y}) = \overline{F}_{V_1} - \overline{s}\,\overline{F}_{V_o} \; .$$

These functions satisfy

$$\begin{cases} \mathcal{L}.G(\overline{X},\overline{Y}) = -\left(\frac{1}{2}\frac{d^2f^2}{d\,F^2} + X^2\,\frac{d^2P}{d\,F^2}\right) G(\overline{X}\,\overline{Y}) - \frac{1}{2}\left(\frac{d\,\delta f^2}{dF}\right) - X^2\,\frac{d\delta P}{dF} \\ \mathcal{L}.g = 0 \end{cases}$$

$$G = g$$

$$\frac{\partial G}{\partial n} - \frac{\partial}{\partial n}\,g = -\overline{X}\,\delta\,B_{m.ie}\,\frac{G}{\overline{X}B_{m_{\overline{\Gamma}_P}}}\left(\frac{1}{2}\,\frac{df^2}{dF} + \overline{X}^2\,\frac{dp}{dF}\right) \quad \text{on } \overline{\Gamma}_P$$

g = constant on Γ_V .

The expressions $\frac{d^2f^2}{dF^2}$ $\frac{d^2P}{dF^2}$ $\frac{d\,\delta f^2}{dF}$ and $\frac{d\,\delta P}{dF}$ are taken with $F = F_o(\overline{X}\,\overline{Y})$

$$\delta I = I - \overline{I}_i = \int_{\overline{\Gamma}_P} \frac{1}{X}\,\frac{\partial g}{dn}\,d\overline{s} - \int_{\overline{\Gamma}_P} \delta\,B_{m.ie}\,d\overline{s} \; ,$$

where I_i is the total current in the approximated configuration defined by system (II) and(I)the desired current. Thus

$$F(X,Y) = \bar{F}_o(X\,Y) + G(X\,Y) \quad ,$$

and in particular the boundary Γ_P is defined with respect to the boundary $\bar{\Gamma}_P$ approximated by the normal component of $\vec{\xi}$

$$\xi_n = \left.\frac{G}{X\,B_m}\right|_{\bar{\Gamma}_P} \quad .$$

The perfection of this numerical program is in progress.

NEIGHBOURING EQUILIBRIA.

The equations satisfied by the functions G and g also allow of the study of the bifurcation in the neighbouring equilibria, these neighbouring equilibria being defined by the new perturbed conditions. These bifurcations are found where certain perturbations no longer correspond to a solution of the equations. They appear instead as the solutions of a problem of homogeneous partial differential equations with given conditions of continuity at the plasma-vacuum interface. They indicate in general the onset of unstable equilibria, whence their interest. In our case, the equilibria remain in symmetry of revolution.

Certain simple cases can as examples be treated analytically. The cylindrical case has been consequently studied for elliptical cross-section and length ℓ which is considered as the limiting case of a torus of the same length. The calculation is carried out in elliptic cylinder coordinates. The limits of stability are determined according to the eccentricity of the ellipse and the proximity of the metal wall. These limits also depend upon the current profile in the plasma. The results show that the instability increases with increasing ellipticity and for current profiles which are more peaked at the centre of the plasma.

REFERENCES.

/1/ - C. MERCIER - H. LUC - Vol.1 "The M.H.D. approach to the problem of plasma confinement in closed magnetic configurations" ; Centre for Inf. and Doc.(CID) Luxembourg May 1974 - EUR.51-27-1A.

/2/ - C. MERCIER - J.C. ADAM - SOUBBARAMAYER - J.L. SOULE - Col. Int. sur les Méthodes de calcul scientifique et technique, IRIA, (1973)
Lecture notes in Mathematics. Springer Verlag (1974). Computing methods in Applied Sciences and Engineering Intern. Symposium-Versailles, Dec.17-21-1973.

/3/ - J.P. BOUJOT - J.P. MORERA - R. TEMAN. Col. Int. sur les Méthodes de calcul scientifique et technique (IRIA - 1973).
Lecture notes in Mathematics - Springer Verlag (1974). Computing methods in Applied Sciences and Engineering - Intern. Symposium - Versailles Dec.17-21-1973.

SUFFICIENT CONDITIONS FOR ABSOLUTE MINIMUM OF THE MAXIMAL FUNCTIONAL IN THE MULTI - CRITERIAL PROBLEM OF OPTIMAL CONTROL

V.V.VELICHENKO
Moscow Phisico-Technical Institute
Moscow, USSR

Introduction

One possible approach to the problem of optimization of a dynamical system

$$x(t) = f(x,u,t) \quad (1)$$

when the simultaneous minimization of the given set Ω of performance criteria-functionals $J_\omega(x,u), \omega \in \Omega$ is required, consists in reducing it to a mono-criterial problem with the single functional

$$J(x,u) = \max_{\omega \in \Omega} \{J_\omega(x,u)\} \quad (2)$$

to be minimized.

The tasks of such a type arise in technical fields (for example, the problem of maximal deviation of the regulated object coordinates minimization) and in the mathematical economics (the example is the problem of minimization of maximal time of production for components of the final product). Some problems with the nondifferentiable performance criteria can be formulated in the same way. For example, the problem of minimization of the functional

$$J(x,u) = |I(x,u)|$$

may be replaced by an equivalent problem of the minimization of the functional

$$J(x,u) = \max\{I(x,u), -I(x,u)\}.$$

The minimax problem considered attracted attention of many authors. R.Bellman suggested treating it by means of dynamical programming [1,2] , A.Ya.Dubovitsky, A.A.Milutin, V.F.Demianov, I.V.Girsanov, V.N.Malozemov, T.K.Vinogradova [3-8] investigated various forms of the necessary optimality conditions, in works [9,10] numerical algorithms for the solution of this problem with the help of the auxiliary functionals were constructed. V.F.Demianov and T.K.Vinogradova [4,7,8] proposed using the necessary optimality conditions for determining the stationary solutions.

In the present report the sufficient optimality conditions for optimization problem with the functional (2) are formulated. Besides the general significance of sufficient conditions for optimization problems, specifically when numerical solutions are interpreted, in the minimax problem considered they are especially interesting when the various necessary conditions (as shown in [7,8] - no equivalent) are discussed. From the results given below it follows that the necessary conditions given here, proposed for the first time by A.Ya.Dubovitsky and A.A.Milutin [3] and including Pontrjagin maximum principle as a main element, are very strong. For formulation of global sufficient conditions with their help they must be supplemented by the conditions for the mutual disposition of extremals in the studied region as a whole, but not for any characteristics of a specific separately taken extremal.

The approach suggested is the extension of the extremals field method used before for the investigation of monocriterial problem [12]. The mathematical tool of the classical calculus of variations is not applicable for treating the problem considered because the functional is not smooth and the controls are restricted. Nevertheless such analogues of an explicit nonlocal relations of the classical calculus of variations as a Hilbert invariant and exact formulas for variations of a functional can be obtained. It is the possibility to write the variation of a functional in the form of exact formula without any assumption of the nearness between the investigated trajectories that permit us to formulate the sufficient conditions for absolute minimum of functional (2) by means of the approach analogous to that of Weierstrass [13].

Statement of the problem

Let us give an explicit formulation of the problem. In (1) x is the n-dimensional phase-coordinate vector and u is the z-dimensional control vector. We shall take the admissible controls to be piecewise-continuous functions $u(t) \in U$. The left end $\{x^0, T_0\}$ of a trajectory $x(t)$ of the system (1) is fixed, the right end $\{x(T), T\}$ in the terminal moment of time must belong to the terminal manifold M, specified by the equations

$$M_j(x,t)=0, \quad j=1,\dots,m\le n. \tag{3}$$

As the performance criteria we shall consider a finite number of functionals which are given as functions of the right end of the trajectory (in such a form integral functional may also be written)

$$\mathcal{J}_\omega(x,u) = \Phi_\omega[x(T),T]. \tag{4}$$

Thus the problem we pose is to choose the control $u(t)\in \mathcal{U}$, that minimizes the functional

$$\mathcal{J}(x,u)=\Phi[x(T),T]=\max_\omega\{\Phi_\omega[x(T),T]\}. \tag{5}$$

The functions $M_j(x,t)$ and $\Phi_\omega(x,t)$ are assumed to be continuous together with their first-order partial derivatives and to have bounded second-order partial derivatives with respect to all arguments, and $f(x,u,t)$ has the same properties with respect to x, u and is continuous **with** respect **to** t (the discontinuous problems also may be treated [12]).

Necessary optimality conditions

Necessary optimality conditions we derive by considering the region of attainability in the $(\ell+m)$ -dimensional vector space of variations of the end-point conditions (3) and the functionals (4) [14,15] . In that way the necessary optimality conditions for the problem discussed are reduced to the necessary minimum conditions for the auxiliary Lagrange functional

$$\psi[x(T),T]=\sum_{\omega=1}^{\ell}\lambda_\omega\Phi_\omega[x(T),T]+\sum_{j=1}^{m}\mu_j M_j[x(T),T] \tag{6}$$

and can therefore be derived with the help of the formula for small variations of a functional [16,12] (this necessary optimality conditions can also be obtained from the results of the work [3]).
The joint use of the results given below with the results of [12, 14,15] gives a possibility to formulate the necessary and sufficient optimality conditions for a broad range of minimax problems, in particular for the problems with nonfixed left end of trajectory and for the problems where the terminal points for functionals (4) is each determined by its own group of conditions of the form (3). We

shall confine ourselves here to the formulation discussed for simplicity.

Let J^* denote the minimal value of the functional (5), and R denote the set of indexes ω for which we have an equality $\Phi_\omega[x(T),T] = J^*$ on the optimal trajectory.

Theorem 1. If the control $u(t)$ and trajectory $x(t)$ are optimal in the problem (1), (3)-(5) then there exist numbers $\lambda_\omega \geq 0$ ($\lambda_\omega = 0$ for $\omega \notin R$) and μ_j not all zero, such that for the vector function $p(t)$ determined by the system of equations

$$\dot{p}(t) = -\nabla_x H(x,p,u,t), \quad H = (p, f(x,u,t)), \quad p(T) = -\nabla_x \Psi[x(T),T] \tag{7}$$

the following conditions are satisfied

$$H(x,p,u,t) = \sup_{v \in U} H(x,p,v,t), \quad T_0 \leq t < T, \tag{8}$$

$$-H[x(T),p(T),u(T-0),T] + \partial\Psi[x(T),T]/\partial T = 0. \tag{9}$$

Any trajectory of the system (1) with the initial point $\{x^\circ, T_0\}$ which satisfies the conditions of the theorem 1 will be denoted $\tilde{x}(t) = \tilde{x}(x^\circ, T_0; t)$ and called an extremal, and corresponding control will be denoted $\tilde{u}(t) = \tilde{u}(x^\circ, T_0; t)$.

Field of extremals

Let a region $A \subset E_{n+1}$ be given. We construct in A the set $\tilde{X}$ of the extremals $\tilde{x}(\xi,\tau;t)$ with the initial points $\{\xi,\tau\} \in A$. Marking all the values concerning extremals with the sign ~ ,we obtain that the extremals of set $\tilde{X}$ determine in A a reference function $\tilde{V}(\xi,\tau) = \Psi[\tilde{x}(\xi,\tau;\tilde{T}(\xi,\tau)), \tilde{T}(\xi,\tau)]$, a synthesis control $\tilde{u}(\xi,\tau) = \tilde{u}(\xi,\tau;\tau)$, a function $\tilde{R}(\xi,\tau)$ which indicates the set of maximal functional from (4), Lagrange multipliers - functions $\tilde{\lambda}_\omega(\xi,\tau)$ and an $(n+1)$-dimensional incline function

$$\tilde{p}(\xi,\tau), \qquad \tilde{H}(\xi,\tau) = (\tilde{p}(\xi,\tau), f(\xi,\tilde{u}(\xi,\tau),\tau)). \tag{10}$$

We say that the set $\tilde{X}$ with the incline function (10) forms an L-continuous field of extremals in A , if there exist constants α and β such that for any two extremals $\tilde{x}'(t) = \tilde{x}(\xi',\tau';t)$ and $\tilde{x}''(t) = \tilde{x}(\xi'',\tau'';t)$, where $|\{\xi',\tau'\} - \{\xi'',\tau''\}| \leq \varepsilon$, the corresponding controls $\tilde{u}'(t)$ and $\tilde{u}''(t)$ and the terminal moments of time $\widetilde{T}'$ and $\widetilde{T}''$ satisfy the conditions (here $\tau^* = max\{\tau', \tau''\}$, $T^* = min\{\widetilde{T}', \widetilde{T}''\}$)

$$\int_{\tau^*}^{T^*} |\widetilde{u}'(t) - \widetilde{u}''(t)| dt \leq \alpha\varepsilon, \qquad |\widetilde{T}' - \widetilde{T}''| \leq \beta\varepsilon.$$

Exact formula for variation of functional

Let $\hat{x}(t)$ be the trajectory of the system (1) with initial and end points $\{\hat{x}(T_0), T_0\} = \{x^0, T_0\}$ and $\{\hat{x}(\hat{T}), \hat{T}\} \in M$ lying entirely in A and corresponding to the admissible control $\hat{u}(t)$. Let

$$\Delta J = \Phi[\hat{x}(\hat{T}), \hat{T}] - \Phi[\widetilde{x}(x^0, T_0; \widetilde{T}), \widetilde{T}]$$

denote the difference between the values of the functional (5) for $\hat{x}(t)$ and for the extremal with the initial point $\{x^0, T_0\}$.

If $\lambda_\omega > 0$ for $\omega \in \widetilde{R}(\xi, \tau)$ for each $\widetilde{x}(\xi, \tau; t) \in \widetilde{X}$ in the conditions of theorem 1, than we can normalize the multipliers λ_ω, μ_j so that to obtain

$$\sum_{\omega \in \widetilde{R}(\xi,\tau)} \widetilde{\lambda}_\omega(\xi, \tau) = 1, \qquad \{\xi, \tau\} \in A. \tag{11}$$

Then from (5), (6) and determination of the reference function $\widetilde{V}(\xi, \tau)$ we have

$$\begin{aligned} J(x, u) &= \Phi[\widetilde{x}(\xi, \tau; \widetilde{T}(\xi, \tau)), \widetilde{T}(\xi, \tau)] = \\ &= \Psi[\widetilde{x}(\xi, \tau; \widetilde{T}(\xi, \tau)), \widetilde{T}(\xi, \tau)] = \widetilde{V}(\xi, \tau). \end{aligned} \tag{12}$$

The representation (12) and the condition (11) permit us to employ the tool of exact formulas for variations of functional for studying the functional in the region A as a whole. For the problem considered all the results of lemma 1 [12] under the conditions of theorem 2 are valid. We shall confine ourselves to the result that is necessary for the discussion of sufficient optimality conditions given below.

Theorem 2. Let the field of extremals in A be L-continuous, the condition (11) be true and $\widetilde{R}(\hat{x}(t), t)$ as a function of t be piecewise-continuous. Then the following formula is true

$$\Delta J = -\int_{T_0}^{\hat{T}} \{H[\hat{x}, \widetilde{p}(\hat{x}, t), \hat{u}(t), t] - H[\hat{x}, \widetilde{p}(\hat{x}, t), \widetilde{u}(\hat{x}, t), t]\} dt. \tag{13}$$

Sufficient optimality conditions

It follows from the formula (13) and maximum condition (8) that under the conditions of theorem 2 for any trajectory $\hat{x}$ of the system (1) lying entirely in A

$$\Phi[\hat{x}(\hat{T}),\hat{T}] \geq \Phi[\tilde{x}(\tilde{T}),\tilde{T}].$$

This inequality proves the following statement.

Theorem 3. Let the field of extremals in A be L-continuous, $\tilde{\lambda}_\omega(x,t)>0$ for $\omega \in \tilde{R}(x,t)$ and $\tilde{R}(x(t),t)$ as a function of t be piecewise-continuous along any admissible trajectory $x(t)$. Then each extremal belonging entirely to A for any of its points in the role of fixed initial data at the left end furnishes in A an absolute minimum to the functional (5) subject to the fulfilment of conditions (3).

Theorem 3 determines the properties of principle under which the global optimality takes place. Its statement is valid also for the piecewise L-continuous fields of extremals. Such generalization of the theorem 3 proved as in [12] has a wide range of applications. If set (4) consists of a unique functional, then the conditions of theorems 1 and 3-4 tranform to the necessary and sufficient conditions [11,12] for mono-criterial problem.

References

1. R.Bellman, "Dynamic Programming", N.Y., 1957.
2. R.Bellman, "Adaptive Control Processes: a Guided Tour", N.Y.,1961.
3. A.Ya.Dubovitsky, A.A.Milutin, Zh. Vychisl.Mat. i Mat.Fiz.,v.5, N 3, 1965.
4. V.F.Demianov, Vestnik Leningradskogo Univers., N 7, 1966.
5. I.V.Girsanov, "Lectures on the Mathematical Theory of the Extremal Problems", M., 1970 (Russian).
6. V.F.Demianov, V.N.Malozemov, "Introduction into Minimax",M.,1972 (Russian).
7. T.K.Vinogradova, V.F.Demianov, Dokl. Akad. Nauk SSSR, v.213, N 3, 1973.
8. --------, Zh. Vychisl. Mat. i Mat. Fiz., v.14, N 1, 1974.
9. V.V.Velichenko, "Numerical method for solving the optimal control problems", dissertation, M., 1966.
10. --------, Kosmicheskie Issledov., v. X, N 5, 1972.
11. L.S.Pontrjagin, V.G.Boltjanskii, R.V.Gamkrelidze, E.F.Mishenko, "The Mathematical Theory of Optimal Processes", N.Y., 1962.
12. V.V.Velichenko, Zh. Vychisl. Mat. i Mat. Fiz.,v. 14, N 1, 1974.
13. G.A.Bliss, "Lectures on the Calculus of Variations", M., 1950.
14. V.V.Velichenko, Dokl. Akad. Nauk SSSR, v. 174, N 5, 1967.
15. --------, Issledovanie Operatsii,Comp.Cent.Akad.Nauk SSSR,N4,1974.
16. L.I.Rozonoer, Avtomat. i Telemech., v.20, N 10, 1959.

STRATIFIED UNIVERSAL MANIFOLDS AND TURNPIKE THEOREMS FOR A CLASS OF OPTIMAL CONTROL PROBLEMS

L.F.Zelikina

Central Economical Mathematical Institute of USSR Academy of Sciences, Moscow, USSR

Consider the following optimal control problem

$$\dot{x} = u\,Q(x) \tag{1}$$

where $x \in R_n^+$, i.e. $x = (x_1, \ldots, x_n)$, $x_i > 0$; u - is n-dimensional control vector belonging to the simplex: $u_i \geq 0$, $\sum_{i=1}^{n} u_i = 1$ for any $1 \leq i \leq n$. Q is a scalar function: $Q(x) \neq 0$ in R_+^n , $Q \in C^2(R_+^n)$. Our goal is to minimize the functional

$$R(x(T), T) \tag{2}$$

where T is the smallest value of t such that $(x(t), t) \in \mathcal{M}$, $\mathcal{M}$ being a given set: $M(x(t), t) = 0$ (called the terminal set).Here R and M are scalar functions: $R, M \in C^1(R_+^{n+1})$. For example, in the time optimal problem of transferring the point from the state (x_0, t_0) to the terminal set $M(x) = 0$ we have to minimize $R(x(T), T) = T - t_0$

A manifold is said to be singular if it is impossible to define the optimal control from the condition of maximum of the Hamiltonian form on this manifold [1] . To know singular manifolds is basic for constructing synthesis in the problems with linear vectogram . A singular manifold is said to be universal if it attracts optimal trajectories. This term was introduced by Isaacs in [2] . The universal manifolds are of great interest among the singular ones.

Isaacs put a question whether a vectogram which is linear in several controls can lead to universal manifolds of codimension more than one, and whether those manifolds are intersection of the universal hypersurfaces (see [2] , ch.7). The synthesis for system

(1) is the affirmative answer to this query.

<u>Definition</u>. A set of measure zero is called a universal set V if for any point $(x,t)\in V$ there exists $T_0>t$ such that for any $T\in(t,T_0)$ there exists neighbourhood of point (x,t) – $U_{x,t}$ such that for any $(x_0,t_0)\in U_{x,t}$ the optimal trajectory starting from the point (x_0,t_0) will belong to V at the moment T.

The part of the state space – $\widetilde{V}$ – and the optimal synthesis in $\widetilde{V}$ is called the universal structure of the optimal synthesis if for any point $(x_0,t_0)\in\widetilde{V}$ the optimal trajectory starting from the point (x_0,t_0) comes up to the universal set, or, put it another way, the optimal structure is the behaviour of optimal trajectories in the region of attraction of the universal set. In our case the universal set is a stratified manifold [3], K-dimensional stratum of which will be called a universal K-dimensional manifold.

<u>Theorem</u> 1

Let the function $Q(x)$ be such that

1) The set $\frac{\partial Q}{\partial x_1}=\frac{\partial Q}{\partial x_2}=\dots=\frac{\partial Q}{\partial x_n}$ is nonempty in R^n_+,

2) On the set $\frac{\partial Q}{\partial x_i}=\frac{\partial Q}{\partial x_j}$ for any i,j $(i\neq j)$ the following condition holds:

$$\frac{\partial}{\partial x_i}\left(\frac{\partial Q}{\partial x_i}-\frac{\partial Q}{\partial x_j}\right)<0\,.$$

Let us denote by $w_{i_1,\dots,i_K}$ the $(K-1)$-vector (see [4], p.22) which is orthogonal to the manifold

$$\frac{\partial Q}{\partial x_{i_1}}=\frac{\partial Q}{\partial x_{i_2}}=\dots=\frac{\partial Q}{\partial x_{i_K}}$$

It is required to meet the following condition:

3) For any K-tuple $(i_1,i_2,\dots,i_K)$, $(K=2,\dots,n)$ the Plukker coordinates of the projection of $w_{i_1,\dots,i_K}$ upon the co-

ordinate plane $(x_{i_1}, \ldots, x_{i_K})$ have the same sign and are not equal to zero. (The $(K-1)$-vector $\mathcal{N}_{i_1,\ldots,i_K}$ is formed with a normal to manifolds

$$\frac{\partial Q}{\partial x_i} = \frac{\partial Q}{\partial x_j}, \quad i \neq j, \quad i,j = i_1, \ldots, i_K)$$

Then the sets

$$\frac{\partial Q}{\partial x_{i_1}} = \frac{\partial Q}{\partial x_{i_2}} = \cdots = \frac{\partial Q}{\partial x_{i_K}} > \frac{\partial Q}{\partial x_e} \tag{3}$$

$$(K = 2, \ldots, n; \quad e \neq i_1, \ldots, i_K)$$

are universal $(n-K+1)$ -dimensional manifolds.

Note, that it follows from conditions (1)-(3) that for any K there exist K -dimensional universal manifolds and that these manifolds have no singularities. In this case the universal structure can be present in the following form. There are C_n^2 $(n-1)$-dimensional manifolds $\frac{\partial Q}{\partial x_i} = \frac{\partial Q}{\partial x_j} > \frac{\partial Q}{\partial x_K}$ (for any $K \neq i, j$) whose intersection is the manifold

$$\frac{\partial Q}{\partial x_1} = \frac{\partial Q}{\partial x_2} = \cdots = \frac{\partial Q}{\partial x_n} \tag{4}$$

These manifolds divide R_+^n into the regions - V^i - where $\frac{\partial Q}{\partial x_i} > \frac{\partial Q}{\partial x_e}$ for any $e \neq i$. In the region V^i the optimal control is:

$u_i = 1, \quad u_e = 0$ for any $e \neq i$

On the manifolds (3) the optimal control u is proportional to the vector whose coordinates are equal to the Plukker coordinates of the projection of $\mathcal{N}_{i_1,\ldots,i_K}$ upon the coordinate plane $(x_{i_1}, \ldots, x_{i_K})$.

In particular, $u_e = 0$ when $e \neq i_1, \ldots, i_K$. Note that this optimal control holds a state point in the universal manifold.

Conditions (1)-(3) are close to the necessary ones when the universal structure of the optimal synthesis has the form described above.

For elucidation let us note that the stratified universal manifold is diffeomorphic to the set (system) of bisector hyperplanes which cut off at the points of intersections of these hyperplanes.

Theorem II

The universal structure of optimal synthesis of the optimal control problem - (1) - is invariant to functionals from the class (2), i.e. if to attain the terminal set the optimal trajectory moves on the universal manifold for a time, then for any sufficiently far terminal set the initial part of the optimal trajectory is independent of a functional from the class (2), and in this case the optimal trajectory attains the universal manifold in the shortest run. If one knows a universal structure of optimal synthesis, then under additional conditions on a function $Q(x)$ one can construct the optimal synthesis in the whole space R^n_+. For this, one has to construct switching surfaces and to find intersection of these surfaces with the stratified universal manifold.

For example, let us consider a time-optimal problem. The terminal set is $x_n = const$. Let the function $Q(x)$ be such that the conditions (1)-(3) and these ones hold:

4) the manifold (4) have no asymptote parallelled to coordinate planes;

5) $\frac{\partial}{\partial x_i}\left(\frac{1}{Q}\frac{\partial Q}{\partial x_i} - \frac{1}{Q}\frac{\partial Q}{\partial x_j}\right) < 0$ in R^n_+ ;

6) $\frac{\partial}{\partial x_i}\left(\frac{1}{Q}\frac{\partial Q}{\partial x_j}\right) < 0$, when $i \neq j$ and $\frac{\partial}{\partial x_i}\left(\frac{1}{Q}\frac{\partial Q}{\partial x_i}\right) \leq 0$;

then a state point either doesn't attain the universal manifold, and in this case there is no more than one switch, or attains the universal manifold in the points of intersection of the latter with the switching surface and in this case it moves on the universal manifold going from strata of less dimension to strata of higher dimension one after another. In the latter case the state point can come to the region V^i at any moment. We can write out explicitly the equations of switching surface. The optimal syntheses for n = 2,3 are given in Fig. 1 and Fig. 2 respectively.

Note that the conditions on a choice of a function $Q(x)$ are not restrictive. For instance, the conditions (1)-(6) are fulfilled for functions $Q(x)=f_1(x_1)\cdot\ldots\cdot f_n(x_n)$ where $f_i(x_i)$ is a monotonic and logarithmically convex function, i.e. $f_i'>0$, $(\ln f_i(x_i))''<0$.

One can obtain the similar results for a case when there are uncontrollable variables in system (1), i.e. for a system

$$\dot{x} = u\,Q(x,y)$$
$$\dot{y}_i = F_i(y)$$

In this case, if $Q(x,y) = Q_1(x)Q_2(y)$, then the universal manifolds are cylindric in y. One can investigate similarly the case in which the manifold (4) is empty, but at least one of the manifolds (3) is nonempty.

The problem (I) arises naturally for an economical problems of optimal allocating of sources among the sectors, producing several factors of production $(x_1,\ldots,x_n)$, where $Q(x)$ is a production function. Theorems I and II have in this case a direct relation to turnpike theorems. The manifold (4) plays a role of von-Neumann's ray of maximal balanced growth.

The calculation of universal manifolds and optimal control on it has made it possible to derive a variety of economic facts of great importance. Namely, the universal manifolds are defined by the condition of equality of relating norm of efficiency of product relative to factors of production. For a Cobb-Duglas's type production functions $Q(x)= x_1^{\alpha_1}\cdot\ldots\cdot x_n^{\alpha_n}$ we have that the optimal controls on the universal manifolds (3) are proportional to α_i - the elasticity of product relative to the factors of production $-x_i$ $(i=i_1,\ldots,i_k)$.
As a consequence of theorem II, this assertion can be de-

rived: the shadow prices (a gradient of a Bellmann function) for a problem of time optimal control are invariant related to a choice of any sufficiently far terminal set. In this case the discounting rate on the universal manifold is equal to the optimal rate of growth and to a norm of efficiency. It is constant only for a first order homogeneous function, i.e. for a case of a constant return to scale.

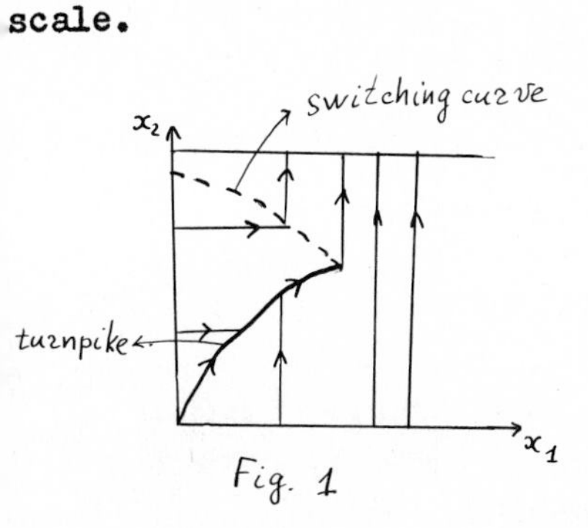

Fig. 1

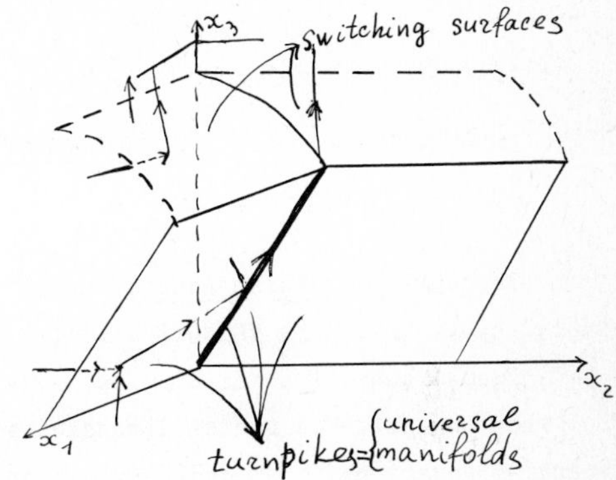

References

[1] L.S. Pontrjagin, V.G. Boltjanskiĭ, R.V. Gamkrelidze and E.F. Miščenko, The mathematical theory of optimal processes, Fizmatgiz, Moscow, 1961; English transl., Wiley, New York, 1962, and Macmillan, New York, 1964.

[2] Isaacs, R., Differential games, Wiley, New-York, 1965, Russian transl., Mir, Moscow, 1967.

[3] Thom, R., Local topological properties of differentiable mappings, Differential Analysis (Papers presented at the Bombay Colloquium, 1964).

[4] Картан Э. Внешние дифференциальные системы и их геометрические приложения. Издательство Московского Университета, 1962.

ON THE NUMERICAL APPROXIMATION OF PROBLEMS OF IMPULSE CONTROLS

J.L. LIONS

COLLEGE DE FRANCE and IRIA-LABORIA

Lecture at I.F.I.P. Meeting, NOVOSIBIRSK, July 1974.

INTRODUCTION

Problems of impulse control arise in a large number of situations in inventory theory (cf. A.F. VEINOTT Jr. [1]).

From a general point of view, they correspond to problems of optimal control when the state is finite dimensionnal (1) and it is given by an Ito's differential equation (or an ordinary differential equation) with jumps and where actually these jumps are indeed the control variable (2).

In a number of recent publications with A. BENSOUSSAN, (cf. bibliography), we have observed that, by using dynamic programming, one is lead, in order to solve these classes of impulse control problems, to partial differential inequalities.

The partial differential inequalities correspond to new free boundary problems and in order to solve these problems (both from a theoretical and a numerical viewpoint), we have introduced with Bensoussan a new tool : the quasi-variational inequalities (Q.V.I. for short).

The aim of this lecture is to recall some general results relative to the Q.V.I., in the evolution and in the stationary cases and to give some ideas

---

(1) If the state is infinite dimensional, one is lead to partial differential inequalities in infinite dimensional spaces.

(2) We have to choose in an optimal manner (if possible) the total number of jumps, the times where the jumps arise, and the "amount" of the jumps (with some constraints on these "amounts").

on the numerical approximation of the solution of Q.V.I. ; we also report briefly on some numerical experiments which have been made, mainly in connection with applications in inventory theory (cf. BENSOUSSAN-GOURSAT-MAAREK[1]), or in energy policies related to optimal use of dams (cf. BRETON-LEGUAY[1]).

The numerical experiments reported in this paper have been conducted by GOURSAT, LEGUAY, MAURIN at IRIA-LABORIA.

Let us notice that the Q.V.I. turn out to be useful in a classical free boundary problem in infiltration theory, (cf. BAIOCCHI[1] ; for earlier work using Variational Inequalities (V.I.), cf. BAIOCCHI-MAGENES[1]).

The Q.V.I. also appear in a natural context in the optimal control of systems governed by V.I. (1); cf. F. MIGNOT[1].

They also appear in the search for Nash points of equilibrium (cf. BENSOUSSAN-LIONS[2]).

The plan is as follows :

---

(1) And therefore in the optimal control of systems whose state is given by the solution of a free boundary problem.

1. - PROBLEMS OF IMPULSE CONTROL

We suppose that the state $y = y(s;v)$ of the system at time s when one applies the control v (to be defined below) is a vector of $\mathbb{R}^n$; we can think of y as being the amount of goods we have in our inventory. We suppose that

$$y(t;v) = x. \tag{1.1}$$

The state equation is an Ito's differential equation with jumps, which corresponds to the following situation : there is a demand, which in the interval t, s is given by

$$D(t,s) = -\mu(s-t) - \sigma(w(s) - w(t)) \tag{1.2}$$

where

$$\mu = \{\mu_1, \ldots, \mu_n\} \in \mathbb{R}^n \text{ (the deterministic part of the demand)} \tag{1.3}$$

and where, to simplify, we assume that

$$\sigma = \sigma \text{ (Identity)}, \tag{1.4}$$

and where $w(t)$ is a Standard Wiener process in $\mathbb{R}^n$ (1).

The control v (or the policy) is of the following form :

$$v = \{\theta^1, \xi^1; \theta^2, \xi^2; \ldots ; \theta^N, \xi^N; N\} \tag{1.5}$$

where

$$t \leqslant \theta^1 < \theta^2 < \ldots < \theta^N \leqslant T = \text{given number, finite or infinite (the horizon of the problem)}, \tag{1.6}$$

$$\xi^i = \text{amount of goods ordered at time } \theta^i, \tag{1.7}$$

$$N = \text{number of orders ;} \tag{1.8}$$

in (1.5) the θ^i, ξ^i, N are at our disposal, with some restrictions on the ξ^i's,

---

(1) We can also take care of the case when we have moreover a Poisson process. Cf. BENSOUSSAN and the A.

that we can express by :

$$\xi^i \in \mu \ , \ \mu = \text{closed set of } \mathbb{R}^n; \tag{1.9}$$

the θ^i, ξ^i and N are random functions.

The state is now given, assuming that the delivery is made without delays [1] by

$$\left| \begin{array}{l} dy = -\mu\, ds - \sigma\, db(s) = \sum_{i=1}^{N} \xi^i \,\delta(s-\theta^i)\, ds \\ \\ y(o) = x. \end{array} \right. \tag{1.10}$$

The pay-off, or cost function, is supposed to be given by

$$J(v) = E\left\{\int_t^T f(y(s;v))ds + kN(v)\right\}, \tag{1.11}$$

where E = expectation.

Remark 1.1.

In the deterministic case, $\sigma = 0$, and one has the same kind of pay-off, without of course taking the expectation in (1.11). ■

In (1.11), f is a given $\geqslant o$ function, which corresponds to a storage cost or to a shortage cost, and k is a given > 0 constant : the set-up cost.

Remark 1.2.

We refer to the bibliography, for much more general ordering costs, cf. in particular, Remark 2.2. Other cases will also be considered below. ■

The problem is now to characterize

$$\inf_v J(v) = u(x,t) \tag{1.12}$$

and to obtain the (or one) optimal policy, if it exists.

---

[1] For the case of deliveries with delays, cf. BENSOUSSAN and the A., in the bibliography and Section 6 below.

2. - FORMULATION IN TERMS OF QUASI-VARIATIONAL INEQUALITIES

One can prove that the function u defined by (1.12) satisfies the following set of inequalities and equalities

$$(2.1)\quad \begin{cases} -\dfrac{\partial u}{\partial t} + \Sigma \mu_i \dfrac{\partial u}{\partial x_i} - \dfrac{\sigma^2}{2} \Delta u - f \leq 0 , \\ u - M(u) \leq 0 \\ (-\dfrac{\partial u}{\partial t} + \sum_{i=1}^{n} \mu_i \dfrac{\partial u}{\partial x_i} - \dfrac{\sigma^2}{2} \Delta u - f)\ (u - M(u)) = 0, \end{cases}$$

where

$$(2.2)\quad Mu(x,t) = k + \inf_{\xi \in \mu} . u(x+\xi, t),$$

and where u is subject to

$$(2.3)\quad u(x,T) = 0.$$

If the state is subject to stay in a set $\bar{\mathcal{O}} \subset \mathbb{R}^n$, where $\mathcal{O}$ is an open set of $\mathbb{R}^n$, assumed to be bounded (this being unessential), then one has to add to (2.1) (where $x \in \mathcal{O}$)(1), the boundary conditions

$$(2.4)\quad \begin{cases} \dfrac{\partial u}{\partial \nu} \leq 0, \quad u - M(u) \leq 0, \quad \dfrac{\partial u}{\partial \nu}\ (u - M(u)) = 0 \\ \text{on } \Gamma = \partial \mathcal{O},\ t < T \end{cases}$$

($\dfrac{\partial}{\partial \nu}$ = normal derivative directed toward the exterior of $\mathcal{O}$). ■

Formulation as a Q.V.I. of evolution.

For $u, v \in H^1(\mathcal{O})$(2), we define

$$(2.5)\quad a(u,v) = \frac{\sigma^2}{2} \Sigma \int_{\mathcal{O}} \frac{\partial u}{\partial x_i} \frac{\partial v}{\partial x_i} dx + \Sigma \int_{\mathcal{O}} \mu_i \frac{\partial u}{\partial x_i} v\, dx.$$

---

(1) In definition (2.2) x and $x + \xi$ should be in $\bar{\mathcal{O}}$.

(2) Sobolev's space of order 1. Cf. SOBOLEV [1] .

For $f,v \in L^2(\mathcal{O})$, $(f,v) = \int_{\mathcal{O}} fv\,dx$ as usual.

It is now a simple matter to show that u is <u>characterized by the</u> Q.V.I. <u>of evolution</u>

$$(2.6)\quad \left| \begin{array}{l} -\left(\frac{\partial u}{\partial t}, v-u\right) + a(u,v-u) - (f,v-u) \geqslant 0 \quad \forall v \in H^1(\mathcal{O}) \\ \text{such that } v \leqslant Mu(x,t), \\ u - Mu \leqslant 0, \\ u(x,T) = 0. \end{array} \right.$$

<u>It is this formulation that we shall use in the numerical approximation.</u> ∎

<u>The stationary case.</u>

Let us consider now the case when

$$T = +\infty$$

and let us assume that $w(t+h)-w(t)$ has a probability law which depends only on h. One considers then the pay-off :

$$(2.7)\quad J(v) = E\left\{\int_0^{\infty} e^{-\alpha s} f(y(s;v))ds + k \sum_{i=1}^{\infty} e^{-\alpha\theta i}\right\}$$

where $\alpha > 0$ is the <u>discount rate</u>.

If we set

$$(2.8)\quad u(x) = \inf.\ J(v),$$

then one can show that u <u>is characterized by the (stationary)</u> Q.V.I.

$$(2.9)\quad a(u,v-u) \geqslant (f;v-u) \qquad \forall v \leqslant M(u), \qquad u \leqslant M(u);$$

let us notice that now $a(u,v)$ is given by :

$$(2.10)\quad a(u,v) = \frac{\sigma^2}{2} \Sigma \int_{\mathcal{O}} \frac{\partial u}{\partial x_i} \frac{\partial v}{\partial x_i}\, dx + \Sigma\, \mu_i \int_{\mathcal{O}} \frac{\partial u}{\partial x_i} v dx + \alpha \int_{\mathcal{O}} uv\, dx.$$ ∎

Remark 2.1.

Assuming that $f \geqslant 0$ belongs to $L^{\infty}(\mathcal{O})$, one can prove that (2.9) admits a unique solution $u \geqslant 0$. (Cf. BENSOUSSAN-GOURSAT-LIONS [1], L. TARTAR [1], Th. LAETSCH [1]).

One can also prove that (2.6) admits a unique solution (cf. BENSOUSSAN-LIONS) by arguments using simultaneously functional analysis and probability theory; a direct proof using only (2.6) is not known yet. ■

Remark 2.2.

The structure of M, as given by (2.2), corresponds to the set up cost k being fixed and independent from the goods which are ordered and of the amount which is ordered. But actually the method is entirely general; to fix ideas, let us assume that we have two goods $(n = 2)$ and that we pay k_i if we order only the ith item $(i=1,2)$ and that we pay $k(< k_1 + k_2)$ if we order simultaneously the two goods. Then :

$$
(2.11)\quad \left|\begin{array}{ll}
M(v) = \inf(M_1(v), M_2(v), M_3(v)) , & \\
M_1(v)(x) = k_1 + \inf. v(x_1+\xi_1, x_2), & \xi_1 \geqslant 0, \\
M_2(v)(x) = k_2 + \inf. v(x_1, x_2+\xi_2), & \xi_2 \geqslant 0, \\
M_3(v)(x) = k + \inf. v(x_1+\xi_1, x_2+\xi_2), & \xi_i \geqslant 0,
\end{array}\right.
$$

and one has the same formulations (and the same results) than above with this new function M.

For an "axiomatic" study of M, we refer to BENSOUSSAN-LIONS, loc. cit., TARTAR [1] . Cf. also JOLY-MOSCO [1]. ■

3. - NUMERICAL APPROXIMATION OF THE STATIONARY CASE (I).

3.1. General remarks.

Let u^o be a given constant, satisfying

$$
(3.1)\qquad u^o \geqslant \frac{1}{\alpha} \sup. f.
$$

Let us introduce λ such that

$$
(3.2)\qquad a(v,v) + \lambda \int_{\mathcal{O}} v^2 dx \geqslant c\|v\|^2_{H^1(\mathcal{O})}, \quad c > 0 , \; v \in H^1(\mathcal{O})
$$

and let us inductively define u^n by the solution of the V.I. (1)

(3.3.)
$$\left| \begin{array}{l} a(u^n, v-u^n) + \lambda(u^n, v-u^n) \geqslant (f+ \lambda u^{n-1}, v-u^n), \\ \forall\ v \leqslant M(u^{n-1}), \quad u^n \leqslant M(u^{n-1}). \end{array} \right.$$

One proves that

(3.4)
$$u^n \downarrow u \quad \text{in } L^p(\mathcal{O}) \text{ strongly and in } H^1(\mathcal{O}) \text{ weakly.}$$ ■

The next step is to <u>approximate</u> u^n ; one uses the standard methods for the numerical approximation of V.I. (Cf. CEA-GLOWINSKI [1] GLOWINSKI-LIONS-TREMOLIERES [1] and the bibliography of these works); in the numerical experiments which have been made so far, one has used the method of <u>finite differences</u>, and not the finite element methods (cf. Remark 3.1 below); let us denote by u_h^n the solution of :

(3.5)
$$\left| \begin{array}{l} a_h(u_h^n, v_h-u_h^n) + \lambda(u_h^n, v_h-u_h^n) \geqslant (f+ \lambda u_h^{n-1}, v_h-u_h^n), \\ \forall\, v_h \leqslant M(u_h^{n-1}), \quad u_h^n \leqslant M(u_h^{n-1}), \end{array} \right.$$

where v_h belongs to the space V_h generated by the characteristic functions of non overlapping cubes corresponding to standard finite differences approximations : $a_h(u_h, v_h)$ denotes the approximation of $a(u,v)$ obtained after replacing $\frac{\partial}{\partial x_i}$ by their approximation ∇_{ih} with "mesh" h ; for the approximation of $\Sigma \mu_i \frac{\partial}{\partial x_i}$ one "follows the characteristics" (which are the trajectories of the state, if $\sigma = 0$).

<u>Remark 3.1.</u>

By virtue of the fact that if $u_h, v_h \in V_h$ as defined above, then $\sup(u_h, v_h) \in V_h$, one can prove that

(3.6)
$$\left| \begin{array}{l} u_h^n \downarrow u_h \text{ (as } n \to \infty) \text{ in } L^p(\mathcal{O}) \text{ strongly,} \\ \nabla_{ih}\, u_h^n \to \nabla_{ih}\, u_h \text{ in } L^2(\mathcal{O}) \text{ weakly} \end{array} \right.$$

where u_h is the solution of the "discrete Q.V.I."

---

(1) Variational Inequality, which admits a unique solution by virtue of LIONS-STAMPACCHIA [1] and of (3.2).

$$(3.7)\quad \left| \begin{array}{l} a_h(u_h, v_h - u_h) \geqslant (f, v_h - u_h), \\ \forall\ v_h \quad V_h, \quad v_h \leqslant M(u_h), \quad u_h \leqslant M(u_h). \end{array} \right.$$

It is an open problem to know if this result still holds true if one replaces finite differences by suitable finite elements. ■

Remark 3.2.

If we start the iterative algorithm (3.3) by

$$u_o = 0$$

and if we denote by u_n the corresponding result, then

$$(3.8)\qquad u_o \leqslant u_1 \leqslant \ldots \leqslant u_{n-1} \leqslant u_n \leqslant \ldots \leqslant u.$$

It is not known if, in general $u_n \to u$, but this seems to be true in all the numerical computations made so far.

Remark 3.3

It has not been proved that $u_h \to u$ as $h \to 0$. But this is conjectured as highly probable and it is "proved" in all the numerical experiments which have been made. The conjecture would be proved if we could prove that

$$(3.9)\qquad u_h \geqslant u_{h'} \quad \text{if } h \geqslant h'$$

which is also a conjecture .... .

Let us notice that all these questions are related to one sided approximations of V.I. : cf. G. Strang [1], U. Mosco and G. Strang [1]. They are also related to the discrete maximum principle, which is a non trivial question for the approximation by finite elements. (cf. Ph. CIARLET and P.A. RAVIART [1]).

3.2. An example

We briefly report on an example of GOURSAT [1] ; the open set $\mathcal{O} \subset R^2$ is given by $\mathcal{O} =]-\frac{1}{2}, 2[\times]-\frac{1}{2}, 2[$;

$$\mu_1 = 12,\ \mu_2 = 8,\ \sigma_1 = 2{,}5\ \sqrt{2}\ 10^{-2}\ \mu_1,\quad \sigma_2 = 3\ \sqrt{2}\ 10^{-2}\ \mu_2 \quad (^1)\ ;$$

M is given by (2.11), where $k_1 = k_2 = 0.7$, $k_3 = 1$;
the function f is given by :

$$f(x) = f_1(x_1) + f_2(x_2),\ f_i(x) = 8\ x \text{ if } x > 0;\ -\ 60\ x \text{ if } x < 0.$$

The function u is minimum at one point denoted by $S = \{x_{1s}, x_{2s}\}$(cf. Fig. 1). It is sufficient to represent the solution for $x_1 \leqslant x_{1s}$, $x_2 \leqslant x_{2s}$ [2].

There is one region (the so-called saturated region) where $u = M(u)$; this region is divided into three parts :

$$(3.10)\quad \left|\begin{array}{l} \mathcal{C}_1 \text{ where } u = M_1(u),\ \mathcal{C}_2 \text{ where } u = M_2(u), \\ \\ \mathcal{C}_3 \text{ where } u = M_3(u)\ ; \end{array}\right.$$

numerical experiments show that (at least in the case under study) these regions do not overlap : they are represented on Fig. 1 ; the "internal" boundary $(s_1) \cup (s) \cup (s_2)$ is <u>a free surface</u>.

One also needs the curves (S_i) where (S_1) for instance denotes the set of minima of the function :

$$x_1 \to u(x_1, x_2)$$

which is enough to consider for $x_2 \in$ projection on the x_2 axis of $\mathcal{C}_1$; (S_2) is defined in a similar manner exchanging the roles of x_1 and of x_2. The curves (S_1) and (S_2) are represented on Fig. 1.

<u>Let us now explain on this example how to derive the optimal policy</u> from the knowledge of u and more precisely of $(s_1) \cup (s) \cup (s_2)$ and of $(S_1) \cup (S_2)$.

----

(1) The operator $-\frac{\sigma^2}{2}\Delta$ is replaced by $-\frac{\sigma_1^2}{2}\frac{\partial^2}{\partial x_1^2} - \frac{\sigma_2^2}{2}\frac{\partial^2}{\partial x_2^2}$ (the variance of the demand depends on the goods that are considered). The values given here correspond to situations arising in practice; cf. BENSOUSSAN-GOURSAT-MAAREK, loc. cit.

(2) But the problem being <u>global</u> in $\mathcal{O}$, one has to compute u everywhere in order to obtain S.

If the inventory $y \notin (s_1) \cup (s) \cup (s_2)$, do not order.

If the inventory reaches (s) at point P, order $\vec{PS}$.

If the inventory reaches (s_1) (resp. (s_2)) at P_1 (resp. P_2), order $\vec{P_1Q_1}$ (resp. $\vec{P_2Q_2}$), i.e. only the first (resp. 2nd) item in the amount P_1Q_1 (resp. P_2Q_2).

To obtain such results takes about 10' on CII 10070.

Remark 3.4.

The numbers σ are "small" against the μ_j's but with sufficient care in the discretization of the first order derivatives, this fact does not lead to severe numerical difficulties. ∎

Remark 3.5.

Specific applications to management problems are given in BENSOUSSAN-GOURSAT-MAAREK [1]. ∎

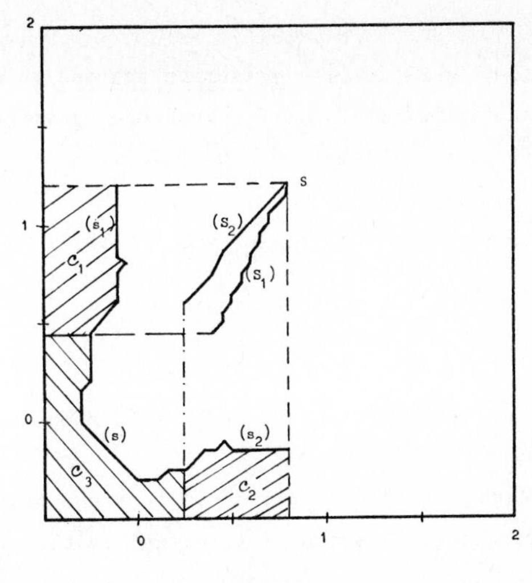

Figure 1.

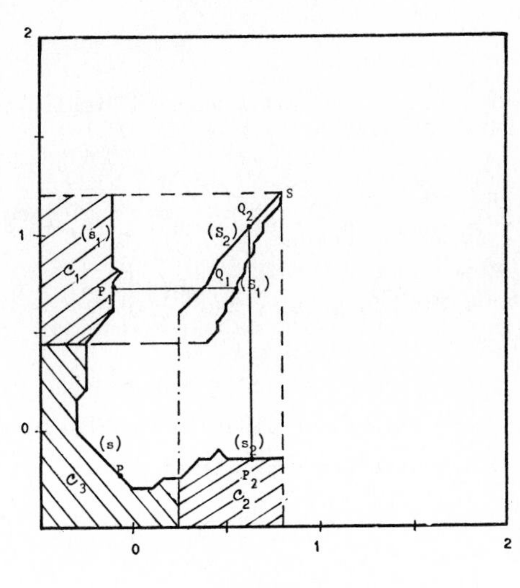

Figure 2.

4. - NUMERICAL APPROXIMATION OF THE EVOLUTION CASE.

4.1. General remarks

It is a simple matter to derive semi-discrete schemes (corresponding to the time discretization) associated to (2.6). For instance, starting with $u^o = 0$, one defines u^n for $n \geqslant 1$ by the solution of the Q.V.I. [1].

$$(4.1)\quad \left| \begin{array}{l} \left(\dfrac{u^n - u^{n-1}}{\Delta t}, v-u^n\right) + a(u^n, v-u^n) \geqslant (f, v-u^n) \\ \\ v \leqslant M(u^n), \quad u^n \leqslant M(u^n); \end{array} \right.$$

in (4.1) we assume that f does not depend on t (cf. BENSOUSSAN-LIONS, loc. cit. for more general cases).

One can prove (BENSOUSSAN, LIONS, loc. cit.) that the function $u_{\Delta t} = u^n$ in $((n-1)\Delta t, n\,\Delta t)$ converges (in a suitable topology) to the solution u of (2.6).

Remark 4.1.

Of course the solution of (4.1) is in turn computed by a space-approximation as in Section 3 - and with analogous open problems. ■

Remark 4.2.

Numerical experiments show good results with an "implicit-explicit" method of the type

$$(4.2)\quad \left| \begin{array}{l} \left(\dfrac{u^n - u^{n-1}}{\Delta t}, v-u^n\right) + a(u^n, v-u^n) \geqslant (f, v-u^n) \\ \\ \forall v \leqslant M(u^{n-1}), \quad u^n \leqslant M(u^{n-1}). \end{array} \right.$$

■

Remark 4.3.

In case (as above) where f does not depend on t, the preceding schemes are also used for solving the stationary problems; convergence of u^n to the state of equilibrium is quite fast. ■

4.2. An example

The example given here is taken from GOURSAT [1]. We consider one item (space dimension = 1) ; μ and σ depend on t [2] :

---

(1) We reverse the order with respect to (2.6), in order to work with the usual notations of numerical analysis; (4.1) is an implicit scheme.

(2) Everything which has been said extends to this case.

(4.3) $$\mu(t) = 5(2.4 + 1.4 \sin \pi t),\ \sigma(t) = 3\sqrt{2}\ 10^{-2}\ \mu(t).$$

The set-up cost depends on the amount which is ordered :

(4.4) $$k(\xi) = k_o + k_1\xi\ ,\qquad k_o = k_1 = 3.10^{-1}\ ;$$

this leads to the same problem than above, with

(4.5) $$Mu(x,t) = \inf_{\xi \geq 0}.\left[k_o + k_1\xi + u(x+\xi,\ t)\right]\ ;$$

we suppose that f depends on x and t :

(4.6) $$f(x,t) = (1 - 0.125t)\ f(x),\ f(x) = 8x \text{ if } x > 0,\ -60\ x \text{ if } x < 0.$$

The set $\mathcal{O}$ is taken to be $]-\frac{1}{2}, 3[$.

The "saturated region" where $u = M(u)$ is shaded on Fig. 3; the boundary (s) is the free surface ; the set of points where $x \to u(x,t)$ is minimum is represented by (S).

When the stock (taking into account the stochastic demand) reaches P (cf. Fig. 3) then we order PQ.

5. - NUMERICAL APPROXIMATION OF THE STATIONARY CASE (II).

5.1. The problem of the dimension

In cases where the space dimension is large (say ≥ 4) one faces the very difficult problem of solving Q.V.I. in high (or very high) dimension.

The only reasonable hope seems to obtain some "sub-optimal" policies of "simple" structure.

In the management literature, it seems that the method usually advocated is essentially to decompose the problem in considering the different goods as being independent.

It is at this point natural to try to extend to Q.V.I. the splitting-up methods ; these methods have been introduced (cf. G.I. MARCHUK [1] , N.N. YANENKO [1] and the bibliography of these books) for the solution of large systems of partial differential equations ; they have been extended, in LIONS-TEMAM [1] to V.I. and in BENSOUSSAN-LIONS-TEMAM [1] to more general variational problems. We are now going to indicate how these methods can be extended to Q.V.I.; one obtains in this manner rather simple algorithms.

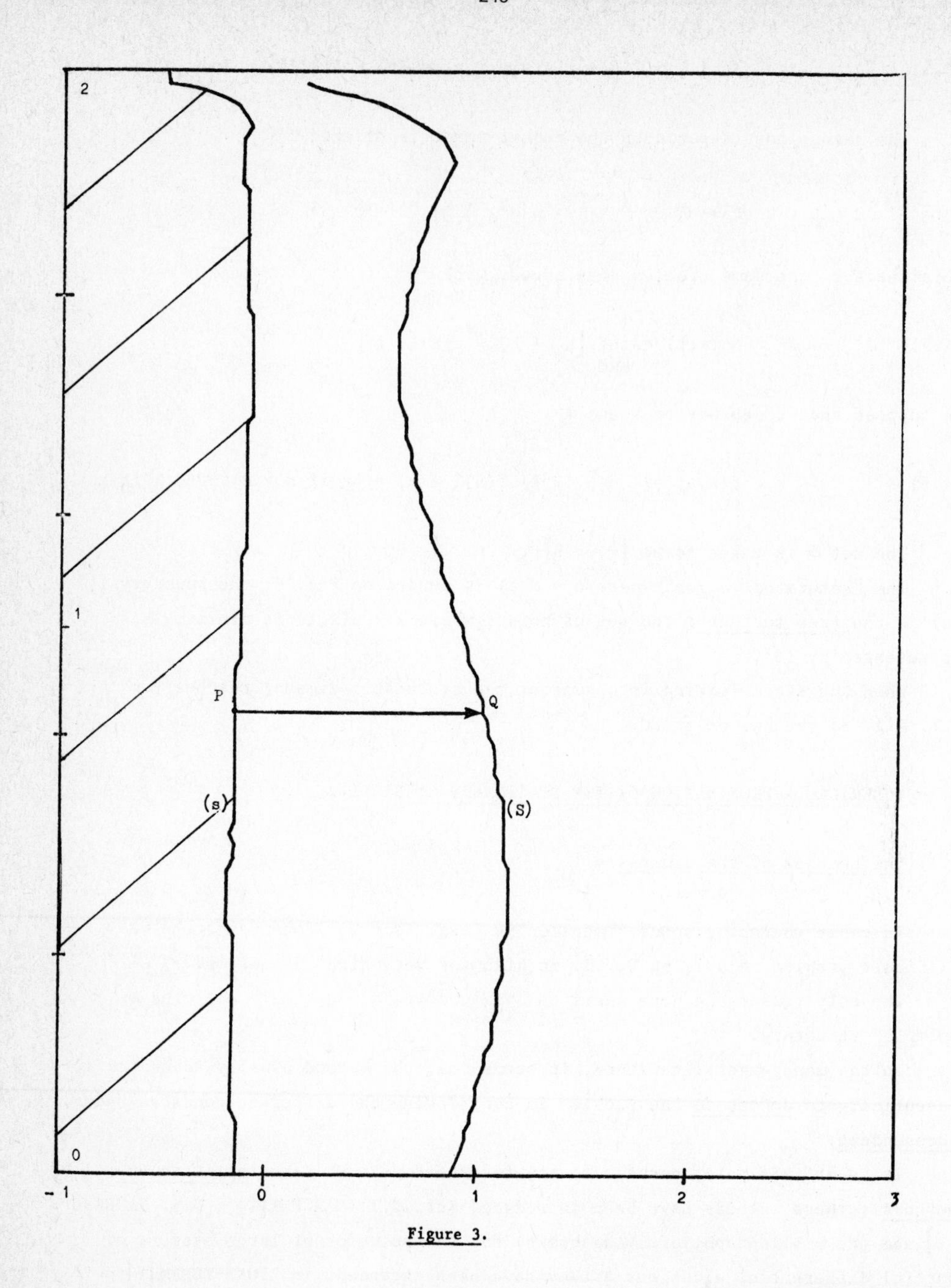

Figure 3.

5.2. Decomposition methods.

We follow here unpublished work of BENSOUSSAN, MAURIN, TEMAM and the A. In order to simplify the exposition (but what follows is entirely general), we suppose that $n = 2$ and that

$$f(x) = f_1(x_1) + f_2(x_2) . \tag{5.1}$$

We consider

$$\left| \begin{array}{l} a(u,v) = a_1(u,v) + a_2(u,v), \\ a_i(u,v) = \dfrac{\sigma_1^2}{2} \displaystyle\int_{\mathcal{O}} \frac{\partial u}{\partial x_i} \frac{\partial v}{\partial x_i} dx + \mu_i \int_{\mathcal{O}} \frac{\partial u}{\partial x_i} v \, dx + \alpha_i \int_{\mathcal{O}} uv \, dx, \\ i = 1,2, \ \alpha_1 + \alpha_2 = \alpha . \end{array} \right. \tag{5.2}$$

We want to solve the stationary problem (2.9).

We introduce $u^o = \max \{\frac{1}{\alpha_1} \sup f_1, \frac{1}{\alpha_2} \sup f_2\}$, and we define u^1 in three steps. The first two steps are done in parallel and are decomposed if $\mathcal{O} = \mathcal{O}_1 \times \mathcal{O}_2$ (which is not restrictive for the applications) : $u^{1/3}$ and $u^{2/3}$ are given by the solution of the Q.V.I.

$$\left| \begin{array}{l} a_1(u^{1/3}, v-u^{1/3}) + \dfrac{1}{\Delta t} (u^{1/3} - u^o, v-u^{1/3}) \geq (f_1, v-u^{1/3}) \\ \forall \ v \leq M_1(u^{1/3}), \ u^{1/3} \leq M_1(u^{1/3}), \end{array} \right. \tag{5.3}$$

$$\left| \begin{array}{l} a_2(u^{2/3}, v-u^{2/3}) + \dfrac{1}{\Delta t} (u^{2/3} - u^o, v-u^{2/3}) \geq (f_2, v-u^{2/3}) \\ \forall \ v \leq M_2(u^{2/3}), \ u^{2/3} \leq M_2(u^{2/3}). \end{array} \right. \tag{5.4}$$

We notice that in (5.3) (resp. in (5.4)) the variable x_2 (resp. x_1) plays the role of a parameter, so that (5.3) and (5.4) are really unidimensional Q.V.I.

We then define (third step) u^1 by :

$$u^1 = \inf (\tfrac{1}{2} (u^{1/3} + u^{2/3}), M (\tfrac{1}{2} (u^{1/3} + u^{2/3}))). \tag{5.5}$$

We inductively proceed :

$$
(5.6)\quad \begin{cases} a_1(u^{n+1/3}, v-u^{n+1/3}) + \frac{1}{\Delta t}(u^{n+1/3} - u^n, v-u^{n+1/3}) \geqslant (f_1, v-u^{n+1/3}) \\ \forall\, v \leqslant M_1(u^{n+1/3}),\ u^{n+1/3} \leqslant M_1(u^{n+1/3}), \end{cases}
$$

$$
(5.7)\quad \begin{cases} a_2(u^{n+2/3}, v-u^{n+2/3}) + \frac{1}{\Delta t}(u^{n+2/3} - u^n, v-u^{n+2/3}) \geqslant (f_2, v-u^{n+2/3}), \\ \forall\, v \leqslant M_2(u^{n+2/3}),\ u^{n+2/3} \leqslant M_2(u^{n+2/3}), \end{cases}
$$

$$
(5.8)\quad u^{n+1} = \inf\left(\tfrac{1}{2}(u^{n+1/3} + u^{n+2/3}),\ M\left(\tfrac{1}{2}(u^{n+1/3} + u^{n+2/3})\right)\right).
$$

One can prove that :

$$
\begin{aligned}
&u^o \geqslant u^{1/3} \geqslant u^{1+1/3} \geqslant \dots \geqslant u^{n-1+1/3} \geqslant u^{n+1/3} \dots \geqslant 0, \\
&u^o \geqslant u^{2/3} \geqslant u^{1+2/3} \geqslant \dots \geqslant u^{n-1+2/3} \geqslant u^{n+2/3} \dots \geqslant 0, \\
&u^o \geqslant u^1 \geqslant u^2 \geqslant \dots \geqslant u^{n-1} \geqslant u^n \geqslant \dots \geqslant 0.
\end{aligned}
$$

<u>It seems highly probable</u> (but it has not been proved) and it is verified on the numerical experiments that

$$
(5.9)\quad u^n \downarrow u.
$$

<u>Remark 5.1</u>

In (5.3) ... (5.7), one can replace $\frac{1}{\Delta t}$ by a parameter ρ_n depending on the order of the iteration; the optimal choice of ρ_n is not known.

<u>Remark 5.2.</u>

If in (5.3) (resp. (5.4)), one replaces $M_1(u^{1/3})$ (resp. $M_2(u^{2/3})$) by $M_1(u^o)$ (resp. $M_2(u^o)$), one obtains V.I. instead of Q.V.I.

The same remark applies when replacing in (5.6) $M_1(u^{n+1/3})$ by $M_1(u^n)$ and in (5.7) $M_2(u^{n+2/3})$ by $M_2(u^n)$.

The numerical experiments which have been made (MAURIN [1]) are satisfactory by both methods.

6. - <u>PROBLEMS WITH DELAYS</u>

Suppose now that when one places an order, the delivery is made **after** a fixed delay h and let us assume that, <u>while expecting a delivery, we do not place new orders</u> (1).

---

(1) Other cases are considered in a series of notes of BENSOUSSAN and the A. to appear.

Then u(x,t) being still defined by (1.12), it is characterized by the Q.V.I.[1]

$$- (\frac{\partial u}{\partial t}, v-u(t)) + a(u(t)), v-u(t)) \geq (f,v-u(t)) \quad \forall v \leq Mu(t-h), \tag{6.1}$$

$$u(t) \leq Mu(t-h), \tag{6.2}$$

$$u(T) = 0. \tag{6.3}$$

It is not difficult to introduce semi-discrete schemes analogous to those introduced in Section 4.

An example of this kind, arising in connection with the optimal use of dams is now under study; cf. BRETON and LEGUAY[1], LEGUAY[1]; one is lead to a system of four Q.V.I. of evolution, with a fixed delay : the numerical results are reported in BRETON and LEGUAY, loc. cit.

7. - A REMARK

It is also possible to apply the methods of Q.V.I. to problems of optimal control with L^1 constraints on the control variable, as those considered by other methods in Cernousko[1], Bratous and Cernousko[1].

----

(2) We write u(t) for $x \to u(x,t)$.

BIBLIOGRAPHY

C. BAIOCCHI [1] C.R. Acad. Sc. Paris, 278, (1974).

C. BAIOCCHI and E. MAGENES [1] These proceedings.

A. BENSOUSSAN and J.L. LIONS [1] C.R. Acad. Sc. Paris :
1) 276 (1973), pp. 1189-1193 ;
2) 276 (1973), pp. 1333-1337 ;
3) 278 (1974), pp. 675-679 ;
4) 278 (1974),
5) 278 (1974).

[2] Lecture in Grenoble. To appear in Springer Lectures Notes.

[3] Applied Mathematics and Optimization, Int. J. (1974)

[4] Ouspechi Mat. Nauk (Dedicated to Prof. Petrowski) (1973).

[5] Book in preparation.

A. BENSOUSSAN, M. GOURSAT and J.L. LIONS [1] C.R. Acad. Sc. Paris, 276 (1973), pp. 1279-1283.

A. BENSOUSSAN, J.L. LIONS and R. TEMAM [1] Cahier IRIA n° 11 June 1972, pp. 5-189.

A. BENSOUSSAN, M. GOURSAT and G. MAAREK [1] Report LABORIA, 1974.

A.S. BRATOUS and F.L. CERNOUSKO [1] Journal of Num. Math. and Math. Physics 14 (1974), pp. 68-78.

A. BRETON and C. LEGUAY [1] Report LABORIA, 1974.

J. CEA and R. GLOWINSKI [1] Méthodes numériques pour l'écoulement laminaire d'un fluide rigide viscoplastique incompressible. Int. J. Comp. Mech. Vol B (1973) pp. 225-255.

F.L. CERNOUSKO [1] Applied Math. and Mech. 35 (1971), 333-342.

Ph. CIARLET and P.A. RAVIART [1] Comp. Methods in Appl. Mech and Eng. 2 (1973), 17-31

R. GLOWINSKI, J.L. LIONS and R. TREMOLIERES [1] Book, Dunod. Paris (1974).

M. GOURSAT [1] Report LABORIA, 1974.

M. JOLY - U. MOSCO [1] C.R. Acad. Sc. Paris, 1974.

Th. LAETSCH [1] To appear.

C. LEGUAY [1] Report LABORIA, 1974.

J.L. LIONS and G. STAMPACCHIA [1] Comm. Pure and Applied Math. XX (1967) pp. 493-519.

J.L. LIONS and R. TEMAM [1] C.R. Acad. Sc. Paris, 263 (1966), pp. 563-565.

G.I. MARCHUK [1] Lecture International Congress of Mathematics, Nice, 1970;

[2] Numerical Methods in Meteorology, Leningrad 1967 (Translation in French, A. Colin, Paris, 1969).

S. MAURIN [1] Report LABORIA, 1974.

F. MIGNOT [1] C.R. Acad. Sc. Paris, 1974.

U. MOSCO and G. STRANG [1] One-sided approximation and variational inequalities.

S.L. SOBOLEV [1] Applications of functional analysis to Mathematical Physics. Leningrad, 1950.

G. STRANG [1] Proceedings of the Conference on Numerical Analysis, Dundee, July 3-7, 1973.

L. TARTAR [1] C.R. Acad. Sc. Paris, 278 (1974).

A. VEINOTT [1] Management Science. 12(11) (1966), pp. 745-777.

N.N. YANENKO [1] Fractional steps Methods. Translation in French, A. Colin (1968).

SATISFICING

by

Roy Radner

Professor of Economics and of Statistics[1]
University of California, Berkeley, California, USA

1. INTRODUCTION AND MOTIVATION

As decision theorists have succeeded in extending their analyses into new domains, and have aspired to new levels of both realism and rigor, they have attempted to apply the rationality postulate to more and more complicated decision problems. In particular, decision theorists have become more concerned with the complexities associated with time, uncertainty, and interpersonal conflict and cooperation; and advances in mathematical theories of optimization, statistical decision-making, and games have provided new concepts and tools for the study of rational behavior in the face of such complexities.

Nevertheless, the very success and expansion of these theories have brought into sharper focus a deep problem for the widespread application of the rationality postulate in decision theory. It is now clear that specialists are far from finding "optimal solutions" to such restricted problems as (1) the management of a network of warehouses under general conditions of uncertain demand, (2) winning a game of chess, or (3) administering a department of mathematics. It is probably not good positive theory to take very seriously an assumption that anyone behaves according to a sequential strategy that maximizes an expected lifetime (or infinite horizon) utility, nor is it good advice to a manager to recommend adoption of the solution of an optimization problem that there is no prospect of solving in the next hundred years.

In other words, decision theory is facing more and more clearly the problem

---

[1]This paper is based on research supported by the National Science Foundation, USA.

of the limits of rationality. I am not speaking here simply of what is often described as the cost of information, but rather of the limited capacities of humans (and machines) for imagination and computation. These limits create theoretical problems on at least two levels. First, there is the profound logical or philosophical problem of defining what one means by "rationality" in the presence of such limits;[2] I shall not discuss this problem here. Second, there is the problem of describing, in terms amenable to theoretical analysis, the different ways humans do behave in complex decision-making situations, and of deducing the consequences of different modes of behavior.

If we are not to discard entirely the rationality postulate in economic theory, then we must elaborate more sophisticated and empirically relevant concepts of rational behavior, which nevertheless retain the important insights provided by the notion of "economic man." Simon has used the term bounded rationality to describe such behavior.[3] I shall not attempt here to give a precise definition of bounded rationality. However, three aspects of bounded rationality do seem important for decision theory: (1) existence of goals, (2) search for improvement, and (3) long-run success.

It is no doubt useful to explain much of economic behavior in terms of "goals" or "motives," and normative economics would appear to be meaningless without reference to goals. On the other hand, an individual economic agent may have "conflicting" goals, and it may be bad psychology in many instances to assume that these conflicts are resolved in terms of a single transitive preference ordering. Such conflicts may be "resolved" in a dynamic way by various mechanisms for switching attention and effort, with results that do not appear to be transitive. (There are, perhaps, useful analogies between individuals with conflicting goals and groups of individuals with conflicting interests.) Also,

---

[2]See, for example, Savage, 1954, pp. 8-17, 59, 83, and Marschak and Radner, 1972, pp. 314-317.

[3]Simon's description is somewhat more general. "Theories that incorporate constraints on the information-processing capacities of the actor may be called theories of bounded rationality." (See H. A. Simon, Ch. 8 of McGuire and Radner, 1972; see, also, Simon, 1959.)

the set of goals may be endogenous, so that, through time, some goals may be dropped and others added to the list.

Even if the theorist draws back from assuming that economic agents behave according to optimal lifetime strategies, it is no doubt useful to postulate that they search for improvements, at least from time to time, and that they take advantage of perceived improvements. How, and under what circumstances, agents search for improvements, and how these improvements are perceived, is, of course, an important subject of study. If repeated improvements can be made in the solution of the same problem, then we have a situation of "expanding rationality." On the other hand, an environment that changes at unpredictable times and in unpredictable directions may make past improvements obsolete, so that the individual is engaged in a race between improvement and obsolescence.

A strategy of search may itself be the object of an improvement effort (as in the planning of research and development), but this leads to a "regression" in the model of decision-making; one eventually reaches a level of behavior at which it is no longer fruitful to assume that the search for improvement is itself being conducted "optimally."

The notion of "adjustment," as it has commonly been used in economic theory, is in the spirit of bounded rationality in the following sense. At a given date the economic agent adopts a particular action (or strategy) that is optimal with respect to the agent's formulation of the decision problem and the agent's "expectations." At the next date, the agent receives new information, which causes him to revise his expectations in _a way that was not anticipated at the previous date_, or even causes him to revise his formulation of the decision problem. This revision of expectations or of problem formulation is to be distinguished from the behavior of a Bayesian statistician with an optimal sequential decision rule, who periodically revises his _a posteriori_ probability distribution on the states of the environment in response to new information, according to a well-defined and completely anticipated (optimal) transformation.

In a similar spirit, a realistic treatment of the search for improvement in a theory of bounded rationality would not follow the present lines of development of the theory of optimal search.[4] Optimal search theory began with a few interesting theorems showing that for some simple search problems the optimal policies could be described in terms of "aspiration levels" and "satisficing." To take a well-known example, suppose that one is searching for larger values in a sequence of independent and identically distributed random variables (with known probability distribution), but there is a constant cost per observation. If one's objective is to maximize the expected value of the difference between the largest value observed and the total cost of observation, then the optimal sequential stopping rule is characterized by an "aspiration level," i.e., there exists a number, the aspiration level, such that one stops searching as soon as one observes a value that is greater than or equal to the aspiration level. However, there are fairly simple (and plausible) examples of search problems in which the optimal policy cannot be characterized by an aspiration level, or even by a rule that determines the aspiration level at each date as a function of the past history of observations. Rather than attempt to characterize optimal search in a greater variety of more and more complicated problems, the theorist following the approach of bounded rationality would observe that aspiration-level and satisficing behavior is common, even in complicated problems, and would endeavor to understand the implications of such behavior in a variety of situations.

In this lecture I shall explore the consequences of satisficing in the context of a simple model of the allocation of an agent's effort to the search for improvement in one or more activities. For any fixed allocation of effort, the performance of each activity is assumed to be a random walk, or more generally, a semimartingale. The expected rate of change per unit time for each activity depends on the effort allocated to it. This expected rate of change is positive if all of the agent's effort is allocated to the activity, and negative if none is. A behavior is a rule that determines, at each date, the current allocation of effort among the activities as a function of the past history of performance

[4]See, for example, MacQueen, 1964, and Rothschild, 1973.

up to that date.

In such a model, performance of the several activities will typically not approach a steady state, even in a stochastic sense, except for very special values of the parameters. In these notes, I examine "long-run success" (i.e., asymptotic performance) with respect to two criteria: (1) the probability of survival, i.e., the probability that performance on one or more activities never falls below certain prescribed levels, and (2) the long-run average rate of growth per unit time.

2. SINGLE OBJECTIVE

2.1. General Formulation of a Satisficing Process

I start with a general formulation of a process of intermittent search for improvement with respect to a single objective. Consider a basic probability space, (X, F, P), where F is a sigma-field of subsets of X, and P is a probability measure on F. Let (F_t), $t = 0,1,2,\ldots$, be an increasing sequence of subfields of F; F_t is to be interpreted as the set of observable events through date t. Let $\{U(t)\}$ be a corresponding sequence of integer-valued random variables on X, such that $U(t)$ is F_t-measurable; $U(t)$ will be called the _performance at t_, relative to a given single objective. Finally, let (T_n), $n = 0,1,2,\ldots$, be a nondecreasing sequence of random times, possibly taking on the value plus infinity, such that $T_n < T_{n+1}$ if T_n is finite; for n odd, T_n is to be interpreted as a date at which a period of search for improvement begins, and T_{n+1} as the date at which that period ends. (A random time T is an integer-valued random variable, possibly equal to plus infinity, such that the event $(T = t)$ is F_t-measurable.) Take $T_o = 0$.

An interval $(T_n \leq t < T_{n+1})$ will be called a _search period_ if n is odd and a _rest period_ if n is even. To capture the idea of intermittent search for improvement I assume: for $T_n \leq t < T_{n+1}$,

$$(2.1)\qquad \begin{aligned} E[U(t+1) \mid F_t] &\geq U(t), \quad \text{if } n \text{ is odd,} \\ E[U(t+1) \mid F_t] &\leq U(t), \quad \text{if } n \text{ is even.} \end{aligned}$$

In other words, $U(t)$ is a submartingale during the search periods, and a supermartingale during the rest periods.

To capture the idea of "satisficing," let $\{S(t)\}$ be a sequence of random variables such that $S(t)$ is F_t-measurable; $S(t)$ is to be interpreted as the "satisfactory level of performance" at date t. The random times T_n are determined by: for n even,

$$(2.2)\qquad \begin{aligned} &T_{n+1} \text{ is the first } t > T_n \text{ such that } U(t) < S(t), \\ &T_{n+2} \text{ is the first } t > T_{n+1} \text{ such that } U(t) \geq S(t); \end{aligned}$$

this is qualified by the convention that, for any n, if T_n is infinite, then so is T_m for every $m > n$.

In the next sections, more specific assumptions will be made about the processes $U(t)$ and $S(t)$.

2.2. A Favorable Satisficing Process

Let $Z(t)$ be the successive increments of the process $U(t)$; thus $Z(t+1) = U(t+1) - U(t)$. Let ξ, η, and β be given positive numbers. For $T_n \leq t < T_{n+1}$, assume:

$$(2.3)\qquad \begin{aligned} &\text{(i) for n even (rest),} \\ &\quad E[Z(t+1) \mid F_t] \leq -\xi , \\ &\quad S(t) = U(T_n) - \beta + 1; \\ &\text{(ii) for n odd (search),} \\ &\quad E[Z(t+1) \mid F_t] \geq \eta , \\ &\quad S(t) = U(T_{n-1}) . \end{aligned}$$

Thus, if a search period ends with $U(T_n) = u$, then the next search period begins as soon as $U(t)$ reaches or falls below $(u - \beta)$, and ends thereafter as soon as $U(t)$ reaches or exceeds u again. During such a search period, u may be called the "aspiration level." For technical reasons, assume further that there is a number b such that

$$(2.3;\ iii)\qquad | Z(t) | \leq b, \quad \text{for all } t.$$

Using an inequality of Freedman, 1973, one can prove:

Proposition 1. The random times T_n have finite expectations; indeed, there are numbers μ_0 and μ_1 such that, for all n,

$$(2.4) \qquad E[T_{n+1} - T_n \mid F_{T_n}] \leq \begin{cases} \mu_0 , & \text{if n is even,} \\ \mu_1 , & \text{if n is odd.} \end{cases}$$

For any nonnegative integer k, let $V_k = U(T_{2k})$. The V_k are the performance levels at which successive search periods end, and each V_k is the aspiration level for the next succeeding search period. It is clear that the V_k form a non-decreasing sequence. If, during search, performance can (with positive probability) increase by more than one unit at a time, then V_k will actually increase from time to time. I shall say that the process is strictly favorable if there is a (strictly) positive number υ such that, for every k,

$$(2.5) \qquad E[V_{k+1} \mid F_{2k}] \geq V_k + \upsilon .$$

Again using Freedman, 1973, one can prove:

Proposition 2. If the process is strictly favorable, then

$$\liminf_{k\to\infty} \frac{V_k}{k} \geq \upsilon \text{, almost surely.}$$

2.3. Random-Walk Search and Rest

In the model of Section 2.2, assume further that, during rest the increments $Z(t+1)$ are independent and identically distributed, with mean $-\xi$, and during search they are also independent and identically distributed, with mean η . In other words, during rest the performance process is a random walk with negative drift, and during search it is a random walk with positive drift. To minimize technical complications, assume further that these random walks are integer-valued and aperiodic.

Let $a(t) = 1$ during search, and 0 during rest. The process

$\{a(t-1), U(t), S(t)\}$ is a Markov chain with countably many states and a single class. Let $D(t) = U(t) - S(t)$. The process $\{a(t-1), D(t)\}$ is also Markovian, with a single class.

Proposition 3. The process $\{a(t-1), D(t)\}$ is positive recurrent. Let $\bar{a}$ denote the long-run frequency with which $a(t) = 1$; and let $\bar{\zeta} = \bar{a}\eta - (1-\bar{a})\xi$; then, almost surely,

$$\lim_{t\to\infty} \frac{U(t)}{t} = \bar{\zeta} . \tag{2.6}$$

If $\bar{\zeta} > 0$, then the process is strictly favorable, in the sense of Proposition 2. In the present case, the sequence (V_k) is a random walk. However, the sequence $U(t)$ is not a random walk, nor even a submartingale. Nevertheless, one can prove for $\{U(t)\}$ the following result.

Proposition 4. If the process is strictly favorable ($\bar{\zeta} > 0$), then there exist positive numbers H and K such that, if $U(0) \equiv u > \beta + b$, then

$$\text{Prob } \{U(t) \leq 0 \text{ for some } t|F_0\} \leq He^{-Ku} .$$

If $\bar{\zeta} = 0$, then the above probability is 1.

Let us say that the process survives if the performance $U(t)$ remains positive for all t. Taken together, Propositions 3 and 4 assert that, for a strictly favorable process, with random-walk rest and search, in the long-run performance increases at a positive average rate per unit time, and the probability of survival approaches unity exponentially as a function of the initial performance level, $U(0)$. This implies further that, if the process has "survived" for a long time, then the performance level is probably very high, and therefore the conditional probability of subsequent survival is close to unity. If the process is not strictly favorable, then the probability of survival is zero.

3. "PUTTING OUT FIRES"

A manager usually supervises more than one activity. For any given level of search effort per unit time, the opportunity cost of searching for improvement in one activity is the neglect of others. Consider a stochastic process $\{U(t),F_t\}$, as in the first paragraph of Section 2.1, but let $U(t)$ be a vector with coordinates $U_i(t)$, $i = 1,\ldots,I$, where $U_i(t)$ is a measure of performance of activity i at date t. An allocation behavior is a sequence, $\{a(t)\}$, where $a(t)$ is an F_t-measurable random vector with coordinates $a_i(t)$, $i = 1,\ldots,I$, such that, for any date t, exactly one coordinate of $a(t)$ is 1, and the other coordinates are 0. If $a_i(t) = 1$, this is interpreted as a search for improvement in activity i at date t.

Concerning the process $U(t)$, I shall make assumptions analogous to those of Section 2.3. As before, let

$$Z(t + 1) = U(t + 1) - U(t). \tag{3.1}$$

For the conditional distribution of $Z(t + 1)$, given F_t, assume:

(3.2a) The distribution of $Z(t + 1)$ depends only on $a(t)$.

(3.2b) $EZ_i(t + 1) = a_i(t)\eta_i - [1 - a_i(t)]\xi_i$, where ξ_i and η_i are given positive parameters.

(3.2c) $\text{Var } Z_i(t + 1) = s_i(a_i[t])$, where $s_i(0)$ and $s_i(1)$ are given positive parameters.

(3.2d) The coordinates of $Z(t + 1)$ are mutually independent.

To minimize technical complications, I also assume:

(3.2e) The coordinates of $Z(t + 1)$ are integer-valued, uniformly bounded by b, and aperiodic.[5]

A common managerial behavior is to pay attention only to those activities that are giving the most trouble; this is colloquially called "putting out fires." Formally, let

$$M(t) = \min_i U_i(t) , \tag{3.3}$$

and define putting out fires by

---

[5] A random variable is aperiodic if 1 is the greatest common divisor of its support.

$$(3.4)\quad \begin{array}{ll} \text{(a)} & \text{if } U_i(t) > M(t)\text{, then } a_i(t) = 0; \\ \text{(b)} & \text{if } U_i(t) = M(t) \text{ and } a_i(t-1) = 1\text{, then } a_i(t) = 1; \\ \text{(c)} & \text{if neither (a) nor (b) holds, then } a_i(t) = 1 \text{ for} \\ & i = \text{the smallest } j \text{ such that } U_j(t) = M(t). \end{array}$$

To compare putting out fires with the satisficing model of Section 2, roughly speaking, the satisfactory level of performance of any activity is here defined to be equal to $M(t) + 1$.

To describe the properties of the performance process under putting out fires, I first define

$$\bar{\zeta} = \Big(1 - \sum_i \frac{\xi_i}{\eta_i + \xi_i}\Big)\Big/\Big(\sum_i \frac{1}{\eta_i + \xi_i}\Big) . \tag{3.5}$$

$$\bar{a}_i = \frac{\bar{\zeta} + \xi_i}{\eta_i + \xi_i}, \quad i = 1,\ldots,I. \tag{3.6}$$

If the limit, as t increases, of $U_i(t)/t$ exists, I shall call this limit the <u>rate of growth of activity i</u>. If $M(t) > 0$ for all t, I shall say that the performance process <u>survives</u>. Define $W(t) = U(t) - M(t)$.

<u>Proposition 3.1</u>. Under putting-out-fires behavior, if $\bar{\zeta} > 0$, then the Markov chain $\{a(t-1), W(t)\}$ is ergodic, and for each activity i,

(a) the long-run frequency with which $a_i(t) = 1$ is almost surely equal to $\bar{a}_i$;

(b) the rate of growth of $U_i(t)$ is almost surely $\bar{\zeta}$ (the same for all activities);

furthermore, if $M(0) > 0$; and if, for every i, $\text{Prob}\{Z_i(t+1) \geq 0 \mid a_i(t) = 0\} > 0$, then

(c) the probability of survival is positive.

In the context of the model defined by (3.2a)-(3.2e) one could explore other allocation behaviors, but the limitation of space does not permit that here. I mention, however, that a necessary and sufficient condition that there exist <u>any</u>

allocation behavior with positive probability of survival is $\bar{\zeta} > 0$. In other words, <u>survival is possible with positive probability if and only if it is possible with putting out fires</u>.

In the special case of two activities (I = 2) the conclusions (a) and (b) of Proposition 3.1 are true also if $\bar{\zeta} \le 0$.

For proofs of the facts mentioned in this section and for an analysis of other allocation behaviors see Radner and Rothschild (1974).

4. BIBLIOGRAPHIC NOTE

The material of this lecture is adapted from Radner, 1973 and 1974, and Radner and Rothschild, 1974. Satisficing plays an important role in the models of stochastic equilibrium and evolution in Winter, 1971, Nelson and Winter, 1972, and Nelson, Winter and Schuette, 1973. Stochastic search for improvement is a key element of a decentralized resource allocation process that converges to Pareto optimal allocations in the presence of nonconvexities, as described in Hurwicz, Radner, and Reiter, 1973. Related stochastic adjustment processes for reaching the core of a game are described in Green, 1970, and in Neuefeind, 1971.

REFERENCES

Freedman, D., "Another Note on the Borel-Cantelli Lemma and the Strong Law," <u>Ann. Probability</u> 1 (1973), <u>no</u>. 6, 910-925.

Green, J., "Some Aspects of the Use of the Core as a Solution Concept in Economic Theory," unpublished thesis, Rochester, 1970; also, <u>Econometrica</u>, forthcoming.

Hurwicz, L., R. Radner, and S. Reiter, "A Stochastic Decentralized Resource Allocation Process," Technical Report No. 20, Center for Research in Management Science, University of California, Berkeley, 1973; also, <u>Econometrica</u>, forthcoming.

MacQueen, J. B., "Optimal Policies for a Class of Search and Evaluation Problems" <u>Management Sci</u>. 10 (1964), <u>no</u>. 4, 746-759.

Marschak, J. and R. Radner, <u>Economic Theory of Teams</u>, Yale University Press,

New Haven, 1972.

McGuire, C. B. and R. Radner (eds.), Decision and Organization, North-Holland Publishing Company, Amsterdam, 1972.

Nelson, R. R. and S. G. Winter, "Toward an Evolutionary Theory of Economic Capabilities," Discussion Paper No. 44, Institute of Public Policy Studies, University of Michigan at Ann Arbor, 1972.

Nelson, R. R., S. G. Winter, and H. L. Schuette, "Technical Change in an Evolutionary Model," Discussion Paper No. 45, Institute of Public Policy Studies, University of Michigan at Ann Arbor, 1973.

Neuefeind, W., "A Tatonnement Process for N-Person Games," CORE Discussion Paper No. 7136, 1971.

Radner, R., "Aspiration, Bounded Rationality, and Control," Presidential Address, Oslo Meeting of the Econometric Society, August 1973 (unpublished manuscript).

Radner, R., "A Behavioral Model of Cost Reduction," Technical Report No. OW-2, Center for Research in Management Science, University of California, Berkeley, 1974; to appear in the Bell J. Econ. and Management Sci., Spring, 1975.

Radner, R. and M. Rothschild, "Notes on the Allocation of Effort," Technical Report No. OW-1, Center for Research in Management Science, University of California, Berkeley, 1974.

Rothschild, M., "Searching for the Lowest Price when the Distribution of Prices is Unknown," J. Political Economy 82 (1974), 689-712.

Savage, L. J., The Foundations of Statistics, Wiley, New York, 1954.

Simon, H. A., "Theories of Decision-Making in Economics and Behavioral Science," Amer. Econ. Rev. 49 (1959), no. 3, 253-283.

Simon, H. A., "Theories of Bounded Rationality," Ch. 8 in C. B. McGuire and R. Radner (eds.), Decision and Organization, North-Holland Publishing Company, Amsterdam, 1972.

Winter, S. G., "Satisficing, Selection, and the Innovating Remnant," Quart. J. Econ. 85 (1971), 237-261.

ON APPROXIMATE SOLUTION OF THE PROBLEM WITH POINT AND BOUNDARY CONTROL

V.V. Alifyorov, Y. Kaimkulov
Automation Institute of Kirghiz SSR
Academy of Sciences, Frunze, USSR

Of obvious interest for practical applications is the investig tion of the problems, where the process control is realized by means of the parameters effecting the plant both at the boundary and at its given points which do not belong to the boundary [I]. In this connectic we consider below an optimal control problem of heat transmission process in a uniform thin bar of a final length. Control is realized by means of a boundary and point parameters. We have found necessary and sufficient optimality condition, which makes possible to write down Fredholm's linear integral equations system of the second order relati to the desired controls. It is proved that this system has a unique solution whose determination is brought to the investigation of the infinite system of linear algebraic equations. We study the question of finding an approximate solution for the problem under consideration an we give the estimates of the errors admissible in this case.

Setting of the problem. Optimality condition

Let function $u(t,x)$, characterizing the controllable plant condi tion in the domain $Q=\{0<x<1,\ 0<t\leq T\}$ satisfy the equation

$$u_t - u_{xx} = p_1(t)\delta(x-x'), \qquad (I)$$

while at the boundary Q let it satisfy additional conditions

$$u(0,x) = u_x(t,0) = 0\ , \quad u_x(t,1) = \alpha[p_2(t) - u(t,1)]\ , \quad \alpha = const>0 \qquad (2)$$

where $p(t)=\{p_1(t), p_2(t)\}$ is controlling vector-function from the space $L_2^2(0,T)$ with the norm

$$\left(\int_0^T [p_1^2(t) + p_2^2(t)]dt\right)^{1/2} < \infty,$$

and $x' \in (0,1)$ is the control action $p_1(t)$ point of application.

In the following each function $p(t) \in L_2^2(0,T)$ will be called an admissible control.

With the fixed admissible control $p = p(t)$ there exists the unique generalized solution $u(t,x) \in W_2^{0,1}(\mathcal{Y})$ of the boundary problem (I)-(2) which satisfies the identity

$$\int_0^1 u(t_1,x)\Phi(t_1,x)dx - \iint_{\mathcal{Y}} [u\Phi_t - u_x\Phi_x]dx\,dt + \alpha\int_0^{t_1} u(t,1)\Phi(t,1)dt =$$

$$= \int_0^1 [p_1(t)\Phi(t,x') + \alpha p_2(t)\Phi(t,1)]dt \quad \forall \Phi(t,x) \in W_2^{1,1}(\mathcal{Y}), \ \mathcal{Y} = \{0 \le x \le 1, 0 \le t \le t_1\}$$

Here t_1 is an arbitrarily fixed time moment from $(0,T)$. This solution is determined by formula [2]

$$u(t,x) = \sum_{n=0}^{\infty} \frac{\cos\lambda_n x}{\omega_n} \int_0^t [p_1(\tau)\cos\lambda_n x' + \alpha p_2(\tau)\cos\lambda_n] e^{\lambda_n^2(\tau - t)} d\tau, \quad (3)$$

where $\{\cos\lambda_n x\}$, λ_n , ω_n are proper functions, values and normalizing multiplier of the boundary problem associated with (I)-(2).

We have to find such admissible control $p^0(t)$, which together with the corresponding generalized solution $u^0(t,x)$ of problem (I)-(2) gives the least possible value to the functional

$$S[p] = \int_0^1 [u(T,x) - u_0(x)]^2 dx + \beta\int_0^T [p_1^2(t) + p_2^2(t)]dt, \ \beta = const > 0, \quad (4)$$

where $u_0(x)$ is the given function from $L_2(0,1)$. The control minimizing S we shall call optimal.

The existence of the unique control optimal on S follows from J.L.Lions' [I] results. Therefore we limit ourselves by its direct estimation. For this purpose we write out optimality conditions.

<u>Theorem 1.</u> In order to make admissible control $p^0(t)$ optimal by S it is necessary and sufficient to fulfil the condition

$$H(v(t,x'), v(t,1), p^0(t)) (=) \sup_p H(v(t,x'), v(t,1), p), \quad (5)$$

where

$$H(v(t,x'), v(t,1), p) = v(t,x')p_1 + \alpha v(t,1)p_2 - \beta(p_1^2 + p_2^2)$$

Symbol (=) denotes equality which holds good almost everywhere while the function $v(t,x)$ belongs to the space $W_2^{1,1}(Q)$ and satisfies the correlations

$$\iint_Q (\varphi v_t - v_x \varphi_x)\,dx\,dt = \alpha \int_0^T v(t,1)\varphi(t,1)\,dt \quad \forall \varphi \in W_2^{0,1}(Q),$$

$$\lim_{t\to T} \int_0^1 \{v(t,x) + 2[u(t,x) - u_0(x)]\}^2 dx = 0.$$

Such function as well as in the case of boundary problem (I)-(2) exists and in a unique way is determined by the formula

$$v(t,x) = 2\sum_{n=0}^{\infty} \frac{\cos\lambda_n x}{\omega_n} \left\{ u_n^o \omega_n - \int_0^T [p_1(\tau)\cos\lambda_n x' + \alpha p_2(\tau)\cos\lambda_n] e^{\lambda_n^2(\tau-T)} d\tau \right\} e^{\lambda_n^2(t-T)}, \tag{6}$$

where u_n^o are Fourier's coefficients of function $u_0(x)$ by the system $\{\cos\lambda_n x\}$.

Provided that there exists a generalized solution of the boundary problem (I)-(2) and the function $v(t,x) \in W_2^{1,1}(Q)$ theorem 1 may be proved by calculating the augmentation of functional S [2].

Since different vector-functions from $L_2^2(0,T)$ are admissible controls, then condition (5) enables us to determine the optimal control by correlations

$$p_1^o(t) = \frac{1}{2\beta} v(t,x'), \qquad p_2^o(t) = \frac{\alpha}{2\beta} v(t,1). \tag{7}$$

Note 1.1. In a similar way may be studied the cases of problems with linear boundary conditions which differ from (2).

Note 1.2. No principal difficulty is caused by the investigation of the case when any finite number of point parameters participate in the process control. In this case correlations (7) are supplemented with ones analogous to the first one from (7).

Optimal control estimation

Here holds good the following

Theorem 2. Optimal control $p^o(t) = \{p_1^o(t), p_2^o(t)\}$ is the unique solution of the integral equations system

$$p^o(t) + \int_0^T K(t,\tau) p^o(\tau)\,d\tau = F(t), \tag{8}$$

where $K(t,\tau)$ is matrix nucleus with elements $K_{ij}(t,\tau)$:

$$K_{11} = \sum_{n=0}^{\infty} \mu_n \cos\lambda_n x' e^{\lambda_n^2(t+\tau-2T)}, \quad K_{12} = K_{21} = \alpha \sum_{n=0} \mu_n \cos\lambda_n \cos\lambda_n x' e^{\lambda_n^2(t+\tau-2T)},$$

$$K_{22} = \alpha^2 \sum_{n=0}^{\infty} \mu_n \cos^2\lambda_n e^{\lambda_n^2(t+\tau-2T)}, \quad F(t) = \{F_1(t), F_2(t)\}, \quad \mu_n = \frac{1}{\beta\omega_n},$$

$$F_1(t)=\frac{1}{\beta}\sum_{n=0}^{\infty}u_n^0\cos\lambda_n x' e^{\lambda_n^2(t-T)},\qquad F_2(t)=\frac{\alpha}{\beta}\sum_{n=0}^{\infty}u_n^0\cos\lambda_n e^{\lambda_n^2(t-T)}.$$

Demonstration. By considering (6) (at $p=p^0(t)$) and (7), we make sure that the optimal control satisfies system (8). Since $\|K\|<\infty$, $\|F\|<\infty$ and $(K\varphi,\varphi)\geqslant 0$*), then according to the integral equations theory, there exists the unique solution for system (8).

In order to find the optimal control $p^0(t)$ we write down system (8) in the form

$$p_1^0(t)=\sum_{n=0}^{\infty}R_n\cos\lambda_n x' e^{\lambda_n^2(t-T)},\qquad p_2^0(t)=\alpha\sum_{n=0}^{\infty}R_n\cos\lambda_n e^{\lambda_n^2(t-T)}, \tag{9}$$

where

$$R_n=\mu_n\Big[u_n^0\omega_n-\cos\lambda_n x'\int_0^T p_1^0(t)e^{\lambda_n^2(t-T)}dt-\alpha\cos\lambda_n\int_0^T p_2^0(t)e^{\lambda_n^2(t-T)}dt\Big].$$

By substituting (9) into (8) and considering the linear independence of the aggregate $\exp\lambda_n^2(t-T)$ we make sure that R_n satisfies the system of linear algebraic equations

$$R_n+\sum_{k=0}^{\infty}K_{nk}R_k=b_n,\quad n=0,1,\ldots, \tag{10}$$

where

$$K_{nk}=\frac{a_{nk}}{\lambda_n^2+\lambda_k^2}\left(1-e^{-(\lambda_n^2+\lambda_k^2)T}\right),\quad a_{nk}=\mu_n(\cos\lambda_n x'\cos\lambda_k x'+\alpha^2\cos\lambda_n\cos\lambda_k),$$

$$b_n=\mu_n\omega_n u_n^0.$$

Equations system (10) has the unique solution with a boundary property in l_2.

Thus, optimal control has the form (9), where R_n is determined in the unique way from system (10).

Assuming in (3) that $p=p^0(t)$, we have

$$u(T,x)=\sum_{n=0}^{\infty}u_n(T)\cos\lambda_n x,\qquad u_n(T)=\beta\sum_{k=0}^{\infty}K_{nk}R_k. \tag{11}$$

Considering (10) and (11) for finding S we obtain the formula

$$S[p^0]=\sum_{n=0}^{\infty}\omega_n\left\{[u_n(T)-u_n^0]^2+[u_n^0-u_n(T)]u_n(T)\right\}=\|u_0(x)\|^2-(u(T,x),u_0(x)).$$

Note 2.1. Analogous results can be obtained when investigating the case when a finite number of pointed parameters participate in the control (see note 1.2.). Here the order of system (8) augments but the properties of nucleous $K(t,x)$ and function $F(t)$ remain unchanged.

---

*) Here and in the following statement $\|\cdot\|$ and $(\cdot\,,\cdot)$ there is a norm and scale product in the space $L_2^2(0,T)$.

Approximate optimal control estimation

It is natural that for algebraic equations system (10) it is impossible to find an accurate solution; therefore, we shall estimate approximately. For this purpose we shall limit ourselves with a finite number of items in all the root determining $K(t,\tau)$ and $F(t)$. As a result we obtain the integral equation

$$p_m^o(t) + \int_0^T K_m(t,\tau) p_m^o(\tau) d\tau = F_m(t). \tag{12}$$

Solution of equation (12) we call m-approximation of optimal control. It is of the form

$$p_{1m}^o(t) = \sum_{n=0}^{m} R_n^m \cos\lambda_n x' e^{\lambda_n^2 (t-T)}, \quad p_{2m}^o(t) = \alpha \sum_{n=0}^{m} R_n^m \cos\lambda_n e^{\lambda_n^2 (t-T)},$$

where R_n^m is determined in the unique way from the equations system

$$R_n^m + \sum_{k=0}^{m} K_{nk} R_k^m = b_n, \qquad n = 0, 1, \dots, m.$$

which corresponds to (10).

Approximate value of S is found by the formula

$$S_m[p_m^o] = \|u_o(x)\|^2 - (u_m(T,x), u_o(x)).$$

Now let us estimate the convergence rate of selected approxima tions to the accurate solution of the problem. From (8) and (12) we ha $p^o - p_m^o = \psi_m - K_m(p^o - p_m^o)$, $\psi_m = F - F_m - (K - K_m)p^o$, where K and K_m are Fredholm's operators in the equations (8) and (12). By multiplying the both parts of the fo equality by $p^o - p_m^o$ with the consideration of non-negativity of opera K_m, we have

$$\|p^o - p_m^o\| \le \|\psi_m\| \tag{13}$$

Similarly we find $\|p^o\| \le \|F\|$. Consequently,

$$\|\psi_m\| \le \|F - F_m\| + \|K - K_m\| \|F\|$$

For estimating the convergence rate of p_m^o to p^o with $m \to \infty$ we estimate the values $\|F - F_m\|$, $\|K - K_m\|$ and $\|F\|$. We have [3]

$$\|F - F_m\| \le \frac{\sqrt{1+\alpha^2}}{\pi\beta} \|u_o(x)\| \left(\sum_{n=m+1}^{\infty} \frac{1}{n^2}\right)^{1/2} = \frac{\sqrt{1+\alpha^2}}{\pi\beta} \|u_o(x)\| \frac{\sqrt{m+2}}{m+1},$$

$$\|K - K_m\| \le \left(\sum_{ij=1}^{2} \int_0^T \int_0^T [K_{ij}(t,\tau) - K_{ij}^m(t,\tau)]^2 d\tau dt\right)^{1/2} \le \frac{1+\alpha^2}{\pi\beta} \frac{m+2}{(m+1)^2},$$

$$\|F\| \leq \frac{\sqrt{1+\alpha^2}}{\beta} \|u_0(x)\| \left(\frac{1}{\lambda^2} + \frac{1}{\pi^2}\sum_{n=1}^{\infty}\frac{1}{n^2}\right)^{1/2} < \frac{\sqrt{1+\alpha^2}}{\beta}\|u_0(x)\|\left(\frac{1}{\lambda_0^2}+\frac{1}{6}\right)^{1/2}$$

Now, considering the obtained inequalities, it follows from (13) hat

$$\|p^0 - p_m^0\| \leq \frac{\sqrt{1+\alpha^2}}{\pi\beta}\|u_0(x)\|\left[\frac{\sqrt{m+2}}{m+1} + \frac{1+\alpha^2}{\beta}\left(\frac{1}{\lambda_0^2}+\frac{1}{6}\right)^{1/2}\frac{m+2}{(m+1)^2}\right].$$

Thus, $\|p^0 - p_m^0\| \to 0$ with $m \to \infty$ not sooner than $\frac{1}{m}$.

For estimating the convergence rate $S_m[p_m^0] \to S[p^0]$ with $m \to \infty$ e have inequality

$$|S[p^0] - S_m[p_m^0]| \leq \mu\left[\|p^0 - p_m^0\|\sum_{n=0}^{\infty}\frac{1}{\lambda^2} + \frac{\|F\|^2}{\pi^2}\frac{m+2}{(m+1)^2}\right], \quad \mu = const > 0.$$

Hence we find that $S_m[p_m^0] \to S[p^0]$ with $m \to \infty$ not sooner than $\frac{1}{m}$.

Note 3.1. The results expounded above represent the generaliza-ion of paper [3].

It should be noted that generally speaking, it seems impossible o essentially improve the estimates obtained by minimization of the unctionals of type (4) by the method expounded above (compare with [3]).

REFERENCES

I. Ж.-Л. ЛИОНС Оптимальное управление системами, описываемыми уравнениями с частными производными. "Мир", М., 1972.

2. А.И. ЕГОРОВ Об условиях оптимальности в одной задаче управ-ения процессом теплопередачи. Ж. вычисл.матем. и матем. физики, т. 12, № 3, 1972.

3. А.И.ЕГОРОВ, Р. РАФАТОВ О приближенном решении одной задачи оптимального управления. Ж. вычисл. матем. и матем. физики, т. 12, № 4, 1972.

A-STABLE METHOD FOR THE SOLUTION OF THE CAUCHY PROBLEM FOR STIFF SYSTEMS OF ORDINARY DIFFERENTIAL EQUATIONS

S.S.ARTEM'EV, G.V.DEMIDOV
Computing Center, Novosibirsk, USSR

For the solution of the Cauchy problem for the system of equations

$$\dot{y} = f(y) \tag{1}$$

there is constructed the Rosenbrock type method accurate to the fifth local order with a single computation of a Jacobian matrix per step of integration. Numerical experiments have shown high efficiency of the proposed method. The following approximation of the exponential function is taken as the basis of the method

$$e^x \approx \varphi_4(x) \equiv 1 + \frac{x}{1-x} - \frac{1}{2}\frac{x^2}{(1-x)^2} + \frac{1}{6}\frac{x^3}{(1-x)^3} + \frac{1}{24}\frac{x^4}{(1-x)^4} \,. \tag{2}$$

From the results of papers [1,2] it immediately follows that

$$|\varphi_4(x)| \leq 1, \quad \text{at} \quad \mathrm{Re}\, x \leq 0 \,. \tag{3}$$

One of the possible versions of the Rosenbrock type formulae based on approximation (2) is of the form

$$y_{n+1} = y_n + \sum_{i=1}^{4} p_i k_i \,, \tag{4}$$

$$k_i = h\,[1 - hf_y]^{-1} f(\eta_i) \,, \tag{5}$$

$$\eta_i = y_n + \sum_{j=1}^{i-1} \beta_{ij} k_j, \quad i = 2,3,4, \qquad (6)$$

$$\eta_1 = y_n,$$

where f_y is the Jacobian matrix of system (1) calculated in the point y_n. Method (4)-(6) is of the fifth local order of accuracy and A-stable provided that the coefficients p_i, β_{ij} satisfy the following system of nonlinear algebraic equations:

$$\sum_{i=1}^{4} p_i = 1, \qquad (7)$$

$$\sum_{i=2}^{4} p_i c_i = -\frac{1}{2}, \qquad (8)$$

$$\sum_{i=2}^{4} p_i c_i^2 = \frac{1}{3}, \qquad (9)$$

$$\sum_{i=2}^{4} p_i c_i^3 = \frac{1}{4}, \qquad (10)$$

$$\beta_{32} c_2 p_3 + (\beta_{42} c_2 + \beta_{43} c_3)\, p_4 = \frac{1}{6}, \qquad (11)$$

$$\beta_{32} c_2^2 p_3 + (\beta_{42} c_2^2 + \beta_{43} c_3^3)\, p_4 = -\frac{1}{4}, \qquad (12)$$

$$\beta_{32} c_2 c_3 p_3 + (\beta_{42} c_2 + \beta_{43} c_3)\, c_4 p_4 = -\frac{5}{24}, \qquad (13)$$

$$\beta_{32}\, \beta_{43} c_2 p_4 = \frac{1}{24}, \qquad (14)$$

$$c_i = \sum_{j=1}^{i-1} \beta_{ij}, \quad i = 2,3,4. \qquad (15)$$

All the solutions (7)-(15) are exhausted by the general solution depending on the two parameters c_2, c_3 and by the singular solution depending on one parameter p_4.

The general solution where $(c_3 \neq c_4)$: c_2, c_3 are parameters, is given by

$$\beta_{32} = \frac{c_3(c_2 - c_3)}{c_2(6 + 4c_2)}, \quad c_4 = -\frac{6 + 7c_2}{8 + 7c_2},$$

$$p_4 = \frac{3 - 4c_2 - c_3(4+6c_2)}{12c_4(c_4 - c_3)(c_4 - c_2)}, \quad p_3 = \frac{1}{c_3(c_3 - c_2)}\left[\frac{1}{3} + \frac{c_2}{2} - p_4c_4(c_4 - c_2)\right],$$

$$p_2 = -\frac{1}{c_2}\left[\frac{1}{2} + c_3p_3 + c_4p_4\right], \quad p_1 = 1 - p_2 - p_3 - p_4,$$

$$\beta_{43} = \frac{1}{24p_1c_2\beta_{32}}, \quad \beta_{42} = \frac{1}{c_2p_4}\left[\frac{1}{6} - p_3c_2\beta_{32} - p_4c_3\beta_{43}\right], \tag{16}$$

$$\beta_{21} = c_2, \quad \beta_{31} = c_3 - \beta_{32}, \quad \beta_{41} = c_4 - \beta_{42} - \beta_{43}.$$

The singular solution ($c_3 = c_4$), where p_4 is a parameter, is given by

$$c_2 = -\frac{16}{7}, \quad c_3 = -\frac{5}{4}, \quad c_4 = -\frac{5}{4}, \quad p_2 = -\frac{7^3}{6\cdot 16\cdot 29},$$

$$p_3 = \frac{17\cdot 16}{3\cdot 5\cdot 29} - p_4, \quad p_1 = 1 - p_2 - p_3 - p_4,$$

$$\beta_{32} = \frac{5\cdot 7\cdot 29}{2\cdot(16)^2\cdot 11}, \quad \beta_{43} = -\frac{4\cdot 11}{3\cdot 5\cdot 29}\cdot\frac{1}{p_4}, \tag{17}$$

$$\beta_{42} = \frac{\left(\frac{1}{6} - \beta_{32}c_2p_3\right) - p_4\beta_{43}c_3}{c_2p_4}, \quad \beta_{21} = c_2,$$

$$\beta_{31} = c_3 - \beta_{32}, \quad \beta_{41} = c_4 - \beta_{42} - \beta_{43}.$$

From this set of solutions we distinguish the variant of the general solution with $c_2 = -1$, $c_3 = 1/2$:

$$c_2 = 1\ , \quad c_3 = \frac{1}{2}\ , \quad c_4 = 1,$$

$$p_1 = \frac{13}{6}\ , \ p_2 = \frac{1}{6}\ , \quad p_4 = -2, \quad p_4 = \frac{2}{3}\ , \tag{18}$$

$$\beta_{21} = -1, \quad \beta_{31} = \frac{1}{8}\ , \quad \beta_{32} = \frac{3}{8}\ , \quad \beta_{41} = \frac{3}{8}\ ,$$

$$\beta_{42} = \frac{19}{24}\ , \quad \beta_{43} = -\frac{1}{6}\ .$$

Method (4)-(6), (18) has the following remarkable properties. In the first place, the domain of influence is reduced to the minimal possible one:

$$|\beta_{ij}| \leq 1, \quad |c_j| \leq 1. \tag{19}$$

In the second place, accumulation of round-off errors characterized by the value ξ :

$$\xi = \sum_{i=1}^{4} |p_i| = 5 \tag{20}$$

is close to the minimal one. If the condition (19) is fulfilled,the minimal $\xi \approx 4.8$, but the coefficients β_{ij}, p_i have a more complicated form. We assume that the fulfilment of condition (19) must ensure high accuracy on smooth slowly changing variables. Tests of the method (4)-(6),(18) on examples of the small stiffness show that it achieves the same accuracy as the Runge-Kutta method with steps two times smaller than those required by the Runge-Kutta method.The global volume of work in comparison with the Runge-Kutta method, is approximately ten times as large. The method suggested will be more effective than the Runge-Kutta method, if stiffness of the system (the relation of the maximal module of eigenvalues of the Jacobian matrix to the minimal one) exceeds one hundred.

Numerical trials of the method on typical test stiff type problems [3] have shown its high efficiency in comparison with the methods of Runge-Kutta, Hamming [4], Brayton, Gustavson, Hachtel [5], and Liniger [6]. We used the standard programs of the "Dubna" monitor system [7] for method of Runge-Kutta and Hamming in numerical experiments, in algorithm [5] changes were introduced in the variation strategy of the step and the order of the method; method [6] was used in the form linearized according to Newton (non-iterative Rosenbrock type algorithm) was used.

References

[1] G.V.Demidov. About one method of constructing stable high order schemes (Russian). Information Bulletin "Chislennye metody mechaniki sploshnoi sredy", t.1, N 6, 1970, Novosibirsk.

[2] V.A.Novikov and G.V.Demidov. A remark on one method of constructing high order schemes (Russian). "Chislennye metody mechaniki sploshnoi sredy", t.3, N 4, 1972, Novosibirsk.

[3] G.Bjurel, G.Dahlquist, B.Lindberg, S.Linde, L.Óden. Survey of stiff ordinary differential equations, The Royal Institute of Technology, Stockholm, Report NA 70.11.

[4] R.V.Hamming. Stable predictor-corrector methods for ordinary differential equations, JACM, 6, 1959, pp 34-47.

[5] R.K.Brayton, F.G.Gustavson, G.D.Hachtel. A new efficient algorithm for solving differential algebraic systems using implicit backward differentiation formulas. Proceedings of the IEEE, volume 60, N 1, January 1972, pp98-108.

[6] W.Liniger. Global accuracy and A-stability of one-and-two-step integration formulae for stiff ordinary differential equations, IBM Rep RC 2396 (1969).

[7] G.L.Mazniy. "Dubna" monitor system (Russian). User's Manual, Dubna, 1971.

SOME METHODS FOR NUMERICAL SOLUTION OF OPTIMAL MODELS IN SPATIAL-PRODUCTION PLANNING

A.E. Bakhtin

Institute of Economics & Industrial Engineering

Siberian Department of the USSR Academy of Sciences

Novosibirsk, USSR

This paper is devoted to problems of numerical solution of some economic-mathematical models used in long-term planning of spatial location of production.

By their structure, models of this kind are of a complicated character and consist as a rule of several closely interconnected submodels. In other words, one has to deal with a system of models. Each model of the system represents certain aspects in a complicated economic system and is meant to clarify certain special problems. The inputs of one model of the system are outputs for other models and affect, therefore, the results obtained with the help of the latter. For this reason, it becomes necessary to have such techniques for the analysis of systems of models and for the synthesis of obtained partial solutions which would allow one to obtain consistent solutions in the optimization system of models under consideration.

The problem of composition is difficult also because the models entering a system are in themselves mathematically difficult objects of investigation:

- they have great dimensionality and a complicated block structure;
- they contain integer conditions or non-linear dependences etc.

Below some examples of models used in long-term spatial-production planning are given and the ideas of the composition of complex solutions are briefly exposed.

Now we shall consider the simplest "production-transportation" model consisting of linear production and transportation submodels.

We introduce the following notation:

i the number of a productive unit located on the territory under consideration, $i = 1, \dots, m$;

κ the number of a product (resource) produced (consumed), $\kappa = 1, \dots, l$;

j the number of the way of production, $j = 1, \dots, n_i$;

$A^i = (a^i_{\kappa j})_{\substack{\kappa=1,\dots,l \\ j=1,\dots,n_i}}$ the matrix of the ways of production of the i-th unit;

$a^i = (a^i_\kappa)_{\kappa=1,\dots,l}$ the vector of constraints for the i-th unit;

$x^i = (x^i_j)_{j=1,\dots,n_i}$ the unknown vector of intensities in the use of production ways by the i-th unit;

$y^i = (y^i_\kappa)_{\kappa=1,\dots,l}$ the unknown vector, the k-th component of which equals the import-export balance of the k-th product for the i-th unit.

The sought vectors x^i, y^i must satisfy balance relations:

$$A^i x^i + y^i = a^i, \qquad x^i \geq 0. \quad (i = 1, \dots, m) \tag{1}$$

We shall designate by P_κ $(\kappa = 1, \dots, l)$ a subset of the set $\{(i,j) : i, j = 1, \dots, m\}$, defining the possible exchange (cooperation) by the k-th item between the units, and by x^κ_{ij}, $(i,j) \in P_\kappa$, an unknown variable equal to the delivery size of the k-th product by the i-th unit to the j-th unit. The unknown x^κ_{ij} must satisfy conditions:

$$\sum_{j:(i,j)\in P_\kappa} x^\kappa_{ji} - \sum_{j:(i,j)\in P_\kappa} x^\kappa_{ij} = y^i_\kappa, \qquad x^\kappa_{ij} \geq 0, \quad i = 1, \dots, m; \quad \kappa = 1, \dots, l. \tag{2}$$

The production-transportation problem is finding such a composition of $(\mathbf{X}, \mathbf{Y}, \mathcal{X})$ $(\mathbf{X} = (x^1, \dots, x^m);$

$$\mathbf{Y} = (y^1, \dots, y^m); \; \mathcal{X} = \{x^\kappa_{ij}\}, \; \kappa = 1, \dots, l, (i,j) \in \mathbf{P}_\kappa),$$

which satisfies the conditions of (1)-(2) and minimizes production--transportation costs

$$\sum_{i=1}^{m} (c^i, x^i) + \sum_{\kappa=1}^{l} \sum_{(i,j)\in \mathbf{P}_\kappa} c^\kappa_{ij} x^\kappa_{ij} \qquad (3)$$

For the search of the optimal composition $(\hat{X}, \hat{Y}, \hat{\mathcal{X}})$ in problem (1)-(3) a finite iterative method /1, 2/ has been used for which a program has been made in the codes of BESM-6 computer and which is used in practical calculations. With regard to such problems sufficient quantity of statistical data has been accumulated.

This method presupposes a mutual exchange of information between the models of the production-transportation system.

In finding production plans x^i and the sizes of y^i the information about reasonable cooperation of units by different kinds of products is used, and in finding the plan of transportations $\mathcal{X}$ the data from production models about y^i are used.

The successive approximation to the optimal composition $(\hat{X}, \hat{Y}, \hat{\mathcal{X}})$ is made either by adding new methods of manufacture and deleting the bad ones or cancelling non-reasonable deliveries; or by introducing new, more effective deliveries to substitute for low-effective deliveries or reducing inefficient production.

In finding the next direction in improvement use is made of productive and transportation evaluations of products in different points (the solutions of dual systems).

This method has been extended to production-transportation problems /3/ in which balance conditions (2) have been replaced by

$$\sum_{j:(i,j)\in \mathbf{P}_\kappa} a^\kappa_{ji} x^\kappa_{ji} - \sum_{j:(i,j)\in \mathbf{P}_\kappa} x^\kappa_{ij} = y^i_\kappa \qquad (2')$$

$$x^\kappa_{ij} \geq 0, \qquad i = 1, \dots, m, \quad \kappa = 1, \dots, l.$$

The above models are linear.

In the practice of long-term planning one can also find such production-transportation models in which the intensities of production ways x_j^i can assume integer values 0 or 1.

Here a situation is discussed where from several ways of a production of unit it is possible to choose only a single method, i.e. to conditions (1)-(3) we add the conditions

$$\sum_{j=1}^{n_i} x_j^i \leq 1, \qquad i = 1, \ldots, m,$$
$$x_j^i = 0 \vee 1, \qquad j = 1, \ldots, n_i . \tag{4}$$

These models are typical of problems about location and specialization of enterprises in an industry or in a group of related industries.

These optimal problems belong to the field of partial integer programming.

The account taken of specific conditions of this problem made it possible in this case also to build up a decomposition solution method acceptable for practical calculations.

The underlying basic ideas were: the implicit enumeration method of E.Balas /4/ and the method of cutting planes /5, 6/.

For certain special cases of models (1)-(4) computational schemes and programs for computers have been made which are employed at this Institute and other institutions in our country. The description and substantiation of the method for the solution of problem (1)-(4) is provided in /7, 8/.

References

1. Bakhtin, A.E., Yu.I.Volkov
   A Technique for Successive Improvement of a Plan for the Solution of One Class of Problems in Linear Programming (Russian).
   Ekonomika i Mat. Metody, v.III, issue 4, 1967.
2. Bakhtin, A.E., E.N.Zvonov
   A Numerical Method for the Calculation of Linear Models in Optimal Spatial-production Planning (Russian).
   Mat. vopr. formirovania ekonom. modelei. Nauka, Novosibirsk, 1973.
3. Bakhtin, A.E., N.F.Pastukhov
   On an Approach to the Solution of Problems in Optimal Spatial-production Planning (Russian).
   Optimiz. planov razvitia i razmeshch. otrasl. prom. Novosibirsk, 1971.
4. Balas, E.
   An Additive Algorithm for Solving Linear Programs with Zero - One Variables.
   Operations Research, 1965, no.13.
5. Balas, E.
   Minimax and Duality for Linear and Nonlinear Mixed-Integer Programming.
   Integer and Nonlinear Programming. Ed. by J.Abadie. North - Holland, Amsterdam, 1970.
6. Benders, J.F.
   Partitioning Procedures for Solving Mixed Variables Programming Problems.
   Numer. Math., 1962, 4, no.3.
7. Bakhtin, A.E., V.K.Korobkov, Z.V.Korobkova
   An Algorithm for Integer Problem Solution to the Choice of Optimal Variants of Development and Location of Plants (Russian).
   Vopr. ĉislen. reŝ. opt. ekon.-mat. zadaĉ. Novosibirsk, 1973.
8. Bakhtin, A.E., A.A.Kolokolov
   A Decomposition Method for the Solution of Integer Production--Transportation Problems (Russian).
   Vopr. ĉisl. reŝ. opt. ekon.-mat. zadaĉ. Novosibirsk, 1973.

An Extension of the Method of Feasible Directions

E. Blum
Universidad Nacional de Ingenieria
Lima/Perú

W. Oettli
Universität Mannheim
68 Mannheim/Germany

In this contribution we are going to discuss the extension of the method of feasible directions [1], [2], [3] to programming problems involving an infinite number of constraints. Problems of this type arise frequently in applications. We shall be working with arbitrary convex approximations instead of with linearizations, simply to emphasize the fact that the feasible direction method belongs to that class of methods where not differentiability but rather convex-likeness of the functions involved is the essential property.

Our programming problem has the following form:

$$\text{(P)} \qquad \min \{F(x) \mid x \in C, f(t,x) \le 0 \ \forall t \in T\} .$$

With $S = \{x \mid x \in C, f(t,x) \le 0 \ \forall t \in T\}$ the admissible domain of (P) we introduce for all $x \in S$ approximations $\Phi(x,\xi), \phi(t,x,\xi)$ for the functions $F(\xi), f(t,\xi)$. We assume that C is a compact convex set of some normed (metrizable) linear space, that T is a compact metric space, and that the functions $F(\xi), f(t,\xi), \Phi(x,\xi), \phi(t,x,\xi)$ are jointly continuous in all their arguments, with $\xi \in C$, $x \in S$, $t \in T$.

We shall be particularly interested in certain elements of S, henceforth denoted by $\hat{x}$, which will be limit points of our iterative procedure. Concerning these points $\hat{x} \in S$ we require in addition that the functions Φ and ϕ are "good" approximations in the sense that

$$|\Phi(\hat{x},\xi) - F(\xi)| \le o(\xi-\hat{x}), |\phi(t,\hat{x},\xi) - f(t,\xi)| \le o(\xi-\hat{x})$$

uniformly for all $t \in T$. Moreover $\Phi(\hat{x},\xi)$ and $\phi(t,\hat{x},\xi)$ have to be *convex* with regard to ξ.

For $\hat{x} \in S$ let us define the set of binding constraints

$$\hat{T} = \{t \in T \mid f(t,\hat{x}) = 0\} ,$$

and consider the following system in ξ:

$$\text{(1)} \qquad \xi \in C, \Phi(\hat{x},\xi) - F(\hat{x}) < 0, \phi(t,\hat{x},\xi) < 0 \ \forall t \in \hat{T} .$$

Under the assumptions made it is not difficult to prove the following

Lemma 1: Let ξ be a solution of (1). Then there exists $x \in [\hat{x},\xi]$ satisfying

$$x \in C, F(x) - F(\hat{x}) < 0, f(t,x) < 0 \ \forall t \in T .$$

From this lemma one obtains immediately the following necessary optimality criterion which may be considered as a generalization of Kolmogorov's criterion for best Chebyshev-approximations.

Theorem 1: If $\hat{x} \in S$ is an optimal solution for the programming problem (P), then system (1) is inconsistent.

We note that the inconsistency of (1) is also a sufficient condition for optimality, if $F(\xi)$ and $f(t,\xi)$ are convex with regard to ξ, and if Slater's assumption is satisfied: there exists $\tilde{x}\in C$ satisfying $f(t,\tilde{x}) < 0 \ \forall t\in\hat{T}$.

Any limit point $\hat{x}$ of the approximation procedure to be described is a stationary point in the sense that it meets the necessary optimality condition of Theorem 1.

Let us now describe the iterative scheme. We choose a *positive* number $\alpha > 0$, and sequences $\theta_k \geq 0, \rho_k \geq 0$ such that $\theta_k \to 0, \sum_k \rho_k < +\infty$. $x^o\in S$ is arbitrary. Given $x^k\in S$ we define x^{k+1} according to the following rules: Let

$$T^k = \{t\in T \,|\, f(t,x^k) \geq -\alpha\} ,$$

$$H^k(\xi) = \max\{\Phi(x^k,\xi) - F(x^k), \phi(t,x^k,\xi) : t\in T^k\} .$$

Let $\xi^k\in C$ be such that

$$H^k(\xi^k) \leq \min \{H^k(\xi) \,|\, \xi\in C\} + \theta_k , \tag{2}$$

and define $x^{k+1}\in\left[x^k,\xi^k\right] \cap S$ such that

$$F(x^{k+1}) \leq \min \{F(x) \,|\, x\in\left[x^k,\xi^k\right] \cap S\} + \rho_k . \tag{3}$$

Obviously x^{k+1} is well defined, and is again in S. Since S is compact, the sequence $\{x^k\}$ has a cluster point $\hat{x}\in S$.

Theorem 2: If $\hat{x}$ is a cluster point of the sequence $\{x^k\}$, then $\hat{x}$ satisfies the necessary optimality criterion of Theorem 1.

Proof: In addition to $H^k(\xi)$ let us define the continuous functions

$$H^\infty(x,\xi) = \max \{\Phi(x,\xi) - F(x), \phi(t,x,\xi) : t\in T\} ,$$

$$\hat{H}(x,\xi) = \max \{\Phi(x,\xi) - F(x), \phi(t,x,\xi) : t\in\hat{T}\} .$$

Since C is compact we can choose a subsequence $x^{\bar{k}}$ such that

$$x^{\bar{k}} \to \hat{x}, \ \xi^{\bar{k}} \to \hat{\xi}\in C .$$

From (2) follows

$$H^k(\xi^k) \leq H^k(\xi) + \theta_k \ \forall\xi\in C .$$

The continuity of $f(t,x)$ over $\hat{T}\times C$, the compactness of $\hat{T}$, and the convergence of $x^{\bar{k}}$ to $\hat{x}$ imply that $\hat{T}\subset T^{\bar{k}}$ for all sufficiently large $\bar{k}$. Also $T^{\bar{k}}\subset T$. Therefore

$$\hat{H}(x^{\bar{k}},\xi^{\bar{k}}) \leq H^\infty(x^{\bar{k}},\xi) + \theta_{\bar{k}} \ \forall\xi\in C$$

for all sufficiently large $\bar{k}$. Passing to the limit we obtain

$$\hat{H}(\hat{x},\hat{\xi}) \leq H^\infty(\hat{x},\xi) \ \forall\xi\in C . \tag{4}$$

From (3) follows

$$F(x^{k+1}) \leq F(x^k) + \rho_k ,$$

thus a fortiori

$$F(\overline{x^{k+1}}) \leq F(\overline{x}^{k+1}) + \sum_{\mu=k+1}^{\infty} \rho_\mu .$$

Again by (3)

$$F(\overline{x^{k+1}}) \leq F(x) + \sum_{\mu=k+1}^{\infty} \rho_\mu \quad \text{for all} \quad x \in \left[\overline{x}^k, \overline{\xi}^k\right] \quad \text{satisfying} \quad \max_{t \in T} f(t,x) \leq 0 .$$

Passing to the limit we obtain by continuity

$$F(\hat{x}) \leq F(x) \quad \text{for all} \quad x \in \left[\hat{x}, \hat{\xi}\right] \quad \text{satisfying} \quad \max_{t \in T} f(t,x) < 0 .$$

This means that the system

$$x \in \left[\hat{x}, \hat{\xi}\right],\ F(x) - F(\hat{x}) < 0,\ f(t,x) < 0\ \forall t \in T \tag{5}$$

is inconsistent. Assume now that (1) has a solution. A slight variant of the proof of Lemma 1 shows that (1) has still a solution if we replace $\hat{T}$ by T. This means there exists $\xi \in C$ satisfying $H^{\infty}(\hat{x},\xi) < 0$. By (4) then $\hat{H}(\hat{x},\hat{\xi}) < 0$. Lemma 1 gives then the existence of x satisfying (5), a contradiction. Thus (1) is inconsistent.

q.e.d.

We may study the rate of convergence of $F(x^k)$ if we require in addition:

$F(\xi), f(t,\xi), \Phi(x,\xi), \phi(t,x,\xi)$ are convex with respect to ξ ;

$\Phi(x,x) = F(x), \phi(t,x,x) = f(t,x)\ \forall x \in S$; the set

$$S_o = \{x \in C \mid f(t,x) \leq 0\ \forall t \in T,\ F(x) \leq F(x^o)\}$$

is bounded; $\exists \tilde{x} \in C$: $f_t(\tilde{x}) < 0\ \forall t \in T$; $\theta_k = 0$ and $\rho_k = 0\ \forall k$.

We use the abbreviations

$$\tau^k = H^k(\xi^k),\ \delta^k = F(x^k) - \hat{F} ,$$

where $\hat{F}$ is the optimal value of (P). Then

$$\tau^k \leq 0,\ \tau^k \to 0;\ \delta^k \geq 0,\ \delta^k \to 0 .$$

We obtain the following results.

<u>Lemma 2:</u> If there exist constants $\mu \geq 0$, $0 < m \leq 1$, such that (i) $\Phi(x,\xi) - \mu|\xi - x|$ is convex with respect to ξ, (ii) $\phi(t,x,\xi) - \mu|\xi - x|^2 \leq f(t,\xi)$, (iii) $\Phi(x,\xi) - (1-m)\mu|\xi - x| \leq F(\xi)$, then $\tau^k \leq \rho(-\delta^k)$ for some $\rho > 0$.

<u>Lemma 3:</u> (a) If $F(\xi) \leq \Phi(x,\xi) + M|\xi - x|^2$, $f(t,\xi) \leq \phi(t,x,\xi) + M|\xi - x|^2$, then $\delta^{k+1} - \delta^k \leq -\gamma(\tau^k)^2$ for some $\gamma > 0$. (b) If, in addition, there exists $\mu > 0$ such that $\Phi(x,\xi) - \mu|\xi - x|^2$ and $\phi(t,x,\xi) - \mu|\xi - x|^2$ are convex with regard to ξ, then $\delta^{k+1} - \delta^k \leq \gamma\tau^k$ for some $\gamma > 0$.

From these follows

<u>Theorem 3:</u> If the assumptions of Lemma 2 and Lemma 3(a) hold, then $\delta^{k+1} \leq$ $\leq (1 - \rho\delta^k)\,\delta^k$ for some $\rho > 0$. If the assumptions of Lemma 2 and Lemma 3(b) hold, then $\delta^{k+1} \leq (1 - \rho)\,\delta^k$ for some $\rho > 0$.

This extends some results of Pironneau - Polak [4]. Proofs are too lengthy to be given here; they will be reproduced elsewhere.

[1] M. Frank, P. Wolfe: An algorithm for quadratic programming. *Naval Res. Logistics Quart.* 3 (1956), 95-110 .

[2] G. Zoutendijk: *Methods of Feasible Directions.* Elsevier, Amsterdam, 1960.

[3] H.P. Künzi, W. Oettli: *Nichtlineare Optimierung* (Lecture Notes in Operations Research and Mathematical Systems, 16), pp. 65-70. Springer, Berlin, 1969.

[4] O. Pironneau, E. Polak: On the rate of convergence of certain methods of centers. *Math. Programming* 2 (1972), 230-257.

A NUMERICAL METHOD FOR SOLVING LINEAR CONTROL PROBLEMS WITH MIXED RESTRICTIONS ON CONTROL AND PHASE COORDINATES

V.I.CHARNY
Institute of Control Sciences, Moscow, USSR

1. The proposed method is a result of the combination of the dual approach and the method of feasible directions. Three types of problems that can be solved by this method are listed below.

1°. Determination of feasible control in the system

$$\frac{dx(t)}{dt} = A(t)\, x(t) + B(t)u(t) + C(t) \,, \tag{1}$$

$$u(t) \geq 0,\; M(t)u(t) \leq N(t)\, x(t) + \rho(t) \,,$$

$$x(0) = x_o,\quad Qx(T) \geq \bar{x} \,.$$

Matrices A, B, M, N, Q, vectors C, ρ, x_o, $\bar{x}$, and terminal time T are given here, with A, B, C, M, N, ρ being piece-wise continuous functions of t ; it is required to find a feasible control $u(t)$ belonging to the class of piece-wise continuous functions of t .

2°. Type (1) problem with delays in differential and finite relations of the system (delays may belong both to phase coordinates and to control).

3°. Finite-dimensional linear programming problem.

2. Let L_n^2 be the space of piece-wise continuous functions of t from L^2 , defined on $[0, T]$ and valued in R^n, $X^{nm} = L_n^2 \times R^m$ - prehilbertian space of pairs $a=\{W_a, V_a\}$ ($W_a \in L_n^2, V_a \in R^m$, $a \in X^{nm}$) with a scalar product $\langle a,b\rangle = \int_0^T (W_a, W_b)dt + (V_a, V_b)$ $(a, b \in X^{nm})$. Problems 1°-3° can be considered as a particular case of functional linear programming problem

$$Lu \leq h \,, \tag{2}$$

in which $u \in X_u$, $h \in X_h$, where $X_u = X^{nm}$, $X_h = X^{kr}$, n, m, k, r are given constants, L is a bounded linear operator transforming X_u into X_h .

The reduction of considered control problems to type (2) problem is a wellknown mode used by various authors, including [1-4] .

3. In application to the problem (2) the numerical method is constructed in the following way.

Let us consider the auxiliary problem

$$\max_{u,\beta} \beta \ : \ Lu \leq h(1+\beta), \quad \beta \leq 0 \tag{3}$$

with a scalar parameter β and apply the method of feasible directions [5,6] to it starting from the feasible solution $u = 0$, $\beta = -1$ of (3). It is not difficult to see that two cases are possible: 1) $\max\beta = 0$; then the first iteration of the method of feasible directions gives the feasible solution of (2) by the formula

$$u = \frac{g}{g_\beta} , \tag{4}$$

where the pair $\{g, g_\beta\}$ defines a feasible direction in the problem (3) for $u=0$, $\beta = -1$ (g, g_β) are defined by relations $Lg \leq hg_\beta, g_\beta > 0$); 2) $\sup \beta < 0$ or $\sup\beta = 0$,but the value $\beta = 0$ in (3) is not attained; in this case there is no feasible direction $\{g, g_\beta\}$ and the problem (2) has no solutions. To determine g, g_β in formula (4) we shall formulate a "best" direction problem [5](for case 1):

$$\max_{g, g_\beta} g_\beta : Lg \leq hg_\beta , \ g_\beta^2 + \langle g, g \rangle = 1. \tag{5}$$

The numerical method for the solution of the problem (2) is based on the application of dual approach [4] to the problem (5) with subsequent utilization of formula (4). In the end this method is reduced to the solution of the problem

$$\min_{\lambda \geq 0} \Phi \ : \ \Phi = (1 + \langle h, \lambda \rangle)^2 + \| L^* \lambda \|^2 , \tag{6}$$

where $\lambda \in X_h$, $\|a\| = \sqrt{\langle a, a \rangle}$, L^* is the operator conjugate with L . Solutions of problems (2) and (6) are related in the following way: 1) let $\lambda \in X_h$ be a solution of (6) and $\min \Phi > 0$; then

$$u = - \frac{L^* \lambda}{1 + \langle h, \lambda \rangle} \tag{7}$$

(the denominator in (7) is not equal to zero, because we can show that $\min \Phi = 1 + \langle h, \lambda \rangle$);

2) let $\inf \Phi = 0$; then the problem (2) is unsolvable;
3) let $\inf \Phi > 0$ be unattainable in X_h ; then we can construct a minimizing sequence $\lambda_n \in X_h$, such that for $u = u_n$, where u_n is defined by substituting $\lambda = \lambda_n$ into (7), the following condition is fulfilled:

$$\lim_{n \to \infty} \| (Lu_n - h)^+ \| = 0, \tag{8}$$

in other words, for u_n positive lacks in (2) tend on the average to zero.

It is not difficult to see that case 2 corresponds to unsolvable conditions in Farkas' lemma.

4. The formulation of problem (6) for problem (1) follows:

$$\min_{\substack{\psi \geq 0, \\ \omega \geq 0, \\ \mu \geq 0}} \Phi : \Phi = \int_0^T (g, g)\, dt + g_\beta^2 , \tag{9}$$

$$g = B^T p - M^T \omega + \psi, \quad g_\beta = 1 + (\mu, \varphi_T) - \int_0^T (\psi, \varphi) dt,$$

$$\frac{dp}{dt} = - A^T p - N^T \omega, \quad p(T) = Q^T \mu.$$

Here $\varphi = N(t)x^o(t) + \rho(t)$, $\varphi_T = Qx^o(T) - \bar{\bar{x}}$,where $x^o(t)$ is the solution of the Cauchy problem:

$$\frac{dx^o(t)}{dt} = A(t)x^o(t) + C(t), \; x^o(0) = x_o .$$

5. The following simple algorithm can be used to solve the problem (6). Let λ_n be n-th approximation to the problem (6) solution. Let u determine λ_n^* by solving the problem

$$\min_{\alpha} \Phi(\lambda_n + \alpha(- \operatorname{grad} \Phi(\lambda_n))^+) \tag{10}$$

$$(\lambda_n^* = \lambda_n + \alpha(- \operatorname{grad} \Phi (\lambda_n))^+ .$$

Then let us find λ_{n+1} by the formula $\lambda_{n+1} = \nu \lambda_n^*$, where ν is the solution of the problem

$$\min_{\nu} \Phi(\nu \lambda_n^*) . \tag{11}$$

The process termination can be controlled by the substitution of $\lambda = \lambda_n$ into (7) and by determination of the system lacks.

Finite formulae can be applied to the one- dimensional problems (10) and (11). The algorithm is proved in [4] . To accelerate its convergence a combined algorithm can be used, in which formulae (10), (11) are used with formulae of the conjugate gradients method.

6. The numerical method permits the following modification. The problem

$$Lu = h, \quad u \geq 0 \tag{12}$$

(the canonical form of the linear programming problem) is considered instead of (2) and the problem

$$\min_{\substack{\omega \in X_h, \\ \lambda \in X_u, \lambda \geq 0}} \Phi : \Phi = (1 + \langle h,\omega \rangle)^2 + \| \lambda - L^* \omega \|^2 \tag{13}$$

is solved instead of (6). The relation of (13) with (12) is analoguos to that of (6) with (2); now instead of (7) the formula

$$u = \frac{\lambda - L^* \omega}{1 + \langle h,\omega \rangle} \tag{14}$$

is used.

The algorithm of consecutive minimization over ω and λ can be used for the solution of problem (13). Besides it is essential that there are no restrictions on ω in (13), and the minimization over λ is possible by finite formulae.

7. The numerical method has been used for the solution of a minimum time problem formulated for the Leontieff type dynamic input-output model (type (1) problem). The relations of this problem [4] are shown below:

$$\begin{gathered} \frac{dV(t)}{dt} = u(t) , \\ u(t) \geq 0, \quad M(t)u(t) \leq V(t) - V^0(t), \\ V(0) = V_0, \quad V(T) \geq \bar{V} , \end{gathered} \tag{15}$$

where matrix M and vectors V^0, V_0, $\bar{V}$ are given. A series of problems of feasible control search with the fixed value of T was solved with different values of T for a minimum time problem ($\min_u T$) . The combined algorithm based on formulae (10),(11) and formulae of the conjugate gradients method was used there. The type (15) problem for a twenty-nine industry model (vector V has 29 components) was solved in less than ten minutes on a third-generation computer (the program was written in ALGOL).

8. As follows from Section 3 the proposed method can be interprete as a feasible directions method in which the direction is determined only once. On the other hand, this method in a way similar to method of penalty functions is reduced to the problem of minimization of a quadratic functional. The basic difference of these methods is in the dependence of $\min \Phi$ on system parameters: in the proposed method $\min \Phi$ has a jump on the boundary of the region of parameters in whi the system has solutions, provided $\|u\|$ doesn't grow to infinity in this region; in analogous case for methods with quadratic functio of penalty the corresponding functional is continuous on the same bou dary. This jump can be used effectively for solving minimum time pr lem: in this case $\min_u T$ must be treated as a limit point of set of values of T, for which the system has a feasible solution.

References

1 N.N.Krasovskii. The control theory of motion. (Russian).Izd. "Nauka", 1968.

2 M.V.Meerov, B.L.Litvak. Optimization of multi-connected control systems (Russian). Izd. "Nauka", 1972.

3 N.V.Gabashvili, N.N.Lominadzé, L.L.Chkhaidzé. An approximate solution of certain optimal control and discrete programming problems (Russian).Tekhnicheskaya Kibernetika, N6, 1972 .

4 V.I.Charny, V.A.Boĭkov. Numerical solution of linear dynamic problems in economic planning (Russian).Preprint, Izd.IAT, 1973

5 G.Hadley. Nonlinear and Dynamic Programming. Addison-Wesley Pub. Co. Inc., Reading, Massachusetts, 1964.

6 A.V.Fiacco, G.P.Mc Cormick. Nonlinear Programming: Sequential Unconstrained Minimization Techniques. New York-London, 1968.

DUAL DIRECTION METHODS FOR FUNCTION MINIMIZATION

Yu.M. Danilin
Institute of Cybernetics Ukrainan
Academy of Sciences, USSR

1°. Let

$$f(x) = \frac{1}{2}(Ax,x) + (b,x) + c,$$

$x \in E^n$, A be a symmetric matrix, and $(Ax,x)>0$, $\forall x \neq 0$, b - n- dimensional vector, c - scalar. Gradient of this function $f'(x) = A(x-x_*)$. Here and further x_* is a point of minimum.

Minimization of the assumed function is equivalent to the solution of the linear equation system

$$(f'(x_i), z_i) = (e_i, x_i - x_*), \qquad 0 \le i \le n-1 \tag{1}$$

or the system of following equations

$$f(x_i) = f(x_*) + \tfrac{1}{2}(f'(x_i), x_i - x_*), \qquad 0 \le i \le n \tag{2}$$

where x_i are arbitrary points; $z_0, \dots, z_{n-1}$ - an arbitrary system of lineary independent vectors; $e_i = f'(x_i) - f'(x_i - z_i) = Az_i$.

The systems (1) and (2) may be written in the form

$$(e_i, x_*) = z^{i+1}, \qquad 0 \le i \le n-1 \tag{3}$$

where

$$z^{i+1} = (e_i, x_i) - (f'(x_i), z_i) \tag{4}$$

or (for the system (2))

$$z^{i+1} = (f'(x_i), x_i) - (f'(y_i), y_i) - 2[f(x_i) - f(y_i)] \tag{5}$$

In the last case $z_i = x_i - y_i$; $e_i = f'(x_i) - f'(y_i)$; points y_i are chosen arbitrarily so as to provide the linear independence of the vectors $z_0, \dots, z_{n-1}$.

The finding of the point x_*, that satisfies the system (3), can be treated as some iterative process of the finding of points

$$\bar{x}_1, \bar{x}_2, \dots, \bar{x}_n$$

satisfying the relations

$$(\bar{x}_{k+1}, e_i) = z^{i+1}, \qquad 0 \le i \le k \le n-1. \tag{6}$$

From the comparison of formulas (4) and (6) it follows that $\bar{x}_n = x_*$.

The finding of the point $\bar{x}_{k+1}$ is realized with recurrent formulas. Let $x_0=\bar{x}_0$ be an arbitrary point, and

$$\bar{x}_{k+1} = \bar{x}_k + p_k , \qquad k = 0,1,\dots, n-1. \tag{7}$$

So from the relations (6) it follows that vector p_k must satisfy such conditions:

$$(p_k, e_i) = 0, \qquad 0 \le i \le k-1 \tag{8}$$

$$(p_k, e_k) = z^{k+1} - (\bar{x}_k, e_k) \tag{9}$$

The last condition may be altered to the form:

$$(p_k, e_k) = -(f'(\bar{x}_k), \tau_k). \tag{10}$$

The choice of vector p_k, satisfying the conditions (8) and (9), permits a wide range of possibilites at $k \le n-2$. Thus, the formula (7) defines actually a wide class of the quadratic algorithms of minimization, in which second derivatives do not participate.

Here we dwell upon the study of the dual direction methods.

If the system of vectors $s_{k+1,i}$, $0 \le i \le k$, is made dual (biortogonal) to the system $e_0,\dots,e_k$, it may be assumed that

$$p_k = \alpha_k s_{k+1,k} \tag{11}$$

where

$$\alpha_k = z^{k+1} - (\bar{x}_k, e_k) = -(f'(\bar{x}_k), \tau_k).$$

So, it appears that

$$\bar{x}_{k+1} = \bar{x}_k + [z^{k+1} - (\bar{x}_k, e_k)]\, s_{k+1,k} \tag{12}$$

Choosing p_k in form (11), the following equality $(f'(\bar{x}_k), \tau_k) = (f'(x_0), \tau_k)$ holds. Consequently

$$\bar{x}_n = x_* = x_0 - \sum_{i=0}^{n-1} (f'(x_0), \tau_i)\, s_{n,i} \tag{13}$$

2°. Let us dwell on the task of minimization of the non-quadratic functions. Let's assume $f(x)$ as a twice continously differentiable function, and $mJ \leq f''(x) \leq MJ, \quad m>0$.

Now we consider the algorithms in the basis of which the formulas (12) and (13) lie.

The method, founded on the application of the formula (13), in its main facilities consists in the following steps. We make the sequence of points

$$x_{k+1} = x_k + \alpha_k \bar{p}_k, \qquad k = n, n+1, \ldots \qquad (14)$$

wherein the vector

$$\bar{p}_k = -\sum_{i=0}^{n-1} \left(f'(x_k), z_{k-i} \right) S_{k+1,k-i} \qquad (15)$$

and as the multiplier α_k, defining the stepsize, we choose the greatest value of the parameter $0 \leq |\alpha| \leq 1$/ i.e. the value obtained by means of fracturing α/, that satisfies unequality

$$f(x_k + \alpha \bar{p}_k) - f(x_k) \leq \varepsilon \alpha \left(f'(x_k), \bar{p}_k \right), \qquad 0 < \varepsilon < \tfrac{1}{2} \qquad (16)$$

At the initial stage of this process some iterations can be realized by the gradient method (that is $\bar{p}_k = -f'(x_k)$) - in the case, when the choice of $\bar{p}_k$ in the form of (15) has resulted in $\left(f'(x_k), \bar{p}_k \right) = 0$.

The vector z_k in the Eq.(15) is determined as $z_k = x_k - y_k$, where the point y_k is found arbitrarily so as to provide the carrying out of the following conditions: a) the vector system $z_k, \ldots, z_{k-n+1}$ must be linearly independent at any k ; b) $\| z_k \| \to 0$ at $k \to \infty$.

The vectors $S_{k+1,k-i}$, $0 \leq i \leq n-1$, form the basis, dual to the basis $e_k, \ldots, e_{k-n+1}$. The finding of the vectors is carried out with the recurrent formulas

$$S_{k+1,k} = \frac{S_{k,k-n}}{(S_{k,k-n}, e_k)}$$

$$S_{k+1,k-j} = S_{k,k-j} - \left(S_{k,k-j}, e_k \right) S_{k+1,k}, \qquad j = 1, \ldots, n-1.$$

If at some value of k it occurs that $(S_{k,k-n}, e_k) = 0$ / it can take place only at the initial stage of the process / we shall choose a new vector z_k and find the corresponding vector e_k. So we can always achieve the fulfilment of the condition $(S_{k,k-n}, e_k) \neq 0$.

Theorem 1. With above-mentioned assumptions for the sequence $\{x_k\}$, determined by formulas $\{(14),(15)\}$, the following statements $f(x_{k+1}) \leq f(x_k)$, $\| x_k - x_* \| \to 0$ are true independently of the choice of initial point x_0 ; what's more, the speed of convergence is superlinear:

$$\| x_{N+\ell} - x_* \| \leq C \lambda_N \lambda_{N+1} \dots \lambda_{N+\ell} \tag{17}$$

Here $\lambda_{N+\ell} < 1$ at any $\ell \geq 0$, $\lambda_i \to 0$ if $i \to \infty$.

If the matrix $f''(x)$ satisfies the Lip. condition, the estimate (17) is defined more precisely as follows:

$$\| x_{k+1} - x_* \| \leq C \| x_{k-n+1} - x_* \| \, \| x_k - x_* \| .$$

The other dual direction algorithms may be found in the following way. The sequence $\{x_k\}$ is formed according to the formula (14), wherein vector $\bar{p}_k = \bar{x}_{k+1} - x_k$ and the point $\bar{x}_{k+1}$ is determined as shown in (12). The parameter α_k is chosen from the condition (16). The different algorithms correspond with the choice of the value of z^{k+1} by the formula (4) or (5). For the sake of brevity we name such algorithms as the methods $\{(14),(4)\}$, $\{(14),(5)\}$. In these algorithms the gradient steps are also possible at the initial stage of the process.

Theorem 2. If the function $f(x)$ and the vector system $z_k, \dots, z_{k-n+1}$ satisfy the above formulated requirements, the sequence $\{x_{\xi n}\}$, $\xi = 0, 1, \dots$, found by method $\{(14), (4)\}$, converges to the point x_* with the superlinear speed independently of the choice of the initial point x_0. And $f(x_{k+1}) \leq f(x_k)$. The analogous properties are peculiar for the method $\{(14), (5)\}$, with the value of the vector z_k constrained as follows:

$$\| z_k \| \leq \min \{ \| x_k - x_{k-1} \|, \| f'(x_k) \| \}$$

The properties of the method $\{(14),(15)\}$ (which was described in the form different from one given here) were studied in the papers [1 - 2] , where the above cited values of the speed of convergence and some other values were derived. The properties of the methods { (14),(4) } and {(14),(5)} were studied in detail in paper [3] . Note that the study of the algorithms alike in their meaning was accomplished in papers [4] and [5].

In conclusion we also notice that the methodology used in item 1° permits creation of conjugate direction methods as well. The theory of convergence of these methods was developed in the paper [6] .

R e f e r e n c e s

1. Yu.M.Danilin, B.N.Pshenichniy, On the Methods of Minimization with the Accelerated Convergence(Russian). Journal "Vychisl. Mat. i Mat. Phys.", 10, N 6, 1970.
2. Yu.M.Danilin, B.N.Pshenichniy, The Estimates for the Speed of Convergence of a Certain Class of Minimization Algorithms (Russian). Journal "Dokl. Akad. Nauk SSSR", t. 213, N 2, 1973.
3. Yu.M.Danilin, On a Certain Class of the Minimization Algorithms with Superlinear Convergence (Russian). Journal "Vychisl. Mat. i Mat. Phys.", 14, N 3, 1974.
4. D.H.Jacobson, Oksman W., An algorithm that minimizes homogeneous functions on N variables in N+2 iterations and rapidly minimizes general functions. J.Math. Anal. Applik., v.38, 1972.
5. H.Y.Huang, Method of Dual matrices for function minimization. JOTA, v. 13, N 5, 1974.
6. Yu.M.Danilin, The Methods of Conjugate Directions for the Minimization Problems Solution (Russian). Journal "Kibernetika", N 5, 1971.

IMPLEMENTATION OF VARIABLE METRIC METHODS FOR CONSTRAINED OPTIMIZATION BASED ON AN AUGMENTED LAGRANGIAN FUNCTIONAL

N. H. Engersbach
DFVLR Institut für
Dynamik der Flugsysteme
8031 Oberpfaffenhofen, F.R.G.

W. A. Gruver*
Institut für Regelungstechnik
Technische Hochschule Darmstadt
6100 Darmstadt, F.R.G.

INTRODUCTION

Penalization and gradient projection are two of the simplest and most useful concepts from nonlinear programming. Both provide a means for extending unconstrained gradient descent techniques to accommodate equality and inequality constraints. It is known that penalty function methods may be used to solve a wide class of problems, even those involving nonconvex constraints. In practice, however, the Hessian matrix of the objective functional becomes ill-conditioned, causing convergence difficulties [1]. On the other hand, gradient projection is a concept that involves linear approximation of the constraints and, therefore, is not inherently suited for nonlinear constraints. In the gradient projection algorithm of Rosen [2], which was originally designed for linear constraints forming a bounded convex region, nonlinear constraints were to be accommodated by a restoration step. This procedure is known to be unsatisfactory if a minimum on the tangent plane approximation lies far from the constraint. The difficulty can be avoided by a simple modification of the gradient projection concept that penalizes violations of the constraints along the search direction [3,4].

Recently, Hestenes and Powell suggested that equality constrained minimization problems be reformulated as the unconstrained minimization of an augmented Lagrangian functional involving the sum of terms that are linear and quadratic in the constraint while using independent updates for the Lagrange multiplier [5,6]. Some extensions of the concept to inequality constrained optimization have also been obtained, including conditions under which the solution of the dual problem agrees with that of the primal [7,8]. An important computational advantage of the augmented Lagrangian concept is that a constrained minimum, if it exists, can be obtained by a finite value of the penalty constant. Computational experience using the technique, although mostl for equality constrained problems, has been encouraging [9,10,11,12,13,14].

This paper presents an extension of the modified gradient projection algorithm [15,16], based on an augmented Lagrangian with quadratic penalization. Numerical results from several problems involving static and dynamic, nonlinear equality and inequality constraints are given.

PROBLEM STATEMENT AND NOTATION

We shall treat the nonlinear programming problem of determining an element x in R^n that minimizes a nonlinear functional $f(x)$ subject to nonlinear equality constraint $g(x)=0$ in R^p. Inequality constraints will be discussed below under Numerical Results. It will be assumed that f and g are second continuously differentiable and that a relative minimum $\hat{x}$ exists. The gradient f_x is an n-tuple, the Jacobian a pxn matrix whose rows are gradients of the elements of g, and g_x^* denotes the transposed matrix. The standard inner product and derived norm on R^n are

$$< u, v >_n = \sum_{i=1}^{n} u_i v_i \ , \qquad \| u \|_n = < u, u >_n^{1/2} .$$

*W. A. Gruver is currently with the Department of Chemical Engineering & Graduate Program in Operations Research at North Carolina State University, Raleigh, North Carolina 27607 USA .

PROJECTION-RESTORATION WITH AN AUGMENTED LAGRANGIAN FUNCTIONAL

Let $p<n$ and the p gradients of the elements of g be linearly independent. Define the augmented Lagrangian functional

$$F(x,\lambda,K) = f(x) + \langle g, \lambda \rangle_p + 0.5\langle g, Kg \rangle_p \tag{1}$$

where λ is the Lagrange multiplier (a p-tuple), and K is a given pxp diagonal matrix of positive constants. A necessary condition that $F(x,\lambda,K)$ have a minimum for fixed K is that the following equations are satisfied: the constraint equation,

$$g(\hat{x}) = 0 \tag{2}$$

and the minimization condition

$$f_x(x) + g_x^*(\hat{x})\hat{\lambda} + g_x^*(\hat{x})Kg(\hat{x}) = 0 \tag{3}$$

An effective approach to the numerical solution of the constrained minimization problem is a two step procedure based on satisfying (2) followed by adjusting x by gradient descent to approximate a solution of (3). The procedure is repeated until a stopping condition is met.

Let the first correction δx, the restoration increment, to an initial guess x be chosen to satisfy (2) to first order,

$$g(x) + g_x(x)\delta x = 0 . \tag{4}$$

From (4), δx may be formally expressed as

$$\delta x = -g_x^+ g \tag{5}$$

where, for brevity, the dependence on x has been dropped, and g_x^+ is a pseudo inverse of g_x. Since $p<n$, there exist an infinite number of solutions to (4) of the form $\delta y + \delta z$ where δy is any solution of (4) and δz is an element in the nullspace of g_x. Uniqueness can be assured in (5) by use of the minimum norm pseudo inverse

$$g_x^+ = g_x^*(g_x g_x^*)^{-1} \tag{6}$$

which selects the smallest value of δx, in the norm on R^n, lying on the intersection of the tangent planes to the constraints. Linear independence of the gradients of the elements of g ensures that $g_x g_x^*$ is invertible. Using (5), an improved value of x is obtained from

$$\bar{x} = x - \hat{\alpha} g_x^*(g_x g_x^*)^{-1} g \tag{7}$$

$$\hat{\alpha} = \operatorname{argmin}\{ \| g(x - \alpha g_x^+ g) \|_p \mid \alpha \text{ in } R^+ \} .$$

The second correction $\delta\bar{x}$, the projection increment, is chosen so that $F(\bar{x},\lambda,K)$ is decreased in the direction of its negative gradient

$$\delta\bar{x} = -F_x(\bar{x},\lambda,K). \tag{8}$$

The multiplier λ needed to compute (8) may be formally expressed, from (3), in terms of a pseudo inverse as

$$\lambda = -(g_x^*)^+ f_x - Kg . \tag{9}$$

For $x \neq \hat{x}$, there is no value of λ in R^p for which (3) is satisfied. However, an approximation is obtained by use of the pseudo inverse

$$(g_x^*)^+ = (g_x g_x^*)^{-1} g_x \tag{10}$$

which selects the least squares value of λ in (9). Using (8), (9) and (10), $\bar{x}$ is updated according to the descent iteration

$$\bar{\bar{x}} = \bar{x} - \hat{\beta} P f_x \tag{11}$$

$$\hat{\beta} = \operatorname{argmin}\{F(\bar{x} - \beta P f_x,\lambda,K) \mid \beta \text{ in } R^+ \}$$

where

$$P = I - g_x^*(g_x g_x^*)^{-1} g_x \tag{12}$$

is an nxn matrix representing the operator that projects the gradient into the manifold formed by the intersection of the constraint tangent planes.

VARIABLE METRIC IMPLEMENTATION

For any symmetric, positive definite matrix G define the weighted inner product and derived norm,

$$< u, v >_G = < u, Gv >_n , \quad \|u\|_G = < u, Gu >_n^{1/2}$$

for u and v in R^n. Let H, the variable metric matrix, be an nxn symmetric, positive definite approximation to the inverse Hessian F_{xx}^{-1}, which is assumed to exist. The variable metric matrix generates the minimum norm pseudo inverse

$$g_x^+ = Hg_x^*(g_xHg_x^*)^{-1}, \tag{13}$$

which minimizes $\|\delta x\|_{H^{-1}}$, and the least squares pseudo inverse

$$(g_x^*)^+ = (g_xHg_x^*)^{-1}g_xH , \tag{14}$$

which minimizes $\|f_x + g_x^*\lambda + g_x^*Kg\|_H^2$.

Using (13) and (14) the restoration and projection increments are,

$$\delta x = -Hg_x^*(g_xHg_x^*)^{-1}g \tag{15}$$

$$\begin{aligned}\delta\bar{x} &= -H(f_x + g_x^*\lambda) \\ &= -HP\ f_x\end{aligned} \tag{16}$$

where

$$P = I - g_x^*(g_xHg_x^*)^{-1}g_xH \tag{17}$$

is an nxn matrix that represents a non-orthogonal projection operator. The variable metric matrix H is updated after each successful projection update according to established formulas [17]. The use of variable metric descent with gradient projection is due to Goldfarb [18] and Kelley and Speyer [4].

For fixed K suppose that a minimum of $F(x,\lambda,K)$ has been obtained. Let the resulting variable metric matrix be $\tilde{H}$. This matrix is now refined according to prescribed changes in the penalty constants $\Delta K = \bar{K} - K$ and the corresponding, although unknown, changes $\Delta\lambda = \bar{\lambda} - \lambda$ in the Lagrange multiplier. The update is based on choosing $\bar{\lambda}$ to minimize the modified residual

$$\|f_x + g_x^*) + g_x^*Kg\|_{\tilde{H}}^2 + \|\bar{\lambda}\|_{(\Delta K)^{-1}}^2 \tag{18}$$

where the second term in (18) is an energy constraint which prevents large changes in the multiplier. If ΔK is nonsingular, the minimum value of $\bar{\lambda}$ can be expressed in closed form as

$$\bar{\lambda} = -(g_x\tilde{H}g_x^* + \Delta K^{-1})^{-1}g_x\tilde{H}(f_x + g_x^*Kg) . \tag{19}$$

This corresponds to an initial projection increment

$$\delta\bar{x}^{(o)} = -H^{(o)}(f_x + g_x^*Kg)$$

$$H^{(o)} \doteq \tilde{H}Q \tag{20}$$

$$Q = I - g_x^*(g_x\tilde{H}g_x^* + \Delta K^{-1})^{-1}g_x\tilde{H} . \tag{21}$$

If one or more constraints are satisfied, (21) cannot be computed because ΔK is singular. In this case, (18) can be minimized sequentially for each nonzero component of $\Delta K = \text{diag}\ (\Delta k_1,\ldots,\Delta k\ )$. The corresponding expressions in place of (20) and (21) are

$$H^{(o)} \doteq \tilde{H}(\sum_{i\varepsilon J} Q_i) , \quad J = \{i \mid 1 \leqslant i \leqslant p,\ \Delta k_i \neq 0\} \tag{22}$$

$$Q_i = I - (\frac{1}{1/\Delta k_i + \|g_{ix}\|_{\tilde{H}}^2})g_{ix}^*g_{ix}\tilde{H} . \tag{23}$$

The above results in (20) to (23) were first obtained by Kelley, et. al. [19] by use of a Schur matrix identity. The derivation of this section, however, provides insight into the update's least squares structure and use with an augmented Lagrangian.

Variable metric implementation of the projection-restoration algorithm requires storage of an nxn symmetric matrix. For problems of large dimension, a less costlier alternative is offered by another class of conjugate direction methods known as conjugate gradient descent. Implementation is identical to that given in the last section except that the gradient F_x is updated before computing the projection increment [3].

NUMERICAL RESULTS

A computer program based on the above method has been written and applied to various problems. The results reported in this section were obtained on a Telefunken TR 440 computer using double precision arithmetic. Convergence of the algorithm was specified by the stopping conditions

$$\|F_x(x,\lambda,K)\|_n < \varepsilon_1 , \qquad \|g(x)\|_p < \varepsilon_2 .$$

Resetting of the variable metric matrix H was enforced after every 2(n-p+1) projection updates. The "complementary DFP" rank two variable metric update was employed [20] together with a gradient linear search by cubic fit for both the restoration and projection increments. The search required, on the average, four to five function evaluations, i.e. evaluation of $F(x,\lambda,K)$ for specified x, λ and K. A more accurate linear search had been used in earlier versions of the program without providing significant improvement.

Inequality constraints

$$g_i(x) \leq 0 , \quad i = p+1,\ldots,p+q$$

have been accommodated by replacing those elements of g in the penalty terms of (1) by $\max\{0,g_i(x)\}$ and by forming the gradients from the ε-active constraints defined by the index set

$$\{i \mid g_i(x) + \varepsilon_i \geq 0, \quad i=p+1,\ldots,p+q\}$$

where ε_i is a small positive constant. This device helps to avoid "jamming" and aids convergence of the algorithm. Further justification is given in the literature [21]. The remaining Kuhn-Tucker condition, that the Lagrange multiplier be non-negative, was not enforced during the descent. In addition, the search was terminated whenever a new constraint is violated.

In the following examples, one cycle is the completion of both a restoration and a projection update. Table 1 lists results for the test problems described below where N_c is the total number of cycles required for convergence within ε_1 and ε_2. N_r and N_p are the total number of function evaluations required for building the restoration and projection increments, respectively. The penalty matrix was set to the identity for the initial cycle and updated once. Following the penalty matrix update, the variable metric matrix was updated using (22) and (23). Convergence tolerences are $\varepsilon_1=10^{-6}$ and $\varepsilon_2=10^{-6}$ for Examples 1 to 4 and $\varepsilon_1=10^{-3}$ and $\varepsilon_2=10^{-6}$ for Examples 5 and 6.

Example 1 [10]. The problem is to minimize the functional

$$f(x) = (x_1-x_2)^2 + (x_2+x_3-2)^2 + (x_4-1)^2 + (x_5-1)^2$$

subject to linear equality constraints

$$g_1(x) = x_1 + 3x_2 = 0$$
$$g_2(x) = x_3 + x_4 - 2x_5 = 0$$
$$g_3(x) = x_2 - x_5 = 0 .$$

The constrained minimum of $f(\hat{x}) = 4.0930$ where

$$\hat{x} = (-.7674,\ .2558,\ .6279,\ -.1163,\ .2558)$$

was obtained from a starting point

$$x = (2,\ 2,\ 2,\ 2,\ 2) .$$

Example 2 [10]. The problem is to minimize the functional

$$f(x) = (x_1-1)^2 + (x_1-x_2)^2 + (x_3-1)^2 + (x_4-1)^4 + (x_5-1)^6$$

subject to nonlinear equality constraints

$$g_1(x) = x_4x_1^2 + \sin(x_4-x_5) - 2\sqrt{2} = 0$$
$$g_2(x) = x_2 + x_3^4x_4^2 - 8 - \sqrt{2} = 0 .$$

A constrained minimum of $f(\hat{x}) = .2415$ where

$$\hat{x} = (1.1661,\ 1.1821,\ 1.3802,\ 1.5060,\ .6109)$$

was obtained from a starting point

$$x = (2,\ 2,\ 2,\ 2,\ 2) .$$

Example 3 [6]. The problem is to minimize the functional

$$f(x) = x_1x_2x_3x_4x_5$$

subject to the nonlinear equality constraints

$$g_1(x) = x_1^2 + x_2^2 + x_3^2 + x_4^2 + x_5^2 - 10 = 0$$
$$g_2(x) = x_2x_3 - 5x_4x_5 = 0$$
$$g_3(x) = x_1^3 + x_2^3 + 1 = 0 .$$

A constrained minimum of $f(\hat{x}) = -2.9197$ where

$$\hat{x} = (-1.7171,\ 1.5957,\ 1.8272,\ .7636,\ .7636)$$

was obtained from a starting point

$$x = (2,\ 2,\ 2,\ 2,\ 2) .$$

Example 4. The problem is to minimize the functional

$$f(x) = x_1^2 + 9x_2^2 + 9x_3^2$$

subject to the mixed constraints

$$g_1(x) = x_1^2 + x_2^2 - 1 = 0$$
$$g_2(x) = x_2 + 0.5 \leq 0 .$$

The constrained minimum of $f(\hat{x}) = 1$ where

$$\hat{x} = (0,\ -1,\ 0)$$

was obtained from a starting point

$$x = (2,\ 2,\ 2) .$$

Example 5. A nontrivial test of the algorithm is provided by a problem involving the optimal positioning of a geostationary satellite [22,15]. We seek a minimum fuel control function and corresponding optimal trajectory subject to dynamic constraints of two body motion, initial state (position and velocity) on a specified transfer orbit, and rendezvous with a specified state on a target orbit. The problem is formulated in a finite dimensional space by requiring control in terms of N impulsive velocity increments which act as equivalent initial conditions for the equations of two body motion. A closed form solution of the resulting initial condition problem and the forward sensitivity matrices due to Goodyear enables accurate evaluation of the constraints and their gradients over the trajectory. The parameter set is the 4N-tuple

$$\pi = (t_1,\ldots,t_N,c_1,\ldots,c_N)$$

consisting of the switching times and velocity increments, respectively. Six equality constraints are obtained from the rendezvous requirement. Additional inequality constraints specify bounds on the switching times and magnitudes of the velocity increments.

In this example, the algorithm is used to compute the positioning control using three impulses (n=12) for the following input data. Transfer orbit: semi major axis, a=25078 km, eccentricity, e=.736047; inclination, i=5°; argument of perigee, ω=180°; longitude of the ascending node, Ω=180°; eccentric anomaly, $E(t_A)$=180°. Target orbit position, $r(t_A)$=(-42164.22,0,0) km and velocity, $v(t_A)$=(0,-3074.65,0) m/sec of the rendezvous point where t_A is the time of apogee passage in the transfer orbit. Position is relative to an earth centered coordinate system with the x-y axis in the plane of the equator. In addition to the six rendezvous (equality) constraints, an inequality constraint was specified by the requirement that the first velocity increment not exceed 1440 m/sec.

Using a starting point of

$$t_1 = 0 \text{ hr} \qquad c_1 = (0, -1420, 0) \text{ m/sec}$$
$$t_2 = 12 \qquad c_2 = (0, -50, 0)$$
$$t_3 = 24 \qquad c_3 = (0, -20, 0) ,$$

the optimal control parameters obtained were

$$\hat{t}_1 = .017 \qquad \hat{c}_1 = (1.94, -1435.6, -111.9)$$
$$\hat{t}_2 = 11.33 \qquad \hat{c}_2 = (17.6, -29.3, 16.7)$$
$$\hat{t}_3 = 22.8 \qquad \hat{c}_3 = (-1.06, -14.5, -8.5) ,$$

corresponding to a minimum fuel cost of 1494.9 m/sec. A more detailed description of this and related examples is given in reference [15].

Example 6. A second application of the algorithm to trajectory optimization is provided by a proposed rendezvous with a comet [23]. For scientific investigation of the comet as it approaches the sun, it is required that the maximum allowable fly-by velocity of the spacecraft be small. By defining the cost functional as the velocity increment required for injection from Earth orbit plus that required for rendezvous, a solution yielding the minimum fuel for the best fly-by is obtained. The impulsive trajectory formulation used in Example 5 is applied and the algorithm is used to compute the optimal parameters using two impulses (n=8) for the following input data. Initial (Earth) orbit: $a=1.5\text{x}10^8$ km, $e=.0167$, $i=0°$, $\omega=101.2208°$, $\Omega=0°$, and the mean anomaly $M(t_a)=-101.2208$ where $t_a=0$ corresponds to an epoch of September 23, 1980. Final orbit: $a=3.3225\text{x}10^8$ km, $e=.8462$, $i=12.35°$, $\omega=185.2°$, $\Omega=334.72°$, $M(t_b)=0$ where $t_b=0$ corresponds to an epoch of December 12, 1980. Six equality constraints are specified by the rendezvous requirement.

Using a starting point of

$$t_1 = 0 \text{ days} \qquad c_1 = (-5.9, -5.9, 0) \text{ km/sec}$$
$$t_2 = 100 \qquad c_2 = (-11, -13, 4) ,$$

the optimal control parameters obtained were

$$\hat{t}_1 = -24.75 \qquad \hat{c}_1 = (-3.005, -8.421, 4.182)$$
$$\hat{t}_2 = 81.16 \qquad \hat{c}_2 = (-1.102, -6.539, -2.416) ,$$

corresponding to a launch date of August 29, 1980 and a flight time of 106 days. The injection velocity is 9.87 km/sec and the fly-by velocity is 7.05 km/sec (Figure 1).

CONCLUSIONS AND COMMENTS

Although the algorithm is heuristic in the sense that convergence criteria are not provided, numerical experiments by the authors indicate that the program is generally more reliable than that employing only gradient projection or penalization alone. The main difficulty to date, and one that is familiar to users of penalty function techniques, is the proper choice and updating of the penalty constants. However, bounds on the magnitudes of the constants are available [11].

The algorithm has also been applied to a problem involving the optimal control of heat distribution at the interface of an inhomogeneous rectangular solid. In view of the large number of parameters resulting from time and spatial discretization of the control function, the gradient of the cost is most efficiently evaluated from the state and adjoint (backward sensitivity) equations. In this case, the Green's function for the system and adjoint are simply related due to the assumed linearity of the model and can be explicitly obtained in terms of a Fourier series.

ACKNOWLEDGEMENT

The work of W. A. Gruver was supported through a Senior Scientist Award administered by the U. S. Special Program of the Alexander von Humboldt Foundation.

REFERENCES

1. Lootsma, F. A.: A Survey of Methods for Solving Constrained Minimization Problems via Unconstrained Minimization, in F. Lootsma (ed), Numerical Methods for Nonlinear Optimization, Academic Press, New York, 313-347 (1972).

2. Rosen, J. B.: The Gradient Projection Method for Nonlinear Programming; Part I: Linear Constraints, J. SIAM, 8, 181-217 (1960); Part II: Nonlinear Constraints, J. SIAM, 9, 414-443 (1961).

3. Miele, A., Huang, H., and Heideman, J.: Sequential Gradient-Restoration Algorithm for the Minimization of Constrained Functions - Ordinary and Conjugate Gradient Versions, J. Optimization Theory and Applications, 4, No 4, 213-242 (1969).

4. Kelley, H. J. and Speyer, J. L.: Accelerated Gradient Projection, in Lectures in Mathematics, 132, Springer Verlag, Berlin-Heidelberg, 151-158 (1970).

5. Hestenes, M. R.: Multiplier and Gradient Methods, in L. Zadeh (ed), Computing Methods in Optimization Problems, 2, Academic Press, New York, 143-163 (1969).

6. Powell, M. J. D.: A Method for Nonlinear Constraints in Minimization Problems, in R. Fletcher (ed), Optimization, Academic Press, New York, 283-298 (1969).

7. Roode, J. D.: Generalized Lagrangian Functions in Mathematical Programming, Thesis, University of Leiden, Netherlands, (1968).

8. Rockafellar, R. T.: Augmented Lagrange Multiplier Functions and Duality in Nonconvex Programming, SIAM J. of Control, to appear.

9. Haarhoff, P. C. and Buys, J. D.: A New Method for the Optimization of a Nonlinear Function Subject to Nonlinear Constraints, Computer J., 13, 178-184 (1970).

10. Miele, A., Cragg, E., Iyer, R., and Levy, A.: Use of the Augmented Penalty Function in Mathematical Programming Problems, Part 1, J. Optimization Theory and Applications, 8, 115-130 (1971).

11. Mårtensson, K.: Methods for Constrained Function Minimization, Report 7107, Div. of Automatic Control, Lund Institute of Technology, Sweden, March 1971.

12. Glad, T.: Lagrange Multiplier Methods for Minimization Under Equality Constraints Report 7323, Div. of Automatic Control, Lund Institute of Technology, Sweden, August 1973.

13. Tripathi, S. S. and Narendra, K. S.: Constrained Optimization Problems Using Multiplier Methods, J. Optimization Theory and Applications, 9, 59-70 (1972).

14. Wierzbicki, A. P.: A Penalty Function Shifting Method in Constrained Static Optimization and its Convergence Properties, Archiwum Automatyki i Telemechaniki, 16, 395-416 (1971).

15. Gruver, W. A. and Engersbach, N. H.: Nonlinear Programming by Projection-Restoration Applied to Optimal Geostationary Satellite Positioning, AIAA Journal, December 1974.

16. Engersbach, N. H. and Gruver, W. A.: Constrained Optimization Based on Generalized Exterior Point Methods, Report IB013-72/10, Deutsche Forschungs- und Versuchsanstalt für Luft- und Raumfahrt, December 1972.

17. Broyden, C. G.: Quasi-Newton Methods, in W. Murray (ed), Numerical Methods for Unconstrained Optimization, Academic Press, 87-106 (1972).

18. Goldfarb, D.: Extension of Davidon's Variable Metric Method to Maximization Under Linear Inequality and Equality Constraints, SIAM J. Applied Math., 17, 739-764, July 1969.

19. Kelley, H. J., Denham, W., Johnson, I., and Wheatley, P.: An Accelerated Gradient Method for Parameter Optimization with Nonlinear Constraints, J. Astronautical Sciences, 13, No 4, 166-169, July-August 1966.

20. Goldfarb, D.: A Family of Variable-Metric Methods Derived by Variational Means, Maths. Computation, 24, 23-26 (1970).

21. Zangwill, W. I.: Nonlinear Programming, Prentice-Hall, Englewood Cliffs, N. J., Chapter 13 (1969).

22. Gruver, W. A. and Engersbach, N.: A Mathematical Programming Approach to the Optimization of Constrained, Impulsive, Minimum-Fuel Trajectories, Report IB013-72/3, Deutsche Forschungs- und Versuchsanstalt für Luft- und Raumfahrt, June 1972.

23. Eckstein, M. C. and Jochim, E. F.: Vorläufige Untersuchung zur Bahnoptimierung für die Helio-C Mission, Report IB522-73/1, Deutsche Forschungs- und Versuchsanstalt für Luft- und Raumfahrt, March 1973.

TABLE 1

CONVERGENCE PROPERTIES OF THE PROJECTION-RESTORATION METHOD

Problem	Cycles N_c	Function Evaluations N_r	N_p	CPU Time[a] sec	Parameters n	Constraints p	q
1	4	20	20	0.668	5	3	
2	11	39	66	1.784	5	2	
3	5	27	25	0.902	5	3	
4	7	36	39	1.350	3	1	1
5	16	61	75	33 [b]	12	6	1
6	15	51	75	20 [b]	8	6	

[a] Telefunken TR 440 Computer (double precision).

[b] Includes evaluation of dynamical system constraints and gradients.

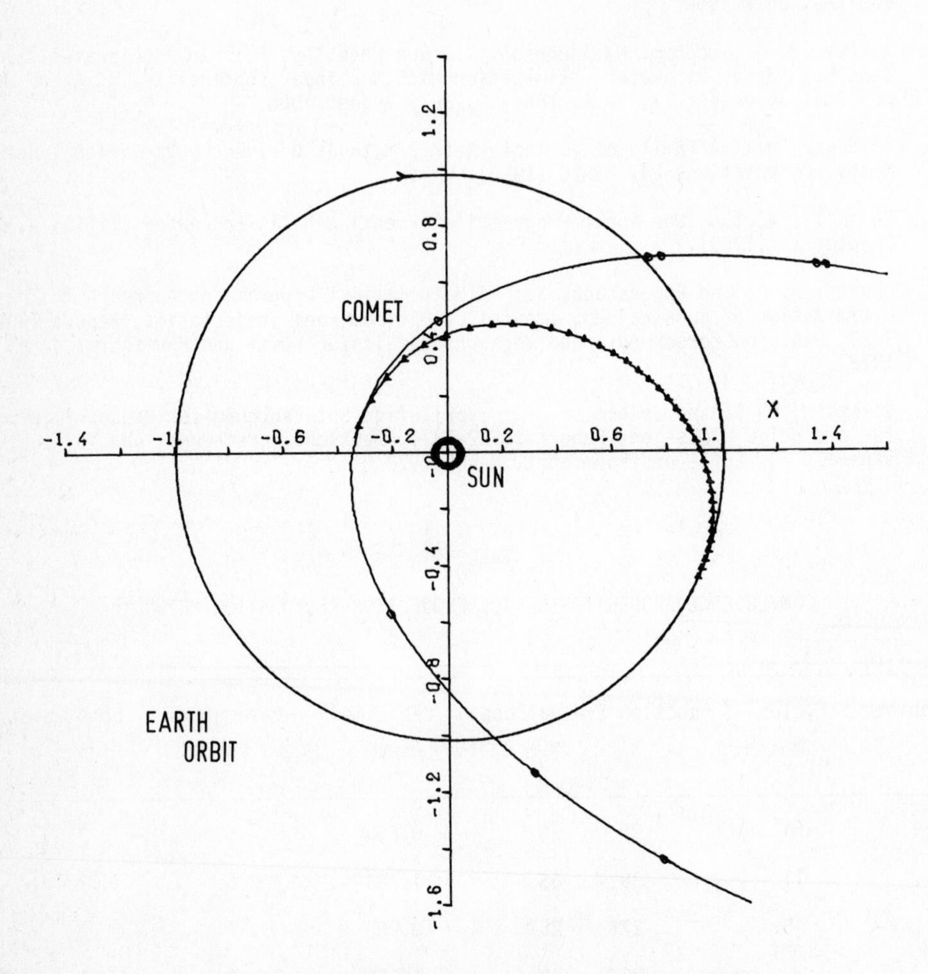

FIGURE 1. RENDEZVOUS COMET ENCKE 1980

LIMIT EXTREMUM PROBLEMS

Yu.M. Ermol'ev, E.A. Nurminskiy
Institute of Cybernetics
Ukrainian Academy of Sciences
252627 Kiev 127 USSR

INTRODUCTION

In applications one has often to tackle extremum problems where the objective function happens to be not strictly fixed as in the general theory of nonlinear programming but changes versus some parameter (time, in particular). That is, instead of $F(x)$ there is a sequence of functions $F^N(x)$ in a certain sense approximating $F(x)$, on the basis of which the extremum of $F(x)$ is to be found. As a rule, one fails to execute the passage to the limit, to find $F(x)$ and then its extremum due to a number of circumstances of which the following might be emphasized:

1. The parameter N corresponds to the discrete time and $F^N(x)$ becomes known at the instant $t=N$ only. In this case the limit passage takes a whole time given for the problem solution.

2. The parameter N is an index of members of the sequence. It may be changed at one's discretion, "frozen", in particular, at some stages of the optimization process, however, the execution of limit passage is technically difficult. Such cases are particularly characteristic of problems of optimizing steady regime of controlled processes when averaged performance figures of the form

$$F(x) = \lim_{N\to\infty} F^N(x) = \lim_{N\to\infty} \frac{1}{N}\sum_{k=1}^{N} g(k,x),$$

$$F(x) = \lim_{N\to\infty} F^N(x) = \lim_{N\to\infty} \sum_{k=1}^{N} \alpha^k g(k,x).$$

are dealt with.

3. The limit passage operation deteriorates some good properties of the function $F^N(x)$ which is characteristic of approximation problems when a function, being due to some reasons "bad", is approximated by a sequence of "good" ones and thus, instead of $F(x)$, it is of advantage to manipulate with the functions $F^N(x)$. An interesting and difficult problem arises here of optimizing the limit function $F(x)$ with the only use of information about members of the sequence $F^N(x)$ approximating $F(x)$. It is important to have in mind that if $F(x)$ is unknown then the examination of only one of the functions $F^N(x)$ in solving the problem approximately does not allow estimation of the accuracy of the obtained approximate solution. Such extremum searching procedures are treated here in which the search of the function $F^N(x)$ extremum is based on the analysis of a sequence of functions $F(x)$. The paper grounds on results of [3].

ALGORITHMS

Consider the extremum problem

$$\min F(x), \; x \in X, \tag{1}$$

where X is a convex closed bounded set, $F(x)$ - a convex but not necessarily continuously differentiable function determined as the limit of a functional sequence:

$$F(x) = \lim_{N \to \infty} F^N(x).$$

The following quite natural algorithm may be offered for the problem solution

$$x^{s+1} = \Pi_X (x^s - \rho_s \hat{F}_x^s(x^s)), \; s = 0, 1, \ldots, \tag{2}$$

where $\rho_s \geq 0$ are step-by-step factors, $\hat{F}_x^s(x^s)$ - a generalized gradient of the convex function $F(x)$, $\Pi_X(\cdot)$ - an operator of projection on the set X. The following theorem holds.

THEOREM 1. Let $F^s(x)$ be convex functions for each s, the sequence $F^s(x)$ uniformly converges on X, and $\rho_s \to 0, \sum \rho_s = \infty$. Then for each convergent subsequence $\{x^{s_k}\}$

$$\lim x^{s_k} = x^* \in X^*,$$

$$\lim F^s(x^s) = F^* = \min \{F(x), x \in X\},$$

where X^* is a set of solutions of problem (1). The requirement for the sequence $F^N(x)$ to converge uniformly is the most essential of the theorem assumptions. However, when functions $F^N(x)$ are convex the uniform convergence follows readily from the point one if some additional fairly weak assumptions are introduced. The rest of conditions do not differ from those of convergence of the known method of generalized gradients, formulated in [1], however, the study of convergence by method (2) with the direct application of the scheme in [1] under assumptions of the theorem is impossible. Theorem 1 is proved in [3] by an approach elaborated in [2]. Of great interest is also a convergence of the stochastic analogue of algorithm (2):

$$x^{s+1}(\omega) = \pi_X(x^s(\omega) - \rho_s \xi^s(x^s, \omega)), \tag{3}$$

where ξ^s is a random vector (a stochastic quasi-gradient) whose conditional expectation

$$E(\xi^s / x^0, \ldots, x^s) = \hat{F}^s_x(x^s). \tag{4}$$

THEOREM 2. Let assumptions of Theorem 1 be satisfied and $\Sigma \rho_s^2 < \infty$. Then algorithm (3) converges in the sense that for almost all ω the limit points of sequence $\{x^s(\omega)\}$ belong to the set X^* and with probability 1 $\lim\limits_s F^s(x^s) = F^*$.

APPLICATIONS

On the method of penalty functions. With the use of the method of penalty functions the problem of minimization of $f^0(x)$ in the domain $\mathcal{D}$ determined by the constraints

$$f^i(x) \le 0, \ i = 1, \ldots, m, \quad x \in X$$

is approximated by the minimization of a function $F(x, c)$ in the domain X so that

$$\min_{x \in \mathcal{D}} f^0(x) = \lim_{c \to c^*} \min_{x \in X} F(x, c)$$

for some C^*, the C^* being often equal to 0, ∞, or must be a sufficiently large number. The C being fixed, the minimization of $F(x,c)$ in the domain X does not yield, generally speaking, the precise solution of the initial problem. However, if such arbitrary sequence $C^N, N=0,1,\ldots$ is chosen that $C^N \to C^*$ and if the limit extremum problem with the function $F^N(x) = F(x, C^N)$ is studied then under appropriate conditions the precise solution will be obtained by method (2). A question is interesting about choosing ways of the sequence that impacts the speed of convergence.

INTERCONNECTED EXTREMUM PROBLEMS

Sometimes there is a set of interconnected extremum problems in which solution of one problem prepares information for solving the others. In problems of vector optimization, for instance, the minimization problem

$$F(x) = \max_{e} \{ f^{e}(x) - \min_{x \in X} f^{e}(x) \}$$

is dealt with when choosing a compromise solution. In this case the problems of minimizing $f^{e}(x)$ prepare information for the basic problem of minimization of the function $F(x)$. Since the solution of each auxiliary problem necessitates an infinite number of iterations, the direct way to calculate $F(x)$ even in a separate point, to say nothing of its extremum search, is a nonconstructive one. In addition, if sequences of points $x^{e}(N)$ such that

$$f^{e}(x^{e}(N)) \to \min_{x \in X} f^{e}(x), \; N \to \infty,$$

are considered together with the limit extremum problem with the function

$$\overset{N}{F}(x) = \max_{e} \{ f^{e}(x)^{e} - f^{e}(x^{e}(N)) \},$$

then, if conditions of Theorem 1 are satisfied, procedure (2) helps us to find the precise minimum of $F(x)$.

OPTIMIZATION OF STEADY REGIMES

The results of Theorem 2 offer quite an effective way of solving function minimization problems of the form

$$F(x) = \lim_{N \to \infty} \frac{1}{N} \sum_{K=1}^{N} g(K, x),$$

It is easily seen that with proper assumptions about the differentiability of functions $g(K,x)$ the stochastic gradient $\xi^{s}(x,\omega)$, satisfying relations (4), can be determined, for instance, as follows:

$$\xi^{s}(x, \omega) = g_{x}(\omega^{N}, x)$$

where ω^{N} is a random variable uniformly distributed on the set $\{1, 2, \ldots \omega\}$ Such a construction allows to calculate at each iteration a derivative of only one member of the increasing sum of terms $\sum_{K}^{N} g(K,x)$. A similar method is also applicable for minimizing

$$F(x) = \lim_{N \to \infty} \sum_{K=1}^{N} \alpha^{K} g(K, x)$$

ON THE RANDOM SEARCH METHOD

When the calculation of gradient of the objective function is complex the following method might prove helpful which also necessitates the solution of limit extremum problems. Instead of the objective function $F(x)$ we consider the function

$$F(x,\alpha) = EF(x - \eta(\alpha)) = \int_{E_n} F(x-y) P(y,\alpha) dy,$$

where the distribution $P(y,\alpha)$ of the random variable $\eta(\alpha)$ for $\alpha \to 0$ concentrates in 0, i.e. $F(x,\alpha) \to F(x), \alpha \to 0$. Then subject to existence of corresponding integrals and $F(x-y) P(y,\alpha) \to 0, y \to 0,$

$$F_x(x,\alpha) = \int_{E_n} F_x(x-y) P(y,\alpha) dy = -\int_{E_n} F(x-y) \frac{P_y(y,\alpha)}{P(y,\alpha)} P(y,\alpha) dy.$$

Thus, the random variable

$$-F(x - \eta(\alpha)) \frac{P_y(\eta(\alpha),\alpha)}{P(\eta(\alpha),\alpha)}$$

for the fixed x coincides on average with the gradient $F(x,\alpha)$. The examination of the sequence $\alpha^s \to 0$ and the limit extremum problem with functions $F^s(x) = F(x,\alpha^s)$ completed, we obtain the possibility to organize, by procedure (3) and for

$$\xi^s(x,\omega) = -F(x - \eta(\alpha^s)) \frac{P_y(\eta(\alpha^s)\alpha^s)}{P(\eta(\alpha^s),\alpha^s)},$$

the iterative process where derivatives of $F(x)$ are not employed.

REFERENCES

1. Yu.M. Ermol'ev, Methods for Solving Nonlinear Extremum Problems, Kibernetika, No. 4, 1966.
2. E.A. Nurminskiy, Convergence Conditions of Nonlinear Programming Algorithms, Kibernetika, No.6, 1972.
3. Yu.M. Ermol'ev, E.A. Nurminskiy, Limit Extremum Problems, Kibernetika, No. 4, 1973.

ALGORITHMS FOR SOLVING NON-LINEAR PROGRAMMING PROBLEMS

Yu.G.Evtushenko

Computing Center USSR. Academy of Sciences. Moscow

1. We consider the following problems

$$\min_{x \in X_1} F(x), \quad X_1 = \{x \in E_n : g(x) = 0,\ h(x) \leq 0,\ 0 \leq x\} \tag{1}$$

$$\min_{x \in X_2} F(x), \quad X_2 = \{x \in E_n : g(x) = 0,\ h(x) \leq 0\} \tag{2}$$

where F, g, h are continuously differentiable functions defined on E_n, Euclidean n space, $x = [x^1, x^2, \ldots, x^n]$ is a point in E_n, functions F, g, h define the mapping $F : E_n \to E_1$, $g : E_n \to E_l$, $h : E_n \to E_c$, $\mathcal{R} = [g, h] \in E_{l+c}$. We define two open sets

$$X_1^0 = \{x : g(x) = 0,\ h(x) < 0,\ x > 0\},\ X_2^0 = \{x : g(x) = 0,\ h(x) < 0\}.$$

$\mathcal{R}_x$ is the $n \times m$ matrix whose ij-th element is $\partial \mathcal{R}^j / \partial x^i$; F_x is the $n \times 1$ column matrix whose ij-th element is $\partial F / \partial x^i$; $\mathcal{D}(x)$, $\mathcal{D}(\sqrt{x})$, $\mathcal{D}(\mathcal{R})$, $\mathcal{D}(\sqrt{-\mathcal{R}})$ are diagonal matrices of an order of $n \times n$, $n \times n$, $m \times m$, $m \times m$ respectively whose j-th diagonal elements are x^j, $\sqrt{x^j}$, $\mathcal{R}^j$, $\sqrt{-\mathcal{R}^j}$ respectively.

For numerical solution of (1) we propose to find the points which are the limits (when $t \to \infty$) of a solution to the following Cauchy problem [1]

$$\dot{x} = -\mathcal{D}(x)[F_x + \mathcal{R}_x v],\ (\dot{\cdot}) = d/dt,\ x(0) = x_0 \in X_1^0 \tag{3}$$

where $v \in E_m$, $m = l + c$ is defined from a linear set of equations

$$A v + \mathcal{R}_x' \mathcal{D}(x) F_x = 0,\ A = \mathcal{R}_x' \mathcal{D}(x) \mathcal{R}_x - \mathcal{D}(\mathcal{R}).$$

We suppose that the sets X_1^0, X_2^0 are non-empty and the set $Z = \{x: F(x) \le F(x_0),\ x \in X_1\}$ is bounded.

Let e^i denote a unit vector whose i-th component is equal to unity.

Definition: The constraint qualification holds at point x, if the vectors $R_x^j(x)$ subject to $R^j(x)=0$, $j \in [1:m]$ and vectors e^i subject to $x^i = 0$ are linearly independent.

Lemma 1: If constraint qualification holds at any point $x \in X_1$, then $A(x)$ is a nonsingular positive definite matrix at any point $x \in X_1$.

The right sides of system (3) are continuous throughout X_1, and there exists a solution to C uchy problem (3). We assume, moreover, that it is unique. Calculating the first derivative of $R(x)$, with respect to system (3), we obtain

$$\dot{R} = -D(R)\vartheta, \quad \dot{F} = -\|D(\sqrt{x})H_x\|^2 - \|D(\sqrt{-R})\vartheta\|^2,$$

$$\text{where } \|z\| = z'z, \quad H_x = F_x + R_x\vartheta.$$

Starting at any point $x_0 \in X_1^0$, a continuous solution $x(x_0,t)$ to (3) exists, and $x(x_0,t)$ remains in X_1^0, for all $t \ge 0$. Let $\tilde{x}$ denote a solution to (3) and V be a neighborhood of $\tilde{x}$.

Theorem 1: Let the set Z be compact, X_1^0, be non-empty, the set of equilibrium points of system (3) be finite on X_1, the constraint qualification hold on the set X_1. Then a solution $x(x_0,t)$ to system (3) converges to the point $\tilde{x}$ for any $x_0 \in V \cap X_1^0$. Moreover, if the function $F(x)$ and the set X_1 are strictly convex, then the solution $x(x_0,t)$ converges to a global solution to (1) for any $x_0 \in X_1^0$.

We can also use a discrete version of (3)

$$x_{s+1} = x_s - \alpha M(x_s)F_x(x_s), \quad M = D(x) - D(x)R_x[R_x'D(x)R_x - D(R)]^{-1}R_x'D(x) \quad (4)$$

We denote $\beta = \max_{x \in Z}\|\vartheta(x)\|$, $a = \max_{x \in Z}\|H_x(x)\|$, $\lambda = \max_{y \in E_n}\max_{x \in Z} y'M(x)y/\|y\|^2$.

We assume that the gradient F_x satisfies Lipshitz condition with constant L on X_1.

Theorem 2: If the assumptions of the previous theorem are satisfied, the function $R(x)$ is linear, $0<\alpha<\min[1/a, 1/b, 2/\lambda L]$ then a solution x_s to (4) converges to $\tilde{x}$ for any $x_0 \in V \cap X_1^0$ and $F(x_{s+1}) \leq F(x_s)$, $x_s \in X_1^0$ for $s=0,1,2,\ldots$

It is possible to change α at each step using a method of steepest descent under the condition $x_s \in X_1^0$. In some particular problems method (3) coincides with the methods proposed in [2,3]

2. Associated with the general non-linear programming problem (2) are verious dual problems. This section develops a new form of a dual problem. Define a modified Lagrangian

$$N(z) = F(x) + \sum_{i=1}^{\ell} p^i g^i(x) + \sum_{i=1}^{c} h^i(x)(w^i)^2$$

where $w \in E_c$, $p \in E_\ell$, $z = (x, p, w) \in E_{n+\ell+c}$.

Consider the max-min problem

$$J = \max_{p \in E_\ell} \max_{w \in E_c} \min_{x \in E_n} N(x, p, w) \tag{5}$$

The necessary conditions for a max-min are given by

$$N_x(z) = N_p(z) = N_w(z) = 0 \tag{6}$$

We shall solve max-min problem (5) instead of (2). The application of a modified Lagranjian w permits us to transform problem (2) to an unconstrained max-min problem. In dealing with N, the multiple w is not constrained to be non-negative, in contrast to an ordinary Lagrangian. Using gradient, max-min algorithms [4,5] yield the following three methods

$$\dot{x} = -N_x, \quad \dot{p} = N_p, \quad \dot{w} = N_w \tag{7}$$

$$\dot{x} = -N_x, \quad \dot{p} = g - g_x' N_{xx}^{-1} N_x, \quad \dot{w} = 2D(w)[h - h_x' N_{xx}^{-1} N_x] \tag{8}$$

$$\dot{x} = -N_x - N_{xx}^{-1}(g_x g + 4h_x D(w) D(w) h), \quad \dot{p} = g, \quad \dot{w} = 2D(w)h \tag{9}$$

Newton's method gives

$$N_{zz}(z)\,\dot{z} = -N_z(z), \tag{10}$$

$$z_{s+1} = z_s - N_{zz}^{-1}(z_s)\,N_z(z_s). \tag{11}$$

Let us denote the points satisfying (6) by $z_* = (x_*, p_*, w_*)$.

Assume now that $N(z)$ is twice continuously differentiable in all arguments in some neighborhood of z_*.

Lemma 2: Let the multiple w_* satisfy the strict complementary slackness conditions, the constraint qualification hold at a feasible point x_*, $N_{xx}(z_*)$ be a positive definite matrix. Then z_* is an isolated local saddle point of $N(z)$ and x_* is a local solution to (2).

We can use the finite difference approximation to (7)-(10). For example, (7) yields

$$x_{s+1} = x_s - \alpha N_x(z_s),\quad p_{s+1} = p_s + \alpha N_p(z_s),\quad w_{s+1} = w_s + \alpha N_w(z_s) \tag{12}$$

where α is the step length.

Theorem 3: If the assumptions of Lemma 2 are satisfied, then the alogorithms (7)-(10) converge locally to z_*.

The finite defference approximations to (7)-(10), similar to (12), converge locally to z_* if α is sufficiently small.

The method (11) converges quadratically to z_*.

3. In this section two modifications of a penalty function algorithm are presented. Consider a max-min problem associated with(2)

$$\max_{0\le\tau\le T}\ \min_{x\in E_n} P(x,\tau),\quad P(x,\tau) = F(x) + \tau\Big[\sum_{i=1}^{l}(g^i)^2 + \sum_{i=1}^{c}\big[h^i + |h^i|\big]^2\Big] \tag{13}$$

The pair $\tau = T$, $x \in Z$ will be a solution to this problem, where

$$Z = \{\bar{x} \in E_n : \min_{x\in E_n} P(x, T) = P(\bar{x}, T)\}.$$

Let X_* denote the set of solutions to problem (2). It is easy to show that if Z, X_* are non-empty, then there exists such a scalar γ that the following inequality holds for any $\tilde{x} \in Z$

$$F(x_*) - \gamma/\tau \leq \mathcal{P}(\tilde{x}, T) \leq F(x_*).$$

If $F(x)$ is bounded below $F(x) \geq \delta$ then $\sum_{i=1}^{\ell} [g^i(x)]^2 + \sum_{i=1}^{c} [h^i(x) + |h^i(x)|]^2 \leq [F(\tilde{x}) - \delta]/T$. By taking T sufficiently large we can thereby find an approximate solution to problem (2) with the required accuracy.

We shall call the constraints essential in problem (2) if the unconstrained infimum of $F(x)$ differs from solution to (2). To solve (13) we shall use a method similar to (3)

$$\dot{x} = -\mathcal{P}_x(x, \tau), \quad \dot{\tau} = \mathcal{P}_\tau(T - \tau), \quad \tau(0) = 0, \quad x(0) = x_0 \qquad (14)$$

<u>Theorem 4:</u> Let F, h be convex, continuously differentiable functions, $g(x)$ be a linear function, Z and X_* be non-empty compact sets, the constraints be essential. Then a solution $x(x_0, t)$ to (14) converges to Z for arbitrary $x_0 \in E_n$.

Another algrithm can be used

$$\dot{x} = -\mathcal{P}_x(x, \tau) \quad \text{where } 0 \leq \tau(t) \leq d\tau/dt, \; x(0) = x_0 \qquad (15)$$

<u>Theorem 5:</u> If the assumptions of the previous theorem are satisfied and X_2^0 is a non-empty set, then a solution $x(x_0, t)$ to (15) converges to X_* for any $x_0 \in E_n$.

References

I.Ю.Г.Евтушенко. Два численных метода решения задач нелинейного программирования.Доклады АН СССР,I974,т.2I5,№I,38-40.

2.Ю.Г.Евтушенко,В.Г.Жадан. Численные методы решения некоторых задач исследования операции. Ж.вычисл.матем. и матем.физ., I973, I3, №3, 583-598.

3.И.И.Дикин. О непрерывных аналогах метода внутренних точек. Управляемые системы,Новосибирск,I97I, № 9.

4.Ю.Г.Евтушенко.Некоторые локальные свойства минимаксных задач. Ж.вычисл.матем.и матем.физ.,I974,I4,№ 3, 669-679.

5. Ю.Г.Евтушенко. Итеративные методы решения минимаксных задач. Ж. вычисл. матем. и матем. физ. 1974, 14, № 5.

STRUCTURAL OPTIMIZATION

C. FLEURY and B. FRAEIJS de VEUBEKE
Aerospace Laboratory
University of Liège
Belgium

PART I : GENERAL CONCEPTS

I. INTRODUCTION

Numerical methods of structural analysis have reached a high standard of efficiency. As a consequence they tend to overgrow their usefulness as numerical checks of stress distribution, amplitudes of displacement, natural frequencies and elastic stability to become adjuvants to design procedures.
The objectives of design vary according to the purpose of the structure. In aerospace engineering the weight is the prominent factor and is often the only goal of numerical optimization studies. In other cases the functional that is subject to minimization is more complex, economical factors of various kind being incorporated with their relative weights into the cost function.
Until recently the minimization was carried out by trial and error, the preliminary design and the modifications introduced after evaluation of a numerical structural analysis being largely based on engineering judgment. Presently there is a tendancy to a more scientific approach in which the changes in design parameters are evaluated on the basis of algorithms. Efficient algorithms are those that tend to bring the functional to its minimum with the smallest number of iterations requiring a subsequent structural reanalysis. Moreover they have to satisfy many kinds od side constraints such as :

- remain within the elastic limits of the material in each structural member under a given set of load distributions;
- keep displacement-type limitations;
- avoid elastic instability;
- keep natural frequencies within prescribed limits;
- keep member sizes above minimum values.

The starting point of such automated Structural Optimization Programs is a given preliminary design. It is therefore difficult to evaluate the cost of optimization procedures, since the computer time devoted to reach a near-optimal stage will heavil depend on the quality of the preliminary design.
In case where the unicity of the optimal solution is not guaranteed a poor preliminary design can even lead to a local and not to the global minimum.

For this reason, while optimization programs will probably remain essential tools, the objectives of optimality will also tend to incorporate the computer in the preliminary design stage. This more direct approach towards an optimal structure is the aim of "Computer aided Design".
It is also a much more ambitious goal and, fortunately perhaps, will never obliterate the exercize of engineering art. It is indeed difficult to conceive a selection by the computer of the best "topology" of structural members to carry the loads according to the purpose of the structure, taking immediatly the effect of side constraints into account. On the other hand, once the topology has been fixed by engineering judgement and experience, we will probably reach the stage where the computer will carry out from there the sizing of the members and even such other alterations in their geometry, permissible under the given topology and external constraints. Whether such ambitious programs will ever become operational within economical limits is a question that only experience will answer.

2. DESIGN VARIABLES

One can divide design variables in groups according to their relative importance. For aerospace structures, with a finite element method idealization, the following groups are proposed :

2.1. Element sizes

They comprize cross-sectional areas of beam, membrane and plate thicknesses ...
The optimization of those variables alone leaves the topology (system of element interconnexions) and other geometrical characteristics (height, length, taper of beams, planforms of membranes and plates ...) unchanged.

2.2. Geometric variables

The choice of geometrical variables may alter the configuration of the structure but not its topology. In the finite element method they correspond to modifications in the nodal coordinates.

2.3. Material properties

The efficiency of the structure can be improved by a change of nature of the material selected for some of its members. For example Young's modulus and material density may be varied but this introduces discrete parameter modifications as opposed to the continuous variations possible in the previous design variables.

2.4. Topology

A change in topology is also, and more fundamentally so, a discrete modification to the structure. For example, a set of members may be replaced by a new one with different elements, differently connected.

The order in which the groups of design variables have been listed is roughly that of increasing complexity in an optimization program and attendant increasing cost.

This consideration has led numerous research teams to limit themselves to the first category. There is also some justification for it in the fact that the general layout of a structure is often dictated by other considerations than a certain definition of optimality. Aerodynamic shape, headroom, access facilities, failsafe design are characteristic examples.

The relative simplicity of dealing with element sizes only is enhanced by the choice of a finite element method for the discretization of the structure. As nodes are kept in place and element interconnexions are invariant, the statics and kinematics of the structure are not modified by alterations in element sizes. This can make a large part of the optimization program a fixed subroutine.

In the sequel we shall deal only with this restricted aspect of optimization.

3. NUMERICAL METHODS OF STRUCTURAL OPTIMIZATION

This section describes briefly two main approaches encountered in structural optimization and discusses their relative capabilities.

3.1. Mathematical programming

In this relatively recent approach, minimum weight design is treated as the mathematical problem of extremizing a cost function in design space. Each dimension of this space is related to one design variable, so that each point corresponds to a possible design. The side-constraints consist of limits to the design variables (element sizes) themselves and to stresses or displacements, the latter constraints being generally functions of the design variables. Symbolically, denoting by A_i (i=1...n) the design variables

$$W = W\,(A_1 \ldots A_n) \text{ min.}$$

$$\underline{A_i} \leqslant A_1 \leqslant \overline{A}_i \qquad i=1\ldots n$$

$$g_j\,(A_1 \ldots A_n) \leqslant \overline{g}_j \qquad j=1\ldots p$$

The cost function and the nature of the second type of constraints determine whether the problem can be treated by linear or non-linear programming. The second case usually prevails for structural optimization.

Drawbacks inherent to the mathematical programming approach appear with large number of design variables as the number of cycles required to get close to the optimum rapidly rises. Each cycle involves a costly stress reanalysis and the computational expenditure rapidly becomes prohibitive. On the other hand the method is very general and reliable. If a solution converges to a local minimum instead of the global minimum required, this can always be checked and the necessary steps be taken to reinitiate the procedure.

3.2. Optimality criteria

Intuitive considerations as to the nature of the optimal design may lead to adopt optimality criteria that are not directly related to the minimization of the given cost function but sometimes constitute a satisfactory approximation to it. They can then provide a basis for the search techniques and lead to simple recursions formulas for redesign. The best known and widely used example of such a procedure is the "fully stressed design" concept. According to it, each component of an optimal structure is stressed to its limit in at least one of the loading conditions.
Convergence to the optimal solution, according to the fully stressed design criterion, is obtained in one iteration for statically determinate structures. In statically determinate cases the internal loads are indeed independent of the design variables and optimality based on fully stressed design coincides with the exact minimal weight criterion if no limitations are put on displacements. In the statically indeterminate case each redesign modifies the internal loading distribution and fully stressed design does not yield the minimum weight but may be considered to approach it satisfactory.
An attractive feature of fully stressed design that explains its relative success is its tendancy to converge in a number of cycles independent of the number of design variables, in contrast to the more rigorous mathematical programming method.
Moreover each redesign cycle is fairly simple.

4. REFERENCES

(1) L.A. SCHMIT and R. H. MALETT
"Structural Synthesis and Design Parameter Hierarchy"
Journal of the Structures Division, ASCE, vol. 89,
August 1963, p. 269.

(2) C.Y. SHEU and W. PRAGER
"Recent Developments in Optimal Structural Design"
Applied Mechanics Review, 21, 1968, p. 985.

(3) R.A. GELLATLY
"The Role of Optimization in the Design of Aircraft Structures"
Proceedings of AGARD Symposium on Structural Optimization,
Istambul, AGARD-CP-36-70, 1970, paper 9.

(4) F.I. NIORDSON and P. PEDERSEN
"A Review of Optimal Structural Design"
DCAMM Rep. 31, Dept. of Solid Mechanics, The Technical
University of Denmark, Copenhagen, 1972.

PART II : AN ALGORITHM FOR MINIMUM WEIGHT DESIGN UNDER A SET OF LOADING MODES WITH CONSTRAINTS ON STRESSES AND DISPLACEMENTS.

1. PROBLEM DEFINITION

The structure is in the linear elastic regime and idealized by finite elements. Under all the specified loading distributions certain constraints on stresses and nodal displacements must be satisfied. The geometry and the material properties are predetermined.
The functional to be minimized

$$W = \sum_i \rho_i L_i A_i \qquad i = 1 \dots n_e \tag{1}$$

is the structural weight, proportional in each element to the material density ρ_i, to the design variable A_i (cross-section of bar , thickness of membrane,...) and a geometrical parameter L_i (length of bar, area of membrane...).

2. DESCRIPTION OF THE CONSTRAINTS

2.1. Production constraints

They place a lower and sometimes an upper limit to the design variables:

$$\underline{A}_i \leq A_i \leq \overline{A}_i \tag{2}$$

2.2. Stress limitations

In bar-type elements the tensile stress limit is determined by elastic the properties of the material; the compressive limit may be reduced to take into consideration, in a simple manner, a safeguard against buckling.
If σ_i is the actual stress in the bar:

$$\underline{\sigma_i} \leqslant \sigma_i \leqslant \overline{\sigma_i} \tag{3}$$

In shear panels one assumes a maximum allowable shear stress, usually governed by buckling considerations ;

$$\tau_i \leqslant \overline{\tau_i} \tag{4}$$

In more general membrane elements, where the three stress components σ_x, σ_y and τ_{xy} play equally important roles, a reference stress related to an elastic limit criterion may be introduced.

2.3. Displacement constraints

They assign upper bounds to generalized displacements. To determine analytical expressions for them in terms of the design variables, the virtual work theorem is used.

If F denotes a vector (column matrix) of externally applied loads,

u the conjugate vector of generalized displacements,

σ the stress vector,

ε the conjugate strain vector,

the virtual work is given by :

$$\Delta = F^T_{(v)}\, u_{(r)} = \int_V \sigma^T_{(r)}\, \varepsilon_{(v)}\, dV = \int_V \sigma^T_{(v)}\, \varepsilon_{(r)}\, dV \tag{6}$$

The subscripts between brackets refer to either a virtual or a real vector, the superscript T denotes transposition. Splitting the integral into the sum of contributions of each finite element :

$$\Delta = \sum_i \int_{V_i} (\sigma^T_{(r)}\, \varepsilon_{(v)})_i\, dV_i \tag{7}$$

According to the finite element theory we have :

$$\int_{V_i} (\sigma^T_{(r)}\, \varepsilon_{(v)})_i\, dV_i = q^T_{(r)i}\, K_i\, q_{(v)i} = q^T_{(r)i}\, g_{(v)i} \tag{8}$$

where q_i is the vector of generalized displacements of element i and g_i its conjugate of generalized loads. K_i is the stiffness matrix of the element.

Let now u_j denote a displacement component of the nodal displacement vector of the structure. Applying a corresponding virtual unit load to the structure, (7) and (8) give :

$$u_j = \sum_i q^T_{(r)i}\, g_{i(j)} \tag{9}$$

where the $g_{i(j)}$ are the corresponding virtual loads generated at each element level. In statically determinate structures the loads $g_{i(j)}$ are uniquely determined by the unit load and (9) turns out to be given in terms of the design variables by ;

$$u_j = \sum_i \frac{c_{ij}}{A_i} \tag{10}$$

where the c_{ij} are constants. In redundant structures those coefficients are themselves implicit functions of the design variables.

3. FORMULATION

Let us begin with the statically determinate case

3.1. Analysis stage

The structure is analyzed under

- the n_r real loading systems of the design specification
- the n_v virtual loading cases connected with each displacement constraint.

3.2. Redesign stage

If the stress constraints are :

$$\sigma_{i\ell} \leq \bar{\sigma}_{i\ell} \qquad i = 1 \ldots n_e \quad \ell = 1 \ldots n_r \tag{11}$$

where $\sigma_{i\ell}$ is the actual stress in element i under the loading case ℓ, the redesign is effected in a single step by :

$$A_i^* = A_i \max_{\ell} \left\{ \frac{\sigma_{i\ell}}{\bar{\sigma}_{i\ell}} \right\} \tag{12}$$

with, in addition, the minimum size requirement :

$$A_i^* \geq \underline{A_i} \tag{13}$$

This method provides a "fully stressed" design, each element reaching its limiting stress (or having its minimal size) under at least one of the loading cases.
If we have displacement constraints, the analysis stage will provide the matrix of c_{ij} coefficients appearing in (10). Taking the A_i^* values appearing in (12) as minimal the problem with the addition of displacement constraints can be stated as follows:

$$W = \sum_i \rho_i L_i A_i \qquad \min$$

under

$$\sum_i \frac{c_{ij}}{A_i} \leq \bar{u}_j \quad j = 1 \ldots n_t \tag{14}$$

$$\bar{A}_i \geq A_i \geq A_i^* \qquad i = 1 \ldots n_e \tag{15}$$

where $\quad n_t = n_r \times n_v \quad .$

Because of the assumption of statical determinacy the formulation is rigorous, the solution unique and only one stress analysis is required.

Real structures, however, are rarely statically determinate. If subjected to different load distributions they are in fact both stiffer and even lighter if proper use is made of the stress cooperation provided by redundancy. But in this case both the c_{ij} and the A_i^* become implicit functions of the design variables. Each change in those will produce new c_{ij} and A_i^* that can only be known exactly through a costly stress reanalysis. The following approach is suggested. The problem as defined by equations (14) and (15) is solved by considering the c_{ij} and A_i^* as constants. The evolved solution for the design variables is inserted in a new stress analysis to provide new c_{ij} and A_i^* values with which to reinitiate problem (14), (15) untill close to convergence.

4. SOLUTION OF THE LINEARIZED PROBLEM (14), (15)

In order to solve this problem, in which the c_{ij} and A_i^* are assumed to be given, it is beneficial to take the reciprocals of sizing variables as new design variables. The recast problem may then be solved by means of the gradient projection method for linear constraints (ref. [6]) adapted to the problem under consideration.

As required by this method, the initial point must be a feasible point, a point lying in the convex region formed by the prescribed constraints.

In any given case, such a point can readily be found by linear scaling of all member sizes, so that a feasible bounded design is generated (one constraint at its critical value, others subcritical). This scaling of all the design variables does not introduce stress redistribution : each stress and each displacement are simply divided by the same scaling factor.

5. APPLICATION OF THE METHOD

Amongst known optimization programs we can mention :

- GELLATLY and BERKE (ref. [2])
- TAIG and KERR (ref. [3]).

As mentioned previously a structural optimization program performs iterative cycling between a structural analysis stage and a redesign stage. The programs under study at the Aerospace Laboratory of the University of Liège are coupled to the extensively developped ASEF code as the analysis module. The structural idealization consists up to now of axial force members and triangular or quadrilateral membrane elements. The degree of displacement polynomials within the elements is allowed to vary from 1 to 3. A symmetry option has been introduced, that constrains members of any specified group to be identical; in this case, the number of design variables is reduced to the number of element groups.

6. EXAMPLES

The method proposed in section 3 has been tested against solutions to classical problems found in the literature.
The two first examples show clearly that the rate of convergence of the redesign procedure is not directly related to the size of the problem under consideration.

6.1. Four-Bar Pyramid (fig. 1)

This very simple structure is subjected to the single loading case given fig.1. Constraints are placed on maximum stress (25000 psi), minimum area (0.1 in^2.) and node displacement in z-direction (0.3 in.). The present results (table I) duplicate those of ref. [2] and [3] . Fig. 4.a shows the strange pattern followed by the iteration procedure : the design seems to converge after 2 cycles and only after several more cycles does the rate of weight reduction accelerate till the final design is generated (after 20 iterations).

6.2. 72- Bar Four Level Tower (fig.3)

This doubly symmetric tower is subjected to two loading cases (table II-a). Symmetry is achieved by use of the input option, which reduces the number of design variables from 72 to 16. The stress limits are again 25000 psi. with 0.1 in^2. minimum area. The displacements of the four uppermost nodes are limited to 0.25 in. in the x and y-directions. In spite of the larger size of the problem, convergence is very rapid and optimal design is reached in only five iterations (see fig. 4.b). The results (table II-b) are the same as those of ref. [3] .

6.3. Cantilever Frame (fig.2)

The 10 bar-truss is subjected to the single loading case indicated on fig. 2. The stress limit in all members is 25000 psi. with 0.1 in^2. minimum area. The node

displacements in y-direction are prescribed to be less than 0.2 in. Table III shows the results obtained from the present method and, for comparison, from other methods (ref. [1], [2], [3]).
In addition, stress constraints have been formulated in a similar way than for displacement constraints:each member stress is linearly expressed in terms of the inverses of the design variables. Corresponding results are given in table III under the title "Experimental Method". This method generates a design weighting 5060.8 lb. which exhibits the following particular characteristics. Member 6 is fully stressed, while being at its minimum area. Furthermore only one displacement constraint is exactly satisfied (node 1) while another prescribed displacement is close to its limiting value (node 3).
For the other designs, shown on table III, these two displacements reach simultaneously their limiting values.

7. CONCLUSION

While using a mathematical programming algorithm, the method that was presented has the convergence characteristics of an optimality criterion approach. Except for the last example (cantilever frame), the same results as those of Taig and Kerr (ref. [3]) have been obtained for each analysis and redesign step. In addition, when there is only one active displacement constraint, the results of Gellatly and Berke (ref. [2]) are also identical to ours. In fact, all these methods are based on the same technique : by means of the virtual work theorem each limited displacement (or linear combination of displacements) is expressed in terms of the design variables. The resulting relations remain exact only in the case of a statically determinate structure; for a redundant structure, they become approximated. The redesign procedures are characterized by the algorithm used in order to resolve the ensuing linearized problem (14) (15).
The present method has the advantage of using a particularly suggestive algorithm : each path up to an "approached" optimum readily shows whether a constraint becomes active or not.
Furthermore each point of this path is a feasible bounded point. That important feature allows the algorithm to be eventually stopped before reaching the optimum, in order to avoid a final divergence due to a too strong internal redundancy.

8. REFERENCES

(1) V. B. VENKAYYA
"Design of Optimum Structures"
Computers and Structures, vol. 1, Nos 1/2, Aug. 1971, p. 265.

(2) R. A. GELLATLY and L. BERKE
"Optimal Structural Design"
USAF Technical Report AFFDL-TR-70-165, February 1971.

(3) I. C. TAIG and R. I. KERR
"Optimization of Aircraft Structures with Multiples Stiffness Requirements"
Second Symposium on Structural Optimization,
AGARD CONFERENCE PROCEEDINGS n° 123, paper 16.

(4) J. M. CHERN and W. PRAGER
"Optimal Design of Beams for Prescribed Compliance under Alternative Loads"
Journal of Optimization Theory and Application, vol. 5, n°6, 1970, p. 424.

(5) J.M. CHERN and W. PRAGER
"Minimum weight Design of Statically Determinate Trusses Subjected to Multiples Constraints".
International Journal of Solids and Structures, vol.7, 1971, p. 931.

(6) J. B. ROSEN
"The Gradient Projection Method for Nonlinear Programming.
Part I. Linear Constraints".
Journal of the Society for Industrial and Applied Mathematics, 8, 1960, p.181.

(7) C. FLEURY
"Methodes numériques d'optimisation des structures"
Rapport SF-19, Laboratoire de Techniques Aéronautiques et Spatiales,
Liège, Belgium.

(8) C. FLEURY
"Applications pratiques des méthodes numériques d'optimisation structurale "
Rapport SF-28 , Laboratoire de Techniques Aéronautiques et Spatiales,
Liège, Belgium.

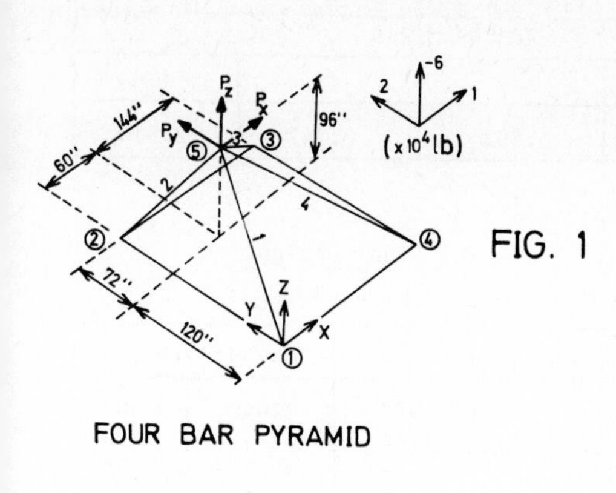

FIG. 1

FOUR BAR PYRAMID

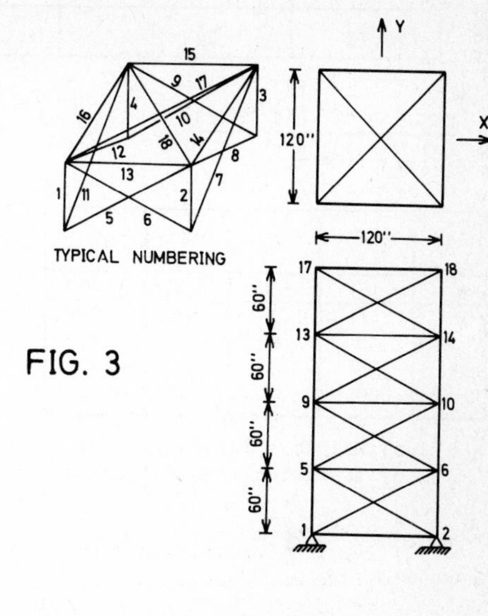

FIG. 3

72 - BAR TOWER

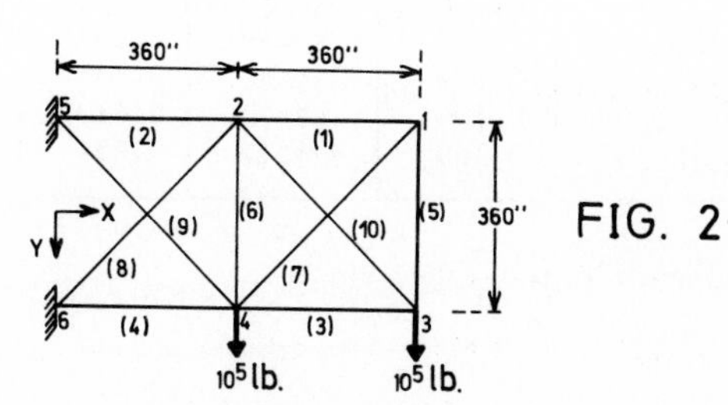

FIG. 2

CANTILEVER FRAME

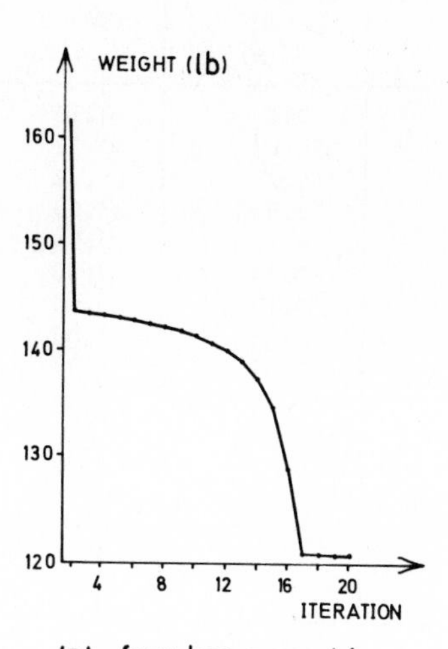

(a) four-bar pyramid

FIG. 4

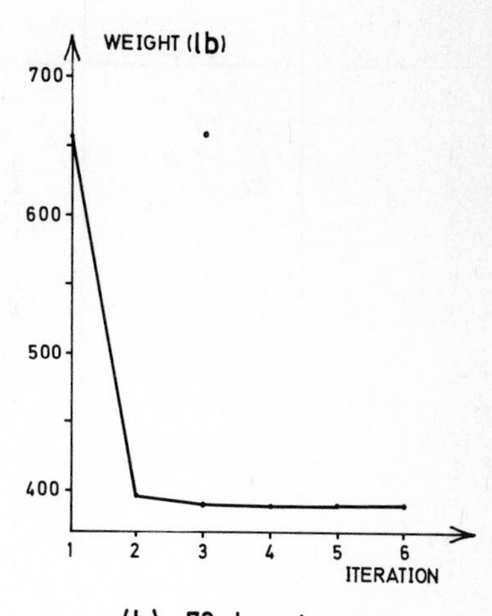

(b) 72-bar tower

ITERATION HISTORY

TABLE I FOUR BAR PYRAMID

WEIGHT	N_o OF ITERATION	MEMBER SIZES			
		1	2	3	4
120.73	22	0.100	3.893	0.747	2.510

TABLE II 72-BAR TOWER

(a) Loading systems (lb)

Load case	Node	X	Y	Z
1	17	5000	5000	-5000
2	17	0	0	-5000
	18	0	0	-5000
	19	0	0	-5000
	20	0	0	-5000

Final Weight : 379.66 lb
No of iteration : 5

(b) Final Design

Member	Size	Member	Size
1-4	1.897	37-40	0.507
5-12	0.516	41-48	0.520
13-16	0.100	49-52	0.100
17,18	0.100	53,54	0.100
22-12	1.280	55-58	0.157
23-30	0.515	59-66	0.536
31-34	0.100	67-70	0.410
35,36	0.100	71,72	0.654

TABLE III CANTILEVER FRAME

MEMBER N_o	REF. [3]	REF. [2]	REF. [1]	PRESENT METHOD (*)	EXP. METHOD
1	0.10	0.10	0.13	0.10	0.10
2	31.98	31.35	30.42	30.73	30.52
3	15.43	15.60	14.90	14.73	15.22
4	22.57	20.03	23.41	23.94	23.20
5	0.57	0.24	0.10	0.10	0.55
6	0.58	0.14	0.10	0.10	0.10
7	0.10	0.10	0.19	0.10	0.10
8	22.76	22.21	21.08	20.95	21.03
9	21.82	22.06	21.08	20.84	21.53
10	6.44	8.35	8.70	8.54	7.46
TOTAL WEIGHT	5167.	5112.17	5084.90	5076.67	5060.85
N_o.OF ITERATIONS	32	19	25	14	19

(*) Taig and Kerr have also obtained this result by fixing member 5 at its minimum area.

ON THE SOLUTION OF A CLASS OF NON LINEAR DIRICHLET PROBLEMS BY A PENALTY-DUALITY METHOD AND FINITE ELEMENTS OF ORDER ONE

R. GLOWINSKI

Université PARIS VI

A. MARROCCO

IRIA-LABORIA

INTRODUCTION.

In this paper, we shall give some results on the approximation and on the numerical solution of some non linear elliptical problems. It is also shown that the iterative method used to solve the approximate problems is also useful for solving other non linear problems arising in mechanics and physics.

1. THE CONTINUOUS PROBLEM.

Let Ω be a bounded open set of $\mathbb{R}^N$, such that its boundary Γ is regular. Let p be such that $1 < p < +\infty$.

We shall denote by V the space $W_o^{1,p}(\Omega)$ whose norm is $\|v\|_1 = (\int_\Omega |\nabla v|^p dx)^{1/p}$.

Let p' be the conjugate of p i.e. $(p-1)(p'-1) = 1$. Let V' be the dual $W^{-1,p'}(\Omega)$ of V and $\|.\|_*$ its norm.

We shall write $\|v\|_s$ instead of $\|v\|_{W^{s,p}(\Omega)}$.

It can be shown (see for example, [1], chapter 2) that the non linear elliptical problem :

$$(1) \qquad -\nabla.(|\nabla u|^{p-2}\nabla u) = f, \quad f \in V'$$

$$(2) \qquad u = o \text{ on } \Gamma$$

has a unique solution and is equivalent to

$$(3) \qquad \int_\Omega |\nabla u|^{p-2}\nabla u.\nabla v \, dx = \langle f,v \rangle \; \forall v \in V, \; u \in V.$$

In (3), $\langle .,. \rangle$ is the bilinear form of the duality between V' and V.

We shall call A the monotonous operator form $V \rightarrow V'$ defined by

$$A(v) = -\nabla(|\nabla v|^{p-2}\nabla v) \; ;$$

if $p = 2$, $A = -\Delta$.

2. THE APPROXIMATED PROBLEM.

For the sake of clarity we shall suppose that Ω is a polyhedral of $\mathbb{R}^2$. Let $\mathcal{T}_h$ be a finite triangulation of Ω such that :

(4) $$T \in \bar{\Omega} \quad \forall T \in \mathcal{T}_h, \qquad \bigcup_{T \in \mathcal{T}_h} T = \bar{\Omega}$$

(5) $$\begin{cases} T \text{ and } T' \in \mathcal{T}_h \Rightarrow T \cap T' = \emptyset \quad \text{or} \\ T \text{ and } T' \text{ have only one common vertex or only one common side.} \end{cases}$$

We choose h egual to the length of the greatest side of the $T \in \mathcal{T}_h$ and we approach V by

(6) $$V_h = \{v_h \mid v_h \in C^\circ(\bar{\Omega}),\ v_h = o \text{ on } \Gamma,\ v_h|_T \in P_1 \quad \forall T \in \mathcal{T}_h\}$$

with P_1 = space of polynomials of order $\leqslant 1$; we have $V_h \subset V$ and we approach (3) by the problem in finite dimension

(7) $$\int_\Omega |\nabla u_h|^{p-2} \nabla u_h . \nabla v_h dx = \langle f, v_h \rangle \quad \forall v_h \in V_h, u_h \in V_h.$$

Problem (7) has a unique solution and the following theorem can be shown

THEOREM 2.1

If the angles of $\mathcal{T}_h$ are bounded from below, uniformly in h, by $\theta_o > o$, we have :

(8) $$\lim \|u_h - u\|_1 = o \text{ when } h \to o \text{ where u is the solution of (3).}$$

3. ESTIMATIONS OF THE ERROR OF APPROXIMATION.

LEMMA 3.1 : We have, $\forall u, v \in V$

(9) $$\langle A(v) - A(u), v-u \rangle \geqslant \alpha \|v-u\|_1^p \text{ if } p \geqslant 2$$

(10) $$\langle A(v) - A(u), v-u \rangle \geqslant \alpha \|v-u\|_1^2 (\|v\|_1 + \|u\|_1)^{p-2} \text{ if } 1 < p \leqslant 2$$

(11) $$\| A(v) - A(u) \|_* \leqslant \beta \|v-u\|_1 (\|v\|_1 + \|u\|_1)^{p-2} \text{ if } p \geqslant 2$$

(12) $$\| A(v) - A(u) \|_* \leqslant \beta \|v-u\|_1^{p-1} \text{ if } 1 < p \leqslant 2$$

with $\alpha, \beta > o$ and independant of u,v.

LEMMA 3.2 : Let u,w be any elements of V. Let u_h, w_h be the solutions of (7) corresponding respectively to f = A(u) and f = A(w). Then :

(13) $$\|u_h\|_1 \leqslant \|u\|_1$$

(14) $$\|w_h - u_h\|_1 \leqslant \left(\frac{\beta}{\alpha}\right)^{p-1} \|w-u\|_1^{\frac{1}{p-1}} (\|w\|_1 + \|u\|_1)^{\frac{p-2}{p-1}} \text{ if } p \geqslant 2$$

(15) $$\|w_h - u_h\|_1 \leqslant \frac{\beta}{\alpha} \|w-u\|_1^{p-1} (\|w\|_1 + \|u\|_1)^{2-p} \text{ if } 1 < p \leqslant 2$$

If Ω is bounded in $\mathbb{R}^2$ and Γ is lipschitz, then $W^{2,p}(\Omega) \subset C^\circ(\bar{\Omega})$ with continuous injection $\forall p$, $1 < p \leqslant +\infty$; from this property, from the results of [2] on the interpolation of differentiable functions and from the above lemmas, we deduce the following theorem :

THEOREM 3.1

Under the hypothesis of theorem 2.1, we have :

(16) $$\|u_h - u\|_1 \leqslant C \|u\|_2^{\frac{1}{p-1}} \|u\|_1^{\frac{p-2}{p-1}} h^{\frac{1}{p-1}} \qquad \forall u \in V \cap W^{2,p}(\Omega),\ p \geqslant 2$$

(17) $$\|u_h - u\|_1 \leqslant C \|u\|_2^{\frac{1}{3-p}} \|u\|_1^{\frac{2-p}{3-p}} h^{\frac{1}{3-p}} \qquad \forall u \in V \cap W^{2,p}(\Omega),\ 1 < p \leqslant 2$$

with C independant of h and u.

From L. TARTAR [3] and from the above results, by non linear interpolation between V and $V \cap W^{2,p}(\Omega)$, we prove the following theorem :

THEOREM 3.2

Under the hypothesis of theorem 2.1, we have for $s \in [1,2]$:

(18) $$\|u_h - u\|_1 \leqslant C \|u\|_1^{\frac{p-2}{p-1}} \|u\|_s^{\frac{1}{p-1}} h^{\frac{s-1}{p-1}} \qquad \forall u \in V \cap W^{s;p}(\Omega),\ p \geqslant 2$$

(19) $$\|u_h - u\|_1 \leqslant C \|u\|_1^{\alpha} \|u\|_s^{\beta} h^{\gamma} \qquad \forall u \in V \cap W^{s,p}(\Omega),\ 1 < p \leqslant 2$$

with C independant of h and u, and in (19) we have

$$\alpha = \frac{(2-p)((2-s)+(s-1)(p-1))}{(2-s) + (s-1)(p-1)(3-p)} \qquad \beta = \frac{p-1}{(2-s)+(s-1)(p-1)(3-p)} \qquad \gamma = (s-1)\beta$$

4. AN ITERATIVE METHOD FOR SOLVING THE APPROXIMATED PROBLEM.

The problems (7) and (20), below, are equivalent.

(20) $$J(u_h) \leq J(v_h)\ \forall v_h \in V_h, u_h \in V_h\ ;\ J(v_h) = \frac{1}{p}\|v_h\|_1^p - \langle f, v_h \rangle.$$

The method of non linear surrelaxation described in [4] is almost inefficient if applied to (20) for $p < 1.5$ and $p \geqslant 10$. The method of auxilatory operator of [5] is suitable for (7) only if p is close to 2.

The remedy is to increase the number of variables while simplifying the non linear structure of (20) by taking $z_h = \nabla v_h$.

Then z_h and v_h are decoupled by penalisation and simultanuous dualisation of $z_h - \nabla v_h = o$ (following a principle due to HESTENES [6]).

Indeed, if a penalisation alone is used, it yields a problem different from the initial problem ; all the less different and the most ill conditioned that the parameter of penalty is small ; if duality alone is used, it yields a problem coercive in z_h and linear, therefore non coercive in v_h.

Let χ_T be the caracteristic function of T,

$$L_h = \{z_h \mid z_h = \sum_{T \in \mathcal{C}_h} z_T \chi_T,\ z_T \in \mathbb{R}^2\}$$

$j(v,z) = \frac{1}{p}\int_\Omega |z|^p dx - \langle f, v \rangle$ with $z \in L^p(\Omega) \times L^p(\Omega)$; (20) is equivalent to

(21) $$j(u_h, y_h) \leqslant j(v_h, z_h) \qquad \forall (v_h, z_h) \in V_h \times L_h,\ \nabla v_h - z_h = o$$

with $Y_h = \nabla u_h$.

By penalisation and dualisation of $\nabla v_h - z_h = o$, we are led to introduce (penalisation) $j_\varepsilon = j + \frac{1}{2\varepsilon}\|z - \nabla v\|^2_{L^2(\Omega)}$ with $\varepsilon > o$, then (dualisation) the Lagrangian

$$\mathcal{L}(v,z;\mu) = j_\varepsilon(v,z) - \int_\Omega \mu \cdot (z - \nabla v)\, dx.$$

Then, we can show ([7], N°28 can also be used) the following :

PROPOSITION 4.1 : $\mathcal{L}$ has a saddle point of the form $(u_h, \nabla u_h ; \lambda_h)$ on $V_h \times L_h \times L_h$; u_h solution of (7).

From this result, we can use the following algorithm for solving (7) :

(22) $\lambda_h^o \in L_h$, given

λ_h^n known, we compute $u_h^n, y_h^n, \lambda_h^{n+1}$ by

(23) $\mathcal{L}(u_h^n, y_h^n ; \lambda_h^n) \leq \mathcal{L}(v_h, z_h; \lambda_h^n) \quad \forall v_h \in V_h, z_h \in L_h \; ; \; u_h^n \in V_h, \; y_h^n \in L_h$

(24) $\lambda_h^{n+1} = \lambda_h^n - \rho_n(y_h^n - \nabla u_h^n), \; \rho_n > o$

THEOREM 4.1

If $o < r_o \leq \rho_n \leq r_1 < \frac{2}{\varepsilon}$, *when* $n \to +\infty$ *and* $\forall \lambda_h^o$, *we have* $u_h^n \to u_h, y_h^n \to \nabla u_h$; u_h *solution of (7).*

COMMENT 4.1 : For μ fixed, $\mathcal{L}$ is strictly convex in (v,z) and quadratic in v ; it implies that (23) can be solved by a modification of a relaxation type on z_h, surrelaxation on v_h, of the standard surrelaxation method on the Dirichlet problem ; the results of [8] apply to this modification which is easy to implement since, for given v_h, μ_h, the minimization in z_h of $\mathcal{L}$ decomposes into Card($\mathcal{T}_h$), easy problems with two variables (this is one of the justification of algorithm (22)-(24)).

COMMENT 4.2 : In some case, algorithm (22)-(24) applied directly to the continuous problem, converges.

COMMENT 4.3 : For p=2, the above method, applied to solve (7), has little interest, since for $\rho_n = 1/\varepsilon$, the sequence (u_h^n) converges in two iterations.

5. APPLICATIONS TO OTHER NON LINEAR PROBLEMS.

With some minor modifications, we can apply the previous method to the following (continuous or approximated) problems :

Elastic-Plastic torsion of a cylindrical beam :

(25) $\text{Min}[\frac{1}{2}\int_\Omega |\nabla v|^2 dx - \int_\Omega fv dx], \; v \in H_o^1(\Omega), |\nabla v| \leq 1$ p.p.

Flow of a Plastic-viscous flow in a pipe :

(26) $\text{Min}[\frac{1}{2}\int_\Omega |\nabla v|^2 dx + g\int_\Omega |\nabla v| dx - \int_\Omega fv dx], \; v \in H_o^1(\Omega)$

Minimal surfaces :

(27) $$\text{Min} \int_\Omega \sqrt{1 + |\nabla v|^2}\,dx, \quad v = g \text{ on } \Gamma .$$

Generally speaking the above method is well adapted to non linear elliptical problems of order 2, when the non linearity is on ∇v ; this is the case of the problem of the Alternator in magneto-static treated in [4] by non linear surrelaxation and of the subsonic flow of a compressible fluid around a profil of R^2, etc...

6. NUMERICAL EXAMPLE.

For $\Omega = \{x \,|\, x_1^2 + x_2^2 < R^2\}$ and $<f,v> = C\int_\Omega v\,dx$, the solution of (3) is given by

$$u(x) = \frac{p-1}{p}\left(\frac{CR}{2}\right)^{\frac{p}{p-1}} R\left(1-\left(\frac{r}{R}\right)^{\frac{p}{p-1}}\right), \text{ with } r = \sqrt{x_1^2 + x_2^2}.$$

By making use of algorithm (22)-(24), we have been able to extend the field (1) of resolution of (1) to $1.1 \leqslant p \leqslant 50$, with computing time of the order of a minute of CII 10070 and for triangulations of about 250 triangles.

(1) limited to $1.5 \leqslant p \leqslant 10$ for non linear S.O.R.

BIBLIOGRAPHY

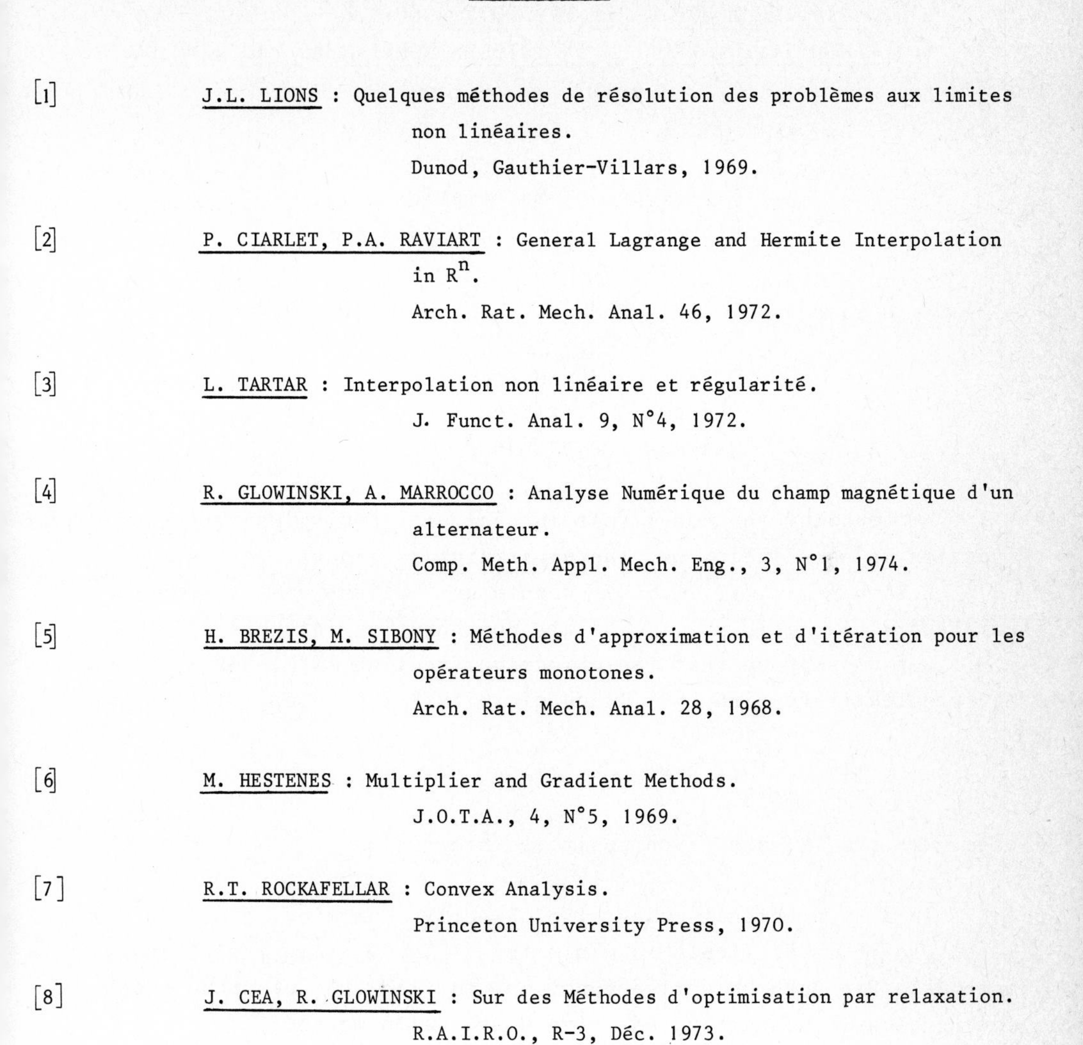

[1] J.L. LIONS : Quelques méthodes de résolution des problèmes aux limites non linéaires.
Dunod, Gauthier-Villars, 1969.

[2] P. CIARLET, P.A. RAVIART : General Lagrange and Hermite Interpolation in R^n.
Arch. Rat. Mech. Anal. 46, 1972.

[3] L. TARTAR : Interpolation non linéaire et régularité.
J. Funct. Anal. 9, N°4, 1972.

[4] R. GLOWINSKI, A. MARROCCO : Analyse Numérique du champ magnétique d'un alternateur.
Comp. Meth. Appl. Mech. Eng., 3, N°1, 1974.

[5] H. BREZIS, M. SIBONY : Méthodes d'approximation et d'itération pour les opérateurs monotones.
Arch. Rat. Mech. Anal. 28, 1968.

[6] M. HESTENES : Multiplier and Gradient Methods.
J.O.T.A., 4, N°5, 1969.

[7] R.T. ROCKAFELLAR : Convex Analysis.
Princeton University Press, 1970.

[8] J. CEA, R. GLOWINSKI : Sur des Méthodes d'optimisation par relaxation.
R.A.I.R.O., R-3, Déc. 1973.

ADAPTIVE MONTE CARLO METHOD FOR SOLVING CONSTRAINED MINIMIZATION PROBLEM IN INTEGER NON-LINEAR PROGRAMMING

Minoru Ichimura
Department of Mathematics
Okayama College of Science
Okayama 700, Japan
and
Kazumasa Wakimoto
Department of Mathematics
Okayama University
Okayama 700, Japan

We consider the non-linear integer programming where the objective function is non-linear, but the constraint is linear.

Let $\mathsf{L} = (\ell_1, \ell_2, \ldots\ldots, \ell_N)$ denote an N-dimensional vector whose components ℓ_i, $i=1,2,\ldots\ldots,N$ are non-negative integers.

Let us suppose that the objective function $f(\mathsf{L})$ is non-linear. The problem is to find L^* which minimizes $f(\mathsf{L})$ subject to

$$(1) \qquad \sum_{i=1}^{N} \ell_i \leqq M,$$

where N and M are given positive integers.

In most practical cases, the objective function $f(\mathsf{L})$ is complicated, so it is difficult to minimize $f(\mathsf{L})$ by using the ordinary methods. For such objective function, to find L^*, we propose the following algorithm, namely, adaptive Monte Carlo method:

Step 1: Generate N-dimensional ramdom vectors of size n satisfying the constraint (1), where an N-dimensional random vector means a selected vector by the equal probability from each of all possible N-dimensional vectors satisfying the constraint (1).

Step 2: Evaluate the objective function $f(\mathsf{L})$ for each random vector and select the vector L_0 which minimizes $f(\mathsf{L})$ in already generated vectors. Set vectors L_0 to initial vector in next step 3.

Step 3: Change the range of components of random vectors by the following procedures:

i) Select the positive integer M' which is less than M.

ii) Generate random vectors, $X^{(k)} = (x_1^{(k)}, x_2^{(k)}, \ldots, x_N^{(k)})$, and $Y^{(k)} = (y_1^{(k)}, y_2^{(k)}, \ldots, y_N^{(k)})$, for $k=1,2,\ldots,m$, whose components satisfy the following constraints:

$$\sum_{i=1}^{N} x_i^{(k)} \leqq M', \quad \sum_{i=1}^{N} y_i^{(k)} \leqq M', \text{ for } k=1,2,\ldots,m,$$

where all components are non-negative integers.

Put $\ell_i^{(k)} = \ell_i^0 + x_i^{(k)} - y_i^{(k)}$, $k=1,2,\ldots,m$, for $i=1, 2,\ldots,N$, where $L_0 = (\ell_1^0, \ell_2^0, \ldots, \ell_N^0)$.

If the vector $(\ell_1^{(k)}, \ell_2^{(k)}, \ldots, \ell_N^{(k)})$ has at least a negative component, reject its vector. Return to step 2.

Example 1.

Define the objective function $f(L)$ as follows:

$$(2) \quad f(L) = \sum_{i=1}^{10} \ell_i^{\,2}$$

where the components of vector L, are non-negative integers and satisfy the following equation:

$$(3) \quad \sum_{i=1}^{10} \ell_i = 100.$$

Step 1: Generate 10-dimensional random vectors of size n satisfying Eq(3) by combinatorial method that presented in Appendix I. In this step, we use the crude Monte Carlo method.

Step 2: Evaluate the objective function $f(L)$ for each random vector and select the vector L_0 which minimizes $f(L)$ in already generated vectors. Set vector L_0 to initial vector in next step 3.

Step 3: Change the range of components of random vectors by the following procedures:

i) Select the positive integer M' which is less than 100. In this example, $M' = 5$ or 1 are used.

ii) Generate random vectors, $X^{(k)} = (x_1^{(k)}, x_2^{(k)}, \ldots, x_{10}^{(k)})$, and $Y^{(k)} = (y_1^{(k)}, y_2^{(k)}, \ldots, y_{10}^{(k)})$, for $k=1,2,\ldots,m$, whose components satisfy the following constraints:

$$\sum_{i=1}^{10} x_i^{(k)} = M' , \qquad \sum_{i=1}^{10} y_i^{(k)} = M', \text{ for } k=1,2,\ldots,m,$$

where all components are non-negative integers.

Put $\ell_i^{(k)} = \ell_i^0 + x_i^{(k)} - y_i^{(k)}$, $k=1,2,\ldots,m$, for $i=1, 2,\ldots,10$, where $L_0 = (\ell_1^0, \ell_2^0, \ldots, \ell_{10}^0)$.

If the vector $(\ell_1^{(k)}, \ell_2^{(k)}, \ldots, \ell_{10}^{(k)})$ has at least a negative component, reject its vector. Return to step 2.

The numerical results are shown in Table 1, and its program in FORTRAN is shown in Appendix II.

Table 1. Numerical results in case of n=10 and m=10

cumulative numbers of generating vector	components of vector L_0	$f(L_0)$	M'
10	(10,23, 7, 2, 5,18, 7,10,17, 1)	1470	
20	(12,21, 7, 3, 5,18, 8,10,15, 1)	1382	5
30	(12,20, 6, 3, 5,17, 8,10,15, 4)	1308	5
40	(11,19, 9, 4, 4,17, 7,10,15, 4)	1274	5
50	(9,19, 9, 6, 5,17, 7, 9,14, 5)	1224	5
60	(8,19,10, 7, 6,17, 8, 7,12, 6)	1192	5
70	(8,17, 9, 8, 7,17, 8, 9,11, 6)	1138	5
80	(9,15, 9, 8, 6,17, 8,11,11, 6)	1118	5
90	(8,14, 9, 7, 9,17,10,11, 9, 6)	1098	5
100	(9,14, 9, 8, 9,16,10,10, 8, 7)	1072	5
110	(9,14, 9, 8, 9,14,10,10, 7,10)	1048	5
120	(8,12,10, 9, 9,14,12,10, 7, 9)	1040	5
130	(9,11,10,10,10,14,12, 8, 7, 9)	1036	5
140	(9,11,10,11,12,11,11,10, 6, 9)	1026	5
150	(10,11,10,10,11,11,11, 9, 7,10)	1014	5
160	(10,11,11,10,10,11,10,10, 7,10)	1012	5
190	(10,11,11, 9,10,10,11,10, 8,10)	1008	5
200	(10,11,11, 9,10,10,11, 9, 9,10)	1006	5
250	(10,10,11,10,10,10,11, 9, 9,10)	1004	1
320	(10,10,11,10,10,10,10, 9,10,10)	1002	1
330	(10,10,10,10,10,10,10,10,10,10)	1000	1

Example 2.

Define the objective function $f(L)$ as follows:

$$(4) \quad f(L) = \sum_{i=1}^{10} \ell_i (1-e^{-\frac{\ell_i}{\alpha_i}})$$

where α_i are constant parameters as shown in Table 2.

Table 2. Constant parameters of objective function

i	1	2	3	4	5	6	7	8	9	10
α_i	6.	7.	10.	18.	19.	23.	25.	26.	30.	47.

Let us find the vector L^* which minimizes the value of objective function $f(L)$ subject to

$$(5) \quad \sum_{i=1}^{10} \ell_i = 100.$$

We take the approach similar to example 1. The numerical results are shown in Table 3. Furthermore, to compare with the crude Monte Carlo method using only the step 1 and step 2, the numerical results of two methods are shown in Fig. 1.

Table 3. Numerical results in case of n=10 and m=10

cumulative numbers of generating vector	components of vector L_0	$f(L_0)$	M'
10	(1, 2, 5, 8,18, 0, 2,19,11,34)	47.396469	
20	(2, 2, 5, 7,17, 0, 2,20,11,34)	47.108032	5
30	(3, 3, 4, 9,15, 2, 3,19, 9,33)	44.611450	5
40	(3, 4, 3, 9,15, 4, 4,19, 9,30)	42.996628	5
50	(3, 5, 5, 9,13, 4, 5,18, 9,29)	41.905899	5
60	(3, 4, 3, 9,12, 7, 7,18, 8,29)	40.623596	5
70	(3, 4, 4, 9,12, 8, 8,18, 8,26)	39.853546	5
80	(3, 3, 5, 7,12, 8, 8,17, 9,28)	39.668411	5
90	(3, 3, 4, 8,13, 7, 8,15,10,29)	39.647949	5
100	(3, 3, 4, 9,11, 6,10,14,11,29)	39.153702	5
120	(3, 3, 5, 8, 9, 6,10,14,13,29)	38.887680	5
130	(3, 2, 4, 9, 9, 7,11,13,14,28)	38.588974	5

140	(3, 3, 4, 8, 7, 8,11,13,14,29)	38.527283	5
150	(2, 2, 5, 9, 9, 9, 9,14,15,26)	38.381607	5
160	(2, 2, 4, 9,10, 9,11,13,15,25)	38.176300	5
180	(2, 4, 5, 8, 8,10,13,12,15,23)	37.931122	5
190	(3, 3, 6, 9, 9, 9,12,12,15,22)	37.920959	5
210	(3, 3, 4,10, 8,10,13,12,15,22)	37.915436	5
220	(3, 4, 4,10, 8,10,12,12,15,22)	37.914230	2
240	(3, 4, 4, 9, 8,11,12,12,15,22)	37.848587	2
250	(3, 4, 5, 8, 8,11,13,12,14,22)	37.841904	2
270	(3, 4, 5, 9, 8,11,12,12,14,22)	37.815933	2
290	(3, 3, 5, 9, 8,11,12,12,14,23)	37.797485	2
310	(3, 3, 4, 9, 9,11,12,12,14,23)	37.795364	2
400	(3, 3, 5, 9, 9,11,11,12,14,23)	37.785004	2

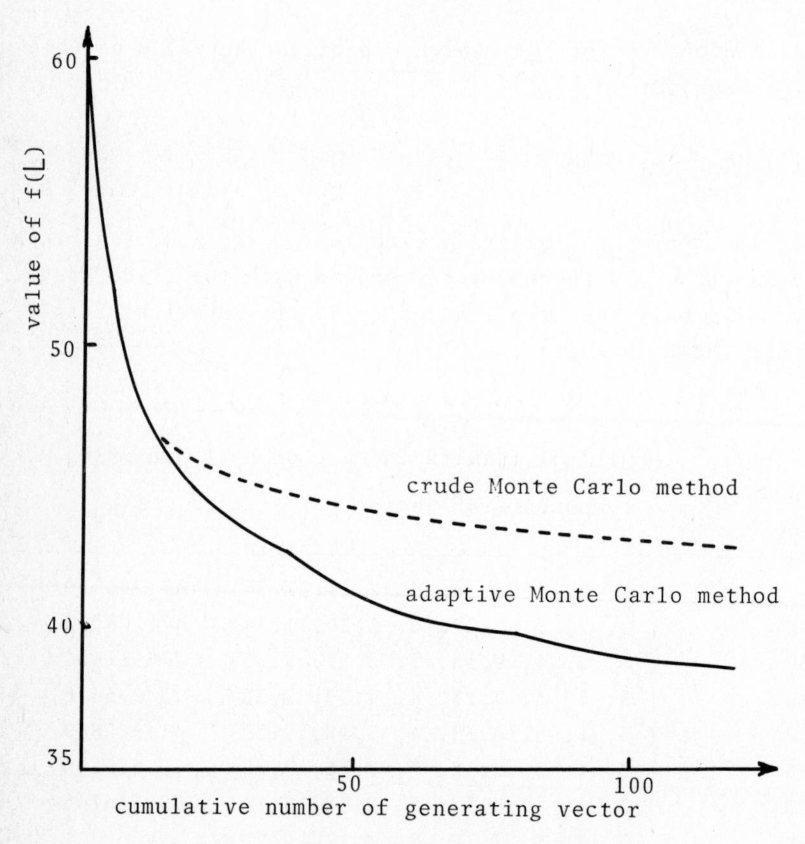

Fig.1 Comparison between adaptive and crude Monte Carlo methods

Acknowledgments.

The authors are grateful to Dr. Koiti Takahasi, the Institute of Statistical Mathematics, Japan. Thanks are due to some members for useful discussions at the Symposium "Information and Statistics" held in Sizuoka University on February, 1973.

Appendix I.

In this appendix, two algorithms are presented for generating an N-dimensional random vector satisfying the constraint (6) as follows:

$$(6) \qquad \sum_{i=1}^{N} \ell_i = M.$$

If an (N+1)-dimensional random vector satisfying the constraint $\sum_{i=1}^{N+1} \ell_i = M$ is generated, it is easy to generate an N-dimensional random vector $\mathsf{L}' = (\ell_1', \ell_2', \ldots, \ell_N')$ satisfying the constraint:

$$(7) \qquad \sum_{i=1}^{N} \ell_i' \leqq M$$

as shown below:

Step 1: Generate an (N+1)-dimensional random vector satisfying the constraint $\sum_{i=1}^{N+1} \ell_i = M$.

Step 2: Put $\ell_i' = \ell_i$, for $i=1,2,\ldots,N$, and $\mathsf{L}' = (\ell_1', \ell_2', \ldots, \ell_N')$.

Algorithm 1.

Step 1: Draw the random numbers of size $N-1$: $n_1, n_2, \ldots, n_{N-1}$ without replacement from the $M+N-1$ consecutive positive integers (from 1 to $M+N-1$) and arrange them in ascending order according to their magnitude:

$n_{(1)} < n_{(2)} < \ldots < n_{(N-1)}$.

Step 2: We get the components of a random vector by the following expression:

$$\ell_i = n_{(i)} - n_{(i-1)} - 1 \ , \quad i=1,2,\ldots,N$$

where $n_{(0)} = 0$ and $n_{(N)} = M+N$.

Algorithm 2.

Step 1: Draw the random numbers of size M: $n_1, n_2, \ldots, n_M$, without replacement from the M+N-1 consecutive positive integers (from 1 to M+N-1) and arrange them in ascending order according to their magnitude: $n_{(1)} < n_{(2)} < \ldots < n_{(M)}$.

Step 2: Put

$k_j = n_{(j)} - (j-1), \ j=1,2,\ldots,M.$

Put

$$\delta_{ij} = \begin{cases} 1, & i = k_j, \\ 0, & \text{otherwise}, \end{cases}$$

for $j=1,2,\ldots,M$, and

$$\ell_i = \sum_{j=1}^{M} \delta_{ij}, \ i=1,2,\ldots,N.$$

Then $(\ell_1, \ell_2, \ldots, \ell_N)$ is an N-dimensional vector satisfying the constraint (6).

We can obviously prove the above algorithm 1 and 2 by repeated combination. The algorithm 2 is efficient when N is large and M is small.

Appendix II.

```
      INTEGER KEKA(10),KST(10),KPA(10),KPB(10),NOKOSU(10),RES(10)
      IXXC=0
      JXXC=0
C.....RESET SUBPROGRAM RAN WHICH GENERATES AN UNIFORMAL RANDOM NUMBER.
      KB=RAN(IXXC,JXXC)
C.....IT = COUNT OF RANDOM VECTORS.
C.....IIT= COUNT OF EVALUATION OF THE OBJECTIVE FUNCTION.
      IT=0
      IIT=0
      KSUM=100
      KOSU=10
      SUMIN=1000000.0
```

```
C.....CRUDE MONTE CARLO METHOD.
      DO 1 J=1,10
      IT=IT+1
      IIT=IIT+1
      CALL RANSUM(KSUM,KOSU,KST,IXXC,JXXC)
      SUM=0.0
      DO 6 I=1,KOSU
      SUM=SUM+KST(I)**2
    6 CONTINUE
      IF(SUM.GE.SUMIN) GO TO 1
      SUMIN=SUM
      DO 8 K=1,KOSU
    8 KEKA(K)=KST(K)
      WRITE(6,101) IT,IIT,(KEKA(I),I=1,KOSU),SUMIN
  101 FORMAT(1H0,'KAISU=',I3,2X,'CT=',I3,5X,'(',10(I3,','),')',2X,F6.0)
    1 CONTINUE
      DO 7 I=1,KOSU
      KST(I)=KEKA(I)
    7 RES(I)=KEKA(I)
C.....ADAPTIVE MONTE CARLO METHOD.
      KSUM=5
      DO 100 KL=1,2
      DO 2 LL=1,20
      DO 3 I=1,10
      CALL RANSUM(KSUM,KOSU,KPA,IXXC,JXXC)
      CALL RANSUM(KSUM,KOSU,KPB,IXXC,JXXC)
      IT=IT+1
      DO 4 J=1,KOSU
      NOKOSU(J)=KST(J)+KPA(J)-KPB(J)
      IF(NOKOSU(J).LT.0) GO TO 3
    4 CONTINUE
      IIT=IIT+1
C.....EVALUATE THE OBJECTIVE FUNCTION.
      SUM=0.0
      DO 20 J=1,KOSU
      SUM=SUM+NOKOSU(J)**2
   20 CONTINUE
      IF(SUM.GE.SUMIN) GO TO 3
      SUMIN=SUM
      WRITE(6,101) IT,IIT,(NOKOSU(K),K=1,KOSU),SUMIN
      DO 21 K=1,KOSU
```

```
 21 RES{K}=NOKOSU{K}
  3 CONTINUE
    DO 22 K=1¬KOSU
 22 KST{K}=RES{K}
  2 CONTINUE
    KSUM=1
100 CONTINUE
    STOP
    END
    SUBROUTINE RANSUM{WA¬KOSU¬KEKA¬IXR¬JXR}
    INTEGER WA¬KOSU¬KR{100}¬KEKA{10}
    KUMI=WA+KOSU
    AK=KUMI
    KR{1}=0
    KP1=KOSU+1
    KR{KP1}=KUMI
    DO 2 J=2¬KOSU
  3 KI=RAN{IXR¬JXR}*AK
    KI=MOD{KI¬KUMI}
    JM1=J-1
    DO 4 L=1¬JM1
    IF{KI.EQ.KR{L}} GO TO 3
    IF{KI.LT.KR{L}} GO TO 12
  4 CONTINUE
    KR{J}=KI
    GO TO 2
 12 ICN=L+1
    KWT=KR{L}
    KR{L}=KI
    DO 13 L=ICN¬J
    KW=KR{L}
    KR{L}=KWT
    KWT=KW
 13 CONTINUE
  2 CONTINUE
    DO 5 J=1¬KOSU
    KEKA{J}=KR{J+1}-KR{J}-1
  5 CONTINUE
    RETURN
    END
```

APPLICATION OF THE QUADRATIC MINIMIZATION METHOD TO THE PROBLEM OF SIMULATED SYSTEM CHARACTERISTICS REPRESENTATION

V.K. Isaev, V.V. Sonin

Moscow Physical-Technical Institute,

Moscow, USSR

The problem of approximation of the experimental function $g(x)$ determined at discrete points $x = x_K$, $K = 1,2,\dots, n$, can be represented as consisting of two steps:

1. The choice of an empirical formula

$$f(x, C_1, C_2, \dots, C_m) \simeq g(x), \quad m < n. \tag{1}$$

2. The search of the constants $C_1, C_2, \dots, C_m$ subject to minimum of discrepancies square sum

$$\Phi(C_1, C_2, \dots, C_m) = \frac{1}{2}\sum_{K=1}^{n}\left[f(x_K, C_1, C_2, \dots, C_m) - g(x_K)\right]^2. \tag{2}$$

The following technique is offered to determine the expression (1). The functions, approximating the experimental data at each of nonintersecting intervals, are constructed. On the basis of these expressions a broken line, whose behaviour qualitatively corres - ponds to the dependence sought for, is generated. Now it is possible to construct a smooth curve arbitrarily closely to the obtained broken line.

Let $H(x)$ be a broken line on $[a,b]$ formed by two smooth arcs, and

$$H(x) = \begin{cases} \varphi_1(x), & \varphi_1(x) \geq \varphi_2(x), \quad a \leq x \leq \xi, \\ \varphi_2(x), & \varphi_2(x) \geq \varphi_1(x), \quad \xi \leq x \leq b. \end{cases} \tag{3}$$

Then for a smooth function

$$\Phi(x,\varepsilon) = \frac{\varphi_1(x) + \varphi_2(x)}{2} + \sqrt{\left[\frac{\varphi_1(x) - \varphi_2(x)}{2}\right]^2 + \varepsilon^2}, \quad \varepsilon > 0, \tag{4}$$

the relation

$$0 < \Phi(x,\varepsilon) - H(x) \leq \varepsilon, \quad \varepsilon > 0, \tag{5}$$

is valid.

Similarly, it is possible to prove that for the functions

$$h(x)=\begin{cases}\varphi_2(x),\ \varphi_1(x)\geqslant\varphi_2(x), & a\leqslant x\leqslant\xi,\\ \varphi_1(x),\ \varphi_2(x)\geqslant\varphi_1(x), & \xi\leqslant x\leqslant b,\end{cases} \tag{6}$$

and

$$\varphi(x,\varepsilon)=\frac{\varphi_1(x)+\varphi_2(x)}{2}-\sqrt{\left[\frac{\varphi_1(x)-\varphi_2(x)}{2}\right]^2+\varepsilon^2},\quad \varepsilon>0, \tag{7}$$

the following expression takes place:

$$0<h(x)-\varphi(x,\varepsilon)\leqslant\varepsilon,\quad \varepsilon>0. \tag{8}$$

Below the function $\Phi(x,\varepsilon)$ is referred to as the upper, and $\varphi(x,\varepsilon)$ - lower envelopes. Suppose that $\varphi_1(x)$ and $\varphi_2(x)$ exist everywhere on $[a,b]$. Despite this fact in general case the construction of the smooth envelope of a multiarc broken line by successive using the expressions (4) and (7) may prove impossible.

This difficulty can be overcome using slight modifications of the described technique and we don't discuss the additional modifications due to limited scope of the paper presented.

Let us consider an example. In fig.1 the dependence $C_{x_0}(M)$ for

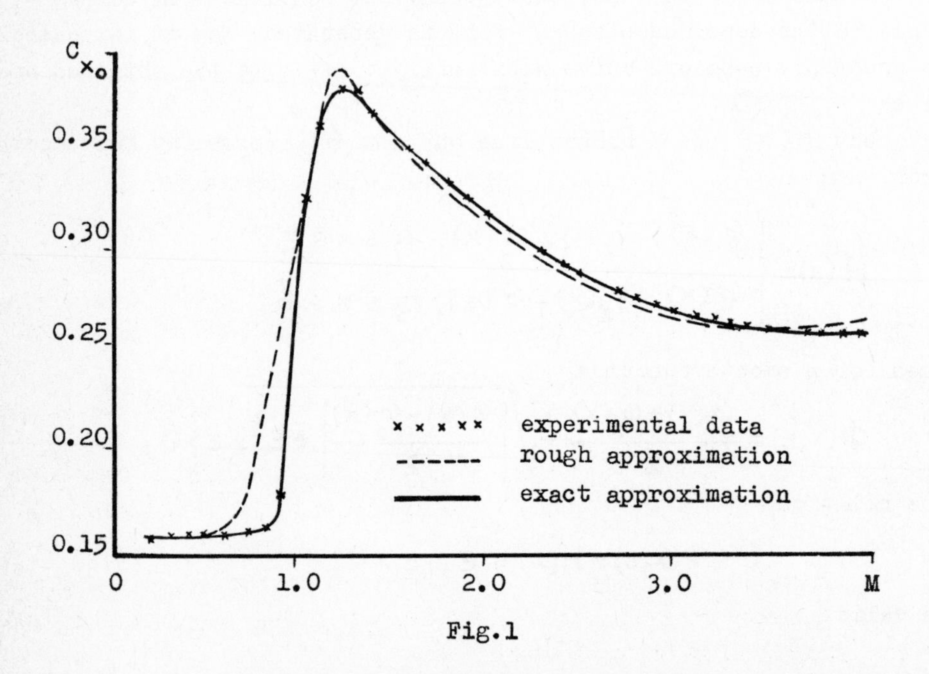

Fig.1

a missile - "the law of 1943" [1] is shown by crosses. The function $C_{x_0}(M)$ acquires practically constant value of 0.157 in the interval (0.1; 0.7). Put in this range of Mach numbers

$$C_{x_0}(M) \simeq \varphi_1(M) = C_1 M + C_2, \quad C_1 = 0, \quad C_2 = 0.157.$$

In the transonic region $C_{x_0}(M)$ increases nearly linear, so for the region $0.7 \leqslant M \leqslant 1.2$ put

$$C_{x_0}(M) \simeq \varphi_2(M) = C_4 M + C_5,$$

where $C_4=0.153$; $C_5=-0.205$. The supersonic part can be represented by the cubic

$$C_{x_0}(M) \simeq \varphi_3(M) = C_7 + C_8 M + C_9 M^2 + C_{10} M^3,$$

$$C_7 = 0.566445, \quad C_8=-0.1758, \quad C_9 = 0.02505, \quad C_{10} = 0.$$

In accordance with the technique described the upper envelope

$$\Phi_{12}(M) = \frac{\varphi_1(M)+\varphi_2(M)}{2} + \sqrt{\left[\frac{\varphi_1(M)-\varphi_2(M)}{2}\right]^2 + C_3^2} \tag{9}$$

($C_3=0.01$), representing $C_{x_0}(M)$ in the range of subsonic and transonic speeds, is constructed, and, finally, the lower envelope of the functions $\Phi_{12}(M)$ and $\varphi_3(M)$, representing $C_{x_0}(M)$ all over the range of $0.1 \leqslant M \leqslant 4$.

$$C_{x_0}(M) = f(M) = \frac{\Phi_{12}(M)+\varphi_3(M)}{2} - \sqrt{\left[\frac{\Phi_{12}(M)-\varphi_3(M)}{2}\right]^2 + C_6^2} \tag{10}$$

($C_6 = 0.01$). In fig.1 the function $f(M)$, corresponding to the given values of $C_1, C_2, \ldots, C_{10}$ is plotted by a dash line. It is shown that generally it correctly reflects the variation of the missile drag coefficient.

Now let us try to find exact values of C_i ($i = 1.2, \ldots, 10$) subject to minimum of square semisum of discrepancies at the points

$$\Phi(C_1, C_2, \ldots, C_{10}) = \frac{1}{2}\sum_{i=1}^{40}\left[f(0.1i;\, C_1, C_2, \ldots, C_{10}) - C_{x_0}(0.1i)\right]^2. \tag{11}$$

It is evident that the expression (11) is not a quadratic function of the parameters $C_1, C_2, \ldots C_{10}$ to be sought . In addition, according to (9) and (10) $\Phi(C_1, C_2, \ldots C_{10})$ is an even function of the parameters C_3 and C_6, so an iterative procedure of the second order is offered for searching the minimum of Φ . A brief description of the algorithm is given below.

Let a certain initial value of C^0 be known. Then any vector C may be presented in the form of

$$C = C^0 + \varepsilon \bar{C}. \tag{12}$$

Here and below usual vectorial-matrix notation is used. When $\varepsilon \bar{C}$ is a sufficiently small vector, $\Phi(C)$ is properly described by its tangential quadratic form

$$\Phi(C) = \Phi(C^0) + \varepsilon \Phi_c^T \bar{C} + \frac{\varepsilon^2}{2} \bar{C}^T \Phi_{cc} \bar{C}. \tag{13}$$

During each iteration the vector $\bar{C}$ is chosen subject to decreasing right-hand part of the expression (13) along it at a sufficiently small ε :

1. Calculate $\Phi(C^0)$, $\Phi_c(C^0)$, $\Phi_{cc}(C^0)$; if there exists the solution of the equation

$$\Phi_{cc} \bar{C} = -\Phi_c, \tag{14}$$

put

$$\bar{C} = -\Phi_{cc}^{-1} \Phi_c \tag{15}$$

and go to 2, otherwise put $C = -\Phi_c$ and go to 4.

2. Calculate $a = [(\Phi_c^T \Phi_c)(\bar{C}^T \bar{C})]^{-1/2} \Phi_c^T \bar{C}$. If $a > 0$, change the sign $\bar{C}$:

$$\bar{C} = \Phi_{cc}^{-1} \Phi_c, \tag{16}$$

as the first order term of (13) is increasing when $\bar{C}$ is calculated according to (15), and go to 5. At $a = 0$ put $\bar{C} = -\Phi_c$ and go to 4. Finally, when $a < 0$ check up the condition

$$\|\bar{C}\| = (\bar{C}^T \bar{C})^{1/2} \leq \alpha, \tag{17}$$

where α is a sufficiently small number (in the example $\alpha = 10^{-6}$). If the condition (17) is fulfilled, go to 3, otherwise - to 5. The fulfilment of (17) means, that the extremum of the quadratic form (13) is at the distance of the order of α from the point C^0 at $\varepsilon = 1$.

3. Print the values of components of the vector C^0, discrepancies

$$\delta_j = f(0.1j, C_1, C_2, \ldots, C_{10}) - C_{x_0}(0.1j),\ j = 1, 2, \ldots, 40, \tag{18}$$

and Silvester's determinants of the matrix Φ_{cc}: $(\Delta_1, \Delta_2, \ldots, \Delta_{10})$, end the calculation. If all the values $\Delta_i > 0$ $(i = 1, 2, \ldots, 10)$, then C^0 is a local minimum of the function Φ . Otherwise it is necessary to recalculate the problem from another initial point or to choose another $\bar{C}$ on the basis of more detailed analysis and proceed with the calculation.

4. Define ε subject to $\varepsilon = (\Phi_c^T \Phi_c)^{-1} D\Phi(C^\circ)$, where D is a specified number (in the example D = 0.1); then go to 6.

5. Put ε = 1 and go to 6.

6. Calculate for the chosen $\overline{C}$

$$\Phi_\varepsilon = \Phi_c^T \overline{C}, \tag{19}$$

$$\Phi_{\varepsilon\varepsilon} = \overline{C}^T \Phi_{cc} \overline{C}, \tag{20}$$

and the value of $\Phi(C^\circ + \varepsilon\overline{C})$ by the expression (13).

7. Define $\Phi(C^0 + \varepsilon\overline{C})$ by the expression (11).

8. If

$$\Phi(C^0) > \Phi(C^0 + \varepsilon\overline{C}), \tag{21}$$

put $C^1 = C^0 + \varepsilon\overline{C}$ and repead 1-6, substituting C^1 for C^0. If (21) is not fulfilled, define $\varepsilon_{new} = \frac{1}{2[\Phi(C^\circ+\varepsilon\overline{C}) - \Phi(C^\circ) - \varepsilon\Phi_\varepsilon]} \cdot (-\Phi_\varepsilon \varepsilon^2)$, recalculate $\Phi(C^0 + \varepsilon_{new}\overline{C})$ according to (11), check up (21) at $\varepsilon = \varepsilon_{new}$ etc until (21) is fulfilled.

The algorithm made it possible to obtain the solution of the set problem for 24 iterations. In fig.1 the final function, corresponding to the parameters values, that minimize Φ, is shown by a solid line. In fig.2 the function $-\lg\Phi$ is plotted versus iteration step number. It should be mentioned, that during iterations 4, 10-16 the tangential quadratic form (13) is nonconvex and $\overline{C}$ is defined by the expression (16).

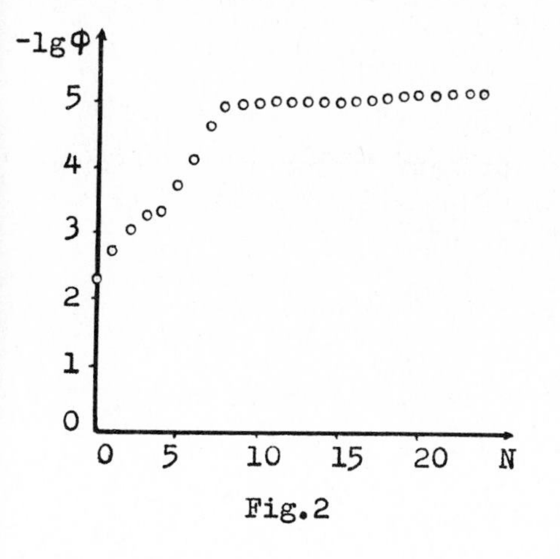

Fig.2

REFERENCE

I. Ф.Р. Гантмахер, Л.М. Левин "Теория полета неуправляемых ракет". Москва, Физматгиз, 1959 г.

MATHEMATICAL PROGRAMMING APPROACH TO A MINIMAX THEOREM OF STATISTICAL DISCRIMINATION APPLICABLE TO PATTERN RECOGNITION

K. ISII, Osaka University

Y. TAGA, Shizuoka University

1. Let $\mathcal{F}_i$'s be two families of all possible p-variate distribution functions with specified mean vectors $\underline{\mu}_i$ and non-degenerate variance-covariance matrices Σ_i, and π_i be prior probability or weight assigned to $\mathcal{F}_i$ for $i=1,2$ ($\pi_1 + \pi_2 = 1$). We are supposed to discriminate whether an observation $\underline{x}$ is from a (true) distribution $F_1 \in \mathcal{F}_1$ or $F_2 \in \mathcal{F}_2$. A randomized decision rule is represented by a pair of functions $\phi_1(\underline{x})$ and $\phi_2(\underline{x}) = 1 - \phi_1(\underline{x})$ ($0 \leqq \phi_1(\underline{x}) \leqq 1$), based on which one decides, with probability $\phi_i(\underline{x})$, that an observed value $\underline{x}$ is a sample from some F_i in $\mathcal{F}_i$ ($i=1,2$). If the pair $F = (F_1, F_2)$ is known, the error probability or classification error for the decision rule $\phi = (\phi_1, \phi_2)$ is clearly given by

$$e(\phi, F) = \pi_1 \int_{R^p} \phi_2(\underline{x})\,dF_1(\underline{x}) + \pi_2 \int_{R^p} \phi_1(\underline{x})\,dF_2(\underline{x}). \tag{1.1}$$

The aim of the present paper is to give the values of $\sup_{F\in\mathcal{F}} \inf_{\phi\in\Phi} e(\phi, F)$ and $\inf_{\phi\in\Phi} \sup_{F\in\mathcal{F}} e(\phi, F)$ together with a saddle point of $e(\phi, F)$, using the mathematical programming method given in one of the authors [3], where Φ denotes the set of all possible classification rule $\phi = (\phi_1, \phi_2)$, and $\mathcal{F} = \mathcal{F}_1 \times \mathcal{F}_2$ is the set of all pairs $F = (F_1, F_2)$ with $F_i \in \mathcal{F}_i$ ($i=1,2$).

2. Some necessary quantities and results used in the main theorems are introduced in the following lemma.

Lemma *Suppose* $1 \leqq \frac{\pi_2}{\pi_1} < 1 + (\underline{\mu}_1 - \underline{\mu}_2)'\Sigma_2^{-1}(\underline{\mu}_1 - \underline{\mu}_2)$. *Then, for every vector* $\underline{x}$ *in* R^p *satisfying* $\frac{\underline{x}'(\underline{\mu}_1 - \underline{\mu}_2)}{\sqrt{\underline{x}'\Sigma_2\underline{x}}} \geqq \frac{\pi_2}{\pi_1} - 1$, *there exists a unique real number* $t = t(\underline{x})$ *which satisfies the equation*

$$(2.1)\qquad \sqrt{\underline{x}'\Sigma_1\underline{x}}\sqrt{\pi_1 t - 1} + \sqrt{\underline{x}'\Sigma_2\underline{x}}\sqrt{\pi_2 t - 1} - \underline{x}'(\underline{\mu}_1 - \underline{\mu}_2) = 0.$$

Further, there exists a vector $\underline{x} = \underline{b}$ *attaining the maximum value, say* t_0, *of* $t(\underline{x})$. *The vector* $\underline{b}$ *is unique up to a positive multiplier, and* $t_0 > \frac{1}{\pi_1}$.

In the following the vector $\underline{b}$ and real number t_0 should be understood to represent those introduced above.

It may be assumed without loss of generality that $\pi_1 \leqq \pi_2$. We have then

Theorem 1 (i) When $1 \leqq \frac{\pi_2}{\pi_1} < 1 + (\underline{\mu}_1 - \underline{\mu}_2)'\Sigma_2^{-1}(\underline{\mu}_1 - \underline{\mu}_2)$, *we have*

$$(2.2)\qquad \max_{F\in\mathcal{F}} \inf_{\phi\in\Phi} e(\phi, F) = \min_{\phi\in\Phi} \sup_{F\in\mathcal{F}} e(\phi, F) = \frac{1}{t_0}.$$

A saddle point (ϕ^*, F^*) *of* $e(\phi, F)$ *is given by any* $F^* = (F_1^*, F_2^*)$ *such that*

$$(2.3)\qquad F_i^* = \frac{1}{\pi_i t_0} G_0 + \left(1 - \frac{1}{\pi_i t_0}\right) G_i \qquad (i=1,2),$$

where G_0 *is the one-point distribution concentrated at* $\underline{m}_0 = \underline{\mu}_1 - \frac{\sqrt{\pi_1 t_0 - 1}}{\sqrt{\underline{b}'\Sigma_1\underline{b}}}\Sigma_1\underline{b}$, G_i *any distribution with mean*

$$\underline{m}_i = \underline{\mu}_i - \frac{(-1)^i \Sigma_i \underline{b}}{\sqrt{\pi_i t_0 - 1}\sqrt{\underline{b}'\Sigma_i\underline{b}}},$$ *variance-covariance matrix*

$\Gamma_i = \frac{\pi_i t_0}{\pi_i t_0 - 1}(\Sigma_i - \frac{1}{\underline{b}'\Sigma_i\underline{b}}\Sigma_i\underline{b}\underline{b}'\Sigma_i)$, *and by any* $\phi^* = (\phi_1^*, \phi_2^*)$ *such that* $0 \leqq \phi_{3-i}^*(\underline{x}) \leqq g_i(\underline{x})$ $(i=1,2)$ *and* $\phi_1^*(\underline{x}) + \phi_2^*(\underline{x}) = 1$, *where* $g_i(\underline{x}) = c_i(\underline{x} - \underline{m}_i)'\underline{b}\underline{b}'(\underline{x} - \underline{m}_i)$ *with*

$$\frac{1}{c_i} = \frac{\pi_i\sqrt{\underline{b}'\Sigma_i\underline{b}}}{\sqrt{\pi_i t_0 - 1}}\left(\frac{\pi_1\sqrt{\underline{b}'\Sigma_1\underline{b}}}{\sqrt{\pi_1 t_0 - 1}} + \frac{\pi_2\sqrt{\underline{b}'\Sigma_2\underline{b}}}{\sqrt{\pi_2 t_0 - 1}}\right)t_0^{\,2}.$$

(ii) *When* $\frac{\pi_2}{\pi_1} \geqq 1 + (\underline{\mu}_1 - \underline{\mu}_2)'\Sigma_2^{-1}(\underline{\mu}_1 - \underline{\mu}_2)$, *we have*

$$\sup_{F\in\mathcal{F}}\inf_{\phi\in\Phi} e(\phi, F) = \min_{\phi\in\Phi}\sup_{F} e(\phi, F) = \pi_1. \tag{2.4}$$

In this case, $\sup_{F\in\mathcal{F}} e(\phi, F)$ *is minimized by* $\phi_1^*(\underline{x}) \equiv 0$ *and* $\phi_2^*(\underline{x}) \equiv 1$, *while a maximizing* F *of* $\inf_{\phi\in\Phi} e(\phi, F)$ *does not always exist. Hence a saddle point does not always exist.*

If we restrict the classification rule to (non-randomized) "linear discrimination", that is, to the case where ϕ_i is the indicator function of a half space (open or closed), we obtain the following theorem. Denote by Φ_0 the set of all linear classification rule ϕ and, in particular, by $\Phi_{\underline{\beta}}$ the set of all ϕ such that ϕ_1 is the indicator function of a half space of the form $\{\underline{x} \mid \underline{\beta}'\underline{x} \geqq c\}$ or $\{\underline{x} \mid \underline{\beta}'\underline{x} > c\}$ (c being arbitrary) for a p-dimensional vector $\underline{\beta}$. Clearly $\Phi_0 = \bigcup_{\underline{\beta}} \Phi_{\underline{\beta}}$. Then we have

Theorem 2 (i) *When* $1 \leqq \frac{\pi_2}{\pi_1} < 1 + (\underline{\mu}_1 - \underline{\mu}_2)'\Sigma_2^{-1}(\underline{\mu}_1 - \underline{\mu}_2)$, *the value of* $\sup_{F\in\mathcal{F}}\inf_{\phi\in\Phi_{\underline{b}}} e(\phi, F)$ *(hence also* $\sup_{F\in\mathcal{F}}\inf_{\phi\in\Phi_0} e(\phi, F)$*) is the same as* $\sup_{F\in\mathcal{F}}\inf_{\phi\in\Phi} e(\phi, F)$ *given in Theorem 1, where* $\underline{b}$ *is the vector defined in the lemma, while* $\inf_{\phi\in\Phi_0}\sup_{F\in\mathcal{F}} e(\phi, F)$ *is in general larger than* $\inf_{\phi\in\Phi}\sup_{F\in\mathcal{F}} e(\phi, F)$.

(ii) *When* $\frac{\pi_2}{\pi_1} > 1 + (\underline{\mu}_1 - \underline{\mu}_2)'\Sigma_2^{-1}(\underline{\mu}_1 - \underline{\mu}_2)$, $\sup_{F\in\mathcal{F}} \inf_{\phi\in\Phi_0} e(\phi, F)$ *and* $\inf_{\phi\in\Phi_0} \sup_{F\in\mathcal{F}} e(\phi, F)$ *coincide with those in the non-restricted case (hence the value is* π_1*).*

Various explicit results are obtained under additional assumptions. Particularly, in the simplest case that $p=1$ and $\pi_1=\pi_2=\frac{1}{2}$, the results coincide with those in Chernoff [1].

3. The formal proofs of Theorems 1 and 2 need not bear any direct reference to the theory of mathematical programming, if once a saddle point (ϕ^*, F^*) has been found. The essentials of our method may lie rather in how to find such a saddle point. For this purpose a mathematical programming approach is useful. For fixed ϕ, the problem to obtain $\sup_{F\in\mathcal{F}} e(\phi, F)$ is regarded as to maximize a linear functional $e(\phi, F)$ in F subject to the linear constraints described in terms of specified $\underline{\mu}_i$ and Σ_i. The assumption of non-degeneracy allows us to make use of the duality theorem given in [3] (the essential part is contained in [2]), and the problem is transformed into a minimization problem. Then $\inf_{\phi\in\Phi} \sup_{F\in\mathcal{F}}$ problem is reduced to a simple minimization problem, and we can obtain the minimizing ϕ^* as well as the minimum value. For this ϕ^* the maximizing $F = F^*$ of $e(\phi^*, F)$ is easily obtained, and the pair (ϕ^*, F^*) thus obtained is introduced in Theorem 1. It remains to verify that (ϕ^*, F^*) is actually a saddle point. Some formal and elementary calculations assure in fact that (ϕ^*, F^*) is a saddle point.

References

[1] Chernoff, H. (1971), "A bound on the classification error for discriminating between populations with specified means and variances," *Studi di probabilità, statistica e ricerca operativa in onore di Giuseppe Pompilj.*

[2] Isii, K. (1964), "Inequalities of the types of Chebyshev and Cramér-Rao and mathematical programming," *Ann. Inst. Statist. Math.*, 16, 277-293.

[3] Isii, K. (1969-), "Lecture Notes on Optimization Theory and its Applications (unpublished)."

PENALTY METHODS AND SOME APPLICATIONS OF MATHEMATICAL PROGRAMMING

A.A. Kaplan

Institute of Mathematics,
Siberian Branch,
U.S.S.R. Academy of Sciences,
Novosibirsk.

I. The problem of minimization of the convex function $f : R^n \to R$ on the convex solid compact set $\Omega \subset R^n$ is under consideration.

Theorem I. Let functions $\varphi_k : R^n \to R$, $k=1,2,\ldots$ be convex in the domain $\Omega_\delta = \{x : \inf_{y\in\Omega} \|x-y\| \leq \delta\}$ at some $\delta > 0$ and quasi-convex on the entire R^n;

I° $\lim_{k\to\infty} \varphi_k(x) = 0$, if $x \in \operatorname{int} \Omega$;

2° $\lim_{k\to\infty} \varphi_k(x) = +\infty$, if $x \bar{\in} \Omega$.

Then functions $F_k(x) = f(x) + \varphi_k(x)$ starting from the certain number achieve their absolute minima; corresponding sequence $\{x^k\}$ of minima is bounded, its any limiting point belongs to Ω and brings the minimum of function f on Ω.

It is necessary to emphasize that in literature the question of convergence of penalty method is usually considered for specific sequences of functions φ_k . The above theorem establishes the conditions of the method's convergence in terms of the structure of penalty functions. In [2] functions of the form $\varphi_k(x) = A_k G(x)$, where $A_k \to \infty$, are investigated and the result depends only on the structural properties of function G .

Theorem 2. Let conditions of theorem I be satisfied and for the boundary points of set Ω in equalities

$$\Phi_K(x) \geq d > 0\ , \quad K = 1,2,\ldots$$

hold at that.

Then if K is sufficiently large, points x^K are located strictly inside Ω.

Let $\Omega = \{x : g^j(x) \leq 0\ , \ j = 1,2,\ldots,m\}$, where $g^j : R^n \to R$ are convex functions. Suppose that f, g^j are differentiable and $\inf_{x \in R^n} \max_{1 \leq i \leq m} g^j(x) < 0$. The following statement is correct.

Theorem 3. Let all functions Φ_K be differentiable and satisfy conditions of theorem I. Let also

$$\nabla \Phi_K(x) = \sum_{j=1}^{m} \Psi_K^j(x) \nabla g^j(x),$$

where functions Ψ_K^j are non-negative and continuous on R^n and for any j the sequence $\{\Psi_K^j(z^K)\}$ vanishes, if $\overline{\lim_{K \to \infty}}\, g^j(z^K) < 0$.

Then the sequence $\{y^K\}$, where $y^K = (y_1^K, \ldots y_m^K)$, $y_j^K = \Psi_K^j(x^K)$ $(j = 1,2,\ldots,m)$ is bounded, every limiting point of the sequence $\{x^K, y^K\}$ is the solution of the dual problem.

The points (x^K, y^K) mentioned in the theorems are feasible for the dual problem. And it means that starting from the certain number at every step of the process bilateral values for the optimal value of the objective function of the initial problem are automatically (calculation of values $\Psi_K^j(x^K)$ will be required only) obtained, if the conditions of theorem 2 are satisfied.

The first theorem of this type was obtained by McCormik and Fiacco [I] for functions $\Phi_K(x) = -r_K \sum_{j=1}^{m} \frac{1}{g^j(x)}$ $(r_K > r_{K+1} > 0\ , \ r_K \to 0)$ and generalized for some special classes of penalty functions in [2].

2. Let $V \subset L_2(Q)$ be Hilbert space and the bilinear form

$$\Phi(u,v) = \int_Q \Big(\sum_{i,j=1}^{n} d_{ij} \frac{\partial u}{\partial x_i} \frac{\partial v}{\partial x_j} + d_0 uv \Big) dx$$

be continious on V. The problem of minimization of the quadratical functional

$$J(u) = \Phi(u,u) - 2 \int_Q fu\, dx$$

on the convex closed set $K \subset V$ is under consideration.

Here $Q \subset R^n$ is bounded domain with piece-wise smooth boundary Γ, $d_{ij} \in C^1(\bar{Q})$, $d_0 \in C(\bar{Q})$, $f \in L_2(Q)$. It is assumed that the quadratic form $\Phi(u,u)$ is positive definite on V.

On the basis of some modification of Ritz's method (and in certain cases, of Courant's method too) the given problem may be approximately expressed as the problem of convex programming. To make this approach efficient the set K has to satisfy the following conditions (!):

(!) Systems of functions

$\{\xi_{ij}\}$, $j=1,2,\ldots,K_i$; $i=1,2,\ldots$; $\xi_{ij}\in V$,
and convex closed sets $\mathcal{N}_\ell \subset R^{K_\ell}$, $\ell=1,2,\ldots$, such that

I) $\sum_{\tau=1}^{K_\ell} a_\tau^\ell \xi_{\ell\tau} \in K$ at all $a^\ell \in \mathcal{N}_\ell$, $\ell=1,2,\ldots$;

2) for any element $v\in K$ for $\varepsilon>0$ it is possible to indicate number T and vectors $a^T\in\mathcal{N}_T$, $a^{T+1}\in\mathcal{N}_{T+1},\ldots$ so that

$$\Big\| \sum_{\tau=1}^{K_{T+i}} a_\tau^{T+i} \xi_{T+i,\tau} - v \Big\|_V \le \varepsilon ;$$

3) for minimization of the convex function of K_ℓ variables on set $\mathcal{N}_\ell$ usual methods of convex programming are applicable, - are known or may be efficiently constructed.

Let us note that the requirement of linear independence of coordinate functions beeing natural for the classic Ritz's scheme may turn rather burdening and complicating the structure of set $\mathcal{N}_\ell$ in this case.

Approximated solution of the initial problem is solved in the following way.

Taking fixed ℓ, we determine element z^ℓ giving the minimum to functional J in the set of elements of the type $u^\ell=\sum_{\tau=1}^{K_\ell} a_\tau^\ell \xi_{\ell\tau}$, where $a^\ell\in\mathcal{N}_\ell$.

Functional

$$\Psi(a^\ell)=J(u^\ell)=\sum_{\tau,\eta=1}^{K_\ell} c_{\tau\eta}\, a_\tau^\ell a_\eta^\ell - 2\sum_{\tau=1}^{K_\ell} f_\tau a_\tau^\ell ,$$

where

$$c_{\tau\eta}=\int_Q \Big(\sum_{i,j=1}^{n} d_{ij}\frac{\partial \xi_{\ell\tau}}{\partial x_i}\frac{\partial \xi_{\ell\eta}}{\partial x_j} + d_0 \xi_{\ell\tau}\xi_{\ell\eta}\Big)dx, \quad f_\tau=\int_Q f\xi_{\ell\tau}\,dx$$

is the sum of a non-negative-definite quadratic form and a linear function depending on K_ℓ variables. Besides $J(u^\ell)\ge \min_{u\in K} J(u)$ at all $a^\ell\in\mathcal{N}_\ell$, so that the unknown element z^ℓ exists and is determined by vector $\bar a^\ell$ giving minimum of Ψ on $\mathcal{N}_\ell$ (such vector $\bar a^\ell$ and hence element z^ℓ are generally speaking not unique).

It's clearly seen that the sequence $\{z^\ell\}$, $\ell=1,2,\ldots$, is a minimizing one in the initial problem which due to the positive definiteness of $\Phi(u,u)$ results in $\|z^\ell - z\|_V \xrightarrow[\ell\to\infty]{} 0$.

With all this to find the approximate solution z^ℓ it is necessary to solve the problem of convex programming in space R^{K_ℓ}.

Three examples of the specific accomplishment of the mentioned scheme are considered below.

It is supposed that $K_1=1, K_2=2,\ldots$, $\xi_{\ell i}=\xi_i$ $(\ell=1,2,\ldots)$ which in case $K=V$ corresponds to the classical scheme of Ritz's method.

First of the problems considered - the problem cf elastic-plasticity - has the following variation wording (more detailed wording of problems under consideration is in [3]) :

$$J(u) - \min$$
$$u \in K \equiv K^1 = \{ v \in \overset{\circ}{W}{}_2^1(Q) \cdot |\nabla v(x)| \leq 1 \text{ at } Q \}.$$

To solve it the complete system of linear-independent functions $\{\xi_k\}$ is introduced in the space $C_0^1(\bar{Q})$. The problem of functional J minimization on the set of linear combinations $u^\ell(x) = \sum_{i=1}^{\ell} a_i \xi_i(x)$ looks like:

$$J(u^\ell) - \min$$

$$|\nabla u^\ell(x)| \leq 1 \text{ at } x \in \bar{Q}.$$

The latter inequality can be put down in the following way:

$$\sum_{k,p=1}^{\ell} a_k a_p \sum_{i=1}^{n} \frac{\partial \xi_k(x)}{\partial x_i} \frac{\partial \xi_p(x)}{\partial x_i} \leq 1 \text{ at } x \in \bar{Q}.$$

To give the set $\mathcal{N}_\ell$ in the above way point 3 of the condition (!)is violated. However due to the choice of the specific penalty function of the penalty method it is possible to avoid the preliminary approximation of set $\mathcal{N}_\ell$.

Namely, under the natural supposition that at certain p

$$\max_{x \in \bar{Q}} \min_{1 \leq i \leq n} \left| \frac{\partial \xi_p(x)}{\partial x_i} \right| > 0,$$

set $\mathcal{N}_\ell$ is a solid compactum it is possible to prove that the sequence of penalty functions

$$\Phi_k(a) = \int_Q e^{A_k(|\nabla u^\ell(x)|^2 - 1)} dx$$

at $A_k \geq 0$, $A_k \to \infty$ satisfies the conditions of theorems I and 2.It is necessary to emphasize that the smoothness degree of an integrated function completely depends on the smoothness of coordinate functions so that with the appropriate choice of the latter ones for calculating $\Phi_k(a)$ cubature formulas of the high degree of accuracy may be applied.

The approach described related to the construction of penalty integral function is universal enough.

In the two variation problems that follow the specificity of set K is essentially taken into consideration which permits us to com to simpler problems of convex programming.

In the second problem

$$V = W_2^1(Q),\ K \equiv K^2 = \{u \in V : u|_\Gamma \geq 0\}.$$

To solve it we shall introduce into consideration the systems of functions $\{\varphi_k\}$ and $\{\psi_k\}$ possessing the following properties:
a) $\psi_k \in C_0^1(\bar{Q})$ and the system $\{\psi_k\}$ is complete in space $C_0^1(\bar{Q})$
b) $\varphi_k \in C^1(\bar{Q})$, $\varphi_k|_\Gamma \geq 0$ and the system $\{\varphi_k\}$ is complete on the boundary in the following sense:for any function $u \in C(\bar{Q})$, $u|_\Gamma \geq 0$ and $\varepsilon > 0$ there exist m and non-negative $\alpha_1, \alpha_2, \dots, \alpha_m$ such that $\| u - \sum_{k=1}^{m} \alpha_k \varphi_k \|_{C(\Gamma)} \leq \varepsilon$ holds.

Let us note that requirements concerning coordinate functions may be weakened by substitution of $C_0^1(\bar{Q})$ by $\overset{\circ}{W}{}_2^1(Q)$, $C^1(\bar{Q})$ by $W_2^1(Q)$.

Fixing t and p we look for the minimum of functional J on the set of linear combinations

$$u^{t,p}(x) = \sum_{i=1}^{t} a_i \psi_i(x) + \sum_{k=1}^{p} b_k \varphi_k(x)$$

under the condition of non-negativity of coefficients b_k.

With all this the sequence of coordinate functions is obtained by mixing sequences $\{\psi_i\}$ and $\{\varphi_i\}$ (for the purpose of simplicity we shall consider $\xi_i = \psi_{i/2}$ if i is even, $\xi_i = \psi_{\frac{i+1}{2}}$ if i is odd) and sets

$$\mathcal{N}_e = \{ a^\ell \in R^\ell : a^\ell_{2\tau-1} \geq 0,\ \tau = 1, 2, \dots [\tfrac{\ell+1}{2}] \}$$

correspond to the sequence $\{\xi_i\}$ formed by it.

It means that to find the approximated solution z^ℓ it will be necessary in this case to solve the simplest problem of quadratic programming. Without loss of generality we may believe that $\bar{Q}$ is located strictly inside unit cube $S = \{x : 0 \leq x_i \leq 1,\ i = 1, 2, \dots, n\}$.

Then functions of the type

$$x_1^{k_1} x_2^{k_2} \dots x_n^{k_n} (1 - x_1)^{z_1} (1 - x_2)^{z_2} \dots (1 - x_n)^{z_n} \qquad (*)$$

with all possible non-negative indices may be taken as φ_i . System $\{\psi_i\}$ is subordinate to standard requirements of Ritz's classic scheme.

In the third problem

$$V = \overset{\circ}{W}{}_2^1(Q),\ K \equiv K^3 = \{u \in V : u(Q) \geq 0\}.$$

The construction of minimization sequence results in the solution of the simplest quadratic programs too. The system of differentiable in Q functions $\{\xi_i\}$, $\xi_i \in C_0(\bar{Q})$, is constructed such that for any non-negative function $u \in C_0^1(\bar{Q})$ and $\varepsilon > 0$ there exist m and non-negative $\alpha_1, \alpha_2, \dots, \alpha_m$ for which

$$\|u-\sum_{i=1}^{m}\alpha_i \xi_i\|_{C(\bar{Q})}\le \varepsilon,\ \|\frac{\partial u}{\partial x_k}-\sum_{i=1}^{m}\alpha_i\frac{\partial \xi_i}{\partial x_k}\|_{L_2(Q)}\le\varepsilon,\ k=1,2,\dots,n$$

holds. Further we set $N_\ell=\{a^\ell\in R^\ell : a^\ell\ge 0\}$. Under the condition $Q\subset S$ as ξ_i functions of type

$$\omega(x)x_1^{k_1}x_2^{k_2}\dots x_n^{k_n}(1-x_1)^{r_1}(1-x_2)^{r_2}\dots(1-x_n)^{r_n} \qquad (**)$$

may be taken where $\omega\in C_0(\bar{Q})$, ω is differentiable and positive on Q.

It is possible to prove that if the boundary Γ is not too bad (i.e. in case of convex or Lipshitz domains) the abovementioned suppositions ensure the correctness of the conditions (!) in all problems.

Theorem 4. The system of functions (*) satisfies condition b). The system of functions (**) is complete in the sense mentioned above.

It is necessary to indicate that for the solution of the last two problems within the framework of the scheme described it is possible to use functions which are usually applied in variation-difference approximation. However, dimension of the problems of quadratic programming obtained will naturally be very large and this may turn out an obstacle for their practical solution.

On the other hand the negative effect related to the bad definiteness of matrices when using Ritz's classic scheme will not evidently be very significant as the conditions of non-negativity which by the meaning of the problem are essential, cut off the vicinity of the absolute minimum of the quadratic functional.

References.

I. Fiacco A., McCormik G. The sequential unconstrained minimization technique for nonlinear programming: A primal dual method. Man. Sci., 10, N 2,1964.

2. Fiacco A., McCormik G. Nonlinear programming. Wiley,1968.

3. Lions J.L. Quelques méthodes de résolution des problèmes aux limites non lineaires. Paris, 1969.

4. Каплан А.А. Характеристические свойства функций штрафа.Докл. АН СССР, т. 210(1973), № 5. (English transl.: A.A.Kaplan.Characteristic properies of penalty functions.Soviet Math.Dokl,3,1973).

NUMERICAL ANALYSIS OF ARTIFICIAL ENZYME MEMBRANE - HYSTERESIS, OSCILLATIONS AND SPONTANEOUS STRUCTURATION

J.P. KERNEVEZ

Département de Mathématiques Appliquées
Université de Technologie de Compiègne
60206 - COMPIEGNE - FRANCE

D. THOMAS

E.R.A. n° 338 du C.N.R.S.
Département de Génie Biologique
Université de Technologie de Compiègne
60206 - COMPIEGNE - FRANCE

1 - INTRODUCTION

Over the past few years there has been an upsurge of interest in partial differential equations with multiple solutions, prompted largely by their probable biological importance, and reflected by a growing number of papers (1). The systems resulting from the coupling between enzyme reactions and diffusion constraints can be ruled by this kind of partial differential equations (2). The binding of enzymes into artificial membranes makes possible the study of the interaction between diffusion and enzyme reaction within a well-defined context (3). This paper deals with the numerical analysis of three experimental systems very recently produced: an hysteresis phenomenon observed by Naparstek et al (4) ; an oscillation phenomenon studied by Naparstek et al (5) and Caplan et al (6) ; an asymetrical structuration shown by Thomas et al (7).

Among the various mechanisms describing information storage in the phase of "short term memory", the most favoured one is based on an all or none process, due to phase transition in an excitable membrane, accompanied with metastable states (8,9). These phase transitions require structural changes at the molecular level, in membrane permeability, or in high molecular weight molecules involved in the process (10).

On the other hand chemical engineers studied intensively cooperative effects, due to non-linear temperature dependance of rate coefficients, and their relation to the stability of chemical reactors. In these works, hysteresis and oscillations resulting, are discussed (11). At the biological point of view, sustained oscillations have been experimentally observed and established beyond doubt for glycolysis by Hess (12) and Betz and Chance (13). Sel'Kov (14) and Higgins (15) have worked out models to represent the observed oscillations. Recently Goldbeter and Lefever (16) have constructed a model explicitly taking into account the allosteric effects which is free from phenomenological factors. According to Prigogine and Nicolis (17) these systems are also likely to occur, under different experimental conditions, in spatially ordered states (18). When associated with diffusion constraints the non-linear enzyme reaction can produce a pattern or structure due to an instability of an homogeneous state.

The experimental systems will be described briefly in order to introduce the equations ruling them. It is possible to get more details from the original papers cited above (4,5,6,7). The existence of one or several solutions will be demonstrated for both transient and steady states. The numerical methods used to solve the described equations will be given together with the numerical results.

2 - STATEMENT OF THE PROBLEMS AND EQUATIONS RULING THE SYSTEMS

2.1. - Hysteresis phenomenon

An enzyme giving an inhibition by excess of substrate is homogeneously distributed inside an artificial membrane (thickness L). The membrane is immersed in a solution of substrate S . The concentration of S is changing slowly enough to enable the quasi steady state hypothesis about the enzyme reaction and the metabolite diffusion which occur inside the membrane. The reference axis is chosen perpendicular to the membrane. S concentration is a function of space and time (x and t).

$$(2.1) \qquad - D_S \frac{\partial^2 [S]}{\partial x^2} + V_M \frac{[S]}{K_M + [S] + \frac{[S]^2}{K_{SS}}} = 0$$

with D_S substrate diffusion coefficient, K_M Michaelis constant and V_M maximum activity by volume unit of membrane.

Let us replace $\frac{x}{L}$ by x and $\frac{[S]}{K_M}$ by s . x and s are now dimensionless.

Equation (2.1) can be written like

$$(2.2) \qquad - \frac{\partial^2 s}{\partial x^2} + \sigma \frac{s}{1+s+as^2} = 0$$

with

$$\sigma = \frac{V_M}{K_M} \frac{L^2}{D} \qquad \text{and} \qquad a = \frac{K_M}{K_{SS}} .$$

Obviously, σ and a are a perfect characterization of the membrane system.

The boundary conditions are:

(2.3) $s(0,t) = s(1,t) = \alpha(t)$ $(\alpha(t) \geqslant 0)$

2.2. - *Oscillation phenomenon*

The system was experimentally built and studied in Harvard by Naparstek et al (4) and a numerical analysis was given by Caplan et al (5). A coating (thickness L) bearing an enzyme activity is along of glass electrode. One side of the coating is impermeable to any molecule. In this system a substrate S (of concentration [S] = s) is transformed into the product H^+ (of concentration $[H^+] = p$), and due to the reaction between H^+ and OH^- a function $a(x,t)$ is introduced in the equations ruling the system:

$$(2.4) \quad \frac{\partial s}{\partial t} - D_S \frac{\partial^2 s}{\partial x^2} + F(s,p) = 0 \quad , \quad \frac{\partial a}{\partial t} - D_H \frac{\partial^2 a}{\partial x^2} - F(s,p) = 0$$

With

$$a = p - \frac{10^{-20}}{p} \quad , \quad F(s,p) = Q_2\, Q_3 \frac{E_0}{Q_2 + Q_3 + Q_3 \dfrac{5.45\ 10^{-5}}{s}}$$

$$Q_2 = \frac{64.5}{1 + 10^{7.29}p + \dfrac{10^{-11.49}}{p}} \quad , \quad Q_3 = \frac{20.2}{1 + 10^{6.92}p}$$

$$D_S = 0.13\ 10^{-5} \quad , \quad D_H = 0.64\ 10^{-5} \quad , \quad L = 10^{-2}$$

The boundary and initial conditions are:

$$(2.5) \quad s(0,t) = 5.5\ 10^{-7} \ ; \ p(0,t) = 10^{-13} \ ; \ \frac{\partial s}{\partial x}(L,t) = \frac{\partial p}{\partial x}(L,t) = 0.$$

$$(2.6) \quad s(x,0) = 5.5\ 10^{-7} \ ; \ p(x,0) = 10^{-13}.$$

2.3. - *Asymetrical structuration*

When pH is taken into account, it is necessary to replace V_M by αV_M, where α is a function of pH $(0 \leqslant \alpha \leqslant 1)$.

Two enzymes E_1 and E_2 are homogeneously immobilized. Reaction E_i is

$$S_i \xrightarrow{E_i} P_i \qquad i = 1,2$$

With an excess of substrates reaction rates are:

$$(2.7) \quad v_i = \alpha_i\, \sigma_i \qquad i = 1,2$$

σ_i being a constant and α_i being a function of pH as indicated in fig. 1.

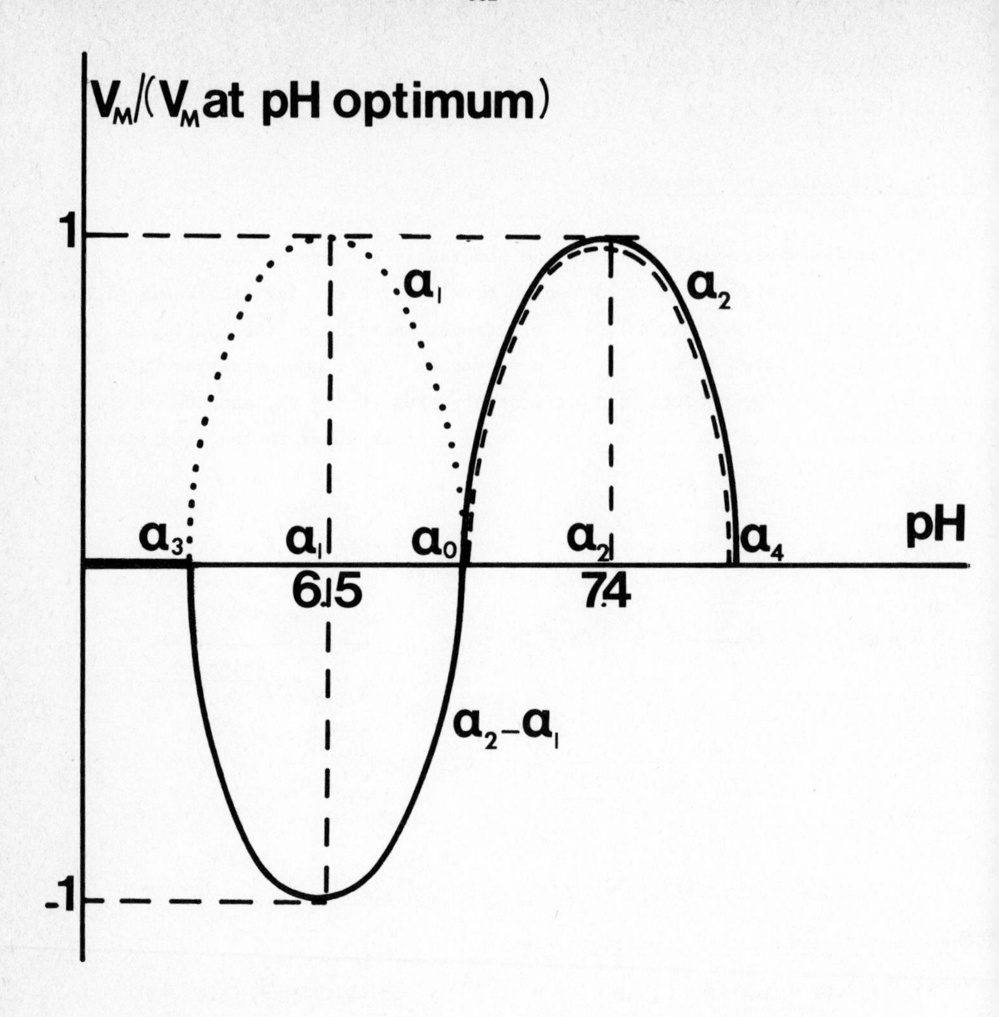

Figure 1. Graphs of α_1 (...) , α_2 (---) and $\alpha_2 - \alpha_1$ (——).

Enzyme E_1 (resp. E_2) has a range of activity at low pH (resp. high pH). It is a catalyst of the transformation of the substrate S_1 (resp. S_2) into the product P_1 (resp. P_2). This product is an acid (resp. a base) so that when it appears the pH is decreasing (resp. increasing). E_i activity is maximum at pH = a_i (i = 1,2).

When the two enzymes are homogeneously immobilized inside a membrane, there is no variation of pH for a pH value a_o.

With a pH value of $a_o + \varepsilon$ in the first compartment and $a_o - \varepsilon$ in the second one there will be amplification of the concentration difference because on both sides there will be an autocatalytic reaction.

Product concentrations are ruled by

(2.8) $$\frac{\partial p_i}{\partial t} - \frac{\partial^2 p_i}{\partial x^2} - \alpha_i \sigma_i = 0 \qquad i = 1,2$$

with $\sigma_1 = \sigma_2 = \sigma$. pH is linked to p_1 and p_2 by

(2.9) $$pH = (pH)_0 + \beta(p_2 - p_1).$$

It is possible to write

(2.10) $$\frac{\partial pH}{\partial t} - \frac{\partial^2 pH}{\partial x^2} = \beta\sigma(\alpha_2 - \alpha_1).$$

The function $(\alpha_2 - \alpha_1)$ is described by curve in Fig. 1.

The boundary and initial conditions are:

(2.11) $pH = a_o - \varepsilon$ at $x = 0$ and $pH = a_o + \varepsilon$ at $x = 1$

(2.12) For $t = 0$ the pH through all the system is equal to a_o.

3 - MATHEMATICAL SECTION

3.1. - Hysteresis phenomenon

Using monotone methods as indicated in Sattinger (19) it is easy to check that

(3.1) $$\begin{cases} -\Delta s + \sigma F(s) = 0 \\ s_{|\Gamma} = 0 \end{cases}$$

has at least one solution.

(Here the equation is written in a domain Ω, open and bounded in R^n ($n = 1,2$ or 3) of boundary Γ, $-\Delta = -\sum_{i=1}^{n} \frac{\partial^2}{\partial x_i^2}$ and $F(s) = s / (1+s+a|s|^2)$.

In fact let $u_o =$ (resp. $v_o = 0$) is an upper (resp. lower) solution of (3.1):

(3.2) $$\begin{cases} -\Delta u_o + \sigma F(u_o) \geqslant 0 \quad , \quad u_{o|\Gamma} = \alpha \\ -\Delta v_o + \sigma F(v_o) = 0 \quad , \quad v_{o|\Gamma} \leqslant \alpha \end{cases}$$

Then we have the following results from Sattinger (19):
There exists a regular solution s of (3.1) such that $v_o \leqslant s \leqslant u_o$.
Now we can prove that there exists values of σ and α for which (3.1) has at least 3 solutions.

Multiple solutions (one dimensional case)

The system (3.1) is now

$$(3.3) \quad \begin{cases} -s'' + \sigma F(s) = 0 \\ s(0) = s(1) = \alpha \end{cases}$$

and we have a first integral

$$(3.4) \quad -s'^2(x) + \sigma G(s(x)) = \sigma G(\mu)$$

Where G is some primitive of F and $\mu = s(\frac{1}{2})$ the minimum of $s(x)$ for $x \quad [0,1]$.

$$(3.5) \quad s'(x) = +\sigma^{1/2}\,(G(s(x)) - G(\mu))^{1/2} \qquad 1/2 \leqslant x \leqslant 1$$

$$(3.6) \quad \sigma^{1/2}(x - 1/2) = \int_\mu^{s(x)} \frac{d\xi}{(G(\xi)-G(\mu))^{1/2}} \qquad 1/2 \leqslant x \leqslant 1$$

and we have the constraint

$$(3.7) \quad \frac{1}{2}\,\sigma^{1/2} = \int_\mu^\alpha \frac{d\xi}{(G(\xi)-G(\mu))^{1/2}}$$

(3.6) will define a solution of (3.3) every time
(3.7) will have a solution μ.

We define f by

$$(3.8) \quad f(\mu) = \int_\mu^\alpha \frac{d\xi}{(G(\xi)-G(\mu))^{1/2}}$$

Then

$$(3.9) \quad f'(\mu) = -\frac{1}{(G(\xi)-G(\mu))^{1/2}} + \left(-\frac{1}{2}\right) \int_\mu^\alpha \frac{F(\xi)-F(\mu)}{(G(\xi)-G(\mu))^{3/2}}\,d\xi$$

our purpose is to show that we can find α and σ such that (3.7) admits 3 different solutions at least.

Let us choose μ^* such that $F'(\mu^*) < 0 \quad (\mu^* > 0)$.

It is easy to see that when $\alpha \to +\infty$ the first term in $f'(\mu^*)$ tends towards 0 and the second one towards a limit $\ell > 0$ $(\ell < +\infty)$.

Taking α sufficiently large we have $f'(\mu^*) > 0$. With this value of α we have necessarily the graph of f like (c) in Fig. 2.

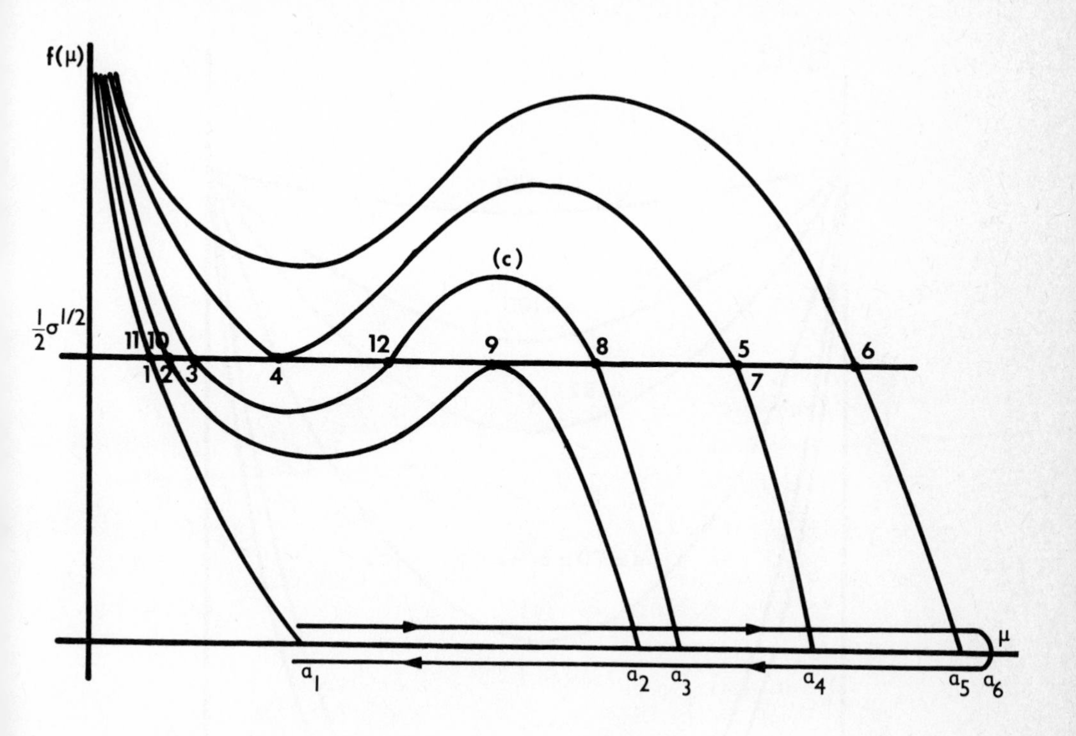

Figure 2. Graphs of f for different values of α.

(In particular because for every σ equation (3.7) has at least one solution).

Now we see on Fig. 2 that according to the value of σ equation (3.7) has 1, 2 or 3 solutions.

Let us take such a value of σ for which (3.7) can have 3 solutions, and assume that α is varying continuously, first increasing from α_1 to α_6, then decreasing to α_1. If we follow the corresponding values of μ, we pass successively by the points 1,2,3,4,5,6,7,8,9,10,11, and there are 2 jumps: first, when we pass through α_4, the point jumps from 4 to 5. Then, when we pass through α_2, the point jumps from 9 to 10.

This corresponds to 2 kinds of profiles of concentration:

- The "low" profiles, for $\alpha_1 < \alpha < \alpha_4$ in the increasing phase and $\alpha_1 < \alpha < \alpha_2$ in the decreasing phase, and
- The "high" profiles, for $\alpha_4 < \alpha$ in the increasing phase and $\alpha > \alpha_2$ in the decreasing phase. (Fig. 3).

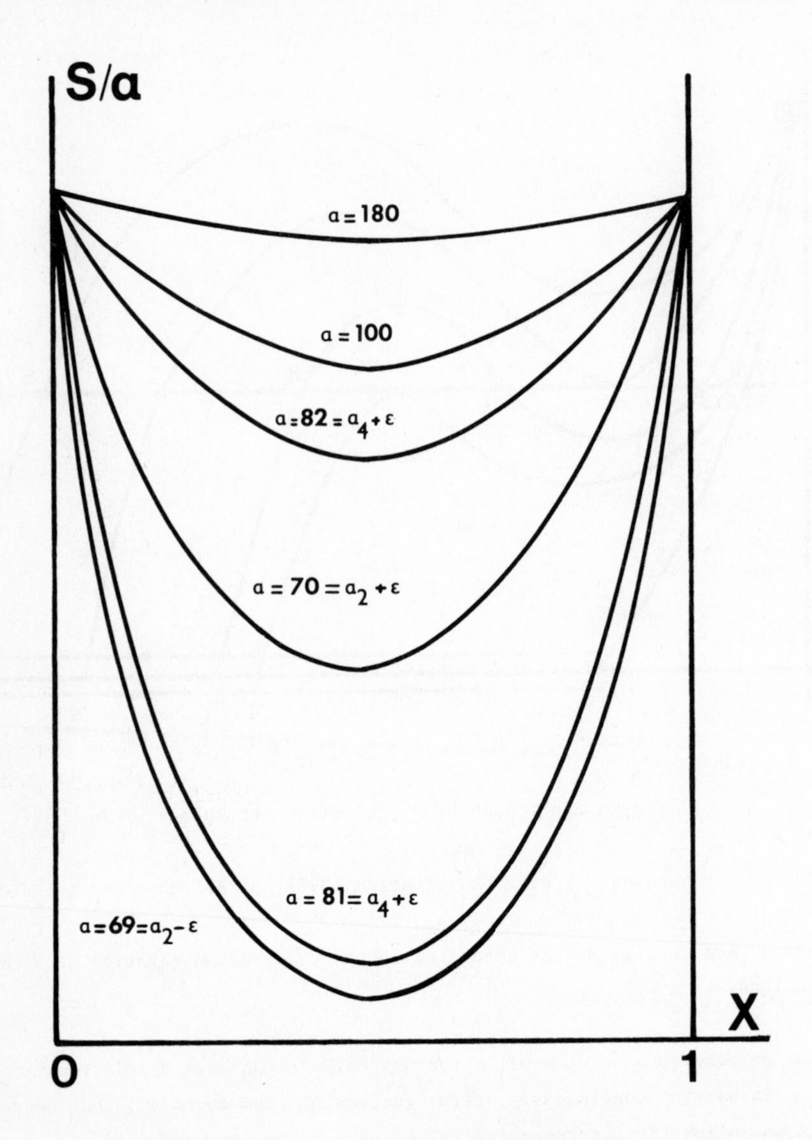

Figure 3. Hysteresis phenomenon: calculated substrate concentration profiles inside an artificial enzyme membrane. These profiles are given for several concentration values outside. The concentration unit in each case is the concentration in the external medium.

3.2. - Oscillation phenomenon

Existence of a solution for equations (2.4) - (2.6) can be proved by standard methods (LIONS (10)): construction of a sequence of Faedo Galerkin approximations, a priori estimates on these approximations, a priori estimates on the time fractionnary derivatives using Fourier transform with respect to time, extraction of a subsequence converging towards a solution of (2.4) - (2.6). The property used in the proof is mainly the boundedness of the function F. In the proof of unicity is used the fact that F is Lipschitz continuous.

3.3. - Spontaneous asymetrical structuration

$$(3.10)\quad \begin{cases} \dfrac{\partial u}{\partial t} - \Delta u + F(u) = 0 \\ u_{|\Gamma} = g(x) \\ u_{|t=0} = u_o(x) \end{cases}$$

F being the full line curve of figure 1. $u_o(x) = a_4$ (resp. $v_o(x) = a_3$) is an upper (resp. lower) solution of

$$(3.11)\quad \begin{cases} -\Delta u + F(u) = 0 \\ u_{|\Gamma} = g(x) \end{cases}$$

provided $a_3 \leqslant g(x) \leqslant a_4$.

The same arguments than in § 3.1 insure that (3.11) has at least one solution. Moreover for any continous $u_o(x)$ with $a_3 \leqslant u_o(x) \leqslant a_4$, we obtain a global regular solution of the initial value problem (3.10), this solution is unique and satisfies $a_3 \leqslant u(x,t) \leqslant a_4$.

4 - NUMERICAL METHODS AND RESULTS

4.1. - Hysteresis phenomenon

For a given boundary value α (2.2) was solved by the following scheme (s_j^k is the approximation of $s(j\,\Delta x)$ at iteration number k and $J\,\Delta x = 1$):

i/ Define $s_j^o = 0$ or according to the solution expected ("low" or "high" profile).

ii/ Define s^{k+1} from s^k by

$$(4.1)\quad \begin{cases} -\dfrac{1}{h^2}\left(s^{k+1}_{j+1} + s^{k+1}_{j-1} - 2s^{k+1}_{j}\right) + \sigma s^{k+1}_{j} \;/\; \left(1 + s^{k}_{j} + a(s^{k}_{j})^2\right) = 0 \\ s^{k+1}_{o} = s^{k+1}_{J} = \alpha \end{cases}$$

iii/ Stop when $\dfrac{\sum_{j=o}^{J} |s^{k+1}_{j} - s^{k}_{j}|}{\sum_{j=o}^{J} |s^{j+1}_{j}|} < \varepsilon$

To solve (3.3), where the boundary value $\alpha(t)$ varies continously with time the initial approximation (s^{o}_{j}) for the solution when $\alpha = \alpha(t)$ was the last approximation s^{k}_{j} for the preceding value of α.

The flux of substrate entering the membrane under quasi-stationary state conditions was calculated as a function of the external substrate concentration for increasing and decreasing values (Figure 4).

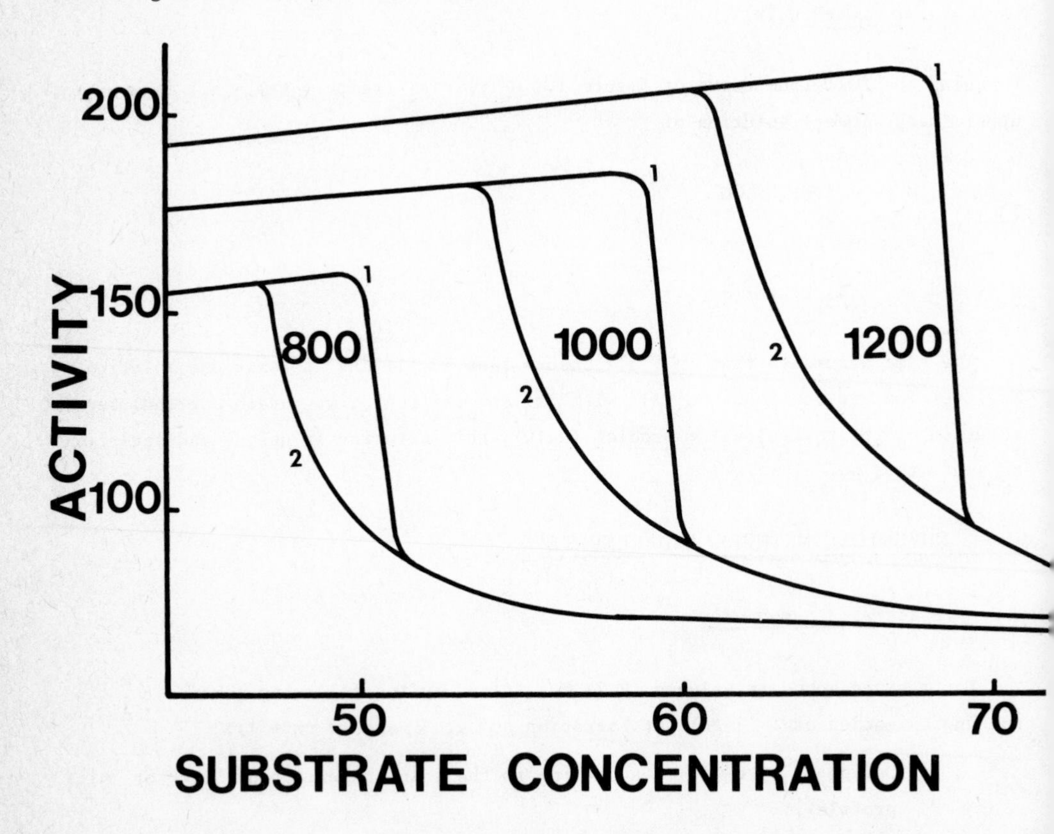

Figure 4. Hysteresis phenomenon: boundary flux as a function of substrate concentration at the boundaries for 3 σ values = 800, 1000 and 1200.

The system parameters were:

$V_M = 4\ 10^{-8}$ mole cm^{-3} sec^{-1} $D = 2.8\ 10^{-8}$ cm^2 sec^{-1}

$K_M = 4\ 10^{-8}$ mole cm^{-3} $e = 4.5\ 10^{-3}$ cm

$K_{SS} = 4\ 10^{-7}$.

On figure 4 there is an evidence for an hysteresis phenomenon, the activity-substrate concentration relationship does not give an unique curve. For increasing substrate concentration values after a critical point there is a strong drop of the enzyme activity. For decreasing values a similar effect is observed. The low and high activity values correspond respectively to high and low concentration profiles of substrate inside the membrane (Fig. 4). Two families of curves are given on Fig. 3, showing the strong variation of concentration level for each critical point. The numerical results are similar to the experimental results presented by Naparstek et al (4).

4.2. - Oscillation phenomenon

The system described by the equations (2.4) to (2.6) was solved numerically by Caplan et al (6). A periodic behavior for the system was shown and a sharp front of substrate concentration in the active layer oscillates between the boundaries ruled respectively by Dirichlet and Neumann conditions.

There is a qualitative agreement between the simulation of Caplan et al (6) and the experimental results of Naparstek et al (7).

4.3. - Spontaneous asymetrical structuration

The system previously described by equations (2.10) to (2.12), was solved numerically by using the explicit method. At time zero the pH is equal to a_o through all the membrane. Asymetrical perturbations were introduced on both boundary values and pH profile in the membrane evolved as shown on Fig. 5. There is an evidence for a spontaneous structuration of the concentration from an homogeneous repartition. This phenomenon is due to the instability of the homogeneous state through the non-linearity and the autocatalysis of the enzyme reactions.

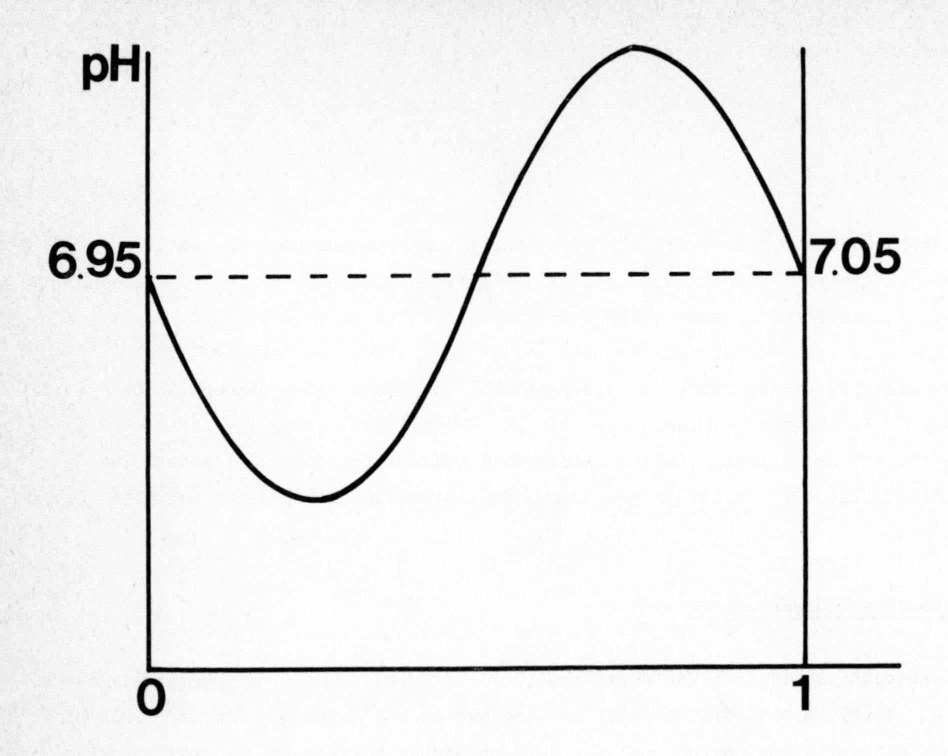

Figure 5. pH profile inside an artificial bienzyme membrane. These profiles are given for different time values.

CONCLUSION

It is interesting at both physical and biological points of view to study instabilities and possibilities of multiple solutions in systems ruled by partial differential equations. In this way, artificial enzyme membranes or at least artificial immobilization of the enzymes could be a mean to study this kind of phenomena, because enzyme kinetics are frequently non-linear or autocatalytic and the systems are ruled by diffusion-reaction coupling, that is to say by partial differential equations.

Moreover it is easily possible to change the boundary conditions, so critical for these kinds of systems.

For artificial enzyme membranes, due to the well-defined context it is possible to write in a simple way equations ruling the systems and to compare calculated and experimental results. This work is also of stage between the classical enzymology in solution and the study of properties of enzymes in very complex distributed biological systems.

REFERENCES

(1) Lecture Notes in Mathematics n° 332, Non Linear Problems in the Physical Sciences and Biology, Springer Verlag 1973.

(2) PRIGOGINE, I (1969). "Structure, Dissipation and Life". In Theoretical Physics and Biology: proceedings, et M. Marois (Amsterdam - North-Holland Co.) p.p. 23-52.

(3) THOMAS, D. and CAPLAN, S.R., in "Membrane Separation Processes", Ed. Meares P. Elsevier Publishing Co. Amsterdam (1974).

(4) NAPARSTEK, A., ROMETTE, J.L., KERNEVEZ, J.P. and THOMAS, D. Nature (1974), 249 , 490

(5) NAPARSTEK, A., THOMAS, D. and CAPLAN, S.R., Biophys. Biochem. acta. (1973), 323 , 643.

(6) CAPLAN, S.R., NAPARSTEK, A. and ZABUSKY, N., (1973), Nature, 245 , 364.

(7) THOMAS, D., GOLDBETER, A. and LEFEVER, R., in press

(8) CHANGEUX, J.P., and THIERY, J., in: "Regulatory Function of Biological Membranes" Vol. 11, Ed. Jänefelt, J., Elsevier Publishing Co. (1968).

(9) KATCHALSKY, A. and OPLATKA, A. (1966), Neurosciences Res. Symp. 1 352 (1966)

(10) KATCHALSKY, A. and SPANGLER, R., Quartely Revs. Biophys., 1 , 127 (1968).

(11) DENBIGH, K.G. and TURNER, J.C.R. in "Chemical Reactor Theory", University Press Cambridge (1971).

(12) HESS, B. (1962) in "Funktionelle und morphologische Organisation der Zelle". Springer Verlag, Berlin.

(13) BETZ, A. and CHANCE, B. (1965). Phase relationship of glycolytic intermediates in yeast cells with oscillatory metabolic control. Archs. Biochem. Biophys., 109 , 585.

(14) SEL'KOV, E.E. (1968), Self-oscillations in glycolysis. Euro. J. Biochem. 4 , 79.

(15) HIGGINS, J. (1964), A chemical mechanism for oscillation of glycolysis intermediates in yeast cells. Proc. Nat. Acad. Sci. U.S.A. 51 , 989.

(16) GOLDBETER, A. and LEFEVER, R. (1972). Dissipative structures for an allosteric model. Application to glycolytic oscillations. Biophys. J. 12 , 302.

(17) PRIGOGINE, I. and NICOLIS, G. (1971). Biological order, structure and instabilities. Quarterly Reviews of Biophysics, 4 , 107.

(18) ZHABOTINSKI, A.M. (1964), Biofizika, 2 , 306.

(19) SATTINGER, D.H. Monotone Methods in Non-linear Elliptic and Parabolic Boundary Value Problems. Indiana Univ. Math. Journal, 21 , 979 - 1000 (1972).

(20) LIONS, J.L. Quelques Méthodes de Résolution des Problèmes aux limites non linéaires, Dunod, Gauthier Villars, Paris, 1969. (English translation by Le Van, Holt, Rinehart and Winston, New-York (1971)).

THE STABILITY OF OPTIMAL VALUES IN PROBLEMS OF DISCRETE PROGRAMMING

V. R. Khachaturov
Computer Center of the Academy
of Sciences of the USSR
Moscow, USSR

Let us consider function $f(X)$, that is given and limited to the set $\mathcal{D}$, consisting of a finite number of elements $X \in \mathcal{D}$.

For an arbitrary $X^0 \in \mathcal{D}$ and any number $R \geq 0$ there exists a non-empty subset $\mathcal{D}(X^0, R) \subset \mathcal{D}$, possessing the following properties:

if $X \in \mathcal{D}(X^0, R)$,

then $|f(X) - f(X^0)| \leq R$;

if $X \notin \mathcal{D}(X^0, R) \quad (X \in \mathcal{D} \setminus \mathcal{D}(X^0, R))$,

then $|f(X) - f(X^0)| > R$.

Let us introduce function $\varphi(X) \equiv f(X) - f(X^0)$ for all $X \in \mathcal{D}$. It is clear that when $X = X^0$, $\varphi(X^0) = 0$. We shall provide several definitions.

Let us designate subset $\mathcal{D}(X^0, R)$ as the region of R-stability for the values $\varphi(X^0) = 0$, and let us say that the value $\varphi(X^0)$ at $\mathcal{D}(X^0, R)$ is R-stable.

The value $\varphi(X^0)$ is absolutely R-stable, if $\varphi(X) = 0$ for all $X \in \mathcal{D}(X^0, R)$.

The value $\varphi(X^0)$ is absolutely stable, if $\varphi(X) = 0$ for all $X \in \mathcal{D}(f(X) = const)$.

The following assertion is evident. If $f(X) \neq const$, then a number R^0 exists, which is the upper boundry of the absolute R-stability of the value $\varphi(X^0)$, i.e., if $0 \leq R < R^0$, then all regions $\mathcal{D}(X^0, R)$ are regions of absolute R-stability, if $R \geq R^0$, then no single region $\mathcal{D}(X^0, R)$ is a region of absolute R-stability.

It is not difficult to note that $\mathcal{D}(X^0, R) = \mathcal{D}(X^0, 0)$ for all $0 \leq R < R^0$. For this reason in order to determine the region of absolute R-stability for any $0 \leq R < R^0$ it is necessary to find all of the solutions for equation $\varphi(X) = 0$ with the condition that $X \in \mathcal{D}$. The solution of such equations, as a rule, is often quite difficult (it is sufficient to recall the Diofan equations).

During the investigation of the stability of value $\varphi(X^0)$ the following problems emerge:

- to determine R^0 – the upper boundry of absolute R-stability;
- for any $R \geq 0$ to evaluate the quantity $|\mathcal{D}(X^0, R)|$ – the number of elements in the set $\mathcal{D}(X^0, R)$;
- for any $R \geq 0$ to construct an algorithm for obtaining all elements of the existing set $\mathcal{D}(X^0, R)$.

In many applications the solutions of these problems are of interest in cases in which the optimal value of the function $f(X)$ is attained at X^0. For the sake of definition let $f(X^0) = \min f(X)$ for $X \in \mathcal{D}$. Then $\varphi(X) \geq 0$ for all $X \in \mathcal{D}$ and $\min \varphi(X) = \varphi(X^0) = 0$.

Let us cite a few examples and classes of functions for which a solution proved to be possible for the problems mentioned above during the investigation of the stability of their optimal values.

1. $X \subset I = \{1, \dots, m\}$, $\mathcal{D} = \{X\}$.

$f(X)$ satisfies the conditions

$$f(X^1) + f(X^2) - f(X^1 \cup X^2) - f(X^1 \cap X^2) \leq 0. \quad (1)$$

For the determination of the minimal value of $f(X^0)$ of such functions let us employ the method of sequential calculation [1].

The function $\varphi(X)$ likewise satisfies conditions (1). For the determination of the region $\mathcal{D}(X^0, R)$ the following generalizations of the rejection rules are employed, the proof of which is given in [2,3].

<u>The first generalized rejection rule</u>. If for X^1 and X^2 the following conditions are fulfilled:

$X^1 \subset X^2$ and $\varphi(X^1) + R < \varphi(X^2)$, then no single $X \supset X^2$ enters into $\mathcal{D}(X^0, R)$.

<u>The second generalized rejection rule</u>. If for X^1 and X^2 the following conditions are fulfilled:

$X^1 \subset X^2$ and $\varphi(X^1) > \varphi(X^2) + R$, then no single

$X \subset X^1$ enters into $\mathcal{D}(X^0, R)$.

The third generalized rejection rule. If for X^1 and X^2 the following conditions are fulfilled: $X^1 \subset X^2$ and

$$\varphi_1(\underline{X}) > \varphi(\tilde{X}) + R \text{ , or } \varphi_2(\underline{X}) > \varphi(\tilde{X}) + R \text{ ,}$$

then no single X of the type $X^1 \subset X \subset X^2$ enters into $\mathcal{D}(X^0, R)$. Here: $\varphi(\underline{X}) = \min_{X^1 \subset X \subset X^2} \varphi(X)$;

$\tilde{X}$ is the arbitrary subset I, in particular it may be $\tilde{X} = X^0$;

$$\varphi_1(X) \equiv \varphi(X^1) - \sum_{i \in X^2 \setminus X^1} [\varphi(X^1) - \varphi(X^1 \cup i)];$$

$$\varphi_2(X^1) = \varphi(X^2) - \sum_{i \in X^2 \setminus X^1} [\varphi(X^2) - \varphi(X^2 \setminus i)].$$

There exists a large class of concrete problems of mathematical programming in which the corresponding functions $\varphi(X)$ satisfy the conditions (1) (for example, different types of problems of distribution [2,4]). For many of these problems effective algorithms have been elaborated for the determination of subset $\mathcal{D}(X^0, R) \subset \mathcal{D}$, that implement the generalized rejection rules [2,5].

2. $X \subset J = \{1, \dots, n\}$, whereupon $|X| = m, m \le n$. $\mathcal{D} = \{X\}$, consequently $|\mathcal{D}| = C_n^m$.

For each $X \in \mathcal{D}$ let us correlate a quadratic system of linear equations:

$$\sum_{K \in X} y_K B_K = B \text{ , where}$$

$$B_K = (a_{K1}, \dots, a_{Km}), \; B = (b_1, \dots, b_m)^T.$$

Let us denote by $\mathcal{D}_S$ the set of all $X \in \mathcal{D}$, for which the corresponding system has a non-negative solution $\{y_K(X)\}, K \in X$.

Let us examine the function

$$f(X) = \begin{cases} \sum_{K \in X} c_K y_K(X), & \text{if } X \in \mathcal{D}_S \\ \infty, & \text{if } X \in \mathcal{D} \setminus \mathcal{D}_S \end{cases}$$

It is evident that the values X^0 and $f(X^0)$ in this case are determined by methods of linear programming. An algorithm has been elaborated [4] for the determination of the regions $\mathcal{D}(X^0, R)$ that is a modification of the algorithm proposed in [6].

3. $X = (x_1, \dots, x_n)$

$x_j \geq 0$ are integer numbers, satisfying the condition

$$\sum_{i=1}^{n} a_i(x_i) \leq b, \quad \mathcal{D} = \{X\}.$$

$$f(X) = -\sum_{i=1}^{n} g_i(x_i)$$

The values X^0 and $f(X^0)$ may be determined by methods of dynamic programming [7]. The determination of region $\mathcal{D}(X^0, R)$ in [4] relies on a modification of the algorithm of Bellman for the case when $g_i(x_i) \geq 0$ and $a_i(x_i)$ are single-valued functions, $a_i(x_i)$ take on integer values in the presence of whole x_i, and $b \geq 0$ are integers.

Thus, for a large class of functions $\varphi(X)$ it turns out to be possible to determine the regions of R-stability of their optimal values. Primarily, this is of great practical significance for the solution of concrete problems of optimal planning - which in practice makes it possible to choose such a solution $\tilde{X} \in \mathcal{D}(X^0, R)$, that satisfies some additional conditions that had not been taken into account in the initial construction of the problem, or else are generally not formalizable. At the same time it is well known that $\varphi(\tilde{X}) > R$, if $\tilde{X} \notin \mathcal{D}(X^0, R)$.

Secondly, these functions $\varphi(X)$ may be used for the determination of optimal values of more "complex" functions $g(X)$. The approximational-combinatorial method for the solution of problems of discrete programming [4, 8], is based on this approach consisting of the following basic elements.

Let the determination of $Y \in \mathcal{D}$ be required, such that

$$g(Y) = \min_{X \in \mathcal{D}} g(X).$$

Let us assume that such a function $f(X)$ is known, for which there exist effective algorithms for the determination of the region $\mathcal{D}(X^0, R)$ and it has been established that $g(Y) \geq f(Y)$. Then a certain value $C \geq f(X^0)$ is chosen and the region $\mathcal{D}(X^0, R)$ is determined for $R = C - f(X^0)$. Then the element $\tilde{Y} \in \mathcal{D}(X^0, R)$ is found such that

$$g(\tilde{Y}) = \min_{X \in \mathcal{D}(X^0, R)} g(X).$$

If $g(\tilde{Y}) \leq C$, then $\tilde{Y} = Y$, $g(\tilde{Y}) = g(Y)$, that is, the problem is solved.

But if $g(\tilde{Y}) > C$, then $\tilde{Y}$ and $g(\tilde{Y})$ are taken as an approximate solution, whereupon $C < g(Y) \leq g(\tilde{Y})$.

Using this method problems of distribution that take into account communication, and territorial-production complexes, the distribution problem with Boolean variables and a series of others were solved [2,4,5,8].

References

1. V. P. Cherenin, "The Solution of Certain Combinatorial Problems of Optimal Planning by the Method of Sequential Calculations", Scientific-methods materials of the economico-mathematical seminar, LEMI AN SSSR, Issue 2, Moscow, Rotaprint, Gipromed, 1962, (Russian).

2. V. R. Khachaturov, Certain Questions and a Supplement to Distribution Problems of the method of Sequential Calculation, Dissertation for the Candidate of physico-mathematical science, Moscow, TsEMI AN SSSR, 1968. (Russian).

3. V. R. Khachaturov, "A Generalization of the Rejection Rules for the Solution of a Certain Class of Combinatorial Problems", Materials of the scientific conference on mathematics and mechanics, AS KazSSR, Alma-Ata, Institute of Economics and AN Kaz SSR, 1968. (Russian).

4. V. R. Khachaturov, "An Approximational-combinatorial Method and Certain of its Applications", Journal of Computer Mathematics and Mathematical Physics, Moscow, 1974 Vol. 14, No 6. (Russian).

5. N. D. Astakhov, V.E. Veselovskii, I. Kh. Sigal, V. R. Khachaturov, "Concerning the Experience in the Solution of Distribution Problems Modified by the Algorithm of Sequential Calculations", The thesis of the papers of the VI All-Union Conference on extremal problems, Tallin, AN EsSSR, 1973, (Russian).

6. K. G. Murty, "Solving the Fixed Charge Problem by Ranking the Extreme Points", Operations Research, 1968, vol. 16, No. 2, p. 268-279.

7. R. Bellman, S. Dreifus, Applied Problems of Dynamic Programming, Moscow, Nauka, 1965. (Russian).

8. V. R. Khachaturov, "Concerning the Approximal-combinatorial Method of Solving Problems of Mathematical Programming", Winter School on mathematical programming in the city of Drogobyche, Issue 3, Moscow, TsEMI AN SSSR, 1970, p. 657-674. (Russian).

Optimal control with minimum problems and variational inequalities

R. Kluge

Central Institute of Mathematics and Mechanics
Academy of Sciences of the GDR
108 Berlin, Mohrenstr. 39

We go to consider here problems of parameter optimization in abstract process relations such as minimum problems and mixed variational inequalities in Banach spaces. Besides we obtain existence results and projected and iterated approximation methods for the solution of these problems which cover, for instance, problems of optimal control, of identification and inverse problems for partial differential equations.

1. Minimum problems

Let V be a real Banach space with norm $\|.\|$, V^* its adjoint space and $(g,u) = g(u)$ for $g \in V^*$, $u \in V$. By the symbols $\rightarrow$, $\rightharpoonup$ we denote strong and weak convergence in V, respectively. The symbol $M[(.)]$ will be used to denote the set of all solutions of the problem represented by the formula $(\cdot)$.

Let $U \subset V$, $U \neq \emptyset$ and $f,h \in (U \rightarrow R^1)$. We consider the problem (1) - (2):

(1) $\quad f(u) = \inf f(v) \;, \quad v \in M\,[(2)]\,,$

(2) $\quad h(v) = \inf h(u) \;, \quad u \in U.$

Existence theorems for (1)-(2) can be obtained from the generalized Weierstraß theorems.

Let $U_n \subset V$, $U_n \neq \emptyset$, $e_n > 0$, $e_n \rightarrow 0$, $f_n, h_n \in (U_n \rightarrow R^1)$, $n=1,2,\ldots.$ As an approximation method for (1)-(2) we consider

(3n) $\quad j_n(w_n) = \inf j_n(u) \;, \; u \in U_n, \; j_n = h_n + e_n \cdot f_n, \; n=1,2,\ldots.$

Let the following assumptions be fulfilled:

I. $w - \overline{\mathrm{Lim}}\, U_n \subset U$.

II. $(f_n, U_n) \rightarrow (f,U)$ upper semi-continuously ($v_n \in U_n, v_n \rightarrow v \in U$ implies $\overline{\lim}\, f_n(v_n) \leq f(v)$) and weakly lower semi-continuously ($v_n \in U_n$, $v_n \rightharpoonup v \in U$ implies $\underline{\lim}\, f_n(v_n) \geq f(v)$).

III. $(h_n, U_n) \rightarrow (h,U)$ upper semi-continuously (u.s.c.) and weakly lower semi-continuously (w.l.s.c.).

Definition 1. The notation "$(f_n,U_n) \rightarrow (f,U)$ with the property (F_+)" means: if $v_n \in U_n$, $v_n \rightharpoonup v \in U$, $\overline{\lim}\, f_n(v_n) \leq f(v)$ then $\|v_n - v\| \rightarrow 0$.

Theorem 1. Let (3n) have at least one solution w_n for every n and let the following assumptions be fulfilled:

1. (1)-(2) has at least one solution w such that there exists a sequence $\{v_n\}$ with the properties $v_n \in U_n$ for $n \geq n_o$, $\|v_n - w\| \to o$ and $\overline{\lim} [(h_n(v_n) - h_n(w_n))/e_n] \leq 0$.
2. One of the following conditions (i),(ii) is fulfilled:
   (i) $U_n \subset E \subset V$, n=1,2,...; E is weakly compact.
   (ii) B is a reflexive space and $\overline{\lim}\, f_n(u_n) = +\infty$ if $u_n \in U_n$ and $\|u_n\| \to +\infty$.

Then $w - \overline{\mathrm{Lim}}\, w_n \neq \emptyset$ and $w - \overline{\mathrm{Lim}}\, w_n \subset M[(1)-(2)]$. If additionally

3. (f_n, U_n) (f, U) with property (F_+)

is fulfilled, then every weakly convergent subsequences of $\{w_n\}$ is also strongly convergent.

Proof: cf. [8]. For special cases see [5-7].

Remarks. 1. Based on the weak* $(B_1$ -) convergence in $B = B_1^*$ an analogous theorem on the convergence of $\{w_n\}$ can be derived [8] . 2. If under the assumptions of Theorem 1 problem (1)-(2) has a unique solution then $w_n \rightharpoonup w$ or $\|w_n - w\| \to o$ holds.

Let be V a Hilbert space H, P(K) the operator of projection from B onto the convex and closed set $K \subset H$, $\{H_n\}$ a sequence of subspaces and $P_n = P(H_n)$. Let for $p > o$ be $[p]$ the smallest integer greater than or equal to p.

Theorem 2. Let (1)-(2) have a unique solution w, (3n) have a solution w_n for every n, $U_n \subset H_n$ be convex and closed; j_n be defined on H and strongly convex with convexity constant $c(j_n)$; the assumptions 1-3 of Theorem 1 be fulfilled. Besides assume that for every j_n the gradient j_n' exists and is Lipschitz-continuous with Lipschitz constant $L(j_n')$. Then the projection-iteration method

$$a_{n+1} = \{P(U_n)P_n [I - t_n j_n']\}^{i_n} a_n, \quad n=1,2,\dots, \; a_1 \in H, \text{ with} \tag{4}$$

$$o < t_n < 16c(j_n)/L(j_n')^2, \quad i_n = [e_o/(1-L_n)], \; e_o > o,$$

$$L_n = \sqrt{1 - 16\, t_n c(j_n) + t_n^2 L(j_n')^2}$$ converges strongly to w in B.

Proof: cf [8]. For special cases see [5-7].

Theorem 3. Let (1) $(M[2] = U)$ have a unique solution w; $U_n \subset H_n$, n=1,2,..., be convex and closed; $U = s - \underline{\mathrm{Lim}}\, U_n$; f_n be defined on H, strongly convex with $c(f_n) \geq c > o$ and have the gradient f_n' with $L(f_n') \leq L$. If $\bigcup_n U_n$ is unbounded, then let $\|f_n'(o)\| \leq q$ and $|f_n(o)| \leq r$.

$$f_n(w_n) = \inf f_n(u) \quad , \quad u \in U_n , \tag{5n}$$

hase a unique solution w_n for every n and the sequence $\{w_n\}$ as well as the sequence $\{a_n\}$ generated by the projection-iteration method.

(6) $\quad a_{n+1} = P(U_n)P_n[I - t_n \cdot f_n']a_n , \; n=1,2,\ldots, \; a_1 \in H ,$

with $0 < \varepsilon_1 \leq t_n \leq \frac{16\, c(f_n)}{L^2} - \varepsilon_2, \; \varepsilon_2 > 0$, converge strongly to w in H.

Proof: cf. [8]. For special cases see [5-7].

Remark. In [8] also the methods (4) and (6) for locally Lipschitz-continuous gradients j_n' and f_n', respectively, are given (see [10]).

2. Generalized trace funtionals

Let also Y be a real Banach space, $X \subset Y$, $F \in (X \times U \to R^1)$, $S \in (U \to X)$. We consider problem (1) in the form

(7) $\quad F(Sw,w) = \inf F(Su,u), \; u \in U.$

If F is w.l.s.c. and S is weakly continuous then $F(S.,.)$ is w.l.s.c. The same is true, if F is (strongly, weakly) - lower semi-continuous (i.e. $x_n \to x$ and $u_n \rightharpoonup u$ implies $\underline{\lim}\, F(x_n,u_n) \geq F(x,u)$) and S is increased continuous (i.e. $u_n \rightharpoonup u$ implies $Su_n \to Su$). In both cases the generalized Weierstraß theorems may be applied. Further existence theorems can be obtained for weakly or increased closed S, for multi-valued $S \in (U \to 2^X)$ and on the base of weak* convergence (cf. [8]).

Let besides be $X_n \subset Y$, $F_n \in (X_n \times U_n \to R^1)$, $S_n \in (U_n \to X_n)$. As an approximation method for (7) we consider (3n) in the form

(8) $\quad F_n(S_n w_n, w_n) = \inf F_n(S_n u,u) \; , \; u \in U_n.$

If $(F_n, X \times U_n) \to (F, X \times U)$ u.s.c. and (strongly, weakly) - l.s.c. [w.l.s.c.] and $(S_n,U_n) \to (S,U)$ increased continuously [continuously and weakly continuously] then $(F_n(S_n.,.), U_n) \to (F(S.,.),U)$ u.s.c. and w.l.s.c. If, in addition, for $x_n \in X_n$ and $x_n \to x \in X$ $[x_n \rightharpoonup x \in X]$ $(F_n(x_n,.)U_n) \to (F(x,.),U)$ with property (F_+) then also $(F_n(S_n.,.),U_n) \to (F(S.,.),U)$ with property (F_+) holds.

So we can apply Theorem 1 (with $f_n = F_n(S_n .,.)$, $h_n \equiv 0$) to (8n).

Let be Y a Hilbert space, $\{Y_n\}$ a sequence of subspaces and $Q_n = P(Y_n)$. Assume $F_n(S_n .,.) = F_1(S_n.) + F_2(.)$, $F_1 \in (Y \to R^1)$ convex, $F_2 \in (V \to R^1)$ strongly convex, U_n, X_n convex, $S_n = \overline{S}_n + \overline{x}_n$, $\overline{S}_n$ linear and bounded, $\overline{x}_n \in X_n$. Then the functional $f_n = F_n(S_n.,.)$ are strongly convex with $c(f_n) = c(F_2)$ and

$$f_n' = F_2' + \overline{S}_n^* Q_n F_1'(S_n.)$$

holds. Under appropriate assumptions the Theorems 2 and 3 may be applied (cf. [8]).

3. Optimization with minimum problems

We consider (7) where $S \in (U \to X)$ is the solution operator of the following problem (9)-(1o):

(9) $k(x,u) = \inf k(y,u)$, $y \in M[(1o)]$,

(1o) $l(y,u) = \inf l(z,u)$, $z \in X$.

Theorem 1 can then be used to investigate the weak or increased continuity of S(cf.[9]). As an approximation method for (7),(9),(1o) we consider (8n) with the solution operator S_n of

(11n) $m_n(x,u) = \inf m_n(y,u)$, $y \in X_n$, $m_n(y,u) = l_n(y,u)+e_n k_n(y,u)$.

In (8n),(11n) as special cases combined Ritz-Ritz (Ritz-projected penalty, Ritz-projected regularization) methods are contained. Theorem 1 gives results on the continuous, weakly continuous or increased continuous convergence of $(S_n,U_n) \rightarrow (S,U)$. Using these results and, once more, Theorem 1 (cf. Part 2) the convergence of solutions of (8n),(11n) to solutions of (7),(9),(1o) can be proved (cf. [9]). In some cases also iteration methods of the form (4) and (6) can be derived ([9]). For applications of the results in problems with partial differential equations see [11] .

4. Mixed variational inequalities

Let be Y a reflexive space; $T,S \in (Y \rightarrow Y^*)$ and $k,l \in (Y \rightarrow R^1)$. We consider the mixed variational inequality

(12) $(Tz,y-z) \geq k(z) - k(y)$, $y \in M[(13)]$,

(13) $(Sy,x-y) \geq k(y) - k(x)$, $x \in X$.

Existence theorems for (12)-(13) can be obtained from the results of BREZIS [1] , BROWDER [2] , KLUGE-BRUCKNER [10] and LIONS [13] on variational inequalities.

Let be $X_n \neq \phi$, convex and closed; T_n, $S_n \in (Y \rightarrow Y^*)$ be monotone and hemicontinuous operators, $R_n = S_n + e_n \cdot T_n$; k_n, $l_n \in (Y \rightarrow R^1)$ be convex and lower semicontinuous functionals and $m_n = l_n + e_n k_n$, $n=1,2,\ldots$.

As an approximation method for (12)-(13) we consider

(14n) $(R_n z_n, x - z_n) \geq m_n(z_n) - m_n(x)$, $x \in X_n$, $n = 1,2,\ldots$.

Let the following assumptions be fulfilled:

IV. $w - \overline{\mathrm{Lim}}\, X_n \subset X$.

V. $X \subset s - \underline{\mathrm{Lim}}\, X_n$ if $S \neq 0$ and $y \in s\text{-}\underline{\mathrm{Lim}}\, X_n$ for any $y \in M[(13)]$ if $S \neq o$.

VI. $(T_n,X_n) \rightarrow (T,X)$ and $(S_n,X_n) \rightarrow (S,X)$ continuously;

$(k_n,X_n) \rightarrow (k,X)$ and $(l_n,X_n) \rightarrow (l,X)$ u.s.c. and w.l.s.c.

Definition 2. The notation "$(T_n,X_n) \rightarrow (T,X)$ with property (S_+)" means: if $y_n \in X_n, y_n \rightharpoonup y \in X$ and $\overline{\lim}\, (T_n y_n, y_n - y) \leq o$ then $\|y_n - y\| \rightarrow o$.

Theorem 4. Let (14n) have at least one solution z_n for every n and let the following assumptions be fulfilled:

1. (12)-(13) has at least one solution z such that there exists a sequence $\{y_n\}$ with the properties: $y_n \in X_n$ for $n \geq n_0$, $\|y_n - z\| \to 0$,
$$\overline{\lim_{i\to+\infty}} \left\{ [(-S_{n_i} y_{n_i}, z_{n_i} - y_{n_i}) + l_{n_i}(y_{n_i}) - l_{n_i}(z_{n_i})]/e_{n_i}\|z_{n_i}\| \right\} \leq 0$$
if $\|z_{n_i}\| \to \infty$ and
$$\overline{\lim_{n\to\infty}} \left\{ [-S_n y_n, z_n - y_n) + l_n(y_n) - l_n(z_n)]/e_n \right\} \leq 0$$
if $\{z_n\}$ is bounded.

2. One of the following conditions (i),(ii) is fulfilled:
   (i) $X_n \subset F \subset Y$, $n = 1,2,\dots$; F is bounded.
   (ii) $$\overline{\lim_{i\to\infty}} \left\{ [(T_{n_i} z_{n_i}, z_{n_i} - y_{n_i}) + f_{n_i}(z_{n_i})]/\|z_{n_i}\| \right\} = +\infty$$
   if $\|z_{n_i}\| \to +\infty$.

Then w-$\underline{\mathrm{Lim}}\ z_n \neq \emptyset$ and w-$\overline{\mathrm{Lim}}\ z_n \subset M[(12)-(13)]$.
If additionally the assumption

3. $(T_n, X_n) \to (T, X)$ with property (S_+) or $(k_n, X_n) \to (k, X)$ with property (F_+)
   is fulfilled, then every weakly convergent subsequence of $\{z_n\}$ is also strongly convergent.

Proof. cf. [8]. For special cases see [4-7].
Remark. Analogous to the theorems 2 and 3 we may give theorems on the strong convergence of projection-iteration methods for (12)-(13) (see [4-8]).

5. Optimization with variational inequalities

We consider (7) where $S \in (U \to X)$ is the solution operator of the following problem

(15) $(Q(z,u), y-z) \geq k(z,u) - k(y,u)$, $y \in M[(16)]$,

(16) $(P(y,u), x-y) \geq l(y,u) - l(x,u)$, $x \in X$.

Theorem 4 can then be used to investigate the weak, strong or increased continuity of S (cf. [9,11]). As an approximation method for (7), (15), (16) we consider (8n) with the solution operator S_n of

(17n) $(R_n(z_n,u), x - z_n) \geq m_n(z_n,u) - m_n(x,u)$, $x \in X_n$, $n = 1,2,\dots$,

$R_n(x,u) = P_n(x,u) + e_n \cdot Q_n(x,u)$, $m_n(x,u) = l_n(x,u) + e_n \cdot k_n(x,u)$.
In (8n),(17n) as special cases combined Ritz-Galerkin (Ritz-projected penalty, Ritz-projected regularization) methods are contained. Theorem 4 gives results on the continuous, weakly continuous or increased continuous convergence of $(S_n, X_n) \to (S, X)$. Using these results and Theorem 1 the convergence of solutions of (8n),(17n) to solutions

of (7), (15), (16) can be proved (cf.[9]). For special cases see KRAUSS [12] and YVON [15]. In some cases also iteration methods of the form (4) and (6) can be derived ([9]).

For applications of the results see [9] and [11] .

R e f e r e n c e s

[1] Brezis, H., Equations et inéquations non linéaires dans les espaces vektoriels en dualité. Ann. Inst. Fourier, Grenoble 18, 115-175 (1968).

[2] Browder, F.E., On the unification of the calculus of variations and the theory of monotone nonlinear operators. Proc. Nat. Acad. Sci. 56, 419-425 (1966).

[3] Kluge, R., Ein Projektions-Iterationsverfahren bei Fixpunktproblemen und Gleichungen mit monotonen Operatoren. Mber. Dt.Akad. Wiss. 11(1969), 599-6o9.

[4] ---, Zur approximativen Lösung nichtlinearer Variationsungleichungen. Mber. Dt. Akad. Wiss. 12 (197o), 12o-134.

[5] ---, Dissertation B. Berlin 197o.

[6] ---, Näherungsverfahren zur approximativen Lösung nichtlinearer Variationsungleichungen. Math. Nachr. 51 (1971), 343-356.

[7] ---, Näherungsverfahren für einige nichtlineare Probleme. Proc. of the Summer School on nonlinear operators. Neuendorf/Hiddensee (GDR)(1972), 133-146. Akademie-Verlag. Berlin 1974.

[8] ---, Variationsungleichungen über Lösungsmengen von Variationsungleichungen. Math. Nachr.

[9] ---, Zur Optimierung in Aufgaben mit Variationsungleichungen. Math. Nachr.

[1o] Kluge, R. und G. Bruckner, Iterationsverfahren für einige nichtlineare Probleme mit Nebenbedingungen. Math. Nachr. 56 (1973), 346-369.

[11] Kluge, R., Krauss, E. und R. Nürnberg, Zur Optimierung in Aufgaben mit Operatorgleichungen und Evolutionsgleichungen. Proc. of a Summer School on nonlinear Operators. Stara Lesná (Czechoslovakia), 1974 .

[12] Krauss, E., Zur Steuerung mit Operatorgleichungen. Proc. of the Summer-School on nonlinear operators. Neuendorf/Hiddensee (GDR) (1972), 169-176. Akademie-Verlag, Berlin 1974.

[13] Lions, J.L., Quelques méthodes de résolution des problémes aux limites non linéaires. Dunod. Gauthier Villars, Paris 1969.

[14] ---, Contrôle optimal de systèmes gouvernés par des équations aux dérivées partielles, Paris, Dunod, Gauthier - Villars, 1968.

[15] Yvon, J.P., These. Etude de quelques problemes de controle pour des systemes distribues. 1973.

DUAL MINIMAX PROBLEMS

P.J. LAURENT
University of Grenoble
B.P. 53 ; 38041 Grenoble, France

In 1972, L. McLinden [3,4] proposed a perturbation method in order to build the dual of a minimax problem. This method is similar to Rockafellar's for dual minimization problems but is much more complicated for several technical reasons. In particular, McLinden permanently works with classes of equivalent convex-concave functionals.

Our aim is to present an equivalent theory of duality for minimax problems, using only the classical duality theory for minimization problems and a notion of partial minimization.

I. NOTATIONS

The notations are essentially the same as in [1,2]. We denote by X and X' two locally convex topological linear spaces in duality ; $< x,x' >$ being the value of the bilinear form at $x \in X$ and $x' \in X'$. In the same way Y, Y' are in duality ; Y_1, Y_1' ; X_2, X_2' ; and so on ... . We consider functionals f defined on X (or X', Y, Y',...) with values in $\overline{\mathbb{R}} = \mathbb{R} \cup \{+\infty\} \cup \{-\infty\}$. The set of functionals f defined on X which are the supremum of a family of continuous affine functionals will be denoted by $\Gamma(X)$. The conjugate of f will be denoted by f^* ; it is an element of $\Gamma(X')$. If a functional f is defined on a product XY, f^* means the conjugate of f with respect to the two variables ; it is an element of $\Gamma(X'Y')$. We will need a notion of partial conjugency ; for example, if f is defined on XY, $p_Y f$ will denote a functional defined on XY' which is obtained by partial conjugency with respect to $y \in Y$:

$$p_Y f(x,y') = \operatorname*{Sup}_{y \in Y} (< y,y' > - f(x,y)) .$$

The conjugency (or partial conjugency) operation will always be applied to convex functionals. Rather to define a notion of conjugency for concave functionals g, we will use the "change of sign operator" θ (such that $\theta g = -g$) and take the conjugate of θg . For example, if $f \in \Gamma(XY)$, $p_Y f$ is concave with respect to $x \in X$ and we can define $p_X\, \theta\, p_Y f$. It is easy to see that :

$$p_X\, \theta\, p_Y f(x',y') = f^*(x',y') = p_Y\, \theta\, p_X f(x',y') .$$

II. DUAL MINIMIZATION PROBLEMS AND DUAL MINIMAX PROBLEMS.

Following Rockafellar [5] , but using notations as in [1,2] , a pair of dual minimization problems is defined by two convex functionals $\varphi \in \Gamma(XY')$ and $\Psi \in \Gamma(X'Y)$ which are mutually conjugate (i.e. such that $\Psi = \varphi^*$ and $\varphi = \Psi^*$).
If we define $f(x) = \varphi(x,y' = 0)$ (*) and $g(y) = \Psi(x' = 0, y)$, the two dual problems are :

$$(P) \qquad \alpha = \operatorname*{Inf}_{x \in X} f(x)$$

$$(Q) \qquad \beta = \operatorname*{Inf}_{y \in Y} g(y)$$

We always have $-\beta \leq \alpha$. The variables y' and x' act as perturbatinn variables. If we consider the two families of minimization problems :

$$(P_{y'}) \qquad h(y') = \operatorname*{Inf}_{x \in X} \varphi(x,y')$$

$$(Q_{x'}) \qquad k(x') = \operatorname*{Inf}_{y \in Y} \Psi(x',y)$$

the problems (P) and (Q) correspond to the value zero of the perturbation variables y' and x'.

The duality between (P) and (Q) could also be defined by the class L of equivalent (see [5]) convex-concave functionals, the extreme elements of which are :

$$\bar{\ell}(x,y) = \theta p_{Y'}\varphi(x,y) = -\operatorname*{Sup}_{y' \in Y'} (< y,y' > - \varphi(x,y'))$$

$$\underline{\ell}(x,y) = p_{X'}\Psi(x,y) = \operatorname*{Sup}_{x' \in X'} (< x,x' > - \Psi(x',y))$$

$$L = \{\ell \,|\, \underline{\ell} \leq \ell \leq \bar{\ell}\}$$

The minimax problem (S) associated to the functionals φ and Ψ consists in finding $[\bar{x},\bar{y}] \in XY$ such that :

$$(S) \qquad \ell(\bar{x},y) \leq \ell(\bar{x},\bar{y}) \leq \ell(x,\bar{y}) \quad , \quad \text{for all } x \in X \text{ and } y \in Y$$

(the solutions of (S) are the same for all $\ell \in L$).

(*) The notation $\varphi(x,y' = 0)$ instead of $\varphi(x,0)$ helps to recall that the variable which is taken equal to zero is $y' \in Y'$. This method will be very useful later for more complicated cases.

It is well-known that the following three propositions are equivalent :

(i) $[\bar{x},\bar{y}]$ is a solution of (S).

(ii) $\bar{x}$ is a solution of (P), $\bar{y}$ is a solution of (Q) and $\alpha = -\beta$.

(iii) $f(\bar{x}) + g(\bar{y}) = 0$.

Thus, hereafter, a minimax problem will rather be given by a pair of mutually conjugate functionals φ and Ψ which are defined on a product of two spaces.

DEFINITION : Consider two minimax problems S_1 (represented by the two functionals φ_{11} and Ψ_{11}, and the associated minimization problems (P_1) and (Q_1)) and S_2 (represented in the same way by φ_{22} , Ψ_{22} , and the associated problems (P_2) and (Q_2)). We will say that S_1 and S_2 are dual if

- P_1 and Q_2 are dual (with respect to φ_{12} and Ψ_{12})
- P_2 and Q_1 are dual (with respect to φ_{21} and Ψ_{21})

This definition can be summarized by the following diagram :

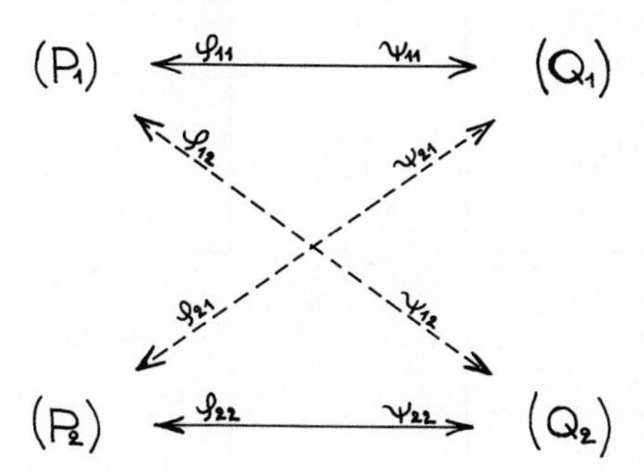

In the next section we will show how this diagram can be easily obtained in the classical framework of duality for two minimization problems, using partial minimization.

III. DUAL MINIMAX PROBLEMS AND PARTIAL MINIMIZATION.

Let (P) and (Q) be two dual minimization problems corresponding to $\varphi \in \Gamma(XY')$ and $\Psi \in \Gamma(X'Y)$ (with $\Psi = \varphi^*$), but suppose that the variable $x \in X$ splits into two variables $x_1 \in X_1$ and $x_2 \in X_2$, i.e. X is the product X_1X_2. In the same way, suppose Y to be the product Y_1Y_2. The spaces $X = X_1X_2$ and $X' = X_1'X_2'$ are in duality and similarly the spaces $Y = Y_1Y_2$ and $Y' = Y_1'Y_2'$ are in duality.

The problem (P) can be written :

$$(P) \quad \alpha = \operatorname*{Inf}_{x_1 \in X_1} f(x_1,x_2)$$

with $f(x_1,x_2) = \varphi(x_1,x_2,y_1' = 0,\ y_2' = 0)$, while the problem (Q) is :

$$(Q) \quad \beta = \operatorname*{Inf}_{\substack{y_1 \in Y_1 \\ y_2 \in Y_2}} g(y_1,y_2)$$

with $g(y_1,y_2) = \Psi(x_1' = 0,\ x_2' = 0,\ y_1,y_2)$.

Now we construct a problem (P_1) which consists in minimizing over X_1 the functional f_1 which is the result of a partial minimization with respect to the variable x_2. Conversely we call (P_2) the problem of minimizing over X_2 the functional f_2 which is the result of a partial minimization with respect to x_1.

In the same way we construct the problems (Q_1) and (Q_2). Let us write the four problems at the vertices of a square :

$$(P_1) : \quad \alpha = \operatorname*{Inf}_{x_1 \in X_1} f_1(x_1) \quad \text{with} \quad f_1(x_1) = \operatorname*{Inf}_{x_2 \in X_2} f(x_1,x_2)$$

$$(Q_1) : \quad \beta = \operatorname*{Inf}_{y_1 \in Y_1} g_1(y_1) \quad \text{with} \quad g_1(y_1) = \operatorname*{Inf}_{y_2 \in Y_2} g(y_1,y_2)$$

$$(P_2) : \quad \alpha = \operatorname*{Inf}_{x_2 \in X_2} f_2(x_2) \quad \text{with} \quad f_2(x_2) = \operatorname*{Inf}_{x_1 \in X_1} f(x_1,x_2)$$

$$(Q_2) : \quad \beta = \operatorname*{Inf}_{y_2 \in Y_2} g_2(y_2) \quad \text{with} \quad g_2(y_2) = \operatorname*{Inf}_{y_1 \in Y_1} g(y_1,y_2)$$

We will show that the duality relations described in the diagram of section II can be easily obtained :

(a) Duality between (P_1) and (Q_1) :

The objective functionals of (P_1) and (Q_1) can be written :

$$f_1(x_1) = \operatorname*{Inf}_{x_2 \in X_2} \varphi(x_1,x_2,y_1' = 0,\ y_2' = 0)$$

$$g_1(y_1) = \operatorname*{Inf}_{y_2 \in Y_2} \psi(x_1' = 0,\ x_2' = 0,\ y_1, y_2)$$

The dual variable of y_1 being y_1', we consider the function of x_1 and y_1' defined by :

$$\varphi_{11}(x_1, y_1') = \operatorname*{Inf}_{x_2 \in X_2} \varphi(x_1, x_2, y_1', y_2' = 0) = \theta p_{X_2} \varphi(x_1, x_2' = 0,\ y_1', y_2' = 0)$$

and similarly :

$$\psi_{11}(x_1', y_1) = \operatorname*{Inf}_{y_2 \in Y_2} \psi(x_1', x_2' = 0, y_1, y_2) = \theta p_{Y_2} \psi(x_1', x_2' = 0,\ y_1, y_2' = 0)$$

The functionals φ_{11} and ψ_{11} are convex ; in general they do not belong to $\Gamma(X_1 Y_1')$ and $\Gamma(X_1' Y_1)$ respectively, and they are not mutually conjugate. But under weak assumptions these properties can be satisfied. Let us consider for example the following two assumptions :

(H_1) for all $x_2' \in X_2'$, the functional

$$[x_1, y_1', y_2'] \to \theta p_{X_2} \varphi(x_1, x_2', y_1', y_2')$$

belongs to $\Gamma(X_1 Y_1' Y_2')$.

(K_1) for all $y_2' \in Y_2'$, the functional

$$[x_1', x_2', y_1] \to \theta p_{Y_2} \psi(x_1', x_2', y_1, y_2')$$

belongs to $\psi(X_1' X_2' Y_1)$.

The assumption (H_1) means that the projection of the epigraph of φ (which is a closed convex set of $X_1 X_2 Y_1' Y_2' \mathbb{R}$) onto the space $X_1 Y_1' Y_2' \mathbb{R}$ is closed. There are many sufficient conditions for obtaining this property.

PROPOSITION :

The assumption (H_1) implies that $\varphi_{11} = \psi_{11}^*$ and the assumption (K_1) that $\psi_{11} = \varphi_{11}^*$.

PROOF : We have :

$$\varphi_{11}^*(x_1', y_1) = (p_{Y_1'} \theta p_{X_1}) \theta p_{X_2} \varphi(x_1', x_2' = 0,\ y_1, y_2' = 0)$$

$$\psi_{11}(x_1', y_1) = \theta p_{Y_2} \psi(x_1', x_2' = 0,\ y_1, y_2' = 0)$$

By assumption (K_1), if we take the conjugate of $\theta p_{Y_2}\Psi$ with respect to the variables $x_1'x_2'y_1$, and then the conjugate with respect to $x_1x_2y_1'$, we obtain $\theta p_{Y_2}\Psi$ itself :

$$(p_{Y_1'}\theta p_{X_1}\theta p_{X_2})(p_{Y_1}\theta p_{X_1'}\theta p_{X_2'})\theta p_{Y_2}\Psi = \theta p_{Y_2}\Psi .$$

But $p_{Y_1}\theta p_{X_1'}\theta p_{X_2'}\theta p_{Y_2}\Psi = \Psi^{*} = \varphi$. It follows that $p_{Y_1'}\theta p_{X_1}\theta p_{X_2}\varphi = \theta p_{Y_2}\Psi$, hence $\varphi_{11}^{*} = \Psi_{11}$

By similar arguments, we prove that assumption (H_1) implies $\Psi_{11}^{*} = \varphi_{11}$.

Thus, under (H_1) and (K_1), the functionals φ_{11} and Ψ_{11} are mutually conjugate and they define a duality between (P_1) and (Q_1). We will call (S_1) the corresponding minimax problem.

ⓑ Duality between (P_2) and (Q_2) :

Proceeding in the same way, we define

$$\varphi_{22}(x_2,y_2') = \operatorname*{Inf}_{x_1\in X_1} \varphi(x_1,x_2,y_1' = 0,y_2') = \theta p_{X_1}\varphi(x_1' = 0,x_2,y_1' = 0,y_2')$$

$$\Psi_{22}(x_2',y_2) = \operatorname*{Inf}_{y_1\in Y_1} \Psi(x_1' = 0,x_2',y_1,y_2) = \theta p_{Y_1}\Psi(x_1' = 0,x_2',y_1' = 0,y_2)$$

and we consider the two assumptions :

(H_2) for all $x_1' \in X_1'$, the functional

$$[x_2,y_1',y_2'] \to \theta p_{X_1}\varphi(x_1',x_2,y_1',y_2')$$

belongs to $\Gamma(X_2Y_1'Y_2')$

(K_2) for all $y_1' \in Y_1'$, the functional

$$[x_1',x_2',y_2] \to \theta p_{Y_1}\Psi(x_1',x_2',y_1',y_2)$$

belongs to $\Gamma(X_1'X_2'Y_2)$.

The assumption (H_2) implies that $\varphi_{22} = \Psi_{22}^{*}$ and (K_2) that $\Psi_{22} = \varphi_{22}^{*}$. With (H_2) and (K_2) the two functionals φ_{22} and Ψ_{22} are mutually conjugate and define a duality between (P_2) and (Q_2). We will call (S_2) the corresponding minimax problem.

ⓒ Duality between (S_1) and (S_2) :

In order to define a duality between (S_1) and (S_2) we have to define

1°/ a duality between the two minimization problems (P_1) and (Q_2) and

2°/ a duality between (P_2) and (Q_1).

Consider the two functionals :

$$\varphi_{12}(x_1,y_2') = \theta p_{X_2}\varphi(x_1,x_2' = 0,y_1' = 0,y_2')$$

$$\Psi_{12}(x_1',y_2) = \theta p_{Y_1}\Psi(x_1',x_2' = 0,y_1' = 0,y_2)$$

One can prove that the assumption (H_1) implies that $\varphi_{12} = \Psi^*_{12}$ and the assumption (K_2) that $\Psi_{12} = \varphi^*_{12}$. Thus the two assumptions (H_1) and (K_2) together imply that φ_{12} and Ψ_{12} are mutually conjugate and define a duality between (P_1) and (Q_2).

Similarly, we define :

$$\varphi_{21}(x_2,y_1') = \theta p_{X_1}\varphi(x_1' = 0,x_2,y_1',y_2' = 0) ,$$

$$\Psi_{21}(x_2',y_1) = \theta p_{Y_2}\Psi(x_1' = 0,x_2',y_1,y_2' = 0) .$$

With the two assumptions (H_2) and (K_1) these two functionals are mutually conjugate and they define a duality between (P_2) and (Q_1).

Finally, with the four assumptions (H_1), (H_2), (K_1) and (K_2), we have obtained two dual minimax problems (S_1), which is the pair (P_1), (Q_1), and (S_2), which is the pair (P_2), (Q_2). The problems (P_1), (P_2), and the problems (Q_1), (Q_2) are obtained by partial minimization of two problems, respectively (P) and (Q), which are dual of each other in the classical sense. This situation is summarized in the following diagram :

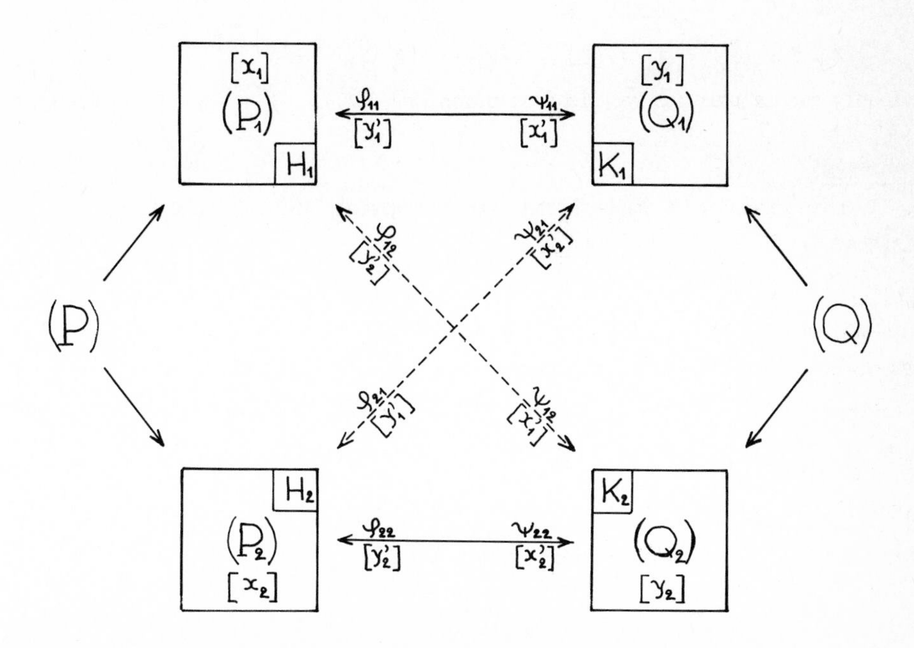

IV. STABILITY AND CHARACTERIZATION OF THE SOLUTIONS .

Using the characterization theorem of section II and the fact that (P_1) and (Q_2) are dual as well as (P_2) and (Q_1), we have the following result :

The pair $[\bar{x}_1,\bar{y}_1]$ is a solution of (S_1) and the pair $[\bar{x}_2,\bar{y}_2]$ is a solution of (S_2) if and only if the following two conditions hold :

$$f_1(\bar{x}_1) + g_2(\bar{y}_2) = 0 \quad \text{and} \quad f_2(\bar{x}_2) + g_1(\bar{y}_1) = 0$$

Using the terminology introduced in [1,2] , the problem (P_1) is stable with respect to the perturbation y_2' if the functional :

$$h_{12}(y_2') = \operatorname*{Inf}_{x_1\in X_1} \varphi_{12}(x_1,y_2')$$

is finite and continuous at $y_2' = 0$.

In the same way, (Q_1) is stable with respect to x_2' if the functional

$$k_{21}(x_2') = \operatorname*{Inf}_{y_1\in Y_1} \Psi_{21}(x_2',y_1)$$

is finite and continuous at $x_2' = 0$.

Speaking of the duality between (S_1) and (S_2), it is logical to say that (S_1) is stable if h_{12} and k_{21} are finite and continuous at $y_2' = 0$ and $x_2' = 0$ respectively.

We have then the following characterization theorem :

If (S_1) is stable, then $[\bar{x}_1,\bar{y}_1]$ is a solution of (S_1), if and only if there exists $[\bar{x}_2,\bar{y}_2] \in X_2\ Y_2$ such that :

$$f_1(\bar{x}_1) + g_2(\bar{y}_2) = 0 \quad \text{and} \quad f_2(\bar{x}_2) + g_1(\bar{y}_1) = 0 .$$

(obviously such a pair $[\bar{x}_2,\bar{y}_2]$ is a solution of (S_2)).

V. EXAMPLES.

In both examples we suppose that the assumptions (H_1),(H_2),(K_1) and (K_2) are satisfied.

Example 1 :

Suppose we have $X_1 = Y_2'$, $X_1' = Y_2$, $X_2 = Y_1'$, $X_2' = Y_1$ and let $\omega_1 \in \Gamma(X_1X_2)$ and $\omega_2 \in \Gamma(X_1X_2)$. Define :

$$\varphi(x_1,x_2,y_1',y_2') = \omega_1(x_1,x_2) + \omega_2(x_1-y_1', x_2-y_2')$$

Then the conjugate of φ is :

$$\Psi(x_1',x_2',y_1,y_2) = \omega_1^*(y_1+x_1', y_2+x_2') + \omega_2^*(-y_1,-y_2)$$

It follows that the minimax problem (S_1) is defined by :

$$\varphi_{11}(x_1,y_1') = \operatorname*{Inf}_{x_2 \in X_2} (\omega_1(x_1,x_2) + \omega_2(x_1-y_1',x_2))$$

$$\Psi_{11}(x_1',y_1) = \operatorname*{Inf}_{y_2 \in Y_2} (\omega_1^*(y_1+x_1',y_2) + \omega_2^*(-y_1,-y_2))$$

and (S_2) by :

$$\varphi_{22}(x_2,y_2') = \operatorname*{Inf}_{x_1 \in X_1} (\omega_1(x_1,x_2) + \omega_2(x_1,x_2-y_2'))$$

$$\Psi_{22}(x_2',y_2) = \operatorname*{Inf}_{y_1 \in Y_1} (\omega_1^*(y_1,y_2+x_2') + \omega_2^*(-y_1,-y_2))$$

(The functionals φ_{12} Ψ_{12} φ_{21} Ψ_{21} are defined in the same way).

The four problems which form (S_1) and (S_2) are :

(P_1) $\alpha = \operatorname*{Inf}_{x_1 \in X_1} f_1(x_1)$ $f_1(x_1) = \operatorname*{Inf}_{x_2 \in X_2} (\omega_1(x_1,x_2)+\omega_2(x_1,x_2))$	(Q_1) $\beta = \operatorname*{Inf}_{y_1 \in Y_1} g_1(y_1)$ $g_1(y_1) = \operatorname*{Inf}_{y_2 \in Y_2} (\omega_1^*(y_1,y_2)+\omega_2^*(-y_1,-y_2))$
(P_2) $\alpha = \operatorname*{Inf}_{x_2 \in X_2} f_2(x_2)$ $f_2(x_2) = \operatorname*{Inf}_{x_1 \in X_1} (\omega_1(x_1,x_2)+\omega_2(x_1,x_2))$	(Q_2) $\beta = \operatorname*{Inf}_{y_2 \in Y_2} g_2(y_2)$ $g_2(y_2) = \operatorname*{Inf}_{y_1 \in Y_1} (\omega_1^*(y_1,y_2)+\omega_2^*(-y_1,-y_2))$

Example 2 :

Let $Y_1 = Y_1' = \mathbb{R}^m$ and $Y_2 = Y_2' = \mathbb{R}^n$. Suppose $\omega \in \Gamma(X_1X_2)$, $f_i \in \Gamma(X_1)$, $i=1,\ldots,m$ and $g_j \in \Gamma(X_2)$, $j=1,\ldots,n$.

Define :

$$\varphi(x_1,x_2,y_1',y_2') = \begin{cases} \omega(x_1,x_2) \text{ if } f_i(x_1)+y_{1i}' \leq 0 , & i=1,\ldots,m \\ \qquad\text{and } g_j(x_2)+y_{2j}' \leq 0 , & j=1,\ldots,n \\ +\infty \quad \text{elsewhere.} \end{cases}$$

The conjugate of φ has the following expression :

$$\Psi(x_1',x_2',y_1,y_2) = \begin{cases} \operatorname*{Sup}_{x_1x_2} (< x_1,x_1' > + < x_2,x_2' > - \omega(x_1,x_2) \\ \qquad - \sum_i y_{1i}\, f_i(x_1) \quad - \sum_j y_{2j}\, g_j(x_2)) \\ \qquad\qquad \text{if } y_{1i} \geq 0,\ i=1,\ldots,m \text{ and } y_{2j} \geq 0,\ j=1,\ldots,n , \\ +\infty \quad \text{elsewhere.} \end{cases}$$

If we define :

$$C_1 = \{x_1 \in X_1 | f_i(x_1) \leq 0 \ , \ i=1,\ldots,m\}$$

$$C_2 = \{x_2 \in X_2 | g_j(x_2) \leq 0 \ , \ j=1,\ldots,n\}$$

the minimax problem (S_1) is given by :

$$\varphi_{11}(x_1,y_1') = \begin{cases} \operatorname*{Inf}_{x_2 \in C_2} \ \omega(x_1,x_2) \text{ if } f_i(x_1) + y_{1i}' \leq 0 \ , \ i=1,\ldots,m \\ +\infty \quad \text{elsewhere} \end{cases}$$

$$\Psi_{11}(x_1',y_1) = \begin{cases} \operatorname*{Inf}_{y_2 \geq 0} \ \operatorname*{Sup}_{\substack{x_1 \in X_1 \\ x_2 \in X_2}} \ (< x_1,x_1' > - \omega(x_1,x_2) - \sum_{i=1}^{m} y_{1i} f_i(x_1) - \sum_{j=1}^{n} y_{2j} g_j(x_2)) \ , \\ \qquad \text{if } y_{1i} \geq 0 \ , \ i=1,\ldots,m \ , \\ +\infty \quad \text{elsewhere} \end{cases}$$

and the minimax problem (S_2) is given by :

$$\varphi_{22}(x_2,y_2') = \begin{cases} \operatorname*{Inf}_{x_1 \in C_1} \ \omega(x_1,x_2) \text{ if } g_j(x_2) + y_{2j}' \leq 0 \ , \ j=1,\ldots,n \ , \\ +\infty \quad \text{elsewhere} \end{cases}$$

$$\Psi_{22}(x_2',y_2) = \begin{cases} \operatorname*{Inf}_{y_1 \geq 0} \ \operatorname*{Sup}_{\substack{x_1 \in X_1 \\ x_2 \in X_2}} \ (< x_2,x_2' > - \omega(x_1,x_2) - \sum_{i=1}^{m} y_{1i} f_i(x_1) - \sum_{j=1}^{n} y_{2j} g_j(x_2)) \ , \\ \qquad \text{if } y_{2j} \geq 0 \ , \ j=1,\ldots,n \ , \\ +\infty \quad \text{elsewhere.} \end{cases}$$

Finally, the four problems which compose (S_1) and (S_2) are :

$(P_1) \quad \alpha = \operatorname*{Inf}_{x_1 \in C_1} \ \omega_1(x_1)$ $\omega_1(x_1) = \operatorname*{Inf}_{x_2 \in C_2} \ \omega(x_1,x_2)$	$(Q_1) \quad \beta = \operatorname*{Inf}_{y_1 \geq 0} \ \eta_1(y_1)$ $\eta_1(y_1) = -\operatorname*{Inf}_{\substack{x_1 \in X_1 \\ x_2 \in C_2}} \ (\omega(x_1,x_2) + \sum_i y_{1i} f_i(x_1))$
$(P_2) \quad \alpha = \operatorname*{Inf}_{x_2 \in C_2} \ \omega_2(x_2)$ $\omega_2(x_2) = \operatorname*{Inf}_{x_1 \in C_1} \ \omega(x_1,x_2)$	$(Q_2) \quad \beta = \operatorname*{Inf}_{y_2 \geq 0} \ \eta_2(y_2)$ $\eta_2(y_2) = -\operatorname*{Inf}_{\substack{x_1 \in C_1 \\ x_2 \in X_2}} \ (\omega(x_1,x_2) + \sum_j y_{2j} g_j(x_2))$

REFERENCES

[1] J.L. JOLY et P.J. LAURENT,
Stability and duality in convex minimization problems. M.R.C. tech. report 1090, Madison (1970). Rev. Franç. d'Inf. et de Rech. Opér., R2 (1971), 3-42.

[2] P.J. LAURENT, Approximation et Optimisation, Hermann (1972).

[3] L. McLINDEN, Minimax problems, saddle functions and duality. M.R.C. tech. report 1190, Madison (1972).

[4] L. McLINDEN, An extension of Fenchel's duality theorem to saddle functions and dual minimax problems. M.R.C. tech. report 1242, Madison (1972).

[5] R.T. ROCKAFELLAR, Convex Analysis, Princeton Univ. Press, 1970.

ON THE TYPE OF A POLYNOMIAL RELATIVE TO A CIRCLE - AN OPEN PROBLEM

JOHN J. H. MILLER,
TRINITY COLLEGE, DUBLIN.

ABSTRACT: This paper is a sequel to [2]. In it we prove a Theorem and Corollary which imply, in particular, the unproved Theorem 4 of that paper. In addition we analyze the reasons for the failure, in certain special cases, of an algebraic algorith for determining the type of a polynomial relative to a circle. For background materia the reader is referred in addition to [1], [3], [4] and the references contained in those papers.

Consider the polynomial with real or complex coefficients

$$f(z) = a_o + a_1 z + \dots + a_n z^n$$

and assume that $a_o \neq 0$ and $a_n \neq 0$. We say that f is of type (p_1, p_2, p_3) relative to a circle in the complex plane if, counting multiplicities, it has p_1 zeros inside, p_2 zeros on, and p_3 zeros outside the circle. To simplify the notation only, we shall assume in what follows that the circle is the unit circle centred at the origin.

We now give some simple definitions. The inverse z^* in the unit circle of a point z is defined by $z^* = 1/\bar{z}$ and that of the polynomial f by $f^*(z) = z^n\overline{f(z^*)}$. It is easy to see that f^* is the polynomial

$$f^*(z) = \bar{a}_n + \bar{a}_{n-1} z + \dots + \bar{a}_o z^n .$$

Furthermore, if the zeros of f are denoted by z_j then the zeros of f^* are z_j^*. Thus f is of type (p_1, p_2, p_3) iff f^* is of type (p_3, p_2, p_1). Clearly z_j lies on the unit circle iff $z_j^* = z_j$, and the zeros common to f and f^* are the zeros z_j lying on the unit circle together with any zero z_j of f whose inverse z_j^* is also a zero of f. If f and f^* have the same zeros and the same set of multiplicities then f is said to be self-inversive. It is not hard to see that a polynomial is a common factor of both f and f^* only if it is a self-inversive polynomial. Such a common factor is called a self-inversive factor of f, and the common factor of f and f^* of maximal degree is called the maximal self-inversive factor of f. It follows that f and f^* have the same maximal self-inversive factor and, in particular, that they have the same zeros on the unit circle.

We now state and prove our main result.

THEOREM: Assume that f is a polynomial of degree n such that $f(0) \neq 0$, $|f^*(0)| - |f(0)| = 0$ and $f^*(0) f(z) - f(0) f^*(z)$ is not identically zero. Then $f^*(0) f(z) - f(0) f^*(z) = z^p g(z)$, where p is some integer satisfying $1 \leqslant p \leqslant [n/2]$ and g is a polynomial of degree $n - 2p$ such that $g(0) \neq 0$. Furthermore g is self-inversive, the maximal self-inversive factor of f is a factor of g and g is of type $(q, n - 2p - 2q, q)$ for some integer q satisfying $0 \leqslant q \leqslant [(n - 2p)/2]$. Moreover, f has at least $p + q$ zeros inside the unit circle, at least $p + q$ zeros outside the unit circle and at most $n - 2p - 2q$ zeros on the unit circle.

PROOF: If $f(z) = a_o + a_1 z + \ldots\ldots + a_n z^n$ then

$$f^*(0) f(z) - f(0) f^*(z) = b_1 z + b_2 z^2 + \ldots\ldots + b_n z^n$$ where for convenience we put

$$b_j = \bar{a}_n a_j - a_o \bar{a}_{n-j}$$

for $j = 0, 1, \ldots, n$. We note that $b_o = 0$ and under the hypotheses of the theorem also $b_n = 0$. Therefore, we have

$$(1) \qquad \bar{b}_{n-j} = - \frac{a_n}{a_o} b_j, \qquad j = 1, \ldots\ldots, n - 1$$

so that

$$(2) \qquad b_j = 0 \text{ iff } b_{n-j} = 0, \quad j = 1, \ldots\ldots, n - 1$$

Furthermore, there is certainly some integer p satisfying $1 \leqslant p \leqslant [n/2]$ such that

$$b_o = \ldots = b_{p-1} = 0, \quad b_p \neq 0$$

and hence by (2) also

$$b_n = \ldots = b_{n-p+1} = 0, \quad b_{n-p} \neq 0.$$

It then follows that

$$(3) \qquad f^*(0) f(z) - f(0) f^*(z) = z^p g(z)$$

where

$$g(z) = b_p + \ldots\ldots + b_{n-p} z^{n-2p}.$$

Moreover, by (1), it is easy to see that

$$g^*(z) = -\frac{a_n}{a_o} g(z)$$

so that g and g^* have the same zeros and hence g is self-inversive. Since the maximal self-inversive factor of f is also a factor of f^*, it follows from (3) that it is also a factor of g. Because g is self-inversive and is of degree $n - 2p$ it must be of type $(q, n - 2p - 2q, q)$ for some q satisfying $0 \leqslant q \leqslant [(n - 2p)/2]$.

To complete the proof of the theorem we put

$$h(z;\lambda) = f^*(0)\, f(z) - \lambda f(0)\, f^*(z) \quad , \quad 0 \leqslant \lambda \leqslant 1.$$

This is a polynomial of degree at most n. The coefficient of z^n is $|f^*(0)|^2 - \lambda|f(0)|^2 = (1 - \lambda)\,|f(0)|^2$, which is non-zero for all λ satisfying $0 \leqslant \lambda < 1$. Thu for all such λ,

$$h^*(z;\lambda) = -\lambda\, \overline{f(0)}\, f(z) + \overline{f^*(0)}\, f^*(z)\,.$$

If α is a common zero of f and f^* then it is clearly also a common zero of h an h^*. Conversely, if α is a common zero of h and h^* then

$$f^*(0)\, f(\alpha) - \lambda\, f(0)\, f^*(\alpha) = 0$$

$$-\lambda\, \overline{f(0)}\, f(\alpha) + \overline{f^*(0)}\, f^*(\alpha) = 0.$$

But the determinant of the coefficient matrix of this system of equations is

$$|f^*(0)|^2 - \lambda^2|f(0)|^2 = (1 - \lambda^2)\,|f(0)|^2$$

which is non-zero for all λ satisfying $0 \leqslant \lambda < 1$. The only solution therefore is th trivial one

$$f(\alpha) = 0, \quad f^*(\alpha) = 0$$

so that α is a common zero of f and f^*. We have thus shown that f and h have the same maximal self-inversive factor for all λ satisfying $0 \leqslant \lambda < 1$.

We now write $f = \phi k$, where ϕ is the maximal self-inversive factor of f. Then $f^* = \phi^* k^*$ and from what we have just proved it follows that

$$h(z;\lambda)/\phi(z) = f^*(0)\, k(z) - \lambda\, f(0)\, k^*(z)$$

has no self-inversive factor for all λ satisfying $0 \leqslant \lambda < 1$. In particular then

h/ϕ has no zeros on the unit circle and is therefore of constant type for all such λ. Since ϕ is independent of λ it follows that h is also of constant type for all such λ.

We look now at the zeros of h for λ arbitrarily close to 1. Writing $\lambda = 1 - \varepsilon$, where $\varepsilon > 0$ is arbitrarily small, we have

$$h(z; 1-\varepsilon) = f^*(0)\, f(z) - f(0)\, f^*(z) + \varepsilon\, f(0)\, f^*(z)$$

$$= z^p\, g(z) + \varepsilon\, f(0)\, f^*(z)\,.$$

Because the zeros of a polynomial are continuous functions of its coefficients, it is easy to see that as $\varepsilon \to 0$ exactly p zeros of h converge to the origin, $n - 2p$ zeros of h converge to the zeros of g and the remaining p zeros of h converge to infinity. But, since h is of constant type for all λ satisfying $0 \leqslant \lambda < 1$, none of these zeros may <u>cross</u> the unit circle as $\varepsilon \to 0$ (some may of course <u>reach</u> the unit circle in the limit). Furthermore, in the limit we have

$$h(z; 1) = z^p\, g(z)$$

which has $p + q$ zeros inside, $n - 2p - 2q$ zeros on and q zeros outside the unit circle. It follows therefore that, for all λ satisfying $0 \leqslant \lambda < 1$, $h(g; \lambda)$ has at least $p + q$ zeros inside, at least $p + q$ zeros outside and at most $n - 2p - 2q$ zeros on the unit circle. In particular, since

$$h(z;0) = f^*(0)\, f(z)$$

the same is true of f. This concludes the proof of the theorem.

Notice that f and g will have the same zeros on the unit circle if no zero of $h(z; 1-\varepsilon)/\phi(z)$ reaches the unit circle in the limit as $\varepsilon \to 0$. In this case the argument in the previous paragraph shows that f has then <u>exactly</u> $p + q$ zeros inside, $n - 2p - 2q$ zeros on and $p + q$ zeros outside the unit circle.

Notice, also, that g will be the maximal self-inversive factor of f if the zeros of $h(z; 1-\varepsilon)/\phi(z)$ neither reach the unit circle nor form self-inversive pairs in the limit as $\varepsilon \to 0$. We have thus established the non-trivial part of the following corollary.

COROLLARY: Suppose that f satisfies the hypotheses of the theorem and that f and g have the same zeros on the unit circle. Then f is of type $(p + q, n - 2p - 2q, p + q)$. If, in addition, g is a factor of f then g is the maximal self-inversive factor of f.

REMARK: The above corollary shows that the structure of the polynomial $f^*(0)\ f(z) - f(0)\ f^*(z) = z^p\ g(z)$ determines the type of f except when g has zeros on the unit circle which either are not zeros of f or are zeros of f but have higher multiplicities.

The observation that, if $f^*(0)\ f(z) - f(0)\ f^*(z) = z^p\ g(z)$ then the corresponding expression for f^* is just $-z^p\ g(z)$, indicates that it is unlikely that the structure of $f^*(0)\ f(z) - f(0)\ f^*(z)$ determines the type of f in the exceptional cases noted in the Remark. For example, if $f(z) = 6z^3 - 31z^2 + 41z - 6$ then $f^*(0)\ f(z) - f(0)\ f^*(z) = 60z\ (z + 1)$, so that $p = 1$, $q = 0$ and $g(z) = z + 1$. We conclude then from the Theorem that f has at least one zero inside the unit circle, at least one zero outside the unit circle and at most one zero on the unit circle. Since the zero of g is not a zero of f, the hypotheses of the Corollary are not fulfilled and we do not know from which side of the unit circle this zero has come. Thus, a priori, the type of f could be either $(2, 0, 1)$ or $(1, 0, 2)$. Now consider the polynomial $e(z) = -6z^3 + 41z^2 - 31z + 6$. Then $e^*(0)\ e(z) - e(0)\ e^*(z) = -60z\ (z + 1)$ so that the type of e could, a priori, also be either $(2, 0, 1)$ or $(1, 0, 2)$. But $e = f^*$ and so e is of one type iff f is of the other.

The following problem therefore remains open: find an efficient algebraic algorithm to determine the type of a polynomial relative to the unit circle, which works for all polynomials satisfying the hypotheses of the Theorem. By efficient we mean that full use should be made of the relations between the coefficients of the polynomial which are implied by the hypotheses of the Theorem (there is an algebraic algorithm (see [3]) which works in all cases, but it is not efficient in the above sense). Such an algorithm could then be implemented algebraically on a computer and it could be used to determine the stability properties of polynomials. It could also be used, in conjunction with Lehmer's algorithm, to isolate and refine with exact error bounds the locations of the zeros of polynomials.

REFERENCES.

1. *JOHN J. H. MILLER,* "On the location of zeros of certain classes of polynomials with applications to numerical analysis" J. Inst. Math. Applic. 8 (1971) 397-406.

2. *JOHN J. H. MILLER,* "On weak stability, stability and the type of a polynomial" Lecture Notes in Mathematics Vol. 228, Ed. John Ll. Morris. Springer-Verlag (1971) 316-320.

3. *JOHN J. H. MILLER,* "Practical algorithms for finding the type of a polynomial". Studies in Numerical Analysis, Ed. B. K. P. Scaife. Academic Press (1974) 253-264.

4. *D. P. MC CARTHY and JOHN J. H. MILLER,* "The Refinement and Implementation of an algorithm for finding the type of a polynomial". Séminaires Analyse Numérique 1972-1973, J. L. Lions, P. A. Raviart Université Paris VI.

ON BAYESIAN METHODS FOR SEEKING THE EXTREMUM

J. Močkus
Institute of Physics and Mathematics
Academy of Sciences Lithuanian SSR
Vilnius, USSR

1. Introduction

Many well known methods for seeking the extremum had been developed on the basis of quadratic approximation.

In some problems of global optimization the function to be minimized can be considered as a realization of some stochastic function. The optimization technique based upon the minimization of the expected deviation from the extremum is called Bayesian.

2. The definition of Bayesian methods

Suppose the function to be minimized is a realization of some stochastic function $f(x) = f(x, \omega), x \in A \subset R^m$ where $\omega \in \Omega$ is some fixed but unknown index.

The probability distribution P on Ω is defined by the equalities

$$P\{\omega : f(x_i, \omega) < y_i, i = 1, \ldots, n, \omega \in \Omega\} = F_{x_1, \ldots, x_n}(y_1, \ldots, y_n), n = 1, 2, \ldots$$

where $F_{x_1, \ldots, x_n}(y_1, \ldots, y_n), x_i \in A, i = 1, \ldots, n$ is the *a priori* probability distribution function.

The observation is evaluation of the function f at some fixed point x_i. The vector

$$z_n = (f(x_i), x_i, i = 1, \ldots, n), n = 1, \ldots, N,$$

contains the information gained in all the observations from 1 to n.

A decision function is the measurable vector- function $d = (d_0, \ldots, d_N)$, which expresses the dependence between the point of the following observation and the results of the previous observations

$$x_{n+1}(d) = d_n(z_n) = x_{n+1}(d, \omega), n = 1, \ldots, N.$$

The decision function d^0 is called the Bayesian method for seeking the minimum, if it minimizes the expected deviation from the extremum

$$E\{f(x_{N+1}(d^{\circ}))-f_0\}=\inf_{d} E\{f(x_{N+1}(d))-f_0\}, \tag{1}$$

where $f_0=\inf\limits_{x\in A} f(x,\omega)$.

The criterion (1) is satisfied under some condition [1] [2] by the solution of the following recurrent equations

$$\begin{aligned} u_N(z_N) &= \inf_{x\in A} E\{f(x)|z_N\}, \\ u_{n-1}(z_{n-1}) &= \inf_{x\in A} E\{u_n(z_{n-1},f(x),x)|z_{n-1}\},\ n=N,\dots,2, \\ u_0 &= \inf_{x\in A} E\{u_1(f(x),x)\}, \end{aligned} \tag{2}$$

where $E\{f(x)|z_N\}$ is a conditional expectation of the stochastic variable $f(x)$ with respect to the stochastic vector z_N.

The equations (2) remain the same in the case of "noisy" observations, when at any fixed point x_i we observe the sum

$$\varphi(x_i)=f(x_i)+\eta_i ,$$

where stochastic variables $\eta_i, i=1,\dots,N$ are independent, the expectation of η_i is zero and variance is fixed. In such a case

$$z_n=(\varphi(x_i),x_i,i=1,\dots,n),\ n=1,\dots,N.$$

3. An illustrative example

Suppose that

$$f(x,\omega)=(x-\omega)^2,\ A=\Omega=[-1,1],\ N=1,$$

and the *a priori* density function

$$p(\omega)=\frac{1}{2},\ \omega\in\Omega,$$

then in accordance with the equations (2) the first observation

$$x_1(d^{\circ})=\pm 1,$$

and the final decision

$$x_2(d^0) = \begin{cases} 1 - \sqrt{f(1)}, & x_1 = 1, \\ -1 + \sqrt{f(-1)}, & x_1 = -1. \end{cases}$$

4. The convergence conditions

In accordance with the definition (1) the Bayesian method depends on the a priori probability distribution P . The conditions when the Bayesian method converges to the minimum of any continuous function are given by

Theorem 1. Assume:

1. the set A is compact, the functions $f(x,\omega)$ and the conditional expectations in (2) are continuous functions of x for all $\omega \in \Omega$, there exists the positive a priori density function and the finite expectation $E\{|f_0|\}$;

2. for all $x \in B(\omega), \delta > 0$ there exists $n_{\delta x}$ such that

$$|F_x(y|z_n) - F_\omega(y)| < \delta, \ n \geq n_{\delta x},$$

3.

$$F_{x_i}(y|z_n) = F_\omega(y), \ i = 1, \dots, n$$

4. for all $x \in A(\omega), y \in R$ there exists δ_{xy} independently of n such that

$$F_x(y|z_n) \geq \delta_{xy},$$

5. $\sup |u_N(z_N)|$ is uniformly integrable with respect to $F_x(y|z_N)$;

then

$$\lim_{N \to \infty} |f(x_{N+1}(d^0,\omega),\omega) - f_0| = 0, \ \omega \in \Omega,$$

where $F_x(y|z_n)$ is the conditional probability distribution function of the stochastic variable $f(x)$ with respect to the stochastic vector z_n ,

$$F_\omega(y) = \begin{cases} 0, & y \leq y_\omega, \\ 1, & y > y_\omega, \end{cases}$$

$y_\omega = f(x_{i(x)}, \omega)$, the index $i(x)$ is defined by the equality

$$r_n = \|x - x_{i(x)}\| = \min_{1 \le i \le n} \|x - x_i\|,$$

and the sets $A(\omega), B(\omega)$ are defined by the conditions

$$A(\omega) = \{x : \lim_{n \to \infty} r_n = \varepsilon, \varepsilon > 0, x \in A\},$$

$$B(\omega) = \{x : \lim_{n \to \infty} r_n = 0, x \in A\},$$

the decision function d^0 corresponds to the solution of the equations (2) when conditional expectations are calculated using the usual formulas of the conditional probability density functions.

5. The one-stage method

One of the simplifications for the solution of the equations (2) is "one-stage" method [1] [3] when at each stage it is assumed that the following observation is the last one. In such a case the sequence of observations is defined by the equations

$$E\{u(z_n, f(x_{n+1}), x_{n+1}) | z_n\} = \min_{x \in A} E\{u(z_n, f(x), x) | z_n\}$$

where

$$u(z_{n+1}) = \min_{x \in A} E\{f(x) | z_{n+1}\}, \; n = 0, \ldots, N.$$

The one-stage Bayesian method converges to the minimum of any continuous function under the conditions of theorem 1.

6. The restricted-memory case

The next simplification for the equations (2) is the restriction of "memory", when one can remember no more than $N_0 < N$ observations [1] [4]. The Bayesian method for seeking the extremum in a special case with $N_0 = 1$ happens to be similar to those of the well known methods of "random search" [5], when the next observation must be on the sphere with the centre at the point of the best observation.

In the case of the restricted memory the Bayesian methods do not necessarily converge.

7. The implementation

The one-stage method had been implemented in the case when the function to be minimized is considered as some realization of the stochastic Gaussian field with the expectation μ , standard deviation σ and the exponential correlation function

$$\rho(x', x'') = \exp\left(-\sqrt{\sum_{i=1}^{m} c_i^2 (x_i' - x_i'')^2}\right),$$

the parameters $\mu, \sigma, c_i, i = 1, \ldots, m$ are estimated by the maximum likelihood method on the basis of some additional observations.

The implementation of one-stage Bayesian method when $f(x)$ is the Viener process is considered in [6].

The development of some system of <u>a priori</u> distributions suitable for different classes of the functions f is probably the most important problem in the application of Bayesian approach to the global optimization.

The Bayesian methods had been applied to some practical problems [7] [8].

References

1. J. Mockus. On a Bayesian method for seeking an extremum. Automatika i vychislitelnaja tekhnika, N 3, 1972.
2. J. Mockus. On the Bayesian methods of optimization. In "Minimization Algorithms and Dynamic Systems" ed. by G. Szego and L. Dixon. North-Holland Publ., Amsterdam (in print).
3. V. Šaltenis. On a method of multiextremal optimization. Automatika i vychislitelnaja tekhnika, N 3, 1973.
4. A. Žilinskas and J. Mockus. On a Bayes method for seeking an extremum. Automatika i vychislitelnaja tekhnika, N 3, 1972.
5. L.A. Rastrigin. The stochastic methods of search. Nauka, Moscow, 1968.
6. A. Žilinskas. One-stage Bayesian method for unidimensional search. Preprint N 663-74. VINITI, Moscow, 1974.
7. L. Telksnys and V. Šaltenis. The application of multiextremal methods to the problem of estimation of parameters of random signals. Proc. 5th Conf. on extremal problems. Gorkij (in print).
8. A. (Ališauskas, A. Lipskis, A. Mačiulis, V. Šaltenis. The application of multiextremal optimization to the planning of experiments for the development of high temperature - resistant polymeric compositions. Proc. 9th Seminar on the stochastic search. Kharkov (in print).

RIEMANNIAN INTEGRAL OF SET-VALUED FUNCTION

E.S.Polovinkin
Moscow State University
Moscow, USSR

Lebesgue integral of a set-valued function is a conception of growing importance for the theories of the optimal control and of the differential games (see for example [1]). For solving concrete problems it will be often useful to reduce a Lebegue set-integral to a riemannien one. In this article we shall find sufficient and necessary conditions of existence of riemannien integral of a set-valued function and prove the equality between riemannien and Lebesgue set-integrals in the case when riemannien set-integral exists.

I. Notations. Definitions. Auxiliary Propositions.

Let R^n be an euclidean n-space, (x,y) - the scalar product of $x,y \in R^n$, $I=[a,b] \subset R^1$, $V=\{x \in R^n \mid \|x\| \leq 1\}$, $W=\{\psi \in R^n \mid \|\psi\|=1\}$. Let A and B be compact subsets of R^n. The Hausdorff distance is definded by formula: $h(A,B)=\min\{\tau \geq 0 \mid A \subset B+\tau V; B \subset A+\tau V\}$. We shall denote by Ω^n the space of all compact subsets of R^n with Hausdorff metric in it. If $\{A_i\} \in \Omega^n$, then we shall define $\sum_{i=1}^{m} A_i = A_1+...+A_m=\{a_1+... ...+a_m \mid a_i \in A_i\}$; if $A \in \Omega^n$ - then $\text{tend}\, A$ is closure of extreme points of A and $|A|=h(A,0)$.

We shall say that $F: I \to \Omega^n$ is set-valued function. The conceptions of limit in the metric space Ω^n (denote $\lim$) and of continuity of function $F: I \to \Omega^n$ are defined as usual.

We shall say that map $F: I \to \Omega^n$ is convex-valued function if all the sets $F(t) \subset R^n$ are convex for every fixed $t \in I$. Support function of the set $A \subset R^n$ is the function $s(\psi, A)=\sup\{(\psi,x) \mid x \in A\}$. We shall consider the support function only for $\psi \in W$.

We shall say that map $F: I \to \Omega^n$ is Lebesgue measurable function on I if for any closed set $D \subset R^n$ the set $\{t \in I \mid F(t) \cap D \neq \emptyset\}$ is Lebesgue measurable set.

Definition 1 Lebesgue integral of the set-valued function $F: I \to \Omega^n$ on I (set-integral) is the set $L\int_I F(t)dt = \{w \in R^n \mid \exists r(t) \in F(t);$ $r(t)$ is Lebesgue measurable function on I and such that $w = L\int_I r(t)dt\}$.

Subdivision of an interval $I=[a,b]$ is finite set of numbers $\omega=\{t_1,t_2,...,t_N\}$ such that $t_1=a$, $t_i<t_{i+1}$, $t_N=b$. Let $\Delta t_i = t_{i+1}-t_i$, $\lambda = \max_{1 \leq i \leq N-1}\{\Delta t_i\}$ be diameter of the subdivision ω . Let $\xi_i \in [t_i, t_{i+1}]$

be an arbitrary point. We shall consider in this article only bounded maps $F: I \to \Omega^n$, that means: $\exists \alpha > 0 : |F(t)| < \alpha \ \forall t \in I$.

Definition 2 Riemannien integral of the bounded set-valued function $F: I \to \Omega^n$ on I is the limit of sets: $R\int_I F(t)dt = \underset{\lambda \to 0}{LIM} \sum F(\xi_i)\Delta t_i$, provided this limit exists.

We shall use following well-known propositions:

Proposition 1. Convex-valued function $F: I \to \Omega^n$ is continuous iff it is bounded and support function $S(\psi, F(t))$ is continuous by t for any $\psi \in W$. ([2]).

Proposition 2. (C-property). Map $F: I \to \Omega^n$ is measurable on I iff for any $\varepsilon > 0$ there exists closed set $I_1 \subset I$ such that Lebesgue measure $m(I \setminus I_1) < \varepsilon$ and $F(t)$ is continuous on I_1. ([3]).

Proposition 3. Let $F: I \to \Omega^n$ be measurable function on I and let there exist the Lebesgue summable function $\mu(t) > 0$ on I such that $|F(t)| \leq \mu(t)$. Then the Lebesgue integral of $F(t)$ is non empty convex compact subset of R^n and $L\int_I F(t)dt = L\int_I coF(t)dt$. (Here "co" means "convex").([4]).

2. The main lemmas.

Lemma 1. The bounded convex-valued function $F: I \to \Omega^n$ is continuous a.e. (almost everywhere) on I iff the function $f(t) = S(\psi, F(t))$ is continuous a.e. on I for any $\psi \in W$.

The proof follows from proposition 1.

Lemma 2. Let the suppositions of proposition 3 be satisfied. Then $S(\psi, L\int_I F(t)dt) = L\int_I S(\psi, F(t))dt$.

Proof. By proposition 3 the integral $L\int_I F(t)dt$ is convex compact, therefore for any $\psi \in W$ there exists vector $x \in L\int_I F(t)dt$ such that $S(\psi, L\int_I F(t)dt) = (\psi, x)$. Then by definition 1 there exists measurable function $r(t) \in F(t)$ such that $x = L\int_I r(t)dt$. Hence

$$S(\psi, L\int_I F(t)dt) = (\psi, L\int_I r(t)dt) = L\int_I (\psi, r(t))dt \leq L\int_I S(\psi, F(t))dt.$$

On the other hand let us consider map $R(t,\psi) = \{x \in F(t) \,|\, S(\psi, F(t)) = (\psi, x)\}$. This map is measurable because $R(t,\psi) = F(t) \cap Q(t,\psi)$ where $Q(t,\psi) = \{x \in R^n \,|\, S(\psi, F(t)) = (\psi, x)\}$. But the map $Q(t,\psi)$ is measurable because $h(Q(t,\psi), Q(t^*,\psi)) = |S(\psi, F(t)) - S(\psi, F(t^*))|$. But if $R(t,\psi)$ is measurable then there exists the measurable function $r(t) \in R(t,\psi)$. Hence we have $S(\psi, F(t)) = (\psi, r(t))$ .Finally

$$L\int_I S(\psi, F(t))dt = L\int_I (\psi, r(t))dt = (\psi, L\int_I r(t)dt) \leq S(\psi, L\int_I F(t)dt).$$

Lemma 3. If convex-valued function $F: I \to \Omega^n$ is continuous a.e. on I then function $\operatorname{tend} F: I \to \Omega^n$ is continuous a.e. on I.

Proof. Let $W(\psi_1,\dots,\psi_k) = \{\psi \in W \mid (\psi,\psi_i)=0;\ 1 \le i \le k\}$. Let us consider a set of mutually orthogonal points $\psi_1,\dots,\psi_n$ from W. We shall define the maps $R(t,\psi_1,\dots,\psi_k)$ in such a way:

$$R(t,\psi_1) = \{x \in F(t) \mid s(\psi_1, F(t)) = (\psi_1, x)\},$$

$$R(t,\psi_1,\psi_2) = \{x \in R(t,\psi_1) \mid s(\psi_2, R(t,\psi_1)) = (\psi_2, x)\},$$

$$\cdots\cdots\cdots$$

$$R(t,\psi_1,\dots,\psi_n) = \{x \in R(t,\psi_1,\dots,\psi_{n-1}) \mid s(\psi_n, R(t,\psi_1,\dots,\psi_{n-1})) = (\psi_n, x)\}.$$

It is easy to show (as in lemma 2) that function $t \to R(t,\psi_1,\dots,\psi_n)$) is continuous a.e. on I and that $\operatorname{tend} F(t) =$

$$= \operatorname{cl} \bigcup_{\psi_1 \in W} \bigcup_{\psi_2 \in W(\psi_1)} \cdots \bigcup_{\psi_n \in W(\psi_1,\dots,\psi_{n-1})} R(t,\psi_1,\dots,\psi_n).$$

Lemma 4. Let function $F: I \to \Omega^n$ be summable on I by Riemann (that is there exists riemannien integral of $F(t)$). Then for any $\psi \in W$ the support function $f(t) = s(\psi, F(t))$ is summable by Riemann and $s(\psi, R\int_I F(t)dt) = R\int_I s(\psi, F(t))dt$.

Proof. It is easy to show that if $\underset{n\to\infty}{LIM}\, P_n = P$, then for any $\psi \in W$ $s(\psi, P) = \lim_{n\to\infty} s(\psi, P_n)$. From here follows

$$s(\psi, R\int_I F(t)dt) = s(\psi, \underset{\lambda\to 0}{LIM} \sum F(\xi_i)\Delta t_i) = \lim_{\lambda\to 0} s(\psi, \sum F(\xi_i)\Delta t_i) =$$

$$= \lim_{\lambda\to 0} \sum s(\psi, F(\xi_i))\Delta t_i = R\int_I s(\psi, F(t))dt.$$

Lemma 5. Let $\{F_\lambda\} \in \Omega^n$, $\lambda > 0$, $F \in \Omega^n$, $|F_\lambda| \le \alpha$. Let sets F_λ and F be convex and for any $\psi \in W$ $\lim_{\lambda\to 0} s(\psi, F_\lambda) = s(\psi, F)$.

Then there exists the limit of sets F_λ in space Ω^n and $\underset{\lambda\to 0}{LIM}\, F_\lambda = F$.

Proof of this lemma is similar to the proof of proposition 1.

Lemma 6. Let $F: I \to \Omega^n$ be convex-valued function on I and for any $\psi \in W$ the support function $f(t) = s(\psi, F(t))$ be summable by Riemann on I. Then $F(t)$ is summable by Riemann and $R\int_I F(t)dt = L\int_I F(t)dt$.

Proof. By lemma 2 we have $s(\psi, L\int_I F(t)dt) = L\int_I s(\psi, F(t))dt =$

$$= R\int_I s(\psi, F(t))dt = \lim_{\lambda\to 0} \sum s(\psi, F(\xi_i))\Delta t_i = \lim_{\lambda\to 0} s(\psi, \sum F(\xi_i)\Delta t_i).$$

From here and lemma 5 follows the proof.

Lemma 7. Let $P \in \Omega^n$. Then $R\int_I P\,dt = (b-a)\,\mathrm{co}\,P$.

Proof. Let $\omega = \{t_1, \dots, t_N\}$ be subdivision of $I = [a, b]$. Consider sets $P_\lambda = \sum_{i=1}^{N-1} (P \cdot \Delta t_i)$. It is easy to show that $P_\lambda \subset (b-a)\,\mathrm{co}\,P$. On the other hand let $x \in (b-a)\,\mathrm{co}\,P$. By Carateodory's theorem there exist some points $x_i \in \mathrm{co}\,P$ and numbers $\mu_i > 0$, $i = 1, \dots, K$; $K \le n+1$, $\sum_{i=1}^{K} \mu_i = 1$, such that $x = (b-a)(\mu_1 x_1 + \dots + \mu_K x_K)$. Let $|P| \le \alpha$. Denote

$$\mu(x) = (b-a) \min\{\mu_i \mid 1 \le i \le K\} \text{ and } \lambda(x) = \min\{\mu(x), \tfrac{\varepsilon}{2n\alpha}\}.$$

It is obvious that $\lambda(x) > 0$ and $x \in P_\lambda + \frac{\varepsilon}{2} V$ for any subdivision with diameter $\lambda \le \lambda(x)$. If $B_\varepsilon(x) = \{y \in R^n \mid \|x-y\| < \frac{\varepsilon}{2}\}$ then $B_\varepsilon(x) \subset P_\lambda + \varepsilon \cdot V$. If we take a finite covering of compact set $(b-a)\,\mathrm{co}\,P$ by neighborhoods $B_\varepsilon(x)$ then we shall find $\lambda_0 > 0$ such that for any subdivision with $\lambda \le \lambda_0$: $(b-a)\,\mathrm{co}\,P \subset P_\lambda + \varepsilon V$. Finally $h(P_\lambda, (b-a)\,\mathrm{co}\,P) < \varepsilon$.

3. The main theorems.

Theorem 1. Let $F: I \to \Omega^n$ be a convex-valued function on I. Map $F: I \to \Omega^n$ is summable by Riemann on I iff $F: I \to \Omega^n$ is continuous a.e. on I. With such conditions riemannien integral of F is equal to Lebesgue integral of F.

Proof. It is well-known that a usual function is summable by Riemann on I iff it is continuous a.e. on I. Therefore the proof is consequence of lemmas 4, 1 and 6.

Theorem 2. If $F: I \to \Omega^n$ is continuous function on I then it is summable by Riemann and $R\int_I F(t)\,dt = R\int_I \mathrm{co}\,F(t)\,dt$.

Proof. By theorem 1 and by continuity of $F(t)$ for any $\varepsilon > 0$ there exists $\delta > 0$ such that for $t, t_* \in I$, $|t - t_*| < \delta$, and for any subdivision $\omega_1 = \{\tau_1, \dots, \tau_N\}$ with diameter δ we have:

$$F(t) \subset F(t_*) + \frac{\varepsilon}{3(b-a)} V \;;\quad h\Big(\sum_{i=1}^{N-1} \mathrm{co}\,F(\gamma_i)\Delta\tau_i,\ R\int_I \mathrm{co}\,F(t)\,dt\Big) < \frac{\varepsilon}{3},$$

where $\gamma_i \in [\tau_i, \tau_{i+1}]$, $\Delta\tau_i = \tau_{i+1} - \tau_i$.

Let us consider subdivision $\omega_2 = \{t_{11}, t_{12}, \dots, t_{1K_1} = t_{21}, t_{22}, \dots, t_{2K_2} = t_{31}, \dots, t_{N1}\}$

with $t_{11}=\tau_1=a$; $t_{21}=\tau_2, \dots ,$ $t_{N1}=\tau_N=b$, and let diameter ω_2 be λ . The integral sum for ω_2 is of a form:

$$\sum_{i=1}^{N-1}\sum_{j=1}^{K_i-1} F(\xi_{ij})\Delta t_{ij} \quad \text{where } \xi_{ij}\in[t_{ij},t_{ij+1}],\ \Delta t_{ij}=t_{ij+1}-t_{ij}.$$

On every segment $[\tau_i , \tau_{i+1}]$ while $\lambda\to 0$ the sum $\sum_{j=1}^{K_i-1} F(\gamma_i)\Delta t_{ij}$ tends to limit which is equal to $R\int_{\tau_i}^{\tau_{i+1}} F(\gamma_i)dt = coF(\gamma_i)\Delta\tau_i$.

Let us choose $\lambda_0>0$ such that for any ω_2 with diameter $\lambda\le\lambda_0$ for every $i=1,\dots,N-1$ we get: $h\left(\sum_{j=1}^{K_i-1}F(\gamma_i)\Delta t_{ij}, coF(\gamma_i)\Delta\tau_i\right)<\frac{\varepsilon}{3(N-1)}$.

From here $h\left(\sum_{i=1}^{N-1}\sum_{j=1}^{K_i-1}F(\gamma_i)\Delta t_{ij}, \sum_{i=1}^{N-1}coF(\gamma_i)\Delta\tau_i\right)<\frac{\varepsilon}{3}$. From $|\gamma_i-\xi_{ij}|<\delta$:

$$h\left(\sum_{i=1}^{N-1}\sum_{j=1}^{K_i-1}F(\xi_{ij})\Delta t_{ij}, \sum_{i=1}^{N-1}\sum_{j=1}^{K_i-1}F(\gamma_i)\Delta t_{ij}\right)\le\sum_{i=1}^{N-1}\sum_{j=1}^{K_i-1}h(F(\xi_{ij}),F(\gamma_i))\Delta t_{ij}<\frac{\varepsilon}{3}.$$

Hence for any $\lambda\le\lambda_0$: $h\left(\sum_{i=1}^{N-1}\sum_{j=1}^{K_i-1}F(\xi_{ij})\Delta t_{ij}, R\int_I coF(t)dt\right)<\varepsilon$.

Theorem 3. If map $F: I\to\Omega^n$ is continuous a.e. on I then it is summable by Riemann and $R\int_I F(t)dt = R\int_I coF(t)dt$.

Proof. Let I_1 be set of points of cease of $F(t)$ on I , then $mI_1=0$. Let $\alpha>0$ be such that $|F(t)|\le\alpha$ for $t\in I$. Let $\varepsilon>0$ and $\delta<\frac{\varepsilon}{6\alpha}$. There exists an open set I_2 such that $mI_2=\delta$ and $I_1\subset I_2\subset I$. For any subdivision ω the integral sum may be decomposed into two parts: $\sum F(\xi_i)\Delta t_i = \sum_I F(\xi_i)\Delta t_i + \sum_{II} F(\xi_i)\Delta t_i$, where $\sum_{II}$ consists of those indexes i for which $[t_i , t_{i+1}]\cap I_2\ne\emptyset$. By theorem 2 we can choose $\lambda_0>0$ such that for any subdivision with diameter $\lambda\le\lambda_0$ we'll have $\sum_{II}\Delta t_i<2\delta$ and $h(\sum_I F(\xi_i)\Delta t_i, R\int_{I\setminus I_2} coF(t)dt)<\frac{\varepsilon}{2}$.

Therefore
$$\begin{aligned} h\left(R\int_I coF(t)dt, \sum F(\xi_i)\Delta t_i\right) &\le h\left(R\int_{I\setminus I_2}coF(t)dt, \sum_I F(\xi_i)\Delta t_i\right) + \\ &+ h\left(R\int_{I_2}coF(t)dt, \sum_{II}F(\xi_i)\Delta t_i\right) < \frac{\varepsilon}{2}+3\alpha\delta<\varepsilon. \end{aligned}$$

Theorem 4. The function $F: I\to\Omega^n$ is summable by Riemann iff the function $coF: I\to\Omega^n$ is continuous a.e. on I . Under this condition the following equalities hold:

$$R\int_I coF(t)dt = R\int_I F(t)dt = R\int_I tend\,coF(t)dt;\quad R\int_I F(t)dt = L\int_I F(t)dt.$$

Proof. Let $coF(t)$ be continuous a.e. on I then $tend\,coF(t)$ is continuous a.e. on I (lemma 3). By theorem 3 and by inclusion: $coF(t)\supset F(t)\supset tend\,coF(t)$ we get the existence of riemannien integral of $F(t)$ and equality of three integrals. On the other hand if $F(t)$ is summable by Riemann then it is easy to show that $co\,F(t)$ is continuous a.e. on I (by lemmas 4 and 1). By equality of three integrals, by theorem 1 and proposition 3 we get:

$$R\int_I F(t)dt = L\int_I F(t)dt.$$

REFERENCES

1. Pontrjagin, L.S., Linear Differential Games. I., D.A.N. SSSR 174. No 6, 1967.
2. Blagodatskikh, V.I., Convexity of Spheres of Achievability Diff. Equations, 8 No 12, 1972, USSR.
3. Plis, A., Remark on Measurable Set-valued Functions. Bull. Acad. Polon. Sci., Sir. Sci. Math. Astr. Phys., 9, 1961.
4. Hermes, H., The Generalized Differential Equation $\dot{x} \in R(t,x)$, Advances in Math., 4, 1970.

CHARACTERISTICS OF SATURATION OF THE CLASS OF CONVEX FUNCTIONS

G.Ŝ.Rubinstein

Institute of Mathematics
Siberian Branch
U.S.S.R. Academy of Sciences
Novosibirsk

I. Formulation of the problem

In many cases real-valued functions are essential only in respect of perfect pre-orders they induce on their domains. Thus we define the following equivalence relation in the space $F(G)$ of all real-valued functions given on one and the same set G :

$$(f \sim g) \Longleftrightarrow (f(x) \leqslant f(y) \Longleftrightarrow g(x) \leqslant g(y)).$$

Equivalence classes F_f corresponding to functions $f \in F(G)$ have the following simple functional description:

$$F_f = \{ u \circ f : \; u \in U(T_f) \},$$

where $U(T_f)$ is the set of all increasing functions defined on the numerical set $T_f = f(G) = \{ f(x) : \; x \in G \}$ and $u \circ f$ is a superposition on u and f , that is function $g(x) = u(f(x)), \; x \in G$.

As usual the saturation of the class $\Phi \subset F(G)$ related to the above-described equivalence is defined as the set

$$U\Phi = \bigcup_{\varphi \in \Phi} F_\varphi = \{ u \circ \varphi : \; \varphi \in \Phi, \; u \in U(T_\varphi) \}.$$

In other words a function $f \in F(G)$ belongs to saturation $U\Phi$ of the class $\Phi \subset F(G)$ if there exists such a function $u \in U(T_f)$ that $u \circ f \in \Phi$.

If the class $\Phi \subset F(G)$ coincides with its saturation $U\Phi$ then it's called saturated.

Hereafter, it will be considered that G is arbitrary (containing more than one point) relatively open convex set in a real vector space, that is such a convex set that for any $x, y \in G$ there exists $\lambda > 0$ for which $y + \lambda(y-x) \in G$.

By $V(G)$ we shall designate the class of convex functions, that is such functions $f \in F(G)$ that

$$f(\lambda x + (1-\lambda)y) \leq \lambda f(x) + (1-\lambda) f(y) \tag{I}$$

holds for arbitrary $x, y \in G$ and $\lambda \in (0,1)$. By $W(G)$ we shall designate the class of quasi-convex functions, that is such functions $f \in F(G)$ that

$$f(\lambda x + (1-\lambda)y) \leq max\{f(x), f(y)\} \tag{2}$$

holds for any $x, y \in G$, $\lambda \in (0,I)$ and the inequality is strict when $f(x) \neq f(y)$.

It is clear that the class of quasi-convex functions is saturated. Hence, it contains saturation $UV(G)$ of more narrow class of convex functions. But there always exist functions $f \in W(G)$ which do not belong to $UV(G)$. In connection with this the American mathematician W.Fenchel stated a well-known problem of characterizing quasi-convex functions that belong to saturation $UV(G)$ of the class of convex functions (see [I] ,p.115-137). The present report is devoted to solving that problem.

2. Auxiliary Functions

Let $F_c(G)$ be a subspace that consists of functions $f \in F(G)$ such that their traces on the cross-cuts of G with any straight line is continuous.

Let us note now that it's sufficient to solve the question on the quasi-convex functions, we are interested in, for more narrow class $W_c(G) = W(G) \cap F_c(G)$.

Indeed, for every function $f \in W(G) \setminus W_c(G)$ it is possible to construct an equivalent function $f_0 = v \circ f$, where $v \in U(T_f)$ coincides with Lebesque measure of set $T_f \cap [t_0, t]$ when $t \in T_f$ is greater than some fixed $t_0 \in T_f$ and if $t \leq t_0$ then it equals Lebesque measure of set $T_f \cap [t, t_0]$ multiplied by - I. If in this case $f_0 \bar{\in} F_c(G)$ then the initial function $f \in W(G)$ does not undoubtedly belong to the set $UV(G)$, we are interested in. Otherwise, the question is to examine function $f_0 \in W_c(G)$.

It's known (see [2]) that if traces of function $f \in F(G)$ on cross-cuts of the set G with any straight line are measurable then inequalities (I) and (2) are implications of related inequalities with some fixed $\lambda \in (0,1)$, for instance, with $\lambda = \frac{1}{2}$. Specifically, it's true for all functions from $UF_c(G)$. And that means we may study question concerning functions from $f \in W_c(G)$ in terms of auxiliary functions

$$\tau_f(t,t') = \sup \{ f(\tfrac{x+y}{2}) : x \in f^{-1}(t), y \in f^{-1}(t') \}$$

defined on $T_f \times T_f$.

Evidently, function $f \in W_c(G)$ belongs to set $UV(G)$ if and only if such a function $u: T_f \to R$ exists that inequalities

$$u(t') - u(t) > 0, \quad u(t) + u(t') - 2u(\tau_f(t,t')) \geq 0 \tag{3}$$

hold for any $t < t'$ from T_f . Under this condition the function $u \in V(T_f)$ is automatically continuous.

All functions $f \in F_c(G)$ evidently have connected ranges. At that if function $f \in W_c(G)$ is not constant it cannot attain its maximum on relatively open convex set G . So, its range T_f either coincides with its open core $\mathring{T}_f$ or contains, besides, one additional point $\theta = \min_{x \in G} f(x)$.

We shall call the point $t^* \in T_f$ regular, the function $f \in W_c(G)$ beeing fixed, if for some $\varepsilon > 0$ there is such a function $u: T \to R$, where $T = T_f \cap (t^* - \varepsilon, t^* + \varepsilon)$, that for any $t < t'$ from T inequalities (3) hold.

It's clear that if the function $f \in W_c(G)$ belongs to $UV(G)$ then all points of the set T_f are regular. The following inverse statement is also correct (its complete proof is given in [3]) :

<u>Theorem I</u>. If all points of set $\mathring{T}_f$ are regular for the function $f \in W_c(G)$ then the function is equivalent to some function $\varphi \in V(G)$.

So, if function $f \in W_c(G)$ does not belong to $U_c V(G) = F_c(G) \cap UV(G)$,then at least one interior point of set T_f is not regular. To complete the characterization of the set $U_c V(G)$ we are to find out necessary and sufficient regularity conditions of interior points of sets T_f corresponding to functions $f \in W_c(G)$.

3. Characteristics of regular points

Let us introduce auxiliary normed linear space of additive functions with finite supports belonging to some bounded connected set $T \subset R$. For the purpose let us assign to every $t \in T$ an additive function $\mu_t : 2^T \to R$ which equals 1 on sets $e \in 2^T$ containing t and equals 0 otherwise. Further let $\Phi_0(2^T)$ be the set of all finite linear combinations of introduced elementary functions with the coefficient sum equal to 0. The norm $\|\cdot\|_v$ in linear space $\Phi_0(2^T)$ is defined as the full variation of corresponding additive functions.

Let us assign now to every function $f \in W_c(G)$ and interval $T = [\alpha, \beta]$ from T_f a cone $K(f,T) \subset \Phi_0(2^T)$ which is the conical hull of the union of two sets:

$$A(f,T) = \{a_{tt'} = \mu_{t'} - \mu_t \; : \; t < t' \text{ from } T\},$$

$$B(f,T) = \{b_{tt'} = \mu_t + \mu_{t'} - 2\mu_{\tau_f(t,t')} : \; t < t' \text{ from } T\}.$$

According to one of non-classical separation theorems (see [4], theorem 9) and some properties derived from proof of theorem I of the preceding paragraph it is possible to demonstrate the validity of the following statements:

<u>Theorem 2</u>. Whatever a function $f \in W_c(G)$ and an interval $T = [\alpha, \beta]$ from T_f are, (*)-closure of the cone $K(f,T)$ coincides with closure of this cone in the topology of normed space $\Phi_0(2^T)$ and, hence, also with its closure in the strongest local-convex topology.

<u>Theorem 3</u>. The interior point t^* of the range T_f of the function $f \in W_c(G)$ is regular if and only if for some $\alpha < t^* < \beta$ from T_f the point $-a_{\alpha t^*}$ does not belong to the closure of the cone $K(f, [\alpha, \beta])$ in one of the three topologies of Theorem 2 (and , therefore, in all the topologies).

<u>Corollary I.</u> The necessary and sufficient regularity condition of inner point t^* of the range T_f of function $f \in W_c(G)$ is the existence of some $\alpha < t^* < \beta$ such that set $T = [\alpha, \beta]$ satisfies the condition

$$\rho_v(-a_{\alpha t^*}, K(f,T)) = \inf_{a \in K(f,T)} \|a + a_{\alpha t^*}\|_v > 0.$$

Corollary 2. If $f \in W_c(G)$ and all points of an interval $[\alpha_0, \beta_0]$ are regular, then for any $t^* \in (\alpha_0, \beta_0)$ inductively defined points

$$\alpha_i = \tau_f(\alpha_{i-1}, t^*), \quad \beta_i = \tau_f(t^*, \beta_{i-1}), \quad i = 1, 2, \dots$$

are such that for some natural number m inequalities

$$\tau_f(\alpha_i, \beta_{i+m}) < t^*$$

hold for all $i = 0, 1, 2, \dots$

With the help of the last corollary we can easily verify, for example, that for $G = (-3, 1)$ function

$$f(x) = \begin{cases} -x-2 & \text{if } -3 < x \leq -1 \\ x^3 & \text{if } -1 < x < 1 \end{cases}$$

from $W_c(G)$ is not equivalent to any convex function, because the point $0 \in \mathring{T}_f$ is not regular.

In conclusion let us mention one circumstance which is often useful for the practical solution of poblems of transformation of quasi-convex functions into the convex ones.

First of all it's clear that if functions $f \in W_c(G_1)$ and $g \in W_c(G_2)$ have the same range and $\tau_f(t,t') \leq \tau_g(t,t')$ is true for any $t < t'$ from $T_f = T_g$, then the function $u \in U(T_g)$ for which $u \circ g \in V(G_2)$, satisfies also the condition $u \circ f \in V(G_1)$. Specifically, it covers the case when $G_2 \subset R$ and function $g \in W_c(G_2)$ is increasing. Then the function $u = g^{-1} \in U(T_g) = U(T_f)$ can be taken as the transforming one for f.

Let us note here that if $T \subset R$ is a connected set without the largest element and a function $\tau : T \times T \to R$ is such that for any $t < t'$ from T

$$\tau(t,t) = t, \quad t < \tau(t,t') < t',$$

then for the existence of an increasing function g satisfying condition $\tau_g = \tau$, it's necessary and sufficient that for any t_1, t_2, t_3 and t_4 from T the following equality takes place

$$\tau(\tau(t_1,t_2), \tau(t_3,t_4)) = \tau(\tau(t_1,t_3), \tau(t_2,t_4)).$$

Besides, the function g we are interested in can be effectively constructed with the help of the function τ.

References.

I. W.Fenchel. Convex cones, set and functions. Princeton,1953.

2. Б.А.Вертгейм и Г.Ш.Рубинштейн. К определению квазивыпуклых функций. Сборник "Математическое программирование",Москва, I966, с.I2I-I34.

3. Г.Ш.Рубинштейн. Характеристика насыщения класса выпуклых функций. Сборник "Оптимизация", вып.9, Новосибирск, I973, с. I65-I80.

4. Г.Ш. Рубинштейн. Теоремы отделимости выпуклых множеств.Сибирский математический журнал, т.5, № 5, I964,с. I098-II24.

A NEW HEURISTIC METHOD FOR GENERAL MIXED INTEGER LINEAR PROGRAMS: A REPORT ON WORK IN PROGRESS

(ABSTRACT)

R.W. Rutledge,
G.P.O. Box 483,
SYDNEY. 2001.
AUSTRALIA.

For linear integer programs, the method to be described starts with the continuous optimum and maintains primal feasibility. Instead of modifying the constraints of the original problem, the method modifies the objective function. A new, non-linear, objective function is set up, based on the original objective function and on two integrality measures. A lattice is constructed in the interior of the constraint polyhedron to enable us to move around inside this region. The simplex method is used to search for the maximum of the derived objective function which is so constructed that a maximum is likely to coincide with an integer solution to the problem. If a maximum is found which is also an integer solution, the parameters of the objective function are changed, and a search for a better integer solution is begun. If a maximum is found which is not an integer solution, the parameters are changed and the search continues. The search stops when no change of parameters will open up new areas of search. The effect of varying the parameters is to stretch or squeeze the derived objective function so that we can escape from local optima that are not integer solutions, and so that we can escape from integer solutions that are not global optima.

Experience with small test problems of proven difficulty (many of them pathological cases) has shown that an appropriate initial choice of parameters, based on the structure of the problem, and automatic adjustment of the parameters during the search, leads quickly to the optimum integer solution. This optimum is often the first integer solution found. The method has not yet been tried on large-scale problems.

It is established that the method generates feasible extreme points amongst which is the optimum integer solution, if there is one. It is also established that there is a path of feasible extreme points between any feasible integer solution and the continuous optimum. The method is guaranteed not to return to an already visited extreme point and to be finite. However, the rules used in setting and varying the parameters have not been proved to lead to the optimum integer solution if one exists. Work is proceeding on this point.

The following account will make the foregoing remarks clearer:

The general mixed integer linear programming problem may be stated as follows:

$$\text{Maximize } F_1 = \sum_{j=1}^{p} \gamma_j \delta_j + \sum_{j=p+1}^{m+n} c_j x_j \qquad (1)$$

subject to:

$$\sum_{j=1}^{p} a_{ij} \delta_j + \sum_{j=p+1}^{m+n} a_{ij} x_j = b_i \qquad i=1,\ldots\ldots\ldots m \quad (2)$$

where the δ_j and the x_j are non-negative variables, the δ_j are to be integers (not necessarily zero or one) and the γ_j, c_j and a_{ij} are constants. There are p integer variables.

Suppose that the maximum of F_1 has been found without the integrality condition for the δ_j and this maximum value of F_1 is Ω. Let us also write

$$F_2 = \sum_{j=1}^{p} f_j \qquad (3)$$

where, using $\{\delta_j\}$ to denote the fractional part of δ_j,

$$f_j = -5.0 \text{ for each integral } \delta_j \text{ (that is for } \{\delta_j\} \geq .9995 \text{ or } \{\delta_j\} \leq .0005), \qquad (4a)$$

$$f_j = -5.0 - \{\delta_j\} \text{ for } 0.0005 < \{\delta_j\} \leq 0.5, \qquad (4b)$$

$$f_j = -6.0 + \{\delta_j\} \text{ for } 0.5 < \{\delta_j\} \leq 0.9995 \qquad (4c)$$

$$F_3 = \text{Number of integer valued } \delta_j \qquad (5)$$

and

$$F_4 = (S + F_2 + ZF_3)/(Q - F_1) \qquad (6)$$

where S,Q and Z are parameters which are calculated and varied as described in the algorithm.

At the s'th iteration a constraint of the form

$$F_1 \geq \Omega - \xi_1 - \xi_2 - \ldots \xi_{s-1} - \xi_s \qquad (7)$$

is added, where ξ_s is chosen so that

(a) each constraint is feasible (i.e. does not violate any of the constraints of the original problem),

(b) at least one of the basic non-integral δ_j (say δ_ρ) becomes integral or one of the integral δ_j changes its value by an integral amount, and

(c) the objective function F_4 increases.

The selection of ξ determines the non-basic variable which is to enter the basis and the δ-variable which is to become integral (say at b_ρ). The tableau is augmented not only by the constraint (7) but by a pseudo-constraint of the form:

$$\delta_\rho \ \{\geqslant \text{ or } \leqslant\} \ b_\rho \tag{8}$$

This requires two new linearly dependent slack variables, say v^+ and v^-, so that the relationship (8) may be expressed as

$$\delta_\rho + v^+ - v^- = b_\rho \tag{8a}$$

These marking "constraints" permit changes in δ_ρ later if this suits us. They do not, of course, genuinely constrain the system. The variables v^+ and v^- are referred to as "marking slacks". The generation of these slacks may be kept under control by removing each of them that becomes basic, together with its non-basic counterpart. This can be done very easily.

At each iteration, the method usually generates a large number of feasible extreme points, using the erasable marking "constraints" referred to above. The effect on the objective function F_4 (and on the subsidiary functions, F_1, F_2 and F_3) of a basis change to a new extreme point is calculated and used as a guide in the choice of a new basis. After the selection, the simplex method is used to transform the matrix.

Before starting the selection process on the first tableau, the problem is scaled by subtracting a constant from the objective function so that we have $\Omega = 3p$, and the parameters are set, as follows:

Define

ξ = decrease in F_1

ψ = increase in F_2

β = increase in F_3

The parameter S is put equal to 5.5p. The parameter Z is set so that for each feasible potential basis change with negative ψ and positive β, the quantity $(\psi + \beta Z)$ is non-negative. If there are no potential basis changes with negative ψ and positive β, Z is put equal to zero. The parameter Q is set so that for each basis change with positive $(\psi + \beta Z)$ and positive ξ the quantity

$$\pi = (Q - \Omega)(\psi + \beta Z) - \xi(S + F_2 + ZF_3)$$

is non negative. If there are no potential changes with positive $(\psi + \beta Z)$ and positive ξ, Q is put equal to $\Omega + 1$.

The criterion for the selection of a new basis is that we take the change giving the maximum of the non-negative π.

The algorithm consists of the following eight steps:

1. Set up the continuous optimal tableau in canonical form.
2. Calculate the feasible basis changes and label these.
3. Calculate the properties of each possible new basis.
4. Set the parameters S,Z and Q of the derived objective function,

   $$F_4 = (S + F_2 + ZF_3)/(Q - F_1)$$

   according to the structure of the problem.
5. Select the change to be made to give a new basis
   (a) if we have an integer solution, and
   (b) otherwise.
6. If no selection can be made, vary the parameters S,Z or Q or permit a selection. If no changes can be made, stop.
7. Transform the tableau.
8. Return to Step 2.

The constraints originally present, together with the lattice generated by the algorithm constitute a set of convex polyhedra whose extreme points are explored by the algorithm. Since there is a finite number of these extreme points, and since the revisiting of a basis is prohibited, the algorithm must, after a finite number of iterations, arrive at a maximum of F_4 with one of the following properties:

(a) it is an integer solution; or
(b) it is not an integer solution, but a change of parameters opens up some more feasible points; or

(c) it is not an integer solution, and no change of parameters will open up more feasible points.

If an integer solution is found, the algorithm is applied to the region between this solution and the continuous optimum. This ensures that we do not finish up with a local optimum.

CLOSED - LOOP DIFFERENTIAL GAMES

N.N. Krasovskii
Institute of Mathematics and Mechanics, Ural
Scientific Center of the Academy of Sciences
of the USSR
S. Kovalevskaja st. 16, Sverdlovsk, USSR

The object of this report is closed-loop differential games which are a formalization of problems in dynamics related to control in presence of uncertainty or conflict. Under the game-theoretic approach to the problem we assume that the uncontrollable factors may appear to be most unfavourable from the point of view of one or the other criterion, important for the evaluation of the process. Thus a game is formed in between our ally, who selects the control in pursuing the one or the other aim and the antagonist who governs the disturbances. The given report contains some conclusions accumulated and summarized by some research members of our Institute [1]. Naturally, our results are closely related to those of other authors. However the framework of this report does not enable us to produce a bibliography.

The model of the differential games under consideration is as follows. Given is a differential equation

$$\dot{x} = f(t, x, u, v), \quad u \in \mathcal{P}, \quad v \in Q, \quad t \geq t_0 \qquad (1)$$

where x is the state space vector; u and v are vectors of control forces for the first and the second player; $\mathcal{P}$ and Q are compacts. Given is a functional

$$\gamma = \varphi(x[t];\ t_0 \leq t) \qquad (2)$$

which evaluates the result of the game.

Here we deal with closed-loop differential games. In assigning his control at current moment of time t the ally may be guided by complete knowledge of realized position $\{t, x\}$ or more generally by the knowledge of the whole history $\{t, x[\tau];\ t_0 \leq \tau \leq t\} = \{t, x_t[\cdot]\}$ $(t \geq t_0)$ of the motion.

The idea of feedback control is simply formalized in the following manner.

The pure strategy $\bar{U}$ for the first player-ally we shall identify with any fixed function $u(t, x) \in \mathcal{P}$. The equation (1), the strategy $\bar{U}$ and the initial position $\{t_*, x_*\}$ generate a bundle of motions $x[t] = x[t, t_*, x_*, \bar{U}]$ $(x[t_*] = x_*)$ each of which is the limit for a certain converging sequence of Euler splines $x_\Delta^{(k)}[t] = x_\Delta[t, t_*, x_*^{(k)}, \bar{U}, v^{(k)}[\cdot]]$ $(k = 1, 2, \ldots)$ with $\delta^{(k)} \to 0$. Here the Euler splines $x_\Delta[t] = x_\Delta[t, t_*, x_*, \bar{U}, v[\cdot]]$ are determined as the solutions of the equation

$$\dot{x}_\Delta[t] = f(t, x_\Delta[t], u(\tau_i, x_\Delta[\tau_i]), v[t]), \quad (\tau_i \leq t < \tau_{i+1}) \qquad (3)$$

where $t_* = \tau_0$, $\tau_i \to \infty$ with $i \to \infty$, $v[t]$ is any measurable function, $v[t] \in Q$ and $\delta = \max_i (\tau_{i+1} - \tau_i)$. The function f in equation (1) is assumed to be continuous, locally Lipschitz in x and to satisfy the conditions for a uniform prolongation of solutions (3). Then the nonempty bundle of motions $\{x[t, t_*, x_*, \bar{U}]\}$ does always exist with no reservations on the functional properties of $u(t, x)$. At the same time one may always return from ideal limit motions $x[t]$ to actually realized Euler splines $x_\Delta[t]$ which yield a good approximation for the ideal motions. The pure strategy reflects the idea of forming the control on the basis of information on the current position $\{t, x[t]\}$ under the condition that the antagonist in forming his control at time t may make use of any conceivable information whatever. In particular, prior to the assignment of his own control force, he may be assumed to know the control force for the ally at same moment of time or even in advance for moments of time in the nearest future.

The mixed strategy $\tilde{U}$ as introduced in the sequel reflects the idea that the ally again forms his control using the information on the current position $\{t, x[t]\}$ but now with condition that precise knowledge of current control force values for the ally is unaccessible for the antagonist. The latter may now but make a guess for a convenient response reaction on his part with the aid of a certain probabilistic technique. Thus a mixed strategy of the first player is identified with the function $\mu(du \mid t, x)$ which assigns for every possible position $\{t, x\}$ a probabilistic measure $\mu(du)$ on $\mathcal{P}$. The motions $x[t] = x[t, t_*, x_*, \tilde{U}]$ (understood to be ideal and deterministic) are determined again by a limit transition but now from Euler broken lines $x_\Delta[t]$

$$\dot{x}_\Delta[t] = \int_P\int_Q f(t, x_\Delta[t], u, v)\,\mu(du \mid \tau_i, x_\Delta[\tau_i])\,\nu_t(dv) \tag{4}$$

$$(\tau_i \leq t < \tau_{i+1})$$

where $\nu_t(dv)$ is a probabilistic measure on Q weakly measurable in t. The transition from deterministic ideal motions $x[t]$ to actually realized stochastic motions $\tilde{x}_\Delta[t]$ is now achieved by means of stochastic control procedures (see as follows at p. 000).

The counterstrategy $U^{(v)}$ is identified with function $u(t, x, v) \in \mathcal{P}$ which is Borel measurable in v. It reflects the idea of forming the control by the ally at every moment of time t on the basis of the realized position $\{t, x[t]\}$ and of the realized at same moment of time t control v for the antagonist. This leads to the so-called informational discrimination of the antagonist. In the case of counterstrategies the motions $x[t, t_*, x_*, U^{(v)}]$ are determined again by a limit transition, but now from Euler splines $x_\Delta[t]$:

$$\dot{x}_\Delta[t] = f(t, x_\Delta[t], u(\tau_i, x_\Delta[\tau_i], v[t]), v[t]) \tag{5}$$

$$(\tau_i \leq t < \tau_{i+1})$$

where $v[t]$ is measurable in t and $v[t] \in Q$. The transition from ideal motions $x[t]$ to those actually realizable $x_\Delta[t]$ is here again achieved by direct return to Euler splines $x_\Delta[t]$ which now again gives a good approximation to the ideal motions. It is noteworthy to say that the concept of counterstrategy is perhaps in its essence the very construction of nearest to control laws in differential games for the ally which are essential in the approach of L.S. Pontriagin and his followers [2]. On the contrary, the main attention in this report will be given to control laws formed in terms of formalized pure or mixed strategies-giving thus a preference to informational discrimination of the ally. It is useful to underline however that with certain conditions presumed (see e.g. as follows at p. 000) the transition from pure to counterstrategies gives no variation in guaranteed result for the game of the ally.

The definitions for strategies V for the second player and of the corresponding motions $x[t]$ and $x_\Delta[t]$ are formed obviously from given definitions by interchange of letters u and v . In passing from information space of positions $\{t, x\}$ to information space of histories $\{t, x_t[\cdot]\}$ all the definitions of above are obviously transformed by direct substitution of history in place of position. In the sequel the terms U and V will denote either pure or mixed strategies or counterstrategies.

The game is determined as a pair of problems each of which is formulated for the player - ally. The problem for the first player is formalized as the one of assigning an optimal minmax strategy U° which satisfies the condition

$$\sup_{x[\cdot]} \gamma(x[\cdot; t_\circ, x_\circ, U^\circ]) = \min_{U} \sup_{x[\cdot]} \gamma(x[\cdot; t_\circ, x_\circ, U]), \tag{6}$$

the problem for the second player is to specify an optimal maxmin strategy V° which satisfies the condition

$$\inf_{x[\cdot]} \gamma(x[\cdot; t_\circ, x_\circ, V^\circ]) = \max_{V} \inf_{x[\cdot]} \gamma(x[\cdot; t_\circ, x_\circ, V]) \tag{7}$$

Equality

$$\inf_{U} \sup_{x[\cdot]} \gamma(x[\cdot]) = \sup_{V} \inf_{x[\cdot]} \gamma(x[\cdot]) = \gamma^\circ \tag{8}$$

being fulfilled, the value γ° will be called the value of the game. If there exists a pair of strategies $\{U^\circ, V^\circ\}$ with equality (3) fulfilled, the given pair will be called the saddle-point or point of equilibrium for the game. A minimaximizing sequence $\{U^{(k)}, V^{(k)}\}$ $(k=1,2,\dots)$ for (8) will be called the ε-equilibrium sequence etc.

The quality of the formalization proposed here may be evaluated at first by verifying the existence of equilibrium and ε-equilibrium situations for more or less simple type of games. The foundation for solving the given question is given by the following theorem on the alternative for a standard approach-evasion game. The game is formed of two problems. Given are closed sets $\mathcal{M}_c$ and $\mathcal{N}_c$ in space $\{t, x\}$, the position $\{t_0, x_0\}$ and the point $\vartheta^* \geq t_0$. Problem of approach for the first player-ally: Specify a strategy U^* which ensures the encounter

$$\{\vartheta, x[\vartheta]\} \in \mathcal{M}_c, \quad \{t, x[t]\} \in \mathcal{N}_c, \quad t_0 \leq t \leq \vartheta \leq \vartheta^* \tag{9}$$

for all motions $x[t] = x[t, t_0, x_0, U^*]$. Problem of evasion for the second player-ally: specify a strategy V^*, which excludes the encounter (9) for all motions $x[t] = x[t, t_0, x_0, V^*]$.

<u>The theorem on the alternative</u>. In the approach-evasion game there either exists a mixed strategy $\tilde{U}^*$ which solves the problem of approach or in contrary there exists a mixed strategy $\tilde{V}^*$ which solves the problem of evasion related even to ε-neighbourhoods $\mathcal{M}_c^{(\varepsilon)}$ and $\mathcal{N}_c^{(\varepsilon)}$ of sets $\mathcal{M}_c$ and $\mathcal{N}_c$. If for any selection of vector s, of t and of x the minor game

$$\min_{u \in \mathcal{P}} \max_{v \in Q} s' f(t, x, u, v) = s' f(t, x, u^*, v^*) \tag{10}$$

does have a saddle-point $\{u^*, v^*\}$, then the alternative is valid in pure strategies $\bar{U}$ and $\bar{V}$. (In our notations the vectors always stand as columns. The upper prime denotes transposition). In general the alternative is always valid in the pairs of classes $\{\{\bar{U}\}\{V^{(u)}\}\}$, $\{\{U^{(v)}\}\{\bar{V}\}\}$.

A similar theorem is true for the information space of histories $\{t, x_t[\cdot]\}$ with functional sets $\mathcal{M}_c$ and $\mathcal{N}_c$ closed in $\{C_{[t_0,t]},\ t_0 \leq t < \infty\}$.

The approach-evasion game constitutes an element in the investigation of the general differential game (6), (7). Consider as an example the game on the minmax-maxmin for the functional

$$\gamma = \min_{t_0 \leq t \leq \vartheta} \rho(x[t], \mathcal{L}) \tag{11}$$

where $\rho(x, \mathcal{L})$ is the euclidean distance of point x from set $\mathcal{L}$ in space $\{x\}$, the moment ϑ being given. Select sets

$$\begin{aligned} \mathcal{N}_c &= [\{t, x\} : t \geq t_0,\ -\infty < x < \infty] \\ \mathcal{M}_c &= [\{t, x\} : t_0 \leq t \leq \vartheta,\ \rho(x, \mathcal{L}) \leq c] \end{aligned} \tag{12}$$

for $c \geq 0$. Assume c_0 is the lower bound over all c for which the first assertion of the theorem on the alternative is true in the class of mixed strategies with $\{t_0, x_0\}$ and $\vartheta^* = \vartheta$ given. Then it is easy to deduce from the alternative that c_0 is the value of the game and that there exists a pair of optimal mixed strategies $\{\tilde{U}^\circ, \tilde{V}^\circ\}$ which constitute the saddle-point. Further assume the saddle-point condition for the minor game (10) is fulfilled. Then for the game under consideration there exists a saddle-point in the class of pure strategies $\{\bar{U}^\circ, \bar{V}^\circ\}$. The optimal strategies $\bar{U}^\circ$ and $\bar{V}^\circ$ ensure the inequalities

$$\begin{aligned} &\min_{t_0 \leq t \leq \vartheta} \rho(x[t, t_0, x_0, \bar{U}^\circ]) \leq c_0 \\ &\rho(x[t, t_0, x_0, \bar{V}^\circ]) \geq c_0, \quad t_0 \leq t \leq \vartheta \end{aligned} \tag{13}$$

This being true for any $\varepsilon > 0$ there exists a $\delta(\varepsilon) > 0$ such that the same strategies when being applied to Euler splines $x_\Delta[t]$ with increment $\delta < \delta(\varepsilon)$ ensure the inequalities

$$\min_{t_0 \leq t \leq \vartheta} \rho(x_\Delta[t, t_0, x_0, \bar{U}^\circ, v[\cdot]]) \leq c_0 + \varepsilon \qquad (14)$$
$$\rho(x_\Delta[t, t_0, x_0, \bar{V}^\circ, u[\cdot]]) \geq c_0 - \varepsilon, \quad t_0 \leq t \leq \vartheta$$

A rather more complicated example is given by the game formed again for the minmax-maxmin of functional (11), but now under condition that moment ϑ is not given but is assigned separately for each motion $x[t]$ with one of the following conditions fulfilled: (1) $\vartheta = \vartheta^*$, (2) when point $x[t]$ runs into the closed domain G , (3) when point $x[t]$ runs into closed set H prior to the motion having attained the open domain $\mathcal{F}$. Making use of a version of the theorem on the alternative for the functional space of histories $x_t[\cdot]$ we are again convinced that there exists a value c_0 of the game. There exists a mixed strategy $\tilde{V}^\circ \div \div \nu^\circ(dv \mid x_t[\cdot])$ here which ensures the inequality $\gamma \geq c_0$ and for any $c^* > c_0$ there exists a mixed strategy $\tilde{U} \div \div \mu^*(du \mid x_t[\cdot])$ which ensures the inequality $\gamma < c^*$.

A similar utilization of the principal element in the form of an appropriate approach-evasion game admits the investigation also of other differential games. Here the theorem on the alternative yields a conclusion that with saddle-point condition for the minor game (10) being fulfilled, the absence of informational discrimination of the antagonist does not worsen the result of the game for the ally.

When describing actual schemes of control one ought to admit the existence of errors (perhaps of minor value) in measurements of the realizations of $x[t]$. Generally speaking the game-theoretic problems of dynamics in adopted formalization are actually improperly posed. Indeed it is possible to demonstrate examples of games with functional (11) when there exists a pure strategy $\bar{V}^\circ$ which guarantees inequalities (13) and (14) with $c_0 > 0$, but which leads to the following phenomena. If when forming the Euler splines $x_\Delta[t]$ one substitutes controls $v^\circ(\tau_i, x^*_\Delta[\tau_i])$ instead of $v^\circ(\tau_i, x_0[\tau_i])$, where measured values $x^*_\Delta[\tau_i]$ deflect from actual values $x_\Delta[\tau_i]$ by minor value $\|x^*_\Delta - x_\Delta\| \leq \eta$

then with $\delta \to 0$, the strategy $\bar{V}^{\circ}$ already does not ensure even the inequality $\rho(x_{\Delta}[t], \mathcal{L}) > 0$. It is of interest to note by the way that in the examples described, the strategy $\bar{V}^{\circ}$ is formed by means of discontinuous functions $v^{\circ}(t, x)$ and there exist no approximations by continuous strategies $\bar{V}^* \div v^*(t, x)$ which could ensure the inequality $\rho(x[t, t_o, x_o, \bar{V}^*], \mathcal{L}) \geq c_o - \varepsilon$, taken even for the ideal motions $x[t, t_o, x_o, \bar{V}^*]$.

Hence a problem of regularizing improperly posed solutions of differential games does arise. The problem is soluble for any approach-evasion game and hence for any game which may be reduced to its elements i.e. to appropriate approach-evasion games. Briefly speaking, the proposed regularized solution, which is said to be the control with leader, is constructed as follows. A computational model system

$$\dot{w} = f^*(t, w, u^*, v^*), \quad u^* \in \mathcal{P}^*, \quad v^* \in Q^* \tag{15}$$

is adjoined to the actual system (1) and a game is formed in between u^* and v^*. Suppose for the given initial position the problem of encounter with set $\mathcal{M}_c$ till moment ϑ^* is solvable for the first player. Then one may specify the three controls $\{u, u^*, v^*\}$ so as to lead the motion $w[t]$ to desired state

$$\{\vartheta, w[\vartheta]\} \in \mathcal{M}_c, \quad \{t, w[t]\} \in \mathcal{N}_c, \quad t_o \leq t \leq \vartheta \leq \vartheta^* \tag{16}$$

and at the same time so as to force the actual motion $x_{\Delta}[t]$ and the motion for the model to track each other. Thus the motion $x_{\Delta}[t]$ is led to a minor ε-neighbourhood of state (16) if only the value $x = x^*_{\Delta}[t]$ introduced into the control law $\bar{U} \div \{u(t, x, w), u^*(t, x, w), v^*(t, x, w)\}$ (or into $U^{(v)} \div \{u(t, x, w, v), u^*(t, x, w), v^*(t, x, w)\}$) does differ from actual value $x_{\Delta}[t]$ by sufficiently small number and the increment δ for Euler splines $x_{\Delta}[t]$ is sufficiently small. The proposition stated here is an accurate mathematical theorem. A similar proposition is also true for the second player's evasion problem with obvious transformation of the procedure and interchange of letters u and v. The scheme of control with a leader results in an interesting interaction of game-type control problems with problems on stability of motion. Here it is also worthwhile to explain the stochastic procedure of control with a leader, which approximates the ideal motion for mixed strategies. Assume each of the players receives his information in the form of a signal $y^{(j)}[t]$ $(j=1,2)$ statistically dependent on the actual motion $x[t]$. Assume the conditional expectations for the deviation of the measured values $\tilde{x}^*_\Delta[t]^{(j)}$ generated by signals $\tilde{y}^{(j)}_t[\cdot]$ with respect to actual values $\tilde{x}_\Delta[t]$ satisfy the inequalities

$$E(\|\tilde{x}^*_\Delta[t]^{(j)} - \tilde{x}_\Delta[t]\| \mid \tilde{x}_\Delta[t]) \le \zeta^{(j)} \qquad (j=1,2) \qquad (17)$$

Assume each player forms his control on the basis of his own strategy $\tilde{U} \div \mu(du \mid t, y^{(1)}_t[\cdot])$ and $\tilde{V} \div \nu(dv \mid t, y^{(2)}_t[\cdot])$ $(y^{(j)}_t[\cdot] = \{y^{(j)}_t[\tau], t_0 \le \tau \le t\})$ within a scheme discrete in time t with his own subdivision $\Delta^{(j)}$ of the time axis t. In forming the realization $\tilde{x}_\Delta[t]$ of the stochastic motion (1) the j-th player being in situation $\{\tau^{(j)}_i, \tilde{y}^{(j)}_{\tau^{(j)}_i}[\cdot]\}$ selects his control realization for the coming time interval as a result of a stochastic test with probabilistic distributions

$\mu(du \mid \tau^{(1)}_i, y^{(1)}_{\tau^{(1)}_i}[\cdot])$ or $\nu(dv \mid \tau^{(2)}_i, y^{(2)}_{\tau^{(2)}_i}[\cdot])$

respectively. In order to found this type of control scheme properly, it is necessary to assume the abilities of the controllers to be in some sense so "flexible" that they would indeed prevail over the informational abilities. These conditions may be expressed in exact form, but the framework of this report does not enable us to do it here. We can say only that the last conditions hold in particular if the information lag exceeds the value of the increment for Euler splines.

Under given conditions the following proposition is true. Assume with $\{t_0, x_0\}$, $\vartheta^* \ge t_0$, $\mathcal{M}_c$ and $\mathcal{N}_c$ given, the problem of approach is soluble in the class of mixed strategies $\bar{U} \div \mu(du \mid t, x)$ within the ideal scheme. Then the procedure of control with leader makes possible the construction of a mixed strategy $\tilde{U} \div \mu(du \mid t, y^{(1)}_t[\cdot])$ which for any $\varepsilon > 0$ and $p^* < 1$ selected in advance ensures, for the stochastic Euler splines $\tilde{x}_\Delta[t]$, the encounter

$$\{\vartheta, \tilde{x}_\Delta[\vartheta]\} \in \mathcal{M}_c^{(\varepsilon)}, \quad \{t, \tilde{x}_\Delta[t]\} \in \mathcal{N}_c^{(\varepsilon)}, \quad t_0 \le t \le \vartheta \le \vartheta^*, \tag{18}$$

with probability $p \geq p^*$ if only the value $\zeta^{(1)} > 0$ in (17) and the increment $\delta^{(1)} = \sup_i (\tau_{i+1}^{(1)} - \tau_i^{(1)})$ for the splines are sufficiently small. On the contrary, assume the problem of evasion is soluble in the class of mixed strategies $\tilde{V} \div \nu(dv \mid t, x)$ within the ideal scheme. Then there exists an $\varepsilon > 0$ such that the procedure of control with leader makes it possible to construct a mixed strategy $\tilde{V} \div \nu(dv \mid y_t^{(2)}[\cdot])$ which for any $p^* < 1$ selected in advance excludes the encounter (18) with probability $p \geq p^*$, if only the values $\zeta^{(2)} > 0$ in (17) and the step $\delta^{(2)}$ are sufficiently small.

As an example of effective construction of solution strategies U and V we here discuss at first the method of extremal aiming of the latter being applied to a game with functional (11) considered for the sake of determinancy in the class of mixed strategies. Assume that at moment t in the course of the game the vector $x[t] = x_*$ is realized. The method of extremal aiming gives a recommendation which is to freeze position $\{t, x_*\}$ and to consider an auxiliary open loop control problem in a fictitious time scale $t \le \tau \le \vartheta$: this is to specify over the set of measures $\mu_\tau(du)$ and $\nu_\tau(dv)$ an optimal pair $\{\mu_\tau^\circ, \nu_\tau^\circ\}$ which ensures for motions

$$\dot{x} = \iint_{\mathcal{P}\,Q} f(\tau, x, u, v)\,\mu_\tau(du)\,\nu_\tau(dv), \quad x(t) = x_* \tag{19}$$

the condition

$$\varepsilon^\circ(t, x_*) = \rho(x^\circ(\tau_0), \mathcal{L}) = \min_\tau \min_{\mu_\tau} \rho(x(\tau), \mathcal{L}) \tag{20}$$

The solution $\{x^\circ(\tau), \mu_\tau^\circ, \nu_\tau^\circ\}$ of the problem satisfies under certain regularity assumptions the minimax condition

$$\iint_{\mathcal{P}Q} s'(\tau) f(\tau, x^{\circ}(\tau), u, v) \mu^{\circ}_{\tau}(du) \nu^{\circ}_{\tau}(dv) =$$

$$= \min_{\mu} \max_{\nu} \iint_{\mathcal{P}Q} s'(\tau) f(\tau, x^{\circ}(\tau), u, v) \mu(du) \nu(dv) \tag{21}$$

which holds almost everywhere and which is an analogy of the maximum principle. Here $s(\tau)$ $(t_* \leq t \leq \tau)$ is a solution of the equation

$$\dot{s} = -\left[\iint_{\mathcal{P}Q} \left\{\frac{\partial f}{\partial x}\right\} \mu^{\circ}_{\tau}(du) \nu^{\circ}_{\tau}(dv)\right] s, \quad s(\tau_{\circ}) = \left[\frac{\partial \rho}{\partial x}\right]_{\{\tau_{\circ}, x^{\circ}(\tau_{\circ})\}} \tag{22}$$

Assume $S(t, x_*, \tau_{\circ})$ is the set of all vectors $s(t)$ in (22) which correspond to all possible solutions of problem (20) for given position $\{t, x_*\}$ with one or the other value of $\tau_{\circ}$ from (20). Then under certain regularity conditions formulated in terms of the properties of $S(t, x, \tau_{\circ})$ one may construct on the basis of sets S the optimal mixed strategies $\tilde{U}^{\circ} \div \mu(du | t, x)$ and $\tilde{V}^{\circ} \div \nu(dv | t, x)$ which ensure (12) with precisely the equality $c_{\circ} = \varepsilon^{\circ}(t_{\circ}, x_{\circ})$ of (20) being true. Thus, the regularity conditions being fulfilled, the control force law when formed by extremal aiming rule in the actual time scale is determined with the aid of a superspeed solution of an auxiliary open loop control problem (20), achieved in a superspeed fictitious time scale $t \leq \tau \leq \leq \vartheta$. The solutions of the auxiliary problems thus act as a sort of a location system, which escorts the actual motion. For the special case of linear equation (1), i.e. with

$$\dot{x} = A(t) x + f(t, u, v) \tag{23}$$

and with convex set $\mathcal{L}$ the computations for the extremal aiming rule are visibly realizable and it is possible to form the control law for the game in the actual time scale, provided a modern computational device of good abilities is available.

We gave a description of an example on the additional open-loop construction for the closed-loop differential game. Such open-loop constructions of extremal aiming may vary greatly depending upon the selected classes of strategies, upon the value to be attained and so on. We now refer to auxiliary constructions of another type, stopping for the sake of determinancy but only on the discussion of a representation of such a class for the case of mixed strategies.

Consider the sets

$$H(t,x) = \bigcap_{\nu} (\bigcap_{\mu} \iint_{\mathcal{P} Q} f(t,x,u,v) \mu(du) \nu(dv) \ . \qquad (24)$$

Assume that in the domain $\{t, x\}$ required, the sets $H(t, x)$ are nonempty. Then any solution $x = w(t)$ of the contingent equation

$$\dot{x} \in H(t, x) \qquad (25)$$

yields for the first player a so-called stable path. If there exists at least one path connecting initial position $\{t_0, x_0\}$ with target $\mathcal{M}_c$, then a mixed strategy $\tilde{U} \div \mu(du \mid t, x)$ may be constructed on the basis of this path which leads all the motions $x[t]$ (the latter being ideal) to set $\mathcal{M}_c$ along the path described.

Further consider the function

$$\varkappa(\ell, t, x) = -\min_{\mu} \max_{\nu} \ell' \iint_{\mathcal{P} Q} f(t, x, u, v) \mu(du) \nu(dv) \qquad (26)$$

With sets $H(t, x)$ (24) nonempty in the domain $\{t, x\}$ of interest and function $\varkappa(\ell, t, x)$ convex in ℓ, the bundle of solutions $x(t, t_0, x_0)$ (25) forms a stable set W for the second player. Therefore it is possible to construct a mixed strategy $\tilde{V} \div \nu(dv \mid t, x)$ for the second player which would hold all the ideal motions $x[t] = x[t; t_0, x_0, \tilde{V}]$ within W. If for $t_0 \leq t \leq \vartheta^*$ the bundle W gives no intersection with $\mathcal{M}_c$ the problem of evasion is then solved.

Let us compare the given conditions with the regularity assumptions of the method of extremal aiming. The general nonlinear case gives a complicated picture. However for the special equation (23) linear in x with $\mathcal{L}$ convex, the answer is simple enough. With $H(t, x)$ (24) nonempty and $\varkappa(\ell, t, x)$ (26) convex, the regularity assumptions for the method of extremal aiming are always fulfilled. A stable path for the first player $x = w^\circ(t) = w^\circ(t, t_\circ, x_\circ)$ $(t_\circ \le t \le \vartheta)$ which is the solution of (25) nearest to $\mathcal{L}$, lies at a distance of

$$\varepsilon = \min_{t_\circ \le t \le \vartheta} \rho(w^\circ(t), \mathcal{L}) = \varepsilon^\circ(t_\circ, x_\circ) \tag{27}$$

from $\mathcal{L}$.

A stable integral manifold $W = \{x = w(t, t_\circ, x_\circ)$, $t_\circ \le t \le \vartheta\}$ for the second player obtained for contingent equation (25) lies at a distance of $\varepsilon = \varepsilon^\circ(t_\circ, u_\circ)$ from $\mathcal{L}$. An opposite assertion is false: the regularity assumptions for the method of extremal aiming generally do not yield the nonemptiness of set $H(t, x)$ or the convexity of function $\varkappa(\ell, t, x)$ (26) in ℓ. Finally we note that in the particular case of an absolutely linear system (23) the conditions on $H(t, x)$ and $\varkappa(\ell, t, x)$ under discussion correspond to the so-called condition of complete sweeping over for the direct method of L.S.Pontriagin [2] which in turn are a generalization of the similarity conditions for linear plants in the pursuit-evasion game [1].

References

1. N.N. Krasovskii. The approach-evasion differential game. I, II, Izv. Akad. Nauk SSSR. Tehn. Kibernet. (Moscow), NN 2, 3, 1973.

2. E.F. Mischenko. The problems of pursuit and evasion from encounter in the theory of differential games. Izv. Akad. Nauk SSSR. Tehn. Kibernet. (Moscow), N 5, 1971.

A PROGAMMED CONSTRUCTION FOR THE POSITIONAL CONTROL

V.D. Batuhtin
Institute of Mathematics and Mechanics, Ural Scientific Center of the Academy of Scienes of the USSR.
S. Kovalevskaja st. 16, Sverdlovsk, USSR.

Let the motion of a competitively controlled system be described by the differential equation

$$dx/dt = f(t, x, u, v), \quad x(t_0) = x_0, \tag{1}$$

where $x \in R^n$ is the phase vector of the system; u and v are the vectors controlling the actions of the players with restrictions $u[t] \in P \subset R^p$, $v[t] \in Q \subset R^q$; P and Q are compacts; the function $f(t,x,u,v)$ is continuous in the totality of the arguments and continuously differentiable in x. In addition, we will assume that the formulated in [1] condition of uniform extendability of the solutions for the equation (1) is fulfilled.

Given are some instant $\vartheta_0 > t_0$, a closed set $T \subset [t_0, \vartheta_0]$, a compact $\mathcal{M} = \{(\vartheta, m): \vartheta \in T,\ m \in \mathcal{M}_\vartheta\}$ in R^{k+1}, where $\mathcal{M}_\vartheta = \{m: (\vartheta, m) \in \mathcal{M}\}$ and a function $\omega(\vartheta, x, m)$ defined on the set $\{(\vartheta, x, m): (\vartheta, m) \in \mathcal{M}, x \in R^n\}$ is continuous in the totality of the arguments and continuously differentiable in x in the domain $\omega_0 < \omega < \omega^0$.

We will call a mixed strategy $\tilde{U} \div \mu_{\{t,x\}}(du)$ of the first player a function $\mu_{\{t,x\}}(du)$ which puts the Borel regular normed measures $\mu(du)$ on P in correspondence to any position $\{t,x\}$. Let us define a motion $x[t; t_0, x_0, \tilde{U}]$ generated by the strategy $\tilde{U}$ as any uniform limit of Euler splines $\tilde{x}_{\Delta^{(k)}}[t]$, for almost all $t \in [\tau_i^{(k)}, \tau_{i+1}^{(k)})$ satisfying the equation

$$d\tilde{x}_{\Delta^{(k)}}[t]/dt = \int_P \int_Q f(t, \tilde{x}_{\Delta^{(k)}}[t], u, v)\, \mu_{\{\tau_i^{(k)}, \tilde{x}_{\Delta^{(k)}}[\tau_i^{(k)}]\}}(du)\, \nu_t(dv),$$

$\tau_{i+1}^{(k)} - \tau_i^{(k)} \le \Delta^{(k)}$, $\Delta^{(k)} \to 0$ for $k \to \infty$, $\nu_t(dv)$ is a Borel regular weak measurable in t on $[t_0, \vartheta_0]$ function normed on Q, that is, a function

$$\alpha(t) = \int_Q \varphi(v) \nu_t(dv)$$

is a Lebesgue measurable function on $[t_0, \vartheta_0)$ for any arbitrary continuous function $\varphi(v) \in C(Q)$. Analogously a mixed strategy $\tilde{V} \div \nu_{\{t,x\}}(dv)$ of the second player and a motion generated by this strategy and also a motion generated by the couple $\{\tilde{U}, \tilde{V}\}$ are defined.

Problem I. For a fixed position $\{t_0, x_0\}$ and a number c it is required to find a mixed strategy $\tilde{U}$ which guarantees the inequality

$$\min_{\vartheta \in T} \min_{m_\vartheta \in \mathcal{M}_\vartheta} \omega(\vartheta, x[\vartheta], m) \leq c$$

for any motion $x[t; t_0, x_0, \tilde{U}]$.

Problem II. For a fixed position $\{t_0, x_0\}$ and a number c it is required to find a mixed strategy $\tilde{V}$ which guarantees the inequality

$$\min_{\vartheta \in T} \min_{m_\vartheta \in \mathcal{M}_\vartheta} \omega(\vartheta, x[\vartheta], m) \geq c$$

for any motion $x[t; t_0, x_0, \tilde{V}]$.

Let us introduce an auxiliary programmed construction for solving these problems. Namely on the space of generalized programmed controls - Borel regular measures $\eta_t = \eta_t(du, dv)$ defined and normed on $P \times Q$ for all $t \in [t_0, \vartheta_0)$ and weak measurable in t on $[t_0, \vartheta_0)$ we will assign a totality of sets called programs.

Then let us define an elementary program $\{\mu_t \times \nu_t^*, [t_*, \vartheta)\}$ on $[t_*, \vartheta)$ as a set of all controls η_t on $[t_*, \vartheta)$ which are represented in the form of the direct product $\eta_t = \mu_t \times \nu_t^*$, where $\mu_t \in \{\mu_t\}$ and ν_t^* are weak measurable on $[t_*, \vartheta)$ Borel regular measures for any $t \in [t_*, \vartheta)$ defined and normed on P and Q respectively.

We will put in correspondence to each position $\{t_*, x_*\}$ $(t_* \in [t_0, \vartheta_0])$ the quantity

$$\tilde{\varepsilon}^0(t_*, x_*) = \min_{\vartheta \in T} \max_{\nu_t} \min_{\mu_t} \min_{m \in \mathcal{M}} \omega(\vartheta, x(\vartheta; t_*, x_*, \mu_{(\cdot)} \times \nu_{(\cdot)}), m) \tag{2}$$

where $x(t) = x(t; t_*, x_*, \mu_{(\cdot)} \times \nu_{(\cdot)})$ is a programmed motion satisfying for almost all $t \in [t_*, \vartheta)$ the equation

$$dx/dt = \iint_{PQ} f(t,x,u,v)\mu_t(du)\nu_t(dv) \qquad (3)$$

Given by(2) optimal programmed controls μ_t°, ν_t° and $m^\circ \in \mathcal{M}$, $\vartheta^\circ \in T$ exist on account of weak compactness in themselves of the elementary programs and the set $\{\nu_t\}_{[t_*,\vartheta)}$ of the controls ν_t on $[t_*, \vartheta)$. Incidentally under the weak convergence of the sequences we understand the convergence in $*$ - weak topology of the sequences of continuous linear functionals defined by the Borel regular measures $\mu^{(k)} = \mu_t^{(k)} \cdot m(dt)$, $\nu^{(k)} = \nu_t^{(k)} \cdot m(dt)$, $m(\cdot)$ - the Lebesgue measure on R^1.

We will say that the elementary program $\{\mu_t \times \nu_t^\circ, [t_*, \vartheta)\}$, where ν_t° is an optimal control for $\{t_*, x_*\}$, is regular in position $\{t_*, x_*\}$, if the problem (2) has an essentially unique solution μ_t° for the fixed control ν_t°, in addition, the minimal point $m^\circ \in \mathcal{M}$ is also unique.

There is valid the following assertion which is an analogy of the maximum principle [2] in the case under study.

Theorem I. Let the regularity condition of the program $\{\mu_t \times \nu_t^\circ, [t_*, \vartheta)\}$ be fulfilled and $\tilde{\varepsilon}^\circ(t_*, x_*) \in (\omega_\circ, \omega^\circ)$. Then for the optimal programmed motion $x^\circ(t) = x(t; t_*, x_*, \mu^\circ_{(\cdot)} \times \nu^\circ_{(\cdot)})$ there is the following minimax condition

$$\iint_{PQ} s^{\circ\prime}(\vartheta,t) f(t,x^\circ(t),u,v)\mu_t^\circ(du)\nu_t^\circ(dv) =$$

$$= \min_\mu \max_\nu \iint_{PQ} s^{\circ\prime}(\vartheta,t) f(t,x^\circ(t),u,v)\mu(du)\nu(dv)$$

for almost all $t \in [t_*, \vartheta]$. Here $s^{\circ\prime}(\vartheta,t) = -[\partial\omega(\vartheta, x^\circ(\vartheta), m^\circ)/\partial x]' \cdot S(\vartheta, t_*, x^\circ(\cdot), \mu^\circ_{(\cdot)} \times \nu^\circ_{(\cdot)})$, $S(t, t_*, x(\cdot), \mu_{(\cdot)} \times \nu_{(\cdot)})$ is the fundamental solution matrix of the first variational approximation equation for equation (3) computed on the motion $x(t) = x(t; t_*, x_*, \mu_{(\cdot)} \times \nu_{(\cdot)})$, the prime denotes transposition.

Let $\nu_t^{(k)} \xrightarrow{w} \nu_t$ and $\{\eta_t\}^*$ be a set of all controls η_t for each of which exists such a sequence $\{\mu_t^{(k)}\}$

that $\mu_t^{(\alpha)} \times \nu_t^{(\kappa)} \xrightarrow{w} \eta_t$. Then the weak closure of the set $\{\eta_t\}^*$ in $*$ - weak topology $C^*([t_*,\vartheta] \times P \times Q)$ we will call the program $\Pi(\nu_t^*)$. We will say that the program $\Pi(\nu_t^\circ)$ is optimal, if a sequence which forms it is maximizing for $\{t_*, x_*\}$. We will call ν_t° a regular control, if for any optimal $\Pi(\nu_t^\circ)$ the minimizing control η_t° in it is unique and also unique is $m^\circ \in \mathcal{M}$. Let us denote by $T^\circ(t_*, x_*)$ the set of the problem's (2) solutions ϑ° and by $\tilde{S}^\circ(t_*, x_*; \vartheta^\circ)$ the set of all vectors $s^\circ(\vartheta^\circ, t)$, $\vartheta^\circ \in T^\circ(t_*, x_*)$ corresponding to all kinds of η_t° for $\{t_*, x_*\}$.

We shall suppose that for any $\{t_*, x_*\}$ $(t_\circ \le t_* \le \vartheta_\circ)$, $t_* \notin T^\circ(t_*, x_*)$, where $\tilde{\varepsilon}^\circ(t_*, x_*) \in (\omega_\circ, \omega^\circ)$ and any Borel regular normed measure $\nu^*(dv)$ on Q there exists an instant $\vartheta^\circ \in T^\circ(t_*, x_*)$ for which two following conditions are fulfilled.

A. Any control ν_t° is regular for $\{t_*, x_*\}$.

B. There exists a Borel regular normed measure $\mu^*(du)$ on P such that for any $s^\circ(\vartheta^\circ, t_*) \in \tilde{S}^\circ(t_*, x_*; \vartheta^\circ)$

$$\int_P \int_Q s^{\circ\prime}(\vartheta^\circ, t_*) f(t_*, x_*, u, v)\, \mu^*(du)\, \nu^*(dv) \le$$

$$\le \min_\mu \max_\nu \int_P \int_Q s^{\circ\prime}(\vartheta^\circ, t_*) f(t_*, x_*, u, v)\, \mu(du)\, \nu(dv).$$

Theorem 2. Let $\tilde{\varepsilon}^\circ(t_\circ, x_\circ) \le c \in (\omega_\circ, \omega^\circ)$ and conditions A, B be fulfilled. Then the mixed strategy $\tilde{U}^{(e)}$ which is extremal [3] to the set $\tilde{W}_c = \{\{t,x\} : \tilde{\varepsilon}^\circ(t,x) \le c\}$ solves Problem I.

Let us denote by $\Sigma(t_*, x_*; \vartheta^\circ)$ the set of all ν_t° for $\{t_*, x_*\}$ and some $\vartheta^\circ \in T^\circ(t_*, x_*)$, $\tilde{S}^\circ(t_*, x_*) = \bigcup_{T^\circ(t_*, x_*)} \tilde{S}^\circ(t_*, x_*; \vartheta^\circ)$. We shall now suppose that two following conditions are fulfilled for any $\{t_*, x_*\} : \tilde{\varepsilon}^\circ(t_*, x_*) \in (\omega_\circ, \omega^\circ)$ instead of A, B.

C. Sets $\Sigma(t_*, x_*; \vartheta^\circ)$ are upper weak semicontinuous in each point $\vartheta^\circ \in T^\circ(t_*, x_*)$.

D. There exists $\nu_*(dv)$ on Q for any $\mu_*(du)$ on P such that for any $s^\circ(\vartheta^\circ, t_*) \in \tilde{S}^\circ(t_*, x_*)$

$$\int_P\int_Q s^{\circ\prime}(\vartheta^\circ, t_*) f(t_*, x_*, u, v)\mu_*(du)\nu_*(dv) \geqslant$$

$$\geqslant \max_{\nu} \min_{\mu} \int_P\int_Q s^{\circ\prime}(\vartheta^\circ, t_*) f(t_*, x_*, u, v)\mu(du)\nu(dv) .$$

Theorem 3. Let $\tilde{\varepsilon}^\circ(t_\circ, x_\circ) \geqslant c \in (\omega_\circ, \omega^\circ)$ and conditions C, D be fulfilled. Then the mixed strategy $\tilde{V}^{(e)}$ which is extremal to the set $\tilde{W}^{(c)} = \{\{t,x\} : \tilde{\varepsilon}^\circ(t,x) \geqslant c\}$ solves Problem II.

With conditions A - D fulfilled simultaneously, the situation of equilibrium takes place.

If the saddle point condition

$$\min_{u\in P} \max_{v\in Q} s' \cdot f(t,x,u,v) = \max_{v\in Q} \min_{u\in P} s' \cdot f(t,x,u,v)$$

for the minor game is fulfilled for any s, t, x, then the problems I, II are solvable in pure strategies.

R e f e r e n c e s

1. V.D. Batuhtin, N.N. Krasovskii. The problem of programmed control on maximin, Izv. AN SSSR, Tehnicheskaja kibernetika, 1972, № 6. (Russian).
2. L.S. Pontrjagin, V.G. Boltjanskii, R.V. Gamkrelidze, E.F. Mistchenko. The mathematical theory of optimal processes, Fizmatgiz, Moscow, 1961. (Russian).
3. N.N. Krasovskii. The differential game of converging - evading, Izv. AN SSSR, Tehnicheskaja kibernetika, 1973, № № 2, 3. (Russian).

AN EXTREMAL CONTROL IN DIFFERENTIAL GAMES

A.G. Chentsov
Institute of Mathematics and Mechanics,
Ural Scientific Center of the Academy of
Sciences of the USSR
S. Kovalevskaja st. 16, Sverdlovsk, USSR

Let us consider a system

$$\frac{dx}{dt} = f(t, x, u, v);\ x \in R^n,\ u \in P \subset R^p,\ v \in Q \subset R^q. \quad (1)$$

Here P and Q are compact sets, $f(\cdot)$ is a continuous function, continuously differentiable in x. It is assumed that the following condition of uniform extendability of the solutions holds: for any bounded set $K \subset R^n$ there exists $\beta = \beta(K) > 0$ such that any solution $x(t) = x(t, t_*, x_*)$ of the equation

$$\frac{dx}{dt} \in \overline{co}\{f : f = f(t, x, u, v),\ u \in P,\ v \in Q\}$$

with the initial condition

$$t_* \in [t_0, \vartheta],\quad x_* \in K \quad \text{is}$$

uniformly bounded in the segment $[t_*, \vartheta]$: $\|x(t)\| \leq \beta$. Here the symbol $\|\cdot\|$ denotes Euclidean norm.

There is given a function $\omega(x, m)$ defined on the space $R^n \times M$ where M is a compact subset of R^m, which is continuous and continuously differentiable in x on domain $\omega \in (\omega_0, \omega^0)$, $\omega_0 < \omega^0$.

The first player by choosing a control $u \in P$, tries to minimize the value of the functional $\rho_M(x[\vartheta]) = \min_M \omega(x[\vartheta], m)$. The second player by choosing a control tries to maximize it.

Similarly to [1], any upper semicontinuous with respect to inclusion in (t, x) function $U(t, x) \subset P$ will be called an admissible strategy U of the first player. The admissible strategy V of the second player can be defined similarly; any upper semicontinuous with respect to inclusion in (t, x, v) function $U_v(t, x, v) \subset P$ will be called a counterstrategy U_v of the first player. We shall say that a function $x[t]$ is the motion generated by a strategy U

(counterstrategy $\mathcal{U}_v$) if it is a solution of the equation

$$\frac{dx}{dt} \in \overline{co}\{f : f = f(t, x, u, v);\ u \in \mathcal{U}(t,x),\ v \in Q\}$$

$$x[t_0] = x_0;$$

$$\frac{dx}{dt} \in \overline{co}\{f : f = f(t, x, u, v);\ u \in \mathcal{U}_v(t,x,v),\ v \in Q\}$$

$$x[t_0] = x_0 .$$

Any solution of the equation

$$\frac{dx}{dt} \in \overline{co}\{f : f = f(t, x, u, v);\ u \in \mathcal{U}(t,x), v \in \mathcal{V}(t,x)\}$$

$$x[t_0] = x_0$$

will be called the motion generated by a pair of strategies $(\mathcal{U}, \mathcal{V})$.

<u>Problem I.</u> There is required to build an optimal counter-strategy $\mathcal{U}_v^\circ$ such that:

$$\max_{\{x[\cdot]|\mathcal{U}_v^\circ\}} \min_{M} \omega(x[\vartheta], m) = \min_{\{\mathcal{U}_v\}} \max_{\{x[\cdot]|\mathcal{U}_v\}} \min_{M} \omega(x[\vartheta], m)$$

And if the condition

$$\min_{P} \max_{Q} s' f(t, x, u, v) = \max_{Q} \min_{P} s' f(t, x, u, v) \quad (2)$$

holds for every $s \in R^n$, $t \in [t_0, \vartheta]$, $x \in R^n$ it is required to find a solution in the class of admissible strategies.

<u>Problem 2.</u> It is required to find a pair $(\mathcal{U}^\circ, \mathcal{V}^\circ)$ of optimal strategies which satisfies

$$\min_{\{\mathcal{U}\}} \max_{\{x[\cdot]|\mathcal{U}\}} \min_{M} \omega(x[\vartheta], m) =$$

$$= \max_{\{x[\cdot]|\mathcal{U}^\circ\}} \min_{M} \omega(x[\vartheta], m) =$$

$$= \min_{\{x[\cdot]|\mathcal{V}^\circ\}} \min_{M} \omega(x[\vartheta], m) =$$

$$= \max_{\{\mathcal{V}\}} \min_{\{x[\cdot]|\mathcal{V}\}} \min_{M} \omega(x[\vartheta], m) .$$

Let us call a set of all regular Borel measures (r.B.m.) $\eta(\cdot)$ on $[t_*,\vartheta]\times P\times Q$, having Lebesgue projection [2, 3] on $[t_*,\vartheta]$, a class $\{H(m(\cdot)),[t_*,\vartheta]\}$ of admissible open-loop controls of the first player on the segment $[t_*,\vartheta]$. A set of all r.B.m. $\nu(\cdot)$ on $[t_*,\vartheta]\times Q$ having Lebesgue projection on $[t_*,\vartheta]$ will be called a class $\{E(m(\cdot)),[t_*,\vartheta]\}$ of admissible open-loop controls of the second player on the segment $[t_*,\vartheta]$. The set of all controls $\eta(\cdot)\in\{H(m(\cdot)),[t_*,\vartheta]\}$ coordinated [2, 3] with open-loop control $\nu(\cdot)$ of the second player we shall call a program $\{\Pi(\nu(\cdot)),[t_*,\vartheta]\}$.

To every open-loop control can be put in correspondence a program motion [2, 3] $\varphi(\cdot,t_*,x_*,\eta(\cdot))$ with initial condition (t_*,x_*). Let $G(\vartheta,t_*,x_*,\nu(\cdot))$ be a set of attainability [1, 3] for a program $\{\Pi(\nu(\cdot)),[t_*,\vartheta]\}$ at the moment ϑ

$$\varepsilon^\circ(t_*,x_*)=\max_{\{E(m(\cdot)),[t_*,\vartheta]\}}\ \min_{G(\vartheta,t_*,x_*,\nu(\cdot))}\ \min_{M}\omega(x,m). \quad (3)$$

Open-loop controls giving minimum and maximum (3) will be called optimal ones.

<u>Theorem I.</u> Let the following condition hold for a position (t_*,x_*) and open-loop control $\nu^*(\cdot)\in\{E(m(\cdot)),[t_*,\vartheta]\}$:

$$\min_{G(\vartheta,t_*,x_*,\nu^*(\cdot))}\ \min_{M}\omega(x,m)\in(\omega_\circ,\omega^\circ).$$

Then the optimal open-loop control $\eta^\circ(\cdot)$ in the program $\{\Pi(\nu^*(\cdot)),[t_*,\vartheta]\}$ satisfies the minimum principle

$$\int_\Delta\int_P\int_Q s_\circ'(t)f(t,\varphi^\circ(t),u,v)\,\eta^\circ(dt\times du\times dv)=$$

$$=\int_\Delta\int_Q \min_P s_\circ'(t)f(t,\varphi^\circ(t),u,v)\,\nu^*(dt\times dv)$$

where Δ is an arbitrary Borel subset of $[t_*,\vartheta]$,

$$\varphi^\circ(t)=\varphi(t,t_*,x_*,\eta^\circ(\cdot)),$$

$$s_\circ'(t)=\left[\frac{\partial}{\partial x}\omega(\varphi^\circ(\vartheta),m^\circ)\right]'S(\vartheta,t,\varphi^\circ(\cdot),\eta^\circ(\cdot)),$$

m° gives a minimum to $\min_M\omega(\varphi^\circ(\vartheta),m)$, $S(\vartheta,t,\varphi^\circ(\cdot),\eta^\circ(\cdot))$ is fundamental solution matrix for a corresponding system in variations.

<u>Theorem 2.</u> Let $\varepsilon^\circ(t_*,x_*)\in(\omega_\circ,\omega^\circ)$ and an open-loop control $\nu^\circ(\cdot)\in\{E(m(\cdot)),[t_*,\vartheta]\}_{opt}$ be such that an optimal control $\eta^\circ(\cdot)$ in the program $\{\Pi(\nu^\circ(\cdot)),[t_*,\vartheta]\}$ be unique, moreover a point $m^\circ\in M$, giving $\min_M\omega(\varphi^\circ(\vartheta),m)$, be also unique. Then for arbitrary Borel subset Δ

$$\int_\Delta \int_P \int_Q s_0'(t) f(t, \varphi^\circ(t), u, v)\, \eta^\circ(dt \times du \times dv) =$$

$$= \int_\Delta \max_Q \min_P s_0'(t) f(t, \varphi^\circ(t), u, v)\, m(dt).$$

Here notations $\varphi^\circ(t)$, $s_0(t)$ have the same meaning as in theorem 1.

For every position (t_*, x_*), $\varepsilon^\circ(t_*, x_*) \in (\omega_0, \omega^\circ)$, denote by $S_0(t_*, x_*)$ a set of all vectors s_0 of the form

$$s_0' = \left[\frac{\partial}{\partial x}\, \omega(\varphi_0(\vartheta), m_0)\right]' S(\vartheta, t_*, \varphi_0(\cdot), \eta_0(\cdot)), \tag{4}$$

where $\varphi_0(\cdot) = \varphi(\cdot, t_*, x_*, \eta_0(\cdot))$, $\eta_0(\cdot) \in \{\Pi(\nu_0(\cdot)), [t_*, \vartheta]\}$ $(\eta^\circ(\cdot), \nu_0(\cdot))$ are optimal for the position (t_*, x_*), m_0 gives $\min_M \omega(\varphi_0(\vartheta), m)$. It will be said that a game is perfectly regular if the set $S_0(t_*, x_*)$ consists of only one vector $s_0 = s_0(t_*, x_*)$ for every position (t_*, x_*), $\varepsilon^\circ(t_*, x_*) \in (\omega_0, \omega^\circ)$.

<u>Lemma I.</u> Let the game be perfectly regular. Then ε° is a continuously differentiable function in domain $\{\varepsilon^\circ \in (\omega_0, \omega^\circ), t \in (t_0, \vartheta)\}$ and

$$\left.\frac{\partial \varepsilon^\circ}{\partial x}\right|_{(t,x)} = s_0(t, x), \quad \left.\frac{\partial \varepsilon^\circ}{\partial t}\right|_{(t,x)} = -\max_Q \min_P s_0'(t, x) f. \tag{5}$$

Extremal strategies under the condition of perfect regularity on domain of differentiability are defined by the relations :

$$U^\circ(t_*, x_*) = \{u^\circ : u^\circ \in P,\ \max_Q s_0'(t_*, x_*) f(t_*, x_*, u^\circ, v) = \min_P \max_Q s_0'(t_*, x_*) f(t_*, x_*, u, v)\},$$

$$V^\circ(t_*, x_*) = \{v^\circ : v^\circ \in Q,\ \min_P s_0'(t_*, x_*) f(t_*, x_*, u, v^\circ) = \max_Q \min_P s_0'(t_*, x_*) f(t_*, x_*, u, v)\},$$

$$U_v^\circ(t_*, x_*, v) = \{u^\circ : u^\circ \in P,\ s_0'(t_*, x_*) f(t_*, x_*, u^\circ, v) = \min_P s_0'(t_*, x_*) f(t_*, x_*, u, v)\}$$

<u>Theorem 3.</u> Let the game be perfectly regular and $\varepsilon^\circ(t_\circ, x_\circ) \in [\omega_\circ, \omega^\circ)$. Then a counterstrategy U°_v (and under condition (2) strategy U°) solves the problem 1. If (2) holds and $\varepsilon^\circ(t_\circ, x_\circ) \in (\omega_\circ, \omega^\circ)$, a pair (U°, V°) solves the problem 2 and $\varepsilon^\circ(t_\circ, x_\circ)$ is the cost of the positional game in pure strategies.

For every optimal for the position (t_*, x_*) open-loop control $v_\circ(\cdot)$ of the second player let us find a set $S_\circ(t_*, x_*, v_\circ(\cdot))$ of all vectors $s_\circ$ determined by (4) with fixed control $v_\circ(\cdot)$. Assume for every position (t_*, x_*) $t_* \in (t_\circ, \vartheta)$ and $\varepsilon^\circ \in (\omega_\circ, \omega^\circ)$ for every optimal open-loop control $v_\circ(\cdot)$ of the second player the set $S_\circ(t_*, x_*, v_\circ(\cdot))$ consists of the unique vector $s_\circ = s_\circ(t_*, x_*, v_\circ(\cdot))$. This condition is fulfilled, for example, when

$$f(t, x, u, v) = A(t)x + f(t, u, v),$$

the set $M \subset R^n$ is convex and closed, and $\omega(x, m) = \|x - m\|$.
<u>Theorem 4.</u> Let $\varepsilon^\circ(t, x)$ be differentialable in the domain $\Gamma = \{t \in (t_\circ, \vartheta), \varepsilon^\circ \in (\omega_\circ, \omega^\circ)\}$. Then for every position $(t_*, x_*) \in \Gamma$ the set $S_\circ(t_*, x_*)$ consists of the unique vector $s_\circ = s_\circ(t_*, x_*)$ and the partial derivatives of $\varepsilon^\circ(t, x)$ satisfy (5).

References

1. N.N. Krasovskii. Game problems on encounter of motions. M., Nauka, 1970. (Russian).
2. V.D. Batuhtin, N.N. Krasovskii. Maximin problem for open-loop control. Izv. AN SSSR. Tehn. kibernet., 1972, № 6. (Russian).
3. A.G. Chentsov. On game problem for open-loop control. Dokl. AN SSSR, 1973, 213, № 2. (Russian).

SOME DIFFERENTIAL GAMES WITH INCOMPLETE INFORMATION

F.L. Chernousko, A.A. Melikyan
Institute for Problems of Mechanics
of the USSR Academy of Sciences
Moscow, USSR

This paper deals with some classes of differential games in which one player has no complete information about the phase coordinates of his partner. The information conditions under consideration allow information delay and (or) gaps of information. The equivalence of these games with incomplete information to certain games with complete information is shown. The solutions of a number of special problems of differential games are given.

I. Formulation of the problem. Let the motion of two controlled objects (players) X and Y on fixed time interval $[t_0, T]$ be described by differential equations and constraints

$$X: \quad \dot{x} = f(t, x, u), \qquad Y: \quad \dot{y} = g(t, y, v),$$
$$u \in U, \qquad v \in V. \tag{I.I}$$

The dimensions of phase-vectors x, y and control-vectors u, v are arbitrary; the vector-functions f, g are given; U, V are bounded closed sets in spaces of vectors u, v respectively. Dots denote the time derivatives.

The information conditions of X are the following. Let $\theta = \theta(t)$ be a continuously differentiable scalar function on $[t_0, T]$, such that $\theta(t) \leqslant t$, $t \in [t_0, T]$. We assume that the following set of quantities becomes known by the player X at every moment $t \in [t_0, T]$: $w = \{t, x(t), y(\theta(t))\}$; i.e. X knows precisely his own phase-vector at every moment of time and receives information about phase-vector of Y with variable time delay $t - \theta(t) \geqslant 0$.

The initial data for (I.I) are of the form

$$x(t_0) = x^\circ, \qquad y(\theta(t_0)) = y^\circ. \tag{I.2}$$

The player X chooses his control vector u at a moment $t \in [t_0, T]$, knowing the information which he obtains at that moment, i.e. he applies feed-back (closed-loop) control strategies of the form $u = u(w)$.

The aim of X is to minimize the functional

$$J = F(x(T), y(T)), \tag{I.3}$$

where F is a given function. The player Y counteracts X and realizes his control in the form of time functions $v(t)$. It is clear that control $v(t)$ must be given on the interval $[\theta(t_0), T]$.

Thus, if X and Y choose certain fixed strategy $\hat{u}(w)$ and control $\hat{v}(t)$, then the value $J(\hat{u}, \hat{v})$ can be calculated, in principle, in the following way. The time function $\hat{v}(t)$ can be substituted in the second equation (I.I) and this equation can be integrated on $[\theta(t_0), T]$ with initial condition (I.2). Substituting the function $y(\theta(t))$ in the strategy $\hat{u}(w)$ and, further, $\hat{u}(w)$ in the first equation (I.I), one can obtain the trajectory $x(t)$, $t \in [t_0, T]$, with initial point (I.2). Hence, the values $y(T)$, $x(T)$ and, therefore, the value of functional (I.3) are known. Note, that to construct the trajectory of X on $[t_0, T]$ one needs the trajectory of Y on interval $[\theta(t_0), \theta(T)]$.

We consider only such functions $u(w)$, $v(t)$, which define a unique absolutely continuous solution of (I.I). Considering the game (I.I)-(I.3) from the point of view of X, we state the following problem of differential game.

Problem I. Find the optimal guaranteeing strategy u^* of the player X, i.e. the minimax strategy, which satisfies the equality

$$J^* = \min_u \sup_v J(u, v) = \sup_v J(u^*, v), \tag{I.4}$$

where min and sup are taken among the above described strategies $u(w)$ and control functions $v(t)$. The information conditions of Y are not essential for problem I.

2. <u>The equivalent game with complete information</u>. Let us show now that the problem I is equivalent to some game with standard information conditions. Consider the dynamic equation of Y, correspondin

to some control $v(t), t \in [\theta(t_0), T]$. Introduce the notation: $z(t) = y(\theta(t))$, $t \in [t_0, T]$. For the derivative of vector-function $z(t)$ we have the equality $\dot{z} = \dot{\theta}\, dy/d\theta$, which we can write by means of equation (I.I) in the form

$$\dot{z} = \dot{\theta} \cdot g(\theta(t), z(t), \tilde{v}(t)), \tag{2.I}$$

where $\tilde{v}(t)$ denotes $v(\theta(t))$. The functional (I.3) for the new game is constructed in the following way. Obtaining at the moment T the information about the phase-vector $y(\theta(T)) = z(T)$, the player X can conclude that the phase-vector $y(T)$ belongs to domain $G(z(T), \theta(T), T)$. Here $G(z, t, T)$ denotes the domain attainable at the moment T for the object Y from (I.I), provided that at the moment t his phase-vector is equal to z. We define now the functional

$$\tilde{J} = \tilde{F}(x(T), z(T)) = \max_{\eta \in G(z(T), \theta(T), T)} F(x(T), \eta). \tag{2.2}$$

We introduce a new game, defined by relations

$$X: \ \dot{x} = f(t, x, u), \qquad Y: \ \dot{z} = \dot{\theta} \cdot g(\theta(t), z, \tilde{v}),$$
$$t \in [t_0, T], \ x(t_0) = x^\circ, \ u \in U, \quad z(t_0) = y^\circ, \ \tilde{v} \in V, \tag{2.3}$$

and by the functional (2.2). The player X in this game can observe set of quantities $\tilde{w} = \{t, x(t), z(t)\}$ at each moment $t \in [t_0, T]$ and applies strategies of the form $u = u(\tilde{w})$. The problem of finding the optimal guaranteeing strategy in the game (2.3), (2.2) with complete information is equivalent to the original problem I. This statement is evident from the previous constructions. Both games have the same optimal value of the minimized functional. For the particular cases of function $\theta(t)$ these problems were considered earlier in the papers [I-5].

If we take $\theta(t) = t - \tau$, $\tau = const.$, we obtain differential games with constant delay of information. The equivalence of such games to certain games with complete information was shown in the paper [I]. In the paper [2] some examples for $\theta(t) = t - \tau$ were considered.

Papers [3-5] are devoted to the games in which the player X observes his partner's phase-vector only on time intervals $[a_i, b_i]$, $i = 1, \ldots, N$, $a_1 = t_0$, $b_N = T$, $b_i < a_{i+1}$, $i = 1, \ldots, N-1$. These games can be considered as a limiting case of the games of section I with the following $\theta(t)$:

$$\theta(t)=t, \quad t\in[a_i, b_i], \quad i=1,\dots,N,$$
$$\theta(t)=b_i, \quad t\in(b_i, a_{i+1}), \quad i=1,\dots,N-1.$$

The intervals $[a_i, b_i]$ can be either fixed or chosen by the first player (the problem of optimal observations) or chosen by the second player (the problem of optimal information noise). In the last two case it is natural to constrain the number or the total time of observation or noises (see ref. [3-5]). Some examples of the problems of these types are given below.

3. Examples. a) Let the relations (I.I) on the interval $[0,T]$ for objects X and Y of the same type have one of two forms

$$\dot{x}=u, \quad x(0)=x^\circ, \qquad \dot{y}=v, \quad y(0)=y^\circ, \tag{3.I}$$
$$\ddot{x}=u, \quad |u|\leqslant\mu, \qquad \ddot{y}=v, \quad |v|\leqslant\nu \tag{3.2}$$
$$x(0)=x^\circ, \quad \dot{x}(0)=\xi^\circ, \qquad y(0)=y^\circ, \quad \dot{y}(0)=\eta^\circ, \quad \mu>\nu>0,$$

i.e. both objects are controlled either by their velocities or by their accelerations. We consider the set of observation moments to consist of N discrete moments of observations a_i : $t_0=a_1\leqslant a_2\leqslant \dots\leqslant a_N=T$. The functional (I.3) is of the form

$$J=|x(T)-y(T)|. \tag{3.3}$$

We consider the problem of finding the distribution of moments a_i^* which is optimal for the player X (in the sense of minimum of (3.3). The equivalent game with complete information in this case is a multistep game, and it can be solved analytically by means of dynamic programming (see ref. [3], [4]). As the result we obtain the following formulas for optimal moments of observations in the games (3.I),(3.2) respectively

$$a_i^*=T\cdot\mathcal{L}_N(k,i), \quad \mathcal{L}_N(k,i)=(1-k^{i-1})/(1-k^{N-1}),$$
$$a_i^*=T\sqrt{\mathcal{L}_N(k,i)}, \quad k=\nu/\mu<1, \quad i=1,\dots,N. \tag{3.4}$$

The formulas for the minimal value of functional (3.3) guaranteed for the player X also were obtained. For instance, in the game (3.I) we have

$$J^* = \max[|x^\circ - y^\circ| - (\mu - \nu)T, \mu T k^{N-1}(1-k)/(1-k^{N-1})].$$

b) Definition. A set Q of moments of observations belonging to the interval $[t_0, T]$ is called a sufficient observation set for a given initial position, if the observations on Q guarantee for the player X the same result (in the sense of the functional (I.3)) as the continuous observations on the interval $[t_0, T]$.

The explicit formulas obtained permit us to find in the problems (3.I), (3.2) the minimal sufficient observation sets. For some initial positions these sets consist of a finite number of observation moments. Note, that a denumerable set of moments of observations a_i^*, $i = 1, 2, \ldots$, obtained from (3.4) by setting $N \to \infty$, with a point of condensation at $t = T$ is a sufficient set for all initial positions.

c) We assume now, that the terminal moment T in the games (3.I), (3.2) is not fixed and is defined by the following capture condition: $|x(T) - y(T)| \leq \ell$, $\ell > 0$. The player X desires to minimize the time of pursuit T. It can be shown that observing the object Y at a finite number of moments the player X can capture Y during the same optimal time as in the corresponding games with information. The minimal sufficient number of moments of observations for (3.I) depends on initial position as follows

$$N(x^\circ, y^\circ) = 1 - [(\ln|x^\circ - y^\circ| - \ln \ell)/\ln k],$$

where the square brackets denote the greatest integer. A similar formula holds for the game (3.2) (see ref. [4]).

d) Let the motion of players X and Y during the fixed time interval $[0, T]$ be described by the following equations

$$\ddot{x} = u, \quad |u| \leq 1, \qquad \dot{y} = v, \quad |v| \leq 1,$$

and the player X minimize the functional (3.3). Using the technique of equivalent games, one can show, that a sufficient observation set for all initial positions has a unique point of condensation which is equal to $t = T - 1$ when $T > 1$ and to $t = 0$, when $T \leq 1$. Considering the game from the point of view of maximizing player Y, we find out that to guarantee for Y the maximum value of functional (3.3) it is necessary and sufficient to observe the position of X

at two moments of time: 0 and $T-1$ ($T>1$) (see ref. [5]).

e) Let the set of moments of observations be chosen by the player Y. It means that the player Y can switch on the noise which eliminates the observations made by the player X. The problem is to find the optimal distribution of intervals of noise (b_i, a_{i+1}), $i=1,\dots,N-1$, during which the observations for X are impossible. The total duration of these intervals is equal to ϑ. A number of specific problems of that type was considered. In these problems it occurred that the optimal set of noise intervals consists only of one interval of the total length ϑ, i.e. the noise is concentrated in one time interval. In different cases this interval is situated at the beginning, at the end of or inside the interval of motion.

f) Consider an example from ref. [2] with $\theta(t)=t-\tau$, τ = const. i.e. the player X receives the information about his partner's phase vector with constant delay τ. Let the motion of players be described by the equation and limitations (3.2) and the game be finished as soon as the condition $|x(T)-y(T)|\leq l$ holds. The player X desires to minimize the time of pursuit T. Using the technique of section I, it was shown that the capture is available for X from any initial position if the inequality $(\mu-\nu)/\mu > \nu\tau^2/(2\cdot l)$ holds. Putting $\tau=0$ we obtain the well-known condition of capture $\mu>\nu$ for the game with complete information. A number of other examples can be found in the references given below.

REFERENCES

I. Chernousko, F.L. On differential games with delay of information. (Russian) Dokl. Akad. Nauk SSSR I88 (I969), 766-769.

2. Sokolov, B.N., Chernousko, F.L. Differential games with delay of information. (Russian) Prikl. Mat. i Mekh. 34 (I970), no.5, 8I2-8I9.

3. Melikyan, A.A., Chernousko, F.L. On differential games with alternating conditions of information. (Russian) Dokl. Akad. Nauk SSSR 203 (I972), 46-49.

4. Melikyan, A.A. On differential games with information gaps. (Russian. Armenian and English summaries) Isv. Akad. Nauk Arm. SSR, Matematika I972, no.4, 300-308.

5. Melikyan, A.A. On minimal observations in one game of approaching. (Russian) Prikl. Mat. i Mekh. 37 (I973), no.3, 426-433.

SOME PROPERTIES OF NONZERO-SUM MULTISTAGE GAMES

Jaroslav DOLEŽAL

Institute of Information Theory and Automation
Czechoslovak Academy of Sciences
180 76 Prague, CSSR

1. INTRODUCTION

This contribution deals with so called multistage games, that is games whose dynamics are governed by vector difference equation. This type of games is studied from the point of view of equilibrium, minimax and noninferior solutions on the open-loop and closed-loop strategy classes.

Previous research in this field was devoted mostly to the case of two-player, zero-sum, multistage games, e.g. see [1] - [4]. Only very few results are known for general, N-player, nonzero-sum, multistage games in comparison with the existing theory of differential games [5] - [7]. General multistage games were extensively studied in the author's thesis [10]. Some results obtained in [10] have been already published, see [8], [9].

Results presented in the following sections are based on [10]. We consider certain type of multistage games in which controls of each payer satisfy given state-dependent constraints. For this multistage games we obtain necessary conditions for equilibrium, minimax and noninferior solutions by direct application of the discrete maximum principle proved in [10]. Applying these necessary conditions to the case of linear multistage games with quadratic cost functionals it was possible to obtain explicit form of all considered solution types.

2. FORMULATION OF MULTISTAGE GAME

Consider the dynamical system described by vector difference equation

$$x_{k+1} = f_k(x_k, u_k^1, \ldots, u_k^N), \qquad k = 0,1,\ldots,K-1, \tag{2.1}$$

where is a given positive integer, $k = 0,1,\ldots,K,$ denotes stage of the system and $f_k: E^n \times E^{m_1} \times \cdots \times E^{m_N} \to E^n,$ $k = 0,1,\ldots,K-1.$ Let the cost functional, which is minimized by player i, $i = 1,\ldots,N$, be given as

$$J_i = g^i(x_K) + \sum_{k=0}^{K-1} h_k^i(x_k, u_k^1, \ldots, u_k^N), \qquad i = 1,\ldots,N. \tag{2.2}$$

Here $g^i: E^n \to E^1;$ $h_k^i: E^n \times E^{m_1} \times \cdots \times E^{m_N} \to E^1,$ $i = 1,\ldots,N;$ $k = 0,1,\ldots,K-1.$
The admissible control vector u_k^i of player i at the stage k, $i = 1,\ldots,N;$ $k = 0,1,\ldots,K-1,$ satisfies constraint

$$u_k^i \in U_k^i(x_k),$$

where

$$U_k^i(x) = \{u^i \mid Q_k^i(x,u^i) = 0,\ q_k^i(x,u^i) \leq 0\},\ i = 1,\ldots,N;\ k = 0,1,\ldots,K-1, \tag{2.3}$$

and $Q_k^i: E^n \times E^{m_i} \to E^{r_k^i};$ $q_k^i: E^n \times E^{m_i} \to E^{s_k^i},$ $i = 1,\ldots,N;$ $k = 0,1,\ldots,K-1.$
The inequality sign for vectors in (2.3) we use in the following sense: Let $a \in E^p$; then $a \leq 0 \iff a_j \leq 0,$ $j = 1,\ldots,p.$

In this paper we shall consider two types of strategies defined further.

<u>Definition 1.</u> Any sequence

$$\{u^i\} \equiv \{u^i_0, u^i_1, \ldots, u^i_{K-1} \mid u^i_k \in U^i_k(x),\ k=0,1,\ldots,K-1\}$$

we shall denote as an admissible open-loop strategy of the i-th player, $i=1,\ldots,N$.

<u>Definition 2.</u> In analogous way, any sequence

$$\{\varphi^i(x)\} \equiv \{\varphi^i_0(x), \varphi^i_1(x), \ldots, \varphi^i_{K-1}(x) \mid \varphi^i_k(x) \in U^i_k(x),\ k=0,1,\ldots,K-1\},$$

where $\varphi^i_k : E^n \to E^{m_i}$, $k=0,1,\ldots,K-1$, we shall denote as an admissible closed-loop strategy of the i-th player, $i=1,\ldots,N$.

To have the problem just stated well-posed, it is also necessary to specify for which solution type necessary conditions should be derived. The following solution types will be considered: equilibrium, minimax and noninferior. For the exact definitions the reader should consult [5].

In general, we make the following assumptions.

<u>Assumption 1.</u> All functions appearing in (2.1) - (2.3) are continuously differentiable in all of their arguments.

Now denote

$$\tilde{h}^i_k(x,u^1,\ldots,u^N) = h^i_k(x,u^1,\ldots,u^N), \qquad i=1,\ldots,N;\ k=0,1,\ldots,K-2;$$

$$\tilde{h}^i_{K-1}(x,u^1,\ldots,u^N) = h^i_{K-1}(x,u^1,\ldots,u^N) + g^i(f_{K-1}(x,u^1,\ldots,u^N)), \quad i=1,\ldots,N.$$

In E^{n+1} consider the sets

$$V^i_k(x,u^1,\ldots,u^{i-1},u^{i+1},\ldots,u^N) = \{(a,v) \mid a\in E^1,\ v\in E^n;\ a=\tilde{h}^i_k(x,u^1,\ldots,u^N),$$
$$v=f_k(x,u^1,\ldots,u^N),\ \forall\, u^i\in U^i_k(x)\},\ i=1,\ldots,N;\ k=0,1,\ldots,K-1. \tag{2.4}$$

Further define in E^{n+1} the convex cone

$$R=\{r \mid r\in E^{n+1},\ r=(\rho,0,\ldots,0),\ \rho\le 0).$$

<u>Assumption 2.</u> For each $i=1,\ldots,N$, the sets $V^i_k(\cdot)$, $k=0,1,\ldots,K-1$, in (2.4) are R-directional convex for any $x\in E^n$, $u^j_k\in E^{m_j}$, $j=1,\ldots,N$, $j\ne i$. For the definition of so called directional convexity see [11].

For an admissible open-loop strategy N-tuple $(\{u^1\},\ldots,\{u^N\})$ and corresponding trajectory $x_0,x_1,\ldots,x_K$, define as $I^i_k(x_k,u^i_k)$, $i=1,\ldots,N$; $k=0,1,\ldots,K-1$, the set of numbers from the set $\{1,2,3,\ldots,s^i_k\}$ denoting those components of $q^i_k(x_k,u^i_k)$, for which equality sign holds. The set $I^i_k(x_k,u^i_k)$ is then called the <u>active set</u> for player i, $i=1,\ldots,N$, at the stage k, $k=0,1,\ldots,K-1$. In a similar way we also define the active set for a closed-loop strategy N-tuple.

As usual, we introduce the Hamiltonian of player i, $i=1,\ldots,N$, at the stage k, $k=0,1,\ldots,K-1$, by formula

$$H^i_{k+1}(x,u^1,\ldots,u^N) = -h^i_k(x,u^1,\ldots,u^N) + \lambda^i_{k+1} f_k(x,u^1,\ldots,u^N),$$

where row-vectors $\lambda^i_{k+1}\in E^n$ will defined later.

3. NECESSARY CONDITIONS FOR EQUILIBRIUM SOLUTION

First let us consider equilibrium solution on the class of open-loop strategies. From its definition we can conclude that each player solves discrete optimization problem supposing, that other players use corresponding equilibrium strategies. Then we can immediately apply general results from [10] to this problem. Formulation from [12] can be also used, because it remains valid under the weaker assumption of R-directional convexity of reachable sets.

Theorem 1. Let the multistage game satisfy Assumptions 1 and 2. Suppose that the admissible open-loop strategy N-tuple $(\{u^{*1}\},\dots,\{u^{*N}\})$ is the equilibrium solution of this game on the class of open-loop strategies. Assume that for $i=1,\dots,N;\ k=0,1,\dots,K-1,$ the vectors (indices ℓ, m denote components of the vector constraints in (2.3))

$$\frac{\partial}{\partial u^i} Q^i_{k\ell}(x^*_k, u^{*i}_k),\quad \ell=1,\dots,r^i_k;\qquad \frac{\partial}{\partial u^i} q^i_{km}(x^*_k, u^{*i}_k),\quad m\in I^i_k(x^*_k,u^{*i}_k),$$

are linearly independent. Here $x^*_0, x^*_1,\dots,x^*_K$ denotes the equilibrium trajectory.

Then for each $i=1,\dots,N$, there exist row-vectors

$$\lambda^i_{k+1}\in E^n,\quad \zeta^i_k\in E^{r^i_k},\quad \xi^i_k\in E^{s^i_k},\quad k=0,1,\dots,K-1,$$

such that the conditions 1) - 3) are satisfied.

1)

$$\lambda^i_k = \frac{\partial}{\partial x} H^i_{k+1}(x^*_k, u^{*1}_k,\dots,u^{*N}_k) + \zeta^i_k \frac{\partial}{\partial x} Q^i_k(x^*_k, u^{*i}_k) + \xi^i_k \frac{\partial}{\partial x} q^i_k(x^*_k, u^{*i}_k),\quad k=0,1,\dots,K-1,$$

where we defined $\lambda^i_0 = 0$ and $\lambda^i_K = -\frac{\partial}{\partial x} g^i(x^*)$.

2)

$$\frac{\partial}{\partial u^i} H^i_{k+1}(x^*_k, u^{*1}_k,\dots,u^{*N}_k) + \zeta^i_k \frac{\partial}{\partial u^i} Q^i_k(x^*_k, u^{*i}_k) + \xi^i_k \frac{\partial}{\partial u^i} q^i_k(x^*_k, u^{*i}_k) = 0,\quad k=0,1,\dots,K-1.$$

3)

$$\xi^i_k \le 0,\qquad \xi^i_k q^i_k(x^*_k, u^{*i}_k) = 0,\quad k=0,1,\dots,K-1.$$

To formulate analogous conditions for the class of closed-loop strategies, we must take into account function dependence given by Def. 2. Thus we obtain

Theorem 2. Consider again the multistage game as in the previous theorem and let the admissible closed-loop strategy N-tuple $(\{\varphi^{*1}(x)\},\dots,\{\varphi^{*N}(x)\})$ be the equilibrium solution on the class of closed-loop strategies. Suppose that all assumptions made in Theorem 1 are satisfied with $u^{*i}_k = \varphi^{*i}_k(x^*_k)$, and moreover, let the functions $\varphi^{*i}_k(x)$, $i=1,\dots,N;\ k=0,1,\dots,K-1,$ be continuously differentiable in the neighbourhood of the trajectory $x^*_0, x^*_1,\dots,x^*_K$.

Then, for $i=1,\dots,N$, there exist row-vectors

$$\lambda^i_{k+1}\in E^n,\quad \zeta^i_k\in E^{r^i_k},\quad \xi^i_k\in E^{s^i_k},\quad k=0,1,\dots,K-1,$$

such that the conditions 1) - 3) are satisfied.

1)

$$\lambda^i_k = \frac{\partial}{\partial x} H^i_{k+1}(x^*_k, \varphi^{*1}_k(x^*_k),\dots,\varphi^{*N}_k(x^*_k)) + \zeta^i_k \frac{\partial}{\partial x} Q^i_k(x^*_k, \varphi^{*i}_k(x^*_k)) + \xi^i_k \frac{\partial}{\partial x} q^i_k(x^*_k, \varphi^{*i}_k(x^*_k)) +$$

$$+ \sum_{j=1}^{N} \Big[\frac{\partial}{\partial u^j} H^i_{k+1}(x^*_k, \varphi^{*1}_k(x^*_k),\dots,\varphi^{*N}_k(x^*_k))\Big]\Big[\frac{\partial}{\partial x}\varphi^{*j}_k(x^*_k)\Big],\qquad k=0,1,\dots,K-1,$$

where we define $\lambda^i_0=0$ and $\lambda^i_K = -\frac{\partial}{\partial x} g^i(x^*_K)$.

2)

$$\frac{\partial}{\partial u^i} H^i_{k+1}(x^*_k, \varphi^{*1}_k(x^*_k),\dots,\varphi^{*N}_k(x^*_k)) + \zeta^i_k \frac{\partial}{\partial u^i} Q^i_k(x^*_k, \varphi^{*i}_k(x^*_k)) + \xi^i_k \frac{\partial}{\partial u^i} q^i_k(x^*_k, \varphi^{*i}_k(x^*_k)) = 0,$$

$$k=0,1,\dots,K-1.$$

3) $\xi^i_k \le 0,\quad \xi^i_k q^i_k(x^*_k, \varphi^{*i}_k(x^*_k)) = 0,\quad k=0,1,\dots,K-1.$

On comparing Theorems 1 and 2 we see that the equilibrium costs differ, in general, as a result of summation term in condition 1) of the last theorem. Because this equation was treated more detailly in [8], we shall not discuss it here further.

4. NECESSARY CONDITIONS FOR OTHER SOLUTIONS

From the definition of the minimax solution we see that, in fact, only two person, zero-sum, multistage game must be solved in which opponent of the i-th player are all remaining players. For such a game both theorems stated in the previous section are identical, i.e. closed-loop solution is synthesis of open-loop problems. Having this in mind, that is $u_k^i = \varphi_k^i(x)$, $i = 1,\dots,N$; $k = 0,1,\dots,K-1$, the following theorem is stated only for the class of the open-loop strategies to make notation simpler.

Theorem 3. Let two-person, zero-sum, multistage game in which results minimax problem of player i, satisfy Assmps. 1 and 2. Suppose that the admissible strategy N-tuple $(\{u'^1\},\dots,\{\bar{u}^i\},\dots,\{u'^N\})$ is the saddle-point of this game (minimax solution). Corresponding trajectory denote $\bar{x}_0, \bar{x}_1, \dots, \bar{x}_K$.

Assume that for $k = 0,1,\dots,K-1$, the vectors

$$\frac{\partial}{\partial u^i} Q_{kl}^i(\bar{x}_k,\bar{u}_k^i),\quad l = 1,\dots,r_k^i;\quad \frac{\partial}{\partial u^i} q_{km}^i(\bar{x}_k,\bar{u}_k^i),\quad m \in I_k^i(\bar{x}_k,\bar{u}_k^i),$$

and the vectors

$$\frac{\partial}{\partial u^j} Q_{kl}^j(\bar{x}_k,u_k'^j),\quad l = 1,\dots,r_k^j;\quad \frac{\partial}{\partial u^j} q_{km}^j(\bar{x}_k,u_k'^j),\quad m \in I_k^j(\bar{x}_k,u_k'^j),$$

are linearly independent ($j = 1,\dots, N, j \neq i$).

Then there exist vectors

$$\bar{\lambda}_{k+1}^i \in E^n,\quad \bar{\xi}_k^j \in E^{r_k^j},\quad \bar{\xi}_k^j \in E^{s_k^j},\quad j = 1,\dots,N;\quad k = 0,1,\dots,K-1,$$

such that conditions 1) - 5) are satisfied.

1)

$$\bar{\lambda}_k^i = \frac{\partial}{\partial x}\bar{H}_{k+1}^i(\bar{x}_k,u_k'^1,\dots,\bar{u}_k^i,\dots,u_k'^N) + \bar{\xi}_k^i\frac{\partial}{\partial x}Q_k^i(\bar{x}_k,\bar{u}_k^i) + \bar{\xi}_k^i\frac{\partial}{\partial x}q_k^i(\bar{x}_k,\bar{u}_k^i)$$

$$- \sum_{\substack{j=1\\ j\neq i}}^{N}\left[\bar{\xi}_k^j\frac{\partial}{\partial x}Q_k^j(\bar{x}_k,u_k'^j) + \bar{\xi}_k^j\frac{\partial}{\partial x}q_k^j(\bar{x}_k,u_k'^j)\right],\qquad k = 0,1,\dots,K-1,$$

where we set $\bar{\lambda}_0^i = 0$ and $\bar{\lambda}_K^i = -\frac{\partial}{\partial x}g^i(\bar{x}_K)$.

Used notation

$$\bar{H}_{k+1}^i(x,u^1,\dots,u^N) = -h_k^i(x,u^1,\dots,u^N) + \bar{\lambda}_{k+1}^i f(x,u^1,\dots,u^N),\quad k = 0,1,\dots,K-1.$$

2)

$$\frac{\partial}{\partial u^j}\bar{H}_{k+1}^i(\bar{x}_k,u_k'^1,\dots,\bar{u}_k^i,\dots,u_k'^N) + \bar{\xi}_k^j\frac{\partial}{\partial u^j}Q_k^j(\bar{x}_k,u_k'^j) + \bar{\xi}_k^j\frac{\partial}{\partial u^j}q_k^j(\bar{x}_k,u_k'^j) = 0,$$
$$j = 1,\dots,N,\ j\neq i;\ k = 0,1,\dots,K-1.$$

3)

$$\frac{\partial}{\partial u^i}\bar{H}_{k+1}^i(\bar{x}_k,u_k'^1,\dots,\bar{u}_k^i,\dots,u_k'^N) + \bar{\xi}_k^i\frac{\partial}{\partial u^i}Q_k^i(\bar{x}_k,\bar{u}_k^i) + \bar{\xi}_k^i\frac{\partial}{\partial u^i}q_k^i(\bar{x}_k,\bar{u}_k^i) = 0,$$
$$k = 0,1,\dots,K-1.$$

4) $\bar{\xi}_k^j \le 0$, $\bar{\xi}_k^j q_k^j(\bar{x}_k,u_k'^j) = 0$, $k = 0,1,\dots,K-1$; $j = 1,\dots,N$, $j \neq i$.

5) $\bar{\xi}_k^i \le 0$, $\bar{\xi}_k^i q_k^i(\bar{x}_k,\bar{u}_k^i) = 0$, $k = 0,1,\dots,K-1$.

The minimax solution is extremly pessimistic, and thus little probable. Moreover, in some well-posed realistic cases we have

$J_i = +\infty$. So we are sometimes forced to construct strategy from other reasons.

Now we shall consider noninferior solution of multistage games. From the definition of noninferior set (see [5]) it follows, that in this case we have only a discrete optimization problem with vector-valued performance index. As controls we can choose any admissible strategy N-tuple. Such optimization problems we treated in [10], [11]. It is evident, that on both, open-loop and closed-loop strategy classes we get same necessary conditions for optimal solution.

First we reformulate Assumption 2. Instead of (2.4) we now define in E^{n+1} the sets

$$\hat{V}_k(x) = \{(a,v) \mid a \in E^N,\ v \in E^n;\ a_j = \tilde{h}^j_k(x,u^1,\dots,u^N),\ j=1,\dots,N,$$
$$v = f_k(x,u^1,\dots,u^N);\quad u^i \in U^i_k(x),\ i=1,\dots,N\},\quad k=0,1,\dots,K-1,$$

and the convex cone

$$\hat{R} = \{r \mid r \in E^{N+n},\ r = (\rho_1,\dots,\rho_N,0,\dots,0);\ \rho_i \le 0,\ i=1,\dots,N\}.$$

Assumption 2a. The sets are R -directional convex for any

Theorem 4. Consider discrete optimization problem with vector valued cost functional $J = (J_1,\dots,J_N)^T$, in which results the problem of finding noninferior solution set for multistage game. Let the Assumps. 1, and 2a be satisfied for this problem. Further suppose that the admissible N-tuple belongs to the noninferior set (denoted $(\{\hat{u}^1\},\dots,\{\hat{u}^N\})$, and denote by $\hat{x}_0,\hat{x}_1,\dots,\hat{x}_K$ corresponding trajectory. Assume that for $i=1,\dots,N$; $k=0,1,\dots,K-1$, the vectors

$$\frac{\partial}{\partial u^i} Q^i_{kl}(\hat{x}_k,\hat{u}^i_k),\quad l=1,\dots,r^i_k;\quad \frac{\partial}{\partial u^i} q^i_{km}(\hat{x}_k,\hat{u}^i_k),\quad m \in I^i_k(\hat{x}_k,\hat{u}^i_k),$$

are linearly independent.

Then there exists a vector $\mu \in E^n$, $\mu^i \ge 0$, $i=1,\dots,N$, $\sum_{i=1}^N \mu^i = 1$, and vectors

$$\hat{\lambda}_{k+1} \in E^n,\quad \hat{\xi}^i_k \in E^{r^i_k},\quad \hat{\xi}^i_k \in E^{s^i_k},\quad k=0,1,\dots,K-1;\quad i=1,\dots,N,$$

such that conditions 1) - 3) are satisfied.

1)

$$\hat{\lambda}_k = \frac{\partial}{\partial x}\hat{H}_{k+1}(\hat{x}_k,\hat{u}^1_k,\dots,\hat{u}^N_k) + \sum_{i=1}^N [\hat{\xi}^i_k \frac{\partial}{\partial x} Q^i_k(\hat{x}_k,\hat{u}^i_k) + \hat{\xi}^i_k \frac{\partial}{\partial x} q^i_k(\hat{x}_k,\hat{u}_k)],\ k=0,1,\dots,K-1,$$

where we define $\hat{\lambda}_0 = 0$ and $\hat{\lambda}_K = -\sum_{i=1}^N \mu^i \frac{\partial}{\partial x} g^i(\hat{x}_K)$.

We used notation

$$\hat{H}_{k+1}(x,u^1,\dots,u^N) = -\sum_{i=1}^N \mu^i h^i_k(x,u^1,\dots,u^N) + \hat{\lambda}_{k+1} f_k(x,u^1,\dots,u^N),\quad k=0,1,\dots,K-1.$$

2)

$$\left.\frac{\partial}{\partial u^i}\hat{H}_{k+1}(\hat{x}_k,\hat{u}^1_k,\dots,\hat{u}^N_k) + \hat{\xi}^i_k \frac{\partial}{\partial u^i} Q^i_k(\hat{x}_k,\hat{u}^i_k) + \hat{\xi}^i_k \frac{\partial}{\partial u^i} q^i_k(\hat{x}_k,\hat{u}^i_k)\right\}\ \begin{matrix} i=1,\dots,N \\ k=0,1,\dots,K-1. \end{matrix}$$

3) $\hat{\xi}^i_k \le 0$, $\quad \xi^i_k q^i_k(\hat{x}_k,\hat{u}^i_k) = 0$,

It is easy to prove that, taking

$$J = \sum_{i=1}^N \mu^i J_i,\quad \mu^i > 0,\quad \sum_{i=1}^N \mu^i = 1,$$

as a scalar cost functional, and solving appropriate optimal control problem, the corresponding strategy N-tuple will be in the noninferior set. Thus solving $(N-1)$-parameter family of scalar optimal control problems we get noninferior set for the game in question except, maybe, for points which are obtained for some $\mu^i = 0$. In general, such points are not necessarily noninferior. Also certain convexity assumptions must be satisfied for such conclusion - see [11]. In a concrete game, just mentioned points must be treated separately in order to obtain the whole noninferior set.

From practical point of view we can such solutions (with some $\mu^i = 0$) simply neglect, because they totally ignore the cost functional of some players. So it is hardly probable that these players will take part in cooperation in such case.

Remark 1. Let $U_k^i \neq U_k^i(x)$, $i = 1, \dots, N$; $k = 0, 1, \dots, K-1$. Except of Theorem 2, all other ones are valid, if in Assump. 1, with respect to u^i, $i = 1, \dots, N$, only continuity of f_k, h_k^i, $i = 1, \dots, N$; $k = 0, 1, \dots, K-1$, is required. Of course, then only maximum condition can be always used -[10].

Remark 2. Now suppose that the initial state x_0 is given. All stated theorems remain valid under the following change. In each theorem in condition 1) delete relation $\lambda_0^i = 0$, which was sooner used to compute x_0 and which is now meaningless.

Remark 3. Maybe, it would be more appropriate to assume $U_k^i \neq U_k^i(X)$ as in Remark 1, when dealing with the open-loop strategy class. Otherwise could happen that if, e.g. one player deviates form his equilibrium strategy, this may result in the inadmissible equilibrium strategies of some other players.

5. LINEAR MULTISTAGE GAMES WITH QUADRATIC COSTS

In this section we apply Theorems 1 - 4 to the special class of multistage game in order to obtain the explicit form for each solution. To avoid rather complicated notation, we suppose that the system is autonomous, i.e. its parameters does not change with k. The same technique is applicable to more complicated cases of these games which were solved in [10]. We shall consider following linear multistage game with quadratic cost functionals (x_0 given)

$$x_{k+1} = A x_k + \sum_{j=1}^{N} B_j u_k^j, \qquad k = 0, 1, \dots, K-1; \tag{5.1}$$

$$J_i = \frac{1}{2} x_K^T S_i x_K + \frac{1}{2} \sum_{k=0}^{K-1} [x_k^T Q_i x_k + \sum_{j=1}^{N} (u_k^j)^T R_{ij} u_k^j], \quad i = 1, \dots, N. \tag{5.2}$$

The dimensions of all variables are the same as in Sec.2. Dimensions of all matrices are then also determined. Without any loss of generality it is assumed, that S_i, Q_i, R_{ij}, $i, j = 1, \dots, N$, are symmetric.
In order to apply Theorems 1 - 4, we make always certain assumptions on mentioned matrices.

Case 1. First let us consider closed-loop equilibrium solution. Assume that the matrices R_{ii}, $i = 1, \dots, N$, are positive definite and the matrices Q_i, S_i, $i = 1, \dots, N$, positive semidefinite. Applying Theorem 2 we get

$$\varphi_k^{*i}(x_k) = R_{ii}^{-1} B_i^T P_{k+1}^i W_k x_k, \qquad i = 1, \dots, N; \quad k = 0, 1, \dots, K-1, \tag{5.3}$$

where symmetric matrices P_k^i, $i = 1, \dots, N$; $k = 1, \dots, K$, are solutions of N coupled discrete matrix Riccati-type equations

$$P_k^i = -Q_i + W_i^T [P_{k+1}^i - \sum_{j=1}^{N} P_{k+1}^j B_j R_{jj}^{-1} B_j^T P_{k+1}^j] W_k, \quad i = 1, \dots, N; \quad k = 1, \dots, K-1, \tag{5.4}$$

with end conditions

$$P_K^i = -S_i, \quad i = 1, \dots, N. \tag{5.5}$$

Here we denoted

$$W_k = [1 - \sum_{j=1}^{N} B_j R_{jj}^{-1} B_j^T P_{k+1}^j]^{-1} A, \quad k = 0, 1, \dots, K-1. \tag{5.6}$$

If we formally consider eq.(5.4) now also for k = 0, we get equilibrium costs for each player

$$J_i^* = -\frac{1}{2} x_0^T P_0^i x_0, \quad i = 1, \cdots, N. \tag{5.7}$$

If the solution can be obtained in just described way, then under made assumptions this solution will be unique.

Case 2. Now let us find open-loop equilibrium solution. The assumptions on various matrices are the same as in Case 1. From (5.2) we conclude the existence of unique equilibrium solution. There exist two equivalent ways to obtain open-loop strategies.

A) Suppose that $(\{u^{*1}\}, \dots, \{u^{*N}\})$ is desired equilibrium solution. From the point of view of player i he only solves optimal control problem with parameters $\{u^{*j}\}$, $j=1,\cdots,N$, $j \neq i$. From Theorem 1 we obtain for player i, $i = 1, \cdots, N$,

$$u_k^{*i} = R_{ii}^{-1} B_i^T [\mathcal{P}_{k+1}^i (\mathcal{W}_k^i x_k + (p_k^i)^T], \qquad k = 0, 1, \dots, K-1, \tag{5.8}$$

where symmetric matrices $\mathcal{P}_k^i$ and row-vectors p_k^i, $k = 1, \cdots, K$, are given by equations

$$\mathcal{P}_k^i = -Q_i + A^T \mathcal{P}_{k+1}^i [1 - B_i R_{ii}^{-1} B_i^T \mathcal{P}_{k+1}^i]^{-1} A, \quad k = 1, \cdots, K-1; \quad \mathcal{P}_K^i = -S_i, \tag{5.9}$$

$$p_k^i = [(w_k^i)^T \mathcal{P}_{k+1}^i + p_{k+1}^i] A, \qquad k = 1, \cdots, K-1; \qquad p_K^i = 0. \tag{5.10}$$

We used notation

$$\left.\begin{aligned} \mathcal{W}_k^i &= [1 - B_i R_{ii}^{-1} B_i^T \mathcal{P}_{k+1}^i]^{-1} A, \\ w_k^i &= [1 - B_i R_{ii}^{-1} B_i^T \mathcal{P}_{k+1}^i]^{-1} [\sum_{j=1; j\neq i}^{N} B_j u_k^{*j} + B_i R_{ii}^{-1} B_i^T (p_{k+1}^i)^T], \end{aligned}\right\} k = 0, 1, \cdots, K-1. \tag{5.11}$$

In fact, we constructed immediately synthesis of open-loop strategies for the i-th player. Eliminating now x_k in (5.8) by successive substitution via (5.1) and doing the same for all $i = 1, \cdots, N$, we see, that we are given by the system of linear algebraic equations for computing u_k^{*i}, $i = 1, \cdots, N$; $k = 0, 1, \cdots, K-1$.
If we consider (5.9) - (5.10) also for $k = 0$, we can write costs of the i-th plyer

$$J_i^* = -[\frac{1}{2} x_0^T \mathcal{P}_0^i x_0 + p_0^i x_0 + q_0^i], \qquad i = 1, \cdots, K, \tag{5.12}$$

where

$$\begin{aligned} q_k^i = {} & q_{k+1}^i + \frac{1}{2} (w_k^i)^T \mathcal{P}_{k+1}^i w_k^i + p_{k+1}^i w_k^i - \frac{1}{2} \sum_{j=1, j\neq i}^{N} (u_k^{*j})^T R_{ij} u_k^{*j} - \frac{1}{2} [(w_k^i)^T \mathcal{P}_{k+1}^i + p_{k+1}^i] \cdot \\ & \cdot B_i R_{ii}^{-1} B_i^T [\mathcal{P}_{k+1}^i w_k^i + (p_{k+1}^i)^T], \quad k = 0, 1, \cdots, K-1; \qquad q_K^i = 0. \end{aligned} \tag{5.13}$$

B) Suppose that each player uses open-loop synthesis (5.8). Then using the same procedure as in Case 1 we obtain

$$u_k^{*i} = R_{ii}^{-1} B_i^T (\Pi_{k+1}^i)^T \Omega_k x_k, \qquad i = 1, \cdots, N; \quad k = 0, 1, \cdots, K-1, \tag{5.14}$$

where, in general, non-symmetric matrices Π_k^i we get from the discrete Riccati-type equations ($i = 1, \cdots, N$; $k = 1, \cdots, K$):

$$\left.\begin{aligned} \Pi_k^i &= -Q_i + \Omega_k^T \Pi_{k+1}^i A, \quad k = 1, \cdots, K-1, \\ \Pi_K^i &= -S_i, \end{aligned}\right\} i = 1, \cdots, N, \tag{5.15}$$

where

$$\Omega_k = [1 - \sum_{j=1}^{N} B_j R_{jj} B_j^T (\Pi_{k+1}^j)^T]^{-1} A, \qquad k = 0, 1, \cdots, K-1. \tag{5.16}$$

Now we do not find an analogy of (5.12).

Comparing Eqs. (5.3) and (5.14) we see that these strategy N-tuples are not necessarily identical.

Case 3. Now let us try to find the minimax strategy for the i-th player. It is convenient to seek directly closed-loop strategy. Matrices $Q_i, S_i, R_{ii}, -R_{ij}, j=1,\dots,N, j\neq i$, are assumed positive definite so as to simplify next calculations. From Theorem 3 we have for minimax strategy $\{\bar{u}^i\}$ of player i and corresponding strategies of opponents $\{u'^j\}, j=1,\dots,N, j\neq i$, the relations

$$\left.\begin{aligned} \bar{u}_k^i &= \bar{\varphi}_k^i(x_k) = R_{ii}^{-1} B_i^T \bar{P}_{k+1}^i \bar{W}_k^i x_k \\ u_k'^j &= \varphi_k'^j(x_k) = -R_{ij}^{-1} B_j^T \bar{P}_{k+1}^i \bar{W}_k^i x_k \end{aligned}\right\} \begin{aligned} &k=0,1,\dots,K-1, \\ &j=1,\dots,N, \ j\neq i, \end{aligned} \tag{5.17}$$

where symmetric matrices $\bar{P}_k^i$ are given by relation

$$\bar{P}_k^i = -Q_i + A^T[(\bar{P}_{k+1}^i)^{-1} - \sum_{j=1}^N B_j R_{ij}^{-1} B_j^T]^{-1} A, \quad k=0,1,\dots,K-1; \quad \bar{P}_K^i = -S_i, \tag{5.18}$$

and

$$\bar{W}_k^i = [1 - \sum_{j=1}^N B_j R_{ij}^{-1} B_j^T \bar{P}_{k+1}^i]^{-1} A, \qquad k=0,1,\dots,K-1. \tag{5.19}$$

Minimax costs for player i are given by formula ($k=0$ in (5.18))

$$\bar{J}_i = -\tfrac{1}{2} x_o^T \bar{P}_o^i x_o. \tag{5.20}$$

Case 4. To find noninferior strategy set we assume matrices $Q_i, S_i, R_{ij}, i,j=1,\dots,N$, positive definite. Then it is possible to obtain noninferior set (with exceptions discussed in Section 4) by solving the optimal control problem with functional

$$J(\mu) = \sum_{i=1}^N \mu^i J_i, \qquad \mu^i > 0, \ i=1,\dots,N, \ \sum_{i=1}^N \mu^i = 1. \tag{5.21}$$

We have denoted $\mu = (\mu^1,\dots,\mu^N)$. By rather straitforward manipulations we obtain for noninferior strategies the relations

$$\hat{u}_k^i = \hat{\varphi}_k^i(x_k) = [\sum_{j=1}^N \mu^j R_{ji}]^{-1} B_i^T \hat{P}_{k+1}(\mu) \hat{W}_k(\mu) x_k, \quad i=1,\dots,N; \ k=0,1,\dots,K-1, \tag{5.22}$$

where

$$\hat{P}_k(\mu) = -\sum_{j=1}^N \mu^j Q_j + A^T[(\hat{P}_{k+1}(\mu))^{-1} - \sum_{i=1}^N B_i [\sum_{j=1}^N \mu^j R_{ji}]^{-1} B_i^T]^{-1} A, \ k=0,1,\dots,K-1; \ \hat{P}_K(\mu) = -\sum_{j=1}^N \mu^j S_j, \tag{5.23}$$

and

$$\hat{W}_k(\mu) = [1 - \sum_{i=1}^N B_i [\sum_{j=1}^N \mu^j R_{ji}]^{-1} B_i^T \hat{P}_{k+1}(\mu)]^{-1} A, \qquad k=0,1,\dots,K-1. \tag{5.24}$$

If we consider Eqs. (5.23) again also for $k=0$ we can get total costs of all players

$$J(\mu) = -\tfrac{1}{2} x_o^T \hat{P}_o(\mu) x_o. \tag{5.25}$$

Each player can compute his own costs using Eq. (5.22).

Remark. From computational point of view it is necessary to assume existence of the all inverses on (5.6), (5.11) and (5.16), in order to apply the scheme described in Cases 1 and 2. On the other hand, the assumptions made in Cases 3 and 4 are sufficient for the existence of inverses in (5.19), (5.24).

6. CONCLUSIONS

For rather general class of N-player, nonzero-sum, multistage games we obtained necessary conditions for equilibrium, minimax and noninferior solution types. We used general version of the discrete maximum principle proved in [10].

As we could expected (by analogy with differential games) also here there exist two different types of equilibrium points depending on the considered strategy class. As a rule, open-loop and closed-loop equlibrium costs are not necessarily identical.

For the class of linear multistage games with quadratic cost functionals we also derived the explicit form of each solution.

REFERENCES

[1] Blaquière A., Gerard F., Leitman G., Quantitative and Qualitative Games, Academic Press, N.Y., 1969.

[2] Wang G., Leitman G., Necessary and Sufficient Conditions for Two-Person, Zero-Sum Multistage Games, JOTA, 4,1969, 145-155.

[3] Propoi A.I., Minimax Control problems with a priori information, Avtom. Telemkh., July 1969, 73-79 (in Russian).

[4] Propoi A.I., Minimax control problems under successively aquiered information, Avtomat. Telemekh., Jan. 1970, 65-75 (in Russian).

[5] Starr A.W., Ho Y.C., Nonzero-sum differential games, JOTA, 3 (1969), 184-206.

[6] Starr A.W., Ho Y.C., Further properties of nonzero-sum differential games, JOTA, 3 (1969), 207-219.

[7] Lukes D.L., Equilibrium Feedback Control in Linear Games with Quadratic Costs, SIAM J. Control, 9 (1971), 234-252.

[8] Doležal J., Open-Loop and Closed-Loop Equilibrium Solutions for Multistage Games, Paper presented on the Conference "Mathematical Questions of Optimal Control", Zakopane, 1974.

[9] Doležal J., Open-Loop Hierarchical Solution for Multistage Games, Paper presented on the FORMATOR Symposium, Prague,1974.

[10] Doležal J., Discrete optimization problems and multistage games,PhD.Thesis, ÚTIA ČSAV, Prague, 1973

[11] DaCunha N.O., Polak E., Constrained minimization under vector-valued criteria in finite dimensional spaces, J. Math. Anal. Appl., 19 (1967), 103-124.

[12] Boltjanskij V.G., Optimal Control of Discrete Systems, Nauka, Moscow, 1973 (in Russian).

EQUILIBRIUM SITUATIONS IN GAMES WITH A HIERARCHICAL STRUCTURE OF THE VECTOR OF CRITERIA

Yu. B. Germeier, I. A. Vatel
Computer Center of the Academy
of Sciences of the USSR
Moscow, USSR

Among models of decision making in n-person games with vector criteria it is possible to distinguish one class of models in which the vector of criteria have a so-called hierarchical structure. Multi-level economic and social systems are used as the basis of such models. Each of these "large" systems are sub-divided into a series of groups, which in their turn are again sub-divided, etc., until some elementary unit remains - a player, constituting the "subject" for making a decision.

It is possible to attribute to each group as a unitary system the propensity to increase certain criteria, and generally speaking, their magnitude depends on the actions of all the members of the given group. At the same time there exists a personal (individual) interest of the given player, whose magnitude, an idealization to a known degree, depends only upon his own actions. These actions for increasing social and personal criteria may be interpreted as a distribution by the player of his resources among the interests of the group, of which he is a member at various levels.

In practice there are many examples of models of such type. Thus, in models of the production sphere three categories of interests may be distinguished - a public interest, the interest of the production unit collective (enterprise, collective farm), and a personal material interest of a hard-working person. In sociological models a system of the interests of an individual, family or social group may be considered. Decision making in the sphere of international relations is connected with an analysis of national, regional and general world interests. If we limit ourselves to the study of models in which the choice of strategy by the players is carried out on a voluntary basis (as, for example, in models of international cooperation pertaining to the protection of the environment), then equilibrium situations may serve as one of the rational principles of decision making. Such models are considered below.

1. A description of the model .

Let n players establish a $(m+1)$-leveled society $(K=\{k: k=1,\dots,m\}$ is the set of index numbers of the levels), that are sub-divided into a series of hierarchically organized groups, the society itself - a zero-level group is determined by the set of the numbers of all the players $s^0=\{i: i=1,\dots,n\}$. The society is sub-divided into groups of the first level, each of which in its turn, into groups of the second level, etc. In this way, each group j of the k -th level may be described by the set s_j^k of the number of players that compose it, whereupon the index assumes the value from the set of the numbers of the group, that exist at a given level k . At the final m -th level each group consists of one player, i.e. $s_i^m = \{i\}$.

A criterion (interest) w_j^k is associated with each group, which it strives to increase. Let us assume, that the magnitude of this criterion depends on the values x_s, that are chosen by only the players of whom it is composed, i.e.

$$w_j^k = f_j^k (x_s \mid s \in s_j^k). \tag{1}$$

Therefore each player i has a $(m+1)$ - dimensional vector of interests with components w_j^k (in addition the numbers of the groups are such that $s_j^k \supset i,\ k \in K$). Each player is regarded to have a propensity to increase the compromise criteria, composed of (1), i.e.

$$w_i = w_i(w_j^k \mid s_j^k \supset i,\ k \in K). \tag{2}$$

It is assumed that the i-th player possesses his own vectorial resource a_i , that he allocates among the interests of the groups at all levels of which he is a member, i.e.,

$$x_i = \{x_i^0, \dots, x_i^m\}$$

and

$$\sum_{k=0}^{m} x_i^k = a_i\ , \ x_i^k \geq 0\ , \ k \in K. \tag{3}$$

Let us adopt the compromise criteria (2) , according to [1] , in the form

$$w_i = \min_{\substack{s_j^k \supset i \\ k \in K}} \lambda_i^k w_j^k = \min_{\substack{s_j^k \supset i \\ k \in K}} \lambda_i^k f_j^k (x_s^k \mid s \in s_j^k). \tag{4}$$

Here λ_i^k are the compromise parameters (weights) that satisfy the conditions

$$\lambda_i^0 = 1\ , \ \lambda_i^k \geq 0\ , \ k \in K \setminus \{0\}\ , \ i \in s^0.$$

It is assumed that the function $f_j^k (x_s^k = 0 \mid s \in s_j^k) = 0$ increases monotonically with the growth of any one of the components of the vector x_s^k and

$$f_j^k(x_s^k = 0 \mid s \in s_j^k) = 0. \qquad (5)$$

2. Equilibrium situations .

One of the widely utilized principles for a joint decision making in n-person games is derived from equilibrium situations [2] . As we shall see below, they possess a series of important properties, in the given model, that speak in favour of their utilization.

Above all we shall be interested in the existence of equilibrium situations in pure strategies. Here the following theorem is relevant [3].

Theorem 1. In a game with scalar resources a_i, in order that the situation $\bar{x} = (\bar{x}_1, \ldots, \bar{x}_n)$ be an equilibrium situation, it is necessary and sufficient, that there exist such an assembly of partitions of the set $K \setminus \{m\}$ into two subsets K_i' and K_i''

$(K_i' \cup K_i'' = K \setminus \{m\},\ K_i' \cap K_i'' = \phi)$, that vector $\bar{x} = (\bar{x}_1, \ldots, \bar{x}_n)$, satisfying (3), is the solution of the following system for all $i \in s^0$:

$$\lambda_i^m f_i^m(\bar{x}_i^m) = \lambda_i^k f_j^k(\bar{x}_s^k \mid s \in s_j^k),\ \bar{x}_i^k \geq 0,\ k \in K_i', \qquad (6)$$

$$\lambda_i^m f_i^m(x_i^m) < \lambda_i^k f_j^k(\bar{x}_s^k \mid s \in s_j^k),\ \bar{x}_i^k = 0,\ k \in K_i'',$$

where for all $k \in K \setminus \{m\}$ the indicies j are determined by the conditions $s_j^k \supset i$.

If the a_i's form a vector, then (6) turn out to be the only necessary conditions for an equilibrium situation. Let us note that in order to prove Theorem 1 $f_i^m(0) = 0$, $i \in s^0$ are the only sufficient conditions; this is also true of $f_j^k(x_s^k = 0 \mid s \in s_j^k) > 0,\ k \in K \setminus \{m\}$.
This fact has a significance for dynamic models where from step to step an accumulation of the values of "social" criteria may take place. It should be emphasized as well that the conditions $f_i^m(0) = 0,\ i \in s^0$ lead to a compulsory setting aside of means for personal criteria, i.e., $\bar{x}_i^m > 0$, $i \in s^0$ and it follows from (6) that in any equilibrium situation, player i measures his results by "a most individual measure" $\lambda_i^m f_i^m(\bar{x}_i^m)$.

Theorem 1 reduces the problem of searching for an equilibrium situation to the solution of a system composed of n systems of equations of the type (3), (6). In this event it is necessary to carefully examine 2^{mn} such systems in order to search for all equilibrium situations.

An important class of societies, for which the analysis is substantially simplified, are the hierarchically coordinated societies, given by the conditions

$$\lambda_i^k = \lambda_j^k \ , \ i \in s_j^k \ , \ k \in K, \tag{7}$$

i.e. all players of group j of the k-th level attach similar weights to the criteria of this group. As has been shown in [3], in an equilibrium situation under conditions (7) for any player i a number τ_i may be found such that

$$K_i' = \{\tau_i, \dots, m-1\}, \ K_i'' = \{0, 1, \dots, \tau_i - 1\},$$

in other words, a non-allocation of means, above all, affects the criteria of rather high levels.

A society is described as harmonious if $K_i'' = \emptyset$ for all $i \in s^0$ in any equilibrium situation, i.e. generally speaking the players distribute their resources among the interests of all levels. It is not difficult to show that a harmonious society is hierarchically coordinated. In this society there exists one equilibrium situation (or several that are equivalent) that is best. Thus, in a harmonious society the interests of all the players, in the sense of choosing an equilibrium situation, coincide. It should be emphasized here that in the case of a non-uniqueness of this best situation, the players should trade information in order to arrive at a joint coordinated choice.

Let us note that if the society is not hierarchically coordinated, then equilibrium situations in pure strategies may not exist.

3. <u>A two-level society</u>.

The situation $m=1$ is a most simple model for the analysis of the problem of the combination of "personal and social interests". Let us write the compromise criteria (2) in the form

$$w_i = \min \{ f^0(x_1, \dots, x_n), \lambda_i f_i(a_i - x_i) \}.$$

For purposes of simplification, here the upper indices are omitted. Let us note that conditions of hierarchical co-ordination (7) are fulfilled.

Without disturbing the community, let us assume that the number of players in the society are ordered in the following manner

$$\lambda_1 f_1(a_1) \geq \lambda_2 f_2(a_2) \geq \dots \geq \lambda_n f_n(a_n) \quad (8)$$

Then the following theorem is true.

<u>Theorem 2</u>. In a two-level society an equilibrium situation $(\bar{x}_1, \dots, \bar{x}_n)$, exists that is best for all (or several equivalent ones); in addition there exists such a number p, that

$$\bar{x}_i \geq 0, \ \lambda_i f_i(a_i - \bar{x}_i) = \bar{w}, \ i = 1, \dots, p \quad (9)$$

$$\bar{x}_i = 0, \ \lambda_i f_i(a_i) < \bar{w}, \quad i = p+1, \dots, n, \quad (10)$$

where $$\bar{w} = f^0(\bar{x}_1, \dots, \bar{x}_p; \ x_{p+1} = \dots = x_n = 0). \quad (11)$$

Players with the numbers $p+1, \dots, n$ may be called "egoists" since they knowingly do not set aside means for social interests. Let us note, that in comparison with non-egoistic players, the egoism of a given player is a result of the small value $\lambda_i f_i(a_i)$ for himself (low opportunities or high personal pretensions). Therefore, one may speak of the egoism of players only in relation to an actual society.

The algorithm for searching for the best equilibrium situation for a two-level society consists of the following: sequentially for $p = 1, \dots, n$ (in correspondence to the ordering of players (8)) a maximization problem $\bar{w}$ (11) is solved under conditions (9) and $0 \leq \bar{x}_i \leq a_i, \ i = 1, \dots, p$. Conditions (10) are verified with the results obtained, and if they have been fulfilled ($p \leq n-1$), then the given situation is an equilibrium one. If $p = n$, then the two-level society is harmonious.

It is not difficult to show, that if the weights are given by the expression

$$\lambda_i = f^0(a_1, \dots, a_n) / f(a_i), \ i \in s^0, \quad (12)$$

then the society is harmonious. Expression (12) may be interpreted as the measuring by each player of the importance of personal and social criteria in proportions equal to the relationship of maximal possible values of these criteria.

As is well known, equilibrium situations are stable, in that sense that it is not advantageous for the player to deviate from the situation if all of the others adhere to it. It turns out that in a two-level society the best equilibrium situation is stable in relation to deviations from it on the part of any aggregate of players (if only they do not form coalitions with side payments). During such a deviation the results decrease at least for one player of the given aggregate. This property may be referred to as strong stability. In particular, it follows from this that the set of the best equilibrium situations are the Pareto set.

4. Linear components.

As an example, let us consider a two-level society with linear components of the criteria vector

$$w_i = \min\left\{\sum_{s=1}^{n} c_s x_s,\ \lambda_i(a_i - x_i)\right\},\ i\in S^0 \qquad (13)$$

where the number of players are ordered in accordance with (8).

Using theorem 2, we obtain that in an equilibrium situation

$$x_i = a_i - \frac{1}{\lambda_i}\sum_{s=1}^{p} c_s a_s \Big/ \left(1 + \sum_{s=1}^{p}\frac{c_s}{\lambda_s}\right), \quad i = 1, \dots, p$$

$$x_i = 0, \qquad i = p+1, \dots, n.$$

Players with numbers $p+1, \dots, n$ are egoists. It may be shown that in order that the society (13) in an equilibrium situation be harmonious, the fulfilment of the following conditions are necessary and sufficient

$$\lambda_n a_n > \sum_{s=1}^{n-1} c_s a_s \Big/ \left(1 + \sum_{s=1}^{n-1}\frac{c_s}{\lambda_s}\right)$$

References

1. Yu. B. Germeier, Introduction to the Theory of Operations Research, Moscow, Nauka, 1971 (In Russian).

2. J. F. Nash, "Non-cooperative Games", Ann. Math., 1951, 54.

3. Yu. B. Germeier, I. A. Vatel, Games with a Hierarchical Vector of Interests, Communications of the Academy of Sciences of the USSR, Series on "Technical Cybernetics", No. 3, 1974 (in Russian).

A CLASS OF LINEAR DIFFERENTIAL EVASION GAMES

N.L.Grigorenko

Moscow State University, Moscow, USSR

In this paper, L.S.Pontrjagin's and E.F.Mischenko's method of evasion [1] - [2] is generalized for linear differential games with bounded resource of control.

Consider the statement of the problem.

A. Let the behaviour of a vector z in n -dimensional Euclidean space R^n be described by the linear vector differential equation

$$\dot{z} = Az - Bu + Cv + a \,, \tag{1}$$

where A is constant quadratic matrix of order n, $u \in R^k$ and $v \in R^e$ are control vectors, B and C are constant matrices of corresponding dimension, $a \in R^n$ is a given vector. The vector $u(t)$ is chosen by the pursuer, and the vector $v(t)$ is chosen by the evader. The vector subspace M of a dimension less than or equal to $n-2$ is given in R^n.

The pursuit is assumed to be completed in the first time t when the point $z(t)$ reaches the set M. The controls $u = u(t)$ and $v = v(t)$ are measurable functions. Assume that the following conditions are valid:

$$v(t) \in Q \quad , \quad u(t) \in P \,, \tag{2}$$

$$\int_0^\infty |v(s)| ds \leq \nu \quad , \quad \int_0^\infty |u(s)| ds \leq \mu \,, \tag{3}$$

$$\int_0^\infty (v(s), v(s)) ds \leq \sigma^2, \quad \int_0^\infty (u(s), u(s)) ds \leq \rho^2, \tag{4}$$

where $P \subset R^k$ and $Q \subset R^e$ are convex compact sets, (b,b) is scalar product of vector b, $\operatorname{Int} Q \ni 0$, $P \ni 0$, $\nu > 0$, $\mu \geq 0$, $\sigma > 0$, $\rho \geq 0$.

The conditions (2)-(4) may be considered from physical point of view as restriction on value, on impulse and on energy of control respectively.

We consider the game (1) from the evader's point of view. It is assumed that evader knows equation of game (1), the type of restrictions on control, the set M, vectors $z(s)$ and $u(s)$ for any t from the interval $\max(0, t-h) \leq s \leq t$, where h is a constant parameter.

The aim of evader is not to attack the set M. We suppose that the control of pursuier is arbitrarily permissible.

We shall give some sufficient conditions that the evader can guarantee the satisfying of condition $z(t) \overline{\in} M$ for all $t \geq 0$ for any initial condition $z_0 \overline{\in} M$.

B. We denote the orthogonal complement of M in the space R^n by L. Further, let W be an arbitrary (so far) vector subspace of the space L. We denote by π the operation of orthogonal projection from R^n onto W. To every point $z \in R^n$ we assign two nonnegative numbers: $z \to (\xi, \eta)$, where ξ is the distance from z to M, and η is the distance from z to L.

C. Conditions of Evasion. There exists a two-dimensional vector subspace W of the space L for which the following conditions hold:

1. There does not exist any fixed one-dimensional vector subspace W^1 of W such that the inclusion

$$\pi e^{\tau A} C Q \subset W^1$$

holds for all small positive values of τ.

2. There exists a constant $\lambda > 1$ such that

$$\lambda \pi e^{\tau A} BP \subset \pi e^{\tau A} CQ$$

for all sufficiently small positive values of τ.

3. $\sigma > \varkappa \rho$, $\nu > \varkappa \mu$, $\varkappa$ is non- negative constant which depends on the game (1)(we'll define constant $\varkappa$ in part E).

It turns out that the following theorem holds.

Theorem of Evasion. If in the game (1) with restrictions (2)-(4) the conditions of evasion are satisfied, then, for any initial value of z_0 which does not belong to M, the evasion game can be played in such a way that the point $z(t)$ will never reach the space M $(0 \leq t < \infty)$ and, moreover, the following estimate of the distance from $z(t)$ to M holds.

D. If, in the game (1), the conditions of evasion C are satisfied, then there exist sequences of positive constants T_n, ε_n and C_n, and a positive integer k, which depends only on the game and does not depend either on its initial condition or on its progress, such that the evasion game can be played in such a way that

1. for $\xi_0 > \varepsilon_1$

$$\xi(t) > \frac{C_n \varepsilon_n^k}{(1+\eta(t))^k},$$

$$T_0 = 0, \quad T_{n-1} \leq t \leq T_n, \quad n = 1, 2, \ldots,$$

2. for $\xi_0 \leq \varepsilon_1$

$$\xi(t) > \frac{C_1 \xi_0^k}{(1+\eta(t))^k}, \quad 0 \leq t \leq T_1,$$

$$\xi(t) > \frac{C_n \varepsilon_n^k}{(1+\eta(t))^k}, \quad T_{n-1} \leq t \leq T_n,$$

$$n = 2, 3, \ldots .$$

E. In this part we shall give the formula for definition of the constant $\varkappa$ and shall show the main distinction of the given method of evasion from L.S.Pontrjagin's method.

Condition of evasion 2 guarantees the following representation of the matrices (for sufficiently small τ) (see [1]):

$\pi e^{\tau A} C = g_\tau \cdot \varphi_\tau \cdot H$ and $\pi e^{\tau A} B = g_\tau \cdot h_\tau$, where g_τ is $n \times \ell$ analytical matrix, the rank of which is equal to the rank of matrix $\pi e^{\tau A} C$ for $\tau \neq 0$, H is $\ell \times \ell$ orthogonal matrix, φ_τ and h_τ are analytical matrices, moreover φ_0 is the matrix of orthogonal projection. The matrix h_τ is defined from these conditions non-uniquely, but the condition of evasion 2 guarantees the upper boundedness of the norm of matrix h_0 . Let C be infinum of the norm of all such matrices. Constant $\varkappa$ is defined by condition

$$\varkappa = C \cdot \sqrt{1 + \left(\max_{w \in \partial \varphi_0 (\frac{1}{\lambda} Q)} \; \min_{\substack{v \in Q \\ \varphi_0(v) = w}} \frac{|w - v|}{|w|} \right)^2} ,$$

where $\partial \varphi_0 (\frac{1}{\lambda} Q)$ is the boundary of set $\varphi_0 (\frac{1}{\lambda} Q)$. We shall give later examples of calculation of constant $\varkappa$ in some given cases.

We shall construct the evasion in game (1) with conditions (2)-(4) on decreasing time interval θ_n , using L.S.Pontrjagin's evasion control (see [1]) $v_n(t) = v_n^1(t) + v_n^2(t)$, where the control $v_n^2(t)$ is used for neutralization of the second player (see §5 [1]) and the control $v_n^1(t)$, $|v_n^1(t)| \leq \rho_n$ is used for doing pass maneuver. We choose the sequences θ_n and ρ_n ($\theta_n = \mathrm{const} \cdot \frac{1}{n}$, $\rho_n = \mathrm{const} \cdot \frac{1}{n}$) in such a way that evasion control be admissible.

F. <u>Example 1.</u> Control example from [2] . In a Euclidean space R^2 , we consider the motions of two points x and y , where x is pursuer and y is the evading object . The process of pursuit is finished when $x=y$. The motions of the points x and y are given by the equations

$$\ddot{x}+\alpha\dot{x}=u, \tag{5}$$

$$\ddot{y}+\beta y=v, \tag{6}$$

where α, β are positive constants, and u and v are control vectors from R^2 satisfying the conditions (2), (3), (4), where P and Q are compact convex subsets of R^2 with $\dim Q=2$, $\operatorname{Int} Q \ni 0$, $P \ni 0$.

It is easy to show, that $h_0=E$, $\varkappa=1$ for game (1) and condition of evasion will be satisfied if there exists such a number $\lambda>1$ that $\lambda P \overset{*}{\subset} Q$ and $\sigma>\rho$, $\nu>\mu$.

Example 2 (given in [3]):

$$\ddot{x}_1=u_1, \quad \dot{y}_1=v_1,$$

$$\dddot{x}_2=u_2, \quad \ddot{y}_2=v_2,$$

where $x_i, y_i, u_i, v_i \in R^1$, $i=1,2$, $x=(x_1,x_2)$ is the pursuer and $y=(y_1,y_2)$ is the evading object, u and v are control vectors satisfying the conditions (2), (3), (4). The process of pursuit is finished when $x_1=y_1$, $x_2=y_2$. It is easy to show that $h_0=0$, $\varkappa=0$ for this game, and the condition of evasion will be satisfied if $\operatorname{Int} Q \ni 0$, $\sigma>0$, $\nu>0$.

R e f e r e n c e s

1. L.S.Pontrjagin, Trudy Steklov Mat. Inst., 112 (1971).

2. L.S.Pontrjagin and E.F.Miščenko, A problem on the escape of one controlled object from another. Dokl. Akad. Nauk SSSR, 189 (1969).

3. M.S.Nikol'skii, About one method of evasion. Dokl. Akad. Nauk SSSR, t. 24, N 2 (1974).

Analytical Study of a Case of the Homicidal Chauffeur Game Problem

Christian Marchal

Onera - 92 32o Chatillon - France

ABSTRACT

This paper is a shorter presentation of the Ref 11 presented at the IFIP conference.

The analytical study of the homicidal chauffeur game problem is developped with the help of the Pontryagin's theory of optimization and its generalization to deterministic two-player zero-sum differential games.

Different phenomena are encountered : universal line, disperal line, barrier, equivocal line etc... and their equations or at least their differential equations are given. The study is not complete because its end is very complicated but is shows at least how a systematic construction of the "separatrix" can solve the problem and help to understand it.

INTRODUCTION

Many people (Ref. 1-7) have studied the homicidal chauffeur game problem as a simple and easily understandable example of pursuit-evasion game. However their results are surprisingly complicated and a systematic analytical study using the generalization of the optimality theory of Pontryagin to deterministic two-player zero-sum differential games may help to understand these results.

1 - THE HOMICIDAL CHAUFFEUR GAME

The evader and the pursuer are on an unlimited plane, the evader has no inertia but a velocity limited to E, the limit velocity of the pursuer is C but he also has a limit radius of curvature equal to R, both players are considered as points and the capture distance is called L.

The performance index is the time of capture t_f and it is assumed that the pursuer only goes at maximum speed in its forward direction.

The simplest set of axes is "the set of axes of the pursuer" in which the pursuer is at the origin and the y axis is in its forward direction. In this set of axes, if x and y are the coordinates of the evader, the equations of motion are :

$$\begin{aligned} dx/dt &= v \cos\Psi - y\Omega \\ dy/dt &= v \sin\Psi - C + x\Omega \end{aligned} \qquad (1)$$

with : v = velocity of the evader : $0 \leq v \leq E$
v, Ψ = control parameters of the evader
C = velocity of the pursuer
Ω = rate of rotation of the pursuer $|\Omega| \leq C/R$
Ω : control parameter of the pursuer

The initial conditions are x_o, y_o, t_o; the playing space is defined by $x^2+y^2 > L^2$ and the "terminal surface" is the cylinder of equation $x^2+y^2 = L^2$ in the x, y, t space.

2 - PRELIMINARY CONSIDERATIONS

2.1 - If $E \geqslant C$ the evader can always escape and the problem has no interest hence we will always assume $E < C$.

2.2 - The problem has an olvious symmetry (+x into -x) and scale considerations on the units of length time show that the problem only depends on the two ratios E/C and L/R and, since $E < C$, we will put : A = Arc sin E/C.

3 - THE PONTRYAGIN'S CONDITIONS

Let us call H the Hamiltonian of the problem and p_x and p_y the adjoint parameters to x and y :

$$H = p_x(v\cos\Psi - y\Omega) + p_y(v\sin\Psi - C + x\Omega) \qquad (2)$$

The Pontryagin's conditions lead to :

3.1 - Optimal controls and the maxi-minimum principle

If p_x and/or $p_y \neq 0$ the maxi-minimization gives for the evader :

$$v = E \quad ; \quad \cos\Psi = p_x / [p_x^2 + p_y^2]^{1/2} \quad ; \quad \sin\Psi = p_y / [p_x^2 + p_y^2]^{1/2} \qquad (3)$$

and for the pursuer :

$$\Omega = \frac{C}{R} \operatorname{sign}(p_x y - p_y x) \qquad (4)$$

3.2 - Final or "transversality" conditions

We will only study a given value of the performance index : $t_f = t_1$ the other results being deduced from this case by a simple translation.

The "terminal surface" $\mathcal{C}$ is then divided into :

$$\left.\begin{array}{llll} \text{The half-cylinder} & \mathcal{C}_+ : & x^2 + y^2 = L^2 & ; t > t_1 \\ \text{The circle} & \mathcal{C}_o : & x^2 + y^2 = L^2 & ; t = t_1 \\ \text{The half-cylinder} & \mathcal{C}_- : & x^2 + y^2 = L^2 & ; t < t_1 \end{array}\right\} \qquad (5)$$

Extremal trajectories of the problem of interest cannot end at $\mathcal{C}_+$: in the vicinity of these trajectories we have necessarily $t_f > t_1$ at least as soon as $t > t_1$

Usual transversality conditions give :

$$\text{At } \mathcal{C}_o : \quad t_f = t_1 \quad ; \quad p_{xf}/x_f = p_{yf}/y_f > 0 \quad ; \quad H_f \leqslant 0 \qquad (6)$$

$$\text{At } \mathcal{C}_- : \quad t_f < t_1 \quad ; \quad p_{xf}/x_f = p_{yf}/y_f > 0 \quad ; \quad H_f = 0 \qquad (7)$$

Conditions (2), (3), (4) give at the final instant :

$$H_f = E\left[p_{xf}^2 + p_{yf}^2\right]^{1/2} - p_{yf} C \qquad (8)$$

Since $p_{xf}/x_f = p_{yf}/y_f > 0$ and $x_f^2 + y_f^2 = L^2$, the condition $H_f \leqslant 0$ is equivalent to :

$$E L - y_f C \leqslant 0 \qquad (9)$$

that is :

$$y_f \geqslant L E / C = L \sin A \qquad (10)$$

Hence the final conditions of extremal trajectories of the problem are (fig. 1) :

Either $t_f = t_1$ and (x_f, y_f) is on the arc ABA' of the figure 1, or $t_f < t_1$ and (x_f, y_f) is at A or at A' (it thus define the "useable part of the terminal surface").

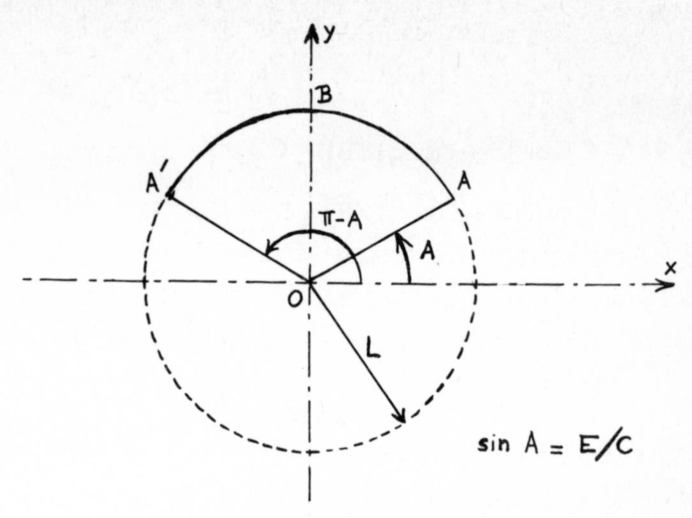

Figure 1 - The final conditions

3.3 - Final "ordinary trajectories of Pontryagin"

Let us integrate back the equations of Pontryagin from the final conditions ; these equations are :

$$dp_x/dt = -\partial H/\partial x = -\Omega p_x \qquad (11)$$

$$dp_y/dt = -\partial H/\partial y = \Omega p_y \qquad (12)$$

$$dH/dt = \partial H/\partial t = 0 \qquad (13)$$

$$dx/dt = Ep_x/[p_x^2 + p_y^2]^{1/2} - y\Omega \qquad (14)$$

$$dy/dt = Ep_y/[p_x^2 + p_y^2]^{1/2} - C + x\Omega \qquad (15)$$

(2), (3), (4) and (13) give :

$$H_f \equiv H = E(p_x^2 + p_y^2) - Cp_y - \frac{C}{R}|p_x y - p_y x| \qquad (16)$$

on the other hand $\Omega = \frac{C}{R}$ sign ($p_x y - p_y x$), let us then study $p_x y - p_y x$:

$$d(p_x y - p_y x)/dt = -Cp_x \qquad (17)$$

since $p_{xf} y_f - p_{yf} x_f = 0$ and since $p_{xf}/x_f > 0$, Ω and $p_x y - p_y x$ are positive in the vicinity of the arc AB of figure 1 and negative near A'B the two cases being symmetrical.

Let us study the first case and let us start from the final point $x_f = L\cos B$, $y_f = L\sin B$, $t_f \leq t_1$ with $A \leq B \leq \pi/2$ if $t_f = t_1$ and $B = A$ if $t_f < t_1$.

Let us choose $p_{xf}^2 + p_{yf}^2 = 1$, we obtain :

$$p_{xf} = \cos B \quad ; \quad p_{yf} = \sin B \qquad (18)$$

and from (11) and (12) :

$$p_x = \cos\left[B + C(t-t_f)/R\right] \quad ; \; p_y = \sin\left[B + C(t-t_f)/R\right] \qquad (19)$$

$$dx/dt = E\cos\left[B + C(t-t_f)/R\right] - Cy/R \qquad (20)$$

$$dy/dt = E\sin\left[B + C(t-t_f)/R\right] - C + Cx/R \qquad (21)$$

with $i=\sqrt{-1}$, (20) and (21) are integrated into :

$$x + iy = R\left[1 - \exp\{iC(t-t_f)/R\}\right] + \left[L+E(t-t_f)\right].\exp\{i\left[B + C(t-t_f)/R\right]\} \qquad (22)$$

these equations are valid only as long as $p_x y - p_y x$ remains positive (wich implies $\Omega = C/R$), i.e. only for :

$$t_f - R(\pi + 2B)/C \leq t \leq t_f \qquad (23)$$

In the x, y plane the different trajectories defined by (22) have a common point D (see fig. 2) for $L + E(t-t_f) = 0$ at $x + iy = R - R\exp\{-iLC/RE\}$; on the other hand the trajectories corresponding to B = A and $t_f \leq t_1$ have the same projection ADin the x, y plane (projection which is an evolute of circle) and this phenomenum corresponds to the phenomenum of barrier discovered by Isaacs (Ref. 1).

3.4 - Singular trajectory

There exists a singular trajectory corresponding to $p_x y - p_y x \equiv 0$, it is easy to study that case :

(17) implies : $\quad p_x \equiv 0 \quad ; \; p_{x_f} = 0$

the final point of the singular trajectory is B (fig. 1 and 2) and then $t_f = t_1$.

(11) and (12) imply then : $\Omega \equiv 0 \quad ; \; p_y \equiv p_{y_f} > 0 \qquad (24)$

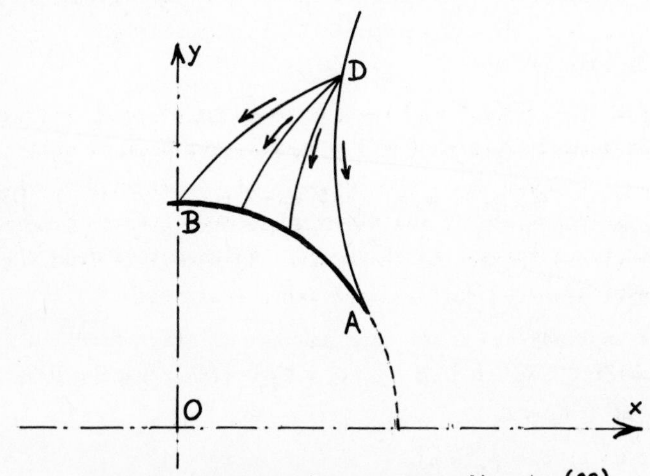

Fig. 2 - Trajectories corresponding to (22)

and from (14) and (15)

$$x \equiv 0 \quad ; \quad y = L + (C-E)(t_1 - t) \qquad (25)$$

The singular trajectory corresponds to a straight race along the y - axis, with its tributaries of the next section it will be a "universal surface" of Isaacs (Ref. 1).

3.5 - The tributaries of the singular trajectory

Let us start from an arbitrary point of the singular trajectory corresponding to $t = t_2 \leqslant t_1$, let us choose $\Omega = C/R$ (or symmetrically $\Omega = -C/R$) and let us integrate the equations (11) to (15) for $t \leqslant t_2$, we obtain (if we choose $p_{yf} = 1$) :

$$p_x = \sin\left[(t_2 - t)\,C/R\right] \quad ; \quad p_y = \cos\left[(t_2 - t)\,C/R\right] \qquad (26)$$

$$x + iy = R + \left[-R + i\{L + C(t_1 - t_2) + E(t - t_1)\}\right].\exp\{-i(t_2 - t)\,C/R\} \qquad (27)$$

3.6 - Thus the Pontryagin's conditions and the symmetry + x into - x lead to four types of extremal trajectories of the problem of interest :

A) The two types the equation of which is written in (22), either for $t_f = t_1$ and $A \leqslant B \leqslant \frac{\pi}{2}$ or for $t_f \leqslant t_1$ and $B = A$, this second type corresponding to the "barriers"

B) The singular trajectory along the y - axis, the equation of which is written in (25).

C) The tributaries of this singular trajectory depending on the parameter t_2 (with $t_2 \leqslant t_1$) and the equation of which, for $t \leqslant t_2$, is written in (27)

Let us note that the trajectories corresponding to the "barriers" goes into the "playing space " in the vicinity of t_f if and only if $E^2/C^2 + L^2/R^2 \leqslant 1$ if not there remain only three types of trajectories given by the ordinary Pontryagin's conditions.

These trajectories can be used to built the "separatrix", i.e. the surface lying between the zone of the playing space $\mathcal{E}$ where $t_f < t_1$ and the zone where $t_f > t_1$, however these trajectories are insufficient, they cannot go for instance inside the loop of the barriers and they have to be completed with the help of the generalization of the optimality theory of Pontryagin to deterministic two-player zero-sum differential games (Ref. 8).

4 - CONSTRUCTION OF THE "SEPARATRIX"

We will only study the case when $\cos A + A\sin A - 1 < L/R \leq \inf[\cos A ; (\pi + 2A)\sin A]$, for instance, for $C = 2E$ and thus $A = \pi/6$ it gives $0.128... < L/R \leq 0.866...$ That case is the case when the two symmetrical barriers exist, don't meet in front of the pursuer (if not it is the "toreador case") and are prolongated by an equivocal line.

The construction of the separatrix (fig. 3) can be done with only Pontryagins trajectories until the point G of the barrier corresponding to $t_f - t = (\pi + 2A) R/C$, that is until the instant t_3 defined by the intersection of the barrier at $(t_f - t) = (\pi + 2A) R/C$ and the proper tributary of the universal line BF. Hence t_3 is the largest instant definded by:

$$t_1 > t_2 > t_3 \qquad (28)$$

$$R + R\exp\{-2iA\} + [(\pi+2A)R\sin A - L]\exp\{-iA\} = R + [-R + i\{L + C(t_1 - t_2) + E(t_3 - t_1)\}]\exp\{iC(t_3 - t_2)/R\} \qquad (29)$$

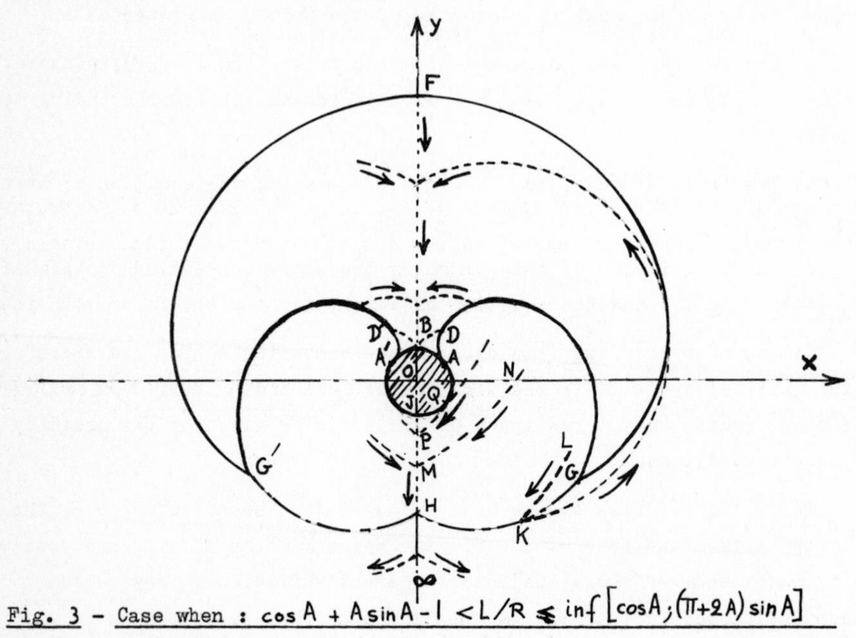

Fig. 3 - Case when : $\cos A + A\sin A - 1 < L/R \leq \inf[\cos A ; (\pi + 2A)\sin A]$

The symmetrical line ADGF G'D'A' (fig. 3) is the intersection of the separatrix with the x, y plane at the instant t_3.

At later instants (for $t_3 < t \leq t_1 - L/E$) the tributaries of the universal line BF and the trajectories of the barrier meet at points of the symmetrical curves DG and D'G' in conditions of a dispersal line, but at G and $t = t_3$

the conditions of the begining of an "equivocal line" are satisfied (Ref. 8) because the conditions (23) end there to be satisfied and the Pontryagin's trajectory of the barrier would continue with $\Omega = -C/R$ instead of $\Omega = +C/R$.

Let us try to obtain the differential equations of this equivocal line GKH (fig. 3) ; along that line the generalized conditions of Pontryagin (Ref. 8) must be satisfied, that is x, y, p_x, p_y and the Hamiltonian H must be such that :

$$\left.\begin{array}{l}
dx/dt = E p_x/[p_x^2 + p_y^2]^{1/2} - y\Omega \\
dy/dt = E p_y/[p_x^2 + p_y^2]^{1/2} - C + x\Omega \\
\Omega = \frac{C}{R}\,\mathrm{sign}\,(p_x y - p_y x) \begin{cases} p_x y - p_y x > 0 \Rightarrow \Omega = C/R \\ p_x y - p_y x = 0 \Rightarrow |\Omega| \leq C/R \\ p_x y - p_y x < 0 \Rightarrow \Omega = -C/R \end{cases} \\
dp_x/dt = -\partial H/\partial x + \lambda(p_x - p_{2x}) = -\Omega p_y + \lambda(p_x - p_{2x}) \\
dp_y/dt = -\partial H/\partial y + \lambda(p_y - p_{2y}) = \Omega p_x + \lambda(p_y - p_{2y}) \\
dH/dt = \partial H/\partial t + \lambda(H - H_2) = \lambda(H - H_2) \\
\lambda \geq 0
\end{array}\right\} (30)$$

p_{2x}, p_{2y} and H_2 being the components of the adjoint vector $\vec{P}$ and the value of the Hamiltonian H along the tributary of the universal line BF at the point x, y, t of interest (which implies of course that the equivocal line remains on the surface defined by the tributaries in the x, y, t space).

Since the equivocal line remains on the surface of the tributaries let us describe it with the help of equation (27) of these tributaries, t_2 being now considered as a function of t and not as a constant as along a given tributary.

The only solution of the system (30) is obtained for $p_x/x = p_y/y < 0$ and thus $p_x y - p_y x = 0$ and it gives with (27) :

$$\left.\begin{array}{l}
d(x+iy)/dt = -E(x+iy)/(x^2+y^2)^{1/2} - iC - i\Omega(x+iy) = \\
= iC\left(1 - \frac{dt_2}{dt}\right)(x+iy-R)/R + i\left(E - C\frac{dt_2}{dt}\right).\exp\{iC(t-t_2)/R\}
\end{array}\right\} (31)$$

The two real quantities Ω and dt_2/dt can be deduced from the first degree and complex equation (31) which is thus the differential equation of the equivocal line GKH.

Along the equivocal line the evader can choose between 2 strategies, either the strategy corresponding to (26) and to the local tributary, or the strategy corresponding to $p_x/x = p_y/y < 0$: the evader runs into the direction of the pursuer and follows the equivocal line until he chooses the other strategy and the pursuer must adapt his own strategy to these modifications.

In the x, y plane the equivocal line reaches either the down part of the circle $x^2+y^2=L^2$ or the negative part of the y axis (as in fig. 3), $H\infty$ is then an ordinary dispersal line and JH is a universal line (where the evader chase the pursuer !) with tributaries such as MN or PQ. There is also tributaries of the equivocal line such as KL starting at K with $p_x/x = p_y/y < 0$.

The analysis can be continued inside the region ADGHJ according to the same principles and leads to many singularities such as the "safe contact motion" (above Q, along the circle of capture), the switch envelopes and the focal lines of Ref. 6 and 7 ; the results are very complicated and numerical computations cannot be avoided.

CONCLUSION

Because of its simplicity the homicidal chauffeur game problem can be studied analytically very far. The notion of "separatrix" (i.e. the surface which, in the playing space, separate the points corresponding to performance indices larger than and smaller than a given value) allows a systematic study of the problem and helps to understant it, for instance the "barriers" correspond to cylindrical parts of the separatrix the generatrices of which are parallel to the t axis.

On the other hand the complexity of the results of this apparently very simple problem emphasizes the need of a systematic way of research and solution of differential game problems.

REFERENCES

[1] ISAACS, R. - "Differential games", Wiley, New-York, (1965).

[2] BRYSON, A.E ; HO, Y.C. - "Applied optimal control"
Blaisdell Publisching Co, Waltham, Mass, (1969).

[3] BREAKWELL, J.V ; MERZ A.W. - "Toward a complete solution of the homicidal chauffeur game"
Proceedings of the first international conference on the theory and applications of differential games. Amherst, Mass, (1969), pp III 1 - III 5.

[4] BREAKWELL, J.V - "Some differential games with interesting discontinuities"
Navy Workshop on Differential games, Aug. 3 (1973, U.S Naval Academy, Annapolis Md.).

[5] BERNHARD P. - "Linear Pursuit-Evasion games and the Isotropic Rocket"
SUDAAR n° 413, Ph D thesis, Dec. 1970, Stanford University Center for Systems Research, Stanford, California.

[6] MERZ A.W. - "The homicidal chauffeur : A differential game"
SUDAAR n° 418, Ph D. thesis, March 1971, Stanford University Center for Systems Research, Stanford California.

[7] MERZ A.W. - "The homicidal chauffeur"
AIAA Journal Vol 12, n° 3, March 1974, pp 259 - 260.

[8] MARCHAL C. - "Generalization of the optimality theory of Pontryagin to deterministic two-player zero-sum differential games".
Communication présentée à la 5ème Conférence IFIP sur les Techniques d'Optimisation - Rome 7.11 mai 1973 - ONERA T.P. n° 1233 (1973).

[9] FRIEDMAN A. - "Differential Games"
Wiley Interscience, a Division of John Wiley and Sons, Inc. - New-York - London - Sydney - Toronto (1971).

[10] KUHN H.W., SZEGO G.P. - "Differential games and related topics".
North Holland Publishing company - Amsterdam - London (1971).

[11] MARCHAL C. - "Analitycal study of the homicidal chauffeur game problem"
Paper presented at the IFIP Technical Conference on Optimization Techniques - Novosibirsk - USSR - 1.7 July 1974 - Also T.P. ONERA n° 1385 (1974).

AN INFORMATIONAL GAME PROBLEM

Yu.S. Osipov
Institute of Mathematics and Mechanics, Ural
Scientific Center of the Academy of Sciences
of the USSR
S. Kovalevskaja st. 16, Sverdlovsk, USSR

The purpose of this report is to give account of one of the possible methods of approaching to the control problems with incomplete information and to demonstrate some results which this method gives. For a certain class of the problems this method seems to be perspective in that it permits one not only to formulate the criteria of solvability of the discussed problems but also to discribe a construction of the controls solving the problem, namely to formulate a generalization of Krasovskii's extremal aiming rule in the theory of positional differential games with complete information. The essence of the method is a substitution of the control problem with incomplete information to more general one in suitable functional space but with complete information. The pay for this substitution is an infinite dimension of a new state space. However the advantage obtained namely the completeness of information and suitable controlled process discription permits overcoming of some serious obstacles. Naturally, such substitution must be invariant in the sense of the initial control problem.

Let us consider the following game problems of convergence.

I. Given is a controlled system

$$\dot{x} = B(t)u - C(t)v + w(t), \quad u \in P, \quad v \in Q, \quad t_0 \leq t \leq \vartheta. \tag{1}$$

Here x is a state vector; u (v) is controlled by the first (second) player; P and Q are compacts. Coefficients in (1) are integrable in Lebesgue sense. The absence of $A(t)x$ from (1) does not diminish the generality of the system [1]. In the space $\{x\}$ given is a convex compact M. The purpose of the first player is to bring $x[t]$ to the set M at the moment $t = \vartheta$. The purpose of the second player is antogonistic. Choosing a control $u[t]$ at the moment t the first player possesses only the information about the convex compact R_t which contains the realized point $x[t]$ of (1). The opponent's behavior is restricted only by the constraint $v[t] \in Q$ and the cases, where the choice of $v[t]$ is based also on the information concerning the control being realized, are not excluded. Let us describe the character of R_t changing. Assume the diameter of R_t is not greater than $\varphi(t)$ where $\varphi(t)$, $t_0 \leq t \leq \vartheta$ is not an increasing function. Further we will distinguish cases (1°) and (2°): for any t_*, $t^* \geq t_*$, R_{t^*} consists only of the points to which it is possible to bring (1) at the moment t^* starting the motion at the moment t_* from the points of R_{t_*} (1°) under the controls being realized in the game on $[t_*, t^*]$, or (2°) under the realized control $u[t]$ and some open-loop measurable control $v(t)$.

Mathematical formalization of this problem is based on one-to-one correspondence between convex compacts and their supporting functionals and is achieved by initial problem absorption in more general control problem. Let $\mathcal{H}_1$ be Lebesgue space of integrable on the ball $E=\{\ell \mid \|\ell\| \leq 1\} \subset \{x\}$ with the square ($\|h\|_1 = (\int_E h^2(\ell)d\ell)^{1/2}$) functions; $\mu(\ell)$, $h_0(\ell)$ -supporting functionals of M, R_{t_0} ; $\mathcal{L} = \{h \in \mathcal{H}_1 \mid h(\ell) \leq \mu(\ell)$ for a.a. $\ell\}$; $d(g) = \operatorname{vraimax}_\ell (g(\ell) + g(-\ell))$; $\mathcal{L}^\varepsilon$ - a closed ε- neighbourhood of $\mathcal{L}$ in $\mathcal{H}_1$. A set of elements $g_t \in \mathcal{H}_1$, $t_* \leq t \leq t^*$, will be called a precision of h_* on $[t_*, t^*]$ if $g_{t_*}(\ell) \leq h_*(\ell)$ for a.a. ℓ, $d(g_t) \leq \varphi(t)$ and for any $t_* \leq t_1 \leq t_2 \leq t^*$ $g_{t_2}(\ell) \leq g_{t_1}(\ell) + \int_{t_1}^{t_2} [f_\xi(\ell) + \ell w(\xi)]d\xi$ for a.a. ℓ.

Here $f_\xi = -\ell C(\xi)v[\xi]$ - in the case (1°) ($v[\xi]$ being realization of v on $[t_*, t^*]$) and $f_\xi = \min \ell C(\xi) v$, $v \in Q$, in the case (2°). A pair $p = \{t, h\}$, $t \in [t_0, \vartheta]$, $h \in \mathcal{H}_1$, will be call a position. A law which puts in correspondence to every p a set $U(p) \subset P$ will be called a strategy U.
Let $p_* = \{t_*, h_*\}$ and Δ - finite separation of $[t_*, \vartheta]$ by the points τ_i ($\tau_0 = t_*$, $\tau_i < \tau_{i+1}$), $\delta(\Delta) = \max_i(\tau_{i+1} - \tau_i)$ be fixed. Then an aggregate of elements $h_t(\ell)_\Delta \in \mathcal{H}_1$, $t_* \leq t \leq \vartheta$, will be called the motion from p_* generated by U if $\forall t$ for almost all ℓ $h_t(\ell) = g_t(\ell) + \int_{t_*}^{t} B(\xi)u[\xi]d\xi$
where $u[t] = u[\tau_i] \in U(\tau_i, h_{\tau_i}(\ell)_\Delta)$, $\tau_i \leq t \leq \tau_{i+1}$; g_t, $t_* \leq t \leq \vartheta$, is a certain precision of h_*. An element g_t describes the deformation and the displacement of R_t under the forces unsubordi-

nated to the first player, moreover a way of g_t forming can make use of any possible information about the system and the control being realized by the first player.

<u>Problem I.</u> Given is a system (1) and a position $p_0 = \{t_0, h^0\}$. It is required to find a strategy U with the property: $\forall \varepsilon > 0\ \exists \delta_0 > 0\ \forall h_t(l)_\Delta = h_t(l; p_0, U)$ with $\delta(\Delta) \leq \delta_0$ $h_\vartheta(l)_\Delta \in \mathcal{L}^\varepsilon$.

<u>Remark I.</u> The strategy U resolving the problem 1 garantees bringing of every compact R_ϑ realized in the game process into a sufficiently small Euclidean neighbourhood of M (while $\delta(\Delta)$ is small). Therefore the problem I can be considered as mathematical formalization of the initial control problem.

Let us discribe the criteria of a solvability of the problem I and the method of the solving strategy construction. Let $\mathcal{H}_1 \supset \supset \mathcal{N}_t \neq \phi$, $t_0 \leq t \leq \vartheta$, and $r(t, h) = \inf \| h - y \|_1$, $y \in \mathcal{N}_t$. Let $D(h, \mathcal{N}_t)$ be a union of all possible minimizing $r(t, h)$ sequences $\{y_k\}$ of the aggregates of all weak in $\mathcal{H}_1$ partial limits sequences $\{\eta_k\}$, where for almost all l $\eta_k(l) = y_k(l) - h(l)$ if $y_k(l) < h(l)$ and $\eta_k(l) = 0$ if $y_k(l) \geq h(l)$.
A strategy $U^e = U^e(t, h) = \{ u^e \mid sB(t)u^e = \max sB(t)u,$

$$u \in P, \quad s = \int_E l g(l) dl, \quad g \in D(h, \mathcal{N}_t) \}$$

will be called an extremal with respect of the sets $\mathcal{N}_t$. A rule G which puts in correspondence to any vector $\{t', t'', h, u(\cdot)\}$, $t'' > t'$, a certain set $G(t', t'', h, u(\cdot)) = \{g_t\}$ of h precisions on $[t', t'')$, moreover in the case (1°), while g_t is being determined, we ought to put $f_\xi = lC(\xi)v$, where $v \in Q$, will be called a counterstrategy of precision.
A set of elements $y_t(l)_\Delta \in \mathcal{H}_1$, $t_* \leq t \leq \vartheta$, will be called the motion from p_* generated by G if $\forall t$ for

$$\text{a. a. } l \qquad y_t(l)_\Delta = g_t(l) + l \int_{t_*}^{t} B(\xi) u(\xi) d\xi$$

where $\{g_t, \tau_i \leq t < \tau_{i+1}\} \subset G(\tau_i, \tau_{i+1}, y_{\tau_i}(l)_\Delta, u(\cdot))$, $y_{\tau_0} = h_*$ and $u(t)$ is any open-loop control. Let Γ_{t_*} be a set of all $h_* \in \mathcal{H}_1$ with the property: $\forall G\ \ \forall \Delta\ \ \exists\, y_t(l)_\Delta$ such that $y_\vartheta(l)_\Delta \in \mathcal{L}^\varepsilon$. Γ_t is analog of the sets of positional absorption[1-3].

<u>Theorem I.</u> The problem I is solvable if and only if $h_0 \in \Gamma_{t_0}$. The solving strategy is a strategy extremal with respect of the sets

Γ_t, $t_0 \leq t \leq \vartheta$

As in the theory of differential games [1,3], there are regular cases [4,5] when Γ_t coincides with the sets of program absorption which can be described effectively.

2. Let us consider now a nonlinear case. Given is a system

(2) $$\dot{x} = f(t, x, u), \quad u \in P, \quad t_0 \leq t \leq \vartheta,$$

where f is continuous and Lipschitzian in x. It is required to bring $x(\vartheta)$ into the compact M by choosing u. Initial state of (2) belongs to a certain compact S_{t_0}. While controlling, S_{t_0} becomes more precise and the results of the precision $S_t \subset S_{t_0}$ at the moment t and the function $h_t(l)$, $l \in S_t$, (3) are known to the first player. The value of this function is a state vector of (2) initiated from $x(t_0) = l$ under realized control $u[\xi]$, $t_0 \leq \xi \leq t$. A way of S_t forming can be arbitrary.

Obviously, the constructions from the point I can't be used here. Let us describe one of the possible ways to investigate this problem. Let $\mathcal{H}_n$ be Lebesgue space of n- dimensional functions integrable with the square on S_{t_0}, $\|h\|_n = (\int_{S_{t_0}} \|h(l)\|^2 dl)^{1/2}$.
A system of compacts of positive measure $S_t \subset S_{t_0}$, $t_* \leq t \leq t^*$, will be called a precision of S on $[t_*, t^*]$ if $S_{t_*} = S$, $S_{t_2} \subset S_{t_1}$, $t_2 > t_1$ and diameter of S_t is less than $\varphi(t)$. A triple $p = \{t, S, h\}$ where $S \subset S_{t_0}$ and h is a contraction on S of an element from $\mathcal{H}_n$ will be called a position. A rule putting in correspondence to any p a set $U(p) \subset P$ will be called a strategy U. A function $x_t(l)_\Delta$, $l \in S_t$, $t_* \leq t \leq \vartheta$, will be called the motion from $p_* = \{t_*, S_*, h_*\}$ generated by U if for any t $x_t(l)_\Delta$ is a contraction on S_t of a certain function from $\mathcal{H}_n$, $x_{t_*}(l) = h_*(l)$ for a.a. $l \in S_{t_*}$ and $\forall t \in (t_*, \vartheta]$ for a.a. $l \in S_{t_*}$, $t \in [t_*, t^*]$ $\dot{x}_t(l) = f(t, x_t(l), u[t])$
where $u[t] = u[\tau_i] \in U(\tau_i, S_{\tau_i}, x_{\tau_i}(l)_\Delta)$, $t \in [\tau_i, \tau_{i+1}) \cap [t_*, t^*]$ and S_t, $t_* \leq t \leq \vartheta$ is a certain precision of S_*. Let $\{m\} = \{m \in \mathcal{H}_n \mid m(l) \in M \text{ a.a. } l\}$ and $\{m\}^\varepsilon$ be a closed ε-neighbourhood of $\{m\}$ in $\mathcal{H}_n$.

<u>Problem 2.</u> Given is a system (2) and a position $p_0 = \{t_0, S_{t_0}, l\}$. It is required to find a strategy U with the property: $\forall \varepsilon > 0\ \exists \delta_0 > 0\ \forall x_t(l)_\Delta = x_t(l, p_0, U)_\Delta$, $\delta(\Delta) \leq \delta_0$, $\exists m \in \{m\}^\varepsilon$ such that $x_\vartheta(l) = m(l)$ for a. a. $l \in S_\vartheta$.

All told in remark I remains true.

Let G be a rule putting in correspondence to every position $\{t', S, h\}$ and to a number $t'' > t'$ an aggregate $G(t', t'', S, h)$ of precisions of S on $[t', t'']$. G will be called a strategy of precision. A function $y_t(\ell)_\Delta$, $\ell \in S_t$, $t_* \leqslant t \leqslant \vartheta$, will be called the motion from $p_* = \{t_*, S_*, h_*\}$ generated by G if $\forall t$ $y_t(\ell)_\Delta$ is a contraction on S_t of a certain function from $\mathcal{H}_n$, $y_{t_*}(\ell) = h_*(\ell)$ for a.a. $\ell \in S_{t_*}$ and $\forall t \in (t_*, \vartheta]$ for a.a. $\ell \in S_{t*}$, $t \in [t_*, t^*]$ $\dot{y}_t(\ell) \in co\{f(t, y_t(\ell)_\Delta, u) \mid u \in P\}$ where $\{S_t, \tau_i \leqslant t < \tau_{i+1}\} \subset G(\tau_i, \tau_{i+1}, S_{\tau_i}, y_{\tau_i}(\ell)_\Delta)$.

Let $\mathcal{K}_{t_*}$ be an aggregate of $h_* \in \mathcal{H}_n$ with the property: $\forall G$ $\forall \Delta$ $\forall S_*$ $\exists y_t(\ell)_\Delta$ $\exists m \in \{m\}$ such that $y_\vartheta(\ell)_\Delta = m(\ell)$ for almost all $\ell \in S_\vartheta$. Let $\mathcal{N}_t \subset \mathcal{H}_n$, $t_0 \leqslant t \leqslant \vartheta$ be non-empty, $r_n(t,g) = \inf \|g - y\|_n$, $y \in \mathcal{N}_t$, and $D(g, \mathcal{N}_t)$ be an aggregate of all weak in $\mathcal{H}_n$ limits of possible minimizing $r_n(t,g)$ sequences. A strategy $U^e = U^e(t, S, h) = \{u^e \mid \int_S (z(\ell) - h(\ell)) f(t, h(\ell), u^e) d\ell = \max_{u \in P} \int_S (z(\ell) - h(\ell)) f(t, h(\ell), u) d\ell,\ z \in D(\tilde{h}, \mathcal{N}_t)\}$ will be called an extremal strategy of the sets $\mathcal{N}_t$. Here $\tilde{h}$ is a continuation of h on S_{t_0}.

Theorem 2. The problem 2 is solvable if and only if $\ell \in \mathcal{K}_{t_0}$. Under this relation an extremal to the sets $\mathcal{K}_t$, $t_0 \leqslant t \leqslant \vartheta$, strategy solves the problem.

$\mathcal{K}_t$ is analog of the sets of positional absorption from [1 - 3]. In the case of linear system following to [1,2] one can distinguish regular cases when $\mathcal{K}_t$'s coincide with the sets of program absorption which have an effective description. The presence of the second player control in (2) does not change the formalization suggested. The changes in the definitions of the motions can be done by analogy with point 1.

References

1. N.N. Krasovskii. Game problems on encounter of motions. "Nauka", Moscow, 1970. (Russian).
2. N.N. Krasovskii. Differential games of convergence-evasion. Izv. AN SSSR. Tehn. kibernet., № 2, 3, 1973. (Russian).
3. Ju.S. Osipov. Differential games with delay. Dokl. AN SSSR, 196, № 4, 1971. (Russian).
4. N.N. Krasovskii, Ju.S. Osipov. Control problem with incomplete information. Izv. AN SSSR. Meh. Tv. Tel., № 4, 1973. (Russian).
5. N.N. Krasovskii, Ju.S. Osipov. On the theory of differential games with incomplete information. Dokl. AN SSSR, 215, № 4, 1974. (Russian).

THE PURSUIT GAME WITH THE INFORMATION LACK OF THE EVADING PLAYER

L.A.Petrosjan
Leningrad State University,
Leningrad, USSR

We consider zero-sum differential game with perscribed duration T. The kinematic equations have the form

$$P: \dot{x} = f(x, u), \quad u \in \mathcal{U} \subset \text{Comp } R^K,$$

$$E: \dot{y} = g(y, v), \quad v \in V \subset \text{Comp } R^e,$$

where $x \in R^n$, $y \in R^n$, $x(0) = x_0$, $y(0) = y_0$. We suppose that for every starting positions x_0, y_0 and every pair of measurable open-loop controls $u(t)$, $v(t)$ there exists a unique solution of the system (1) with the initial conditions x_0, y_0.

The state of information is defined in the following way. The player E (evader) at each instant $t \in [0, T]$ knows the time t, the initial position of P (pursuier), x_0, and his own position $y(t)$. When $t \in [0, \ell]$, P at each moment t knows his position $x(t)$, the time t and the initial state of the player E at $t=0$, y_0; when $t \in [\ell, T]$, P at each moment t knows his position $x(t)$, the time t and the position of player E at moment $t-\ell$, $y(t-\ell)$., $\ell > 0)$.

The payoff of the player E is defined as

$$M(x(T), y(T)),$$

where $M(x, y)$ is a given continuous function on $R^n \times R^n$. The game is supposed to be zero-sum.

<u>Pure strategies.</u> Under pure strategy of the player in the considered game we shall understand the so called piecewise control strategies (PCS). Under the (PCS) of the player E, $v(\cdot)$, we shall understand the pair $\{\tau, b\}$, where τ is a finite decomposition of the time interval $[0, T]$, $0 = t_1 \leq t_2 \leq \cdots \leq t_s = T$ and b-mapping which assigns to every state of information at the moment $t_k \in \tau$, t_k, x_0, $y(t_k)$, a measurable open-loop

control $v(t)$ defined on the time interval $[t_k, t_{k+1})$. Under the (PCS) of the player P, $u(\cdot)$, we shall understand the pair $\{\sigma, a\}$ where σ is a finite decomposition of the time interval $[0,T]$, $0 = t'_1 \leq t'_2 \leq \cdots \leq t'_q = T$ and a-mapping which assigns to every state of information at the moments $0 \leq t'_k \leq \ell$ $(t'_k \in \sigma)$, t'_k , y_0 , $x(t'_k)$, a measurable open-loop control $u(t)$ defined on the time interval $[t'_k, t'_{k+1})$; and at moments $\ell < t'_k \leq T$ $(t'_k \in \sigma)$, it assigns to the t'_k , $y(t'_k - \ell)$, $x(t'_k)$ a measurable open-loop control $u(t)$ defined on the time interval $[t'_k, t'_{k+1})$.

Every pair of (PCS) $u(\cdot)$, $v(\cdot)$ and the initial conditions x_0 , y_0 uniquely determine the trajectories $x(t), y(t)$ as solutions of (1), $t \in [0,T]$, and the payoff $M(x(T), y(T))$.

Thus we can define the pay-off function, as functional of pure strategies (PCS) in the following way

$$K(x_0, y_0; u(\cdot), v(\cdot)) = M(x(T), y(T)), \qquad (3)$$

where $x(t)$, $y(t)$ are the trajectories corresponding to the strategy pair $u(\cdot)$, $v(\cdot)$ and initial conditions x_0 , y_0 .

The game under consideration is one with incomplete information for both players, as we well know from the general game theory, usually

$$\begin{aligned} &\operatorname{Sup}_{v(\cdot)} \operatorname{Inf}_{u(\cdot)} K(x_0, y_0; u(\cdot), v(\cdot)) \neq \\ &\neq \operatorname{Inf}_{u(\cdot)} \operatorname{Sup}_{v(\cdot)} K(x_0, y_0; u(\cdot), v(\cdot)) . \end{aligned} \qquad (4)$$

In case (4) holds it is difficult to speak about the solution of the game in any sense. So we have to follow the von Neumann's approach (see [1]) and introduce the mixed strategies in hope of finding the saddle point in an enlarged class.

Mixed behaviour piecewise control strategies (MB PCS). Under the (MB PCS) of the player E, $\nu(\cdot)$, we shall understand the pair $\{\tau, d\}$, where τ is a finite decomposition of the time interval $[0, T]$, $0 = t_1 \leq t_2 \leq \cdots \leq t_s = T$, and b is a mapping which assigns to every state of information at moments $t_k \in \tau$, t_k, x_0, $y(t_k)$, a probability measure ν concentrated on a finite set of open-loop measurable controls $v(t)$, $t \in [t_k, t_{k+1})$. Under the (MB PCS) of the player P, $\mu(\cdot)$, we shall understand the pair $\{\sigma, c\}$, where σ is a finite decomposition of the time interval $[0, T]$, $0 = t_1' \leq t_2' \leq \cdots \leq t_q' = T$, and c is a mapping which assigns to every state of information at the moments $0 \leq t_k' \leq \ell$ $(t_k' \in \sigma)$, t_k', y_0, $x(t_k')$ a probability measure μ concentrated on a finite set of open-loop measurable controls $u(t)$, $t \in [t_k', t_{k+1}')$; and at the moments $\ell \leq t_k' \leq T$ $(t_k' \in \sigma)$ it assigns to the t_k', $y(t_k' - \ell)$, $x(t_k')$ a probability measure μ concentrated on a finite set of open-loop measurable controls $u(t)$, $t \in [t_k', t_{k+1}')$.

Every strategy pair of (MB PCS) $\mu(\cdot)$, $\nu(\cdot)$ defines random trajectories $x(t)$, $y(t)$ from the initial position x_0, y_0. Thus the payoff $M(x(T), y(T))$ becomes random reachable and we have to consider its mathematical expectation. The latter is uniquely determined by the initial conditions x_0, y_0 and (MB PCS) strategy pair $\mu(\cdot)$, $\nu(\cdot)$. We shall write it as a functional of $\mu(\cdot)$, $\nu(\cdot)$

$$E(x_0, y_0; \mu(\cdot), \nu(\cdot)) = \operatorname{Exp} M(x(T), y(T)), \tag{5}$$

when the expectation is taken by the probability measure over the trajectories $x(t), y(t)$ $(x(0) = x_0, y(0) = y_0)$ corresponding to the (MB PCS) strategy pair $\mu(\cdot)$, $\nu(\cdot)$.

We shall derive later some sufficient condition under which the equation (4) holds in the class of (MB PCS) strategies.

An auxiliary zero-sum game Γ_y. Let $C_P^t(x)$, $C_E^t(y)$ be reachable sets of positions for the players P and E from the starting positions x, y by the moment t. We shall consider a simultaneous game Γ_y, $y \in C_E^{T-\ell}(y_0)$ over the sets of strategies $C_P^T(x_0)$, $C_E^{\ell}(y)$. The game proceedes as follows. The players P and E choose simultaneously and independently of each other the points $\xi \in C_P^T(x_0)$ and corresponding ly $\eta \in C_E^e(y)$. The payoff of player E is defined as $M(\xi,\eta)$. If we suppose the compactness of the sets $C_P^T(x_0)$ and $C_E^t(y)$, the game Γ_y for every $y \in C_E^{T-e}(y_0)$ (sec [1] has the saddle point in mixed strategies, that means in the class of probability measures over the sets $C_P^t(x_0)$, $C_E^e(y)$ (the payoff $M(\xi,\eta)$ is assumed to be continuous).

We shall pose the following conditions on the class of games $\Gamma_y, y \in C_E^{T-e}(y_0)$.

1. For every $\varepsilon > 0$, there exists such N , that in the game Γ_y , P has an ε -optimal mixed strategy μ_ε , which prescribes equal probabilities $1/N$, to N points $\xi_i(y) \in C_P^T(x_0)$ and the number of the points N does not depend on y , when $y \in C_E^{T-e}(y_0)$.

2. Let $y(t)$ be any motion of E on the time interval $t_1 \leq t \leq t_2$, then there exists such N nonintersecting trajectories $\xi_i[y(t-e)] = \xi_i(t)$, that $\xi_i(y(t-e)) \in C_P^T(x_0)$, where every $\xi_i[y(t-e)]$ is a spectrum point of the strategy μ_ε , which is ε -optimal in the game $\Gamma_{y(t-e)}$.

Let $V(y)$ be the value of the game Γ_y and

$$V(\bar{y}) = \max_{y \in C_E^{T-e}(y_0)} V(y) .$$

Now we can describe the construction of the ε -optimal (MB PCS) for both players in the previous game.

Theorem. Let the sets $C_P^T(x_0)$, $C_E^e(y)$, be compact for each $y \in C_E^{T-e}(y_0)$ and the conditions 1, 2 be satisfied. Suppose that for every $\varepsilon_1 > 0$ player P can guarantee ε_1 -capture at the moment T with any of the points ξ_i moving along the trajectories $\xi_i[y(t-e)] = \xi_i(t)$, when E moves along $y(t)$.

Then the value of the game is equal to $V(\bar{y})$ (see (6)) . The ε -optimal (MB PCS) for E includes the open-loop control transition to $\bar{y}$ on the time interval $t \in [0, T-e]$, and further transfer to any point $y(T) \in C_E^e(\bar{y})$, which occurs after the realisation of the random device according to the $\varepsilon/3$ - optimal mixed strategy of the player E in the game $\Gamma_{\bar{y}}$ at the moment $t = T - \ell$.

The ε -optimal (MB PCS) for P randomly chooses at $t=0$ with the probability $1/N$ any of the points ξ_i contained in the spectrum of his $\varepsilon/3$ optimal strategy in the game Γ_{y_0} and prescribes the pursuit of this point to guarantee the $\varepsilon/3$ capture with it at the terminal moment T.

When $M(\xi, \eta) = \rho(\xi, \eta)$, where ρ is an euclidean distance the value of the game, strategies mentioned in the theorem have an interesting geometric interpretation. The value $V(y)$ of Γ_y is equal to the radius $R(y)$ of the minimal sphere $S(y)$ which contains the set $C_E^\ell(y)$. The value of the previous game $V(\bar{y})$ is equal to the maximal radius $R(\bar{y}) = \max_{y \in C_E^{T-\ell}(y_0)} R(y)$.
The optimal (NB PCS) of player P is pure and consists in the pursuing of the centre $O(y)$ of the minimal sphere $S(y)$.

The optimal strategy of the player E is (MB PCS). On the time interval $[0, T-\ell)$ he moves to the point $\bar{y}$, for wich the radius $R(y)$ of minimal sphere containing the set $C_E^\ell(y)$ reaches its maximal value. Let us consider now the auxilary game $\Gamma_{\bar{y}}$. The payoff function in this game is convex, so the maximising player has an optimal mixed strategy which prescribes positive probabilities to no more than $n+1$ points of the set $C_E^\ell(\bar{y})$, where n is a dimension of the space R^n. One can prove that these points lie on the boundary of the minimal sphere containing $C_E^\ell(\bar{y})$. We denote them $\eta_1, \cdots, \eta_{n+1}$. Let $O(\bar{y})$ be the center of this minimal sphere, then there exist such

$$\lambda_i, \quad \lambda_i \geqslant 0, \quad i = 1, \ldots, n+1,$$

$$\sum_{i=1}^{n+1} \lambda_i = 1,$$

that

$$\sum_{i=1}^{n+1} \lambda_i \eta_i = O(\bar{y})$$

At the moment $t = T - \ell$, E chooses with the probability λ_i, $i = 1, \ldots, n+1$, the direction to one of the points η_i, $i = 1, \ldots, n+1$ and on the interval $(T-\ell, T]$ moves to reach it at moment T.

REFERENCES

1 . Karlin S. Mathematical methods and theory of games, programming and economics, Pergamon Press, 1953.

ON CONSTRUCTING INVARIANT SETS IN LINEAR DIFFERENTIAL GAMES

György Sonnevend, senior research fellow, Computing and Automation Institute of the Hungarian Academy of Sciences,
Budapest, Hungary

1. Introduction

In the theory of differential games considerable success was achieved for the cases of linear games of pursuit and evasion.

In this contribution we investigate another game problem: how to find /extremal/ invariant sets, or in other words, stable sets of preference. Before giving exact definitions let us consider two simple examples of the type of problems we are going to study. Suppose there are given two controllable objects, the crocodile, x, and the boy, y,

$$\ddot{x} = u, \quad \|u\| \leq \rho, \quad \dot{y} = v, \quad \|v\| \leq \sigma, \quad x, y \in R^k \tag{1}$$

Do numbers $\mathcal{H}_1, \mathcal{H}_2$ exist such that the conflicting players can quarantee, each, by some behaviour, which do not anticipates future, $\mathcal{H}_2 \geq \|x(t)-y(t)\| \geq \mathcal{H}_1$, at least for all large enough t, where $\mathcal{H}_2$ is minimized by the crocodile, $\mathcal{H}_1$ maximized by the boy?

Here we propose some methods which might be useful in dealing with similar problems in linear games. The main idea is to periodize the control process: to return, after time intervals of length t_0, to a suitable chosen /parametrized/ set K_0. The chief tools of our construction are the tools of the theory of linear differential games of pursuit, introduced by L.S.Pontrjagin [1] . The formulation of our problem as well as our results are in close connection with and were inspired by the evasion problem in the treatment of L.S.Pontrjagin, [2].

2. Definition of the problem

A linear differential game is a control process, idealized by the equation

$$\dot{z} = Cz + u - v, \quad z \in R^n, \quad u \in P, \quad v \in Q, \quad z(0) = z_o, \tag{2}$$

where C is a quadratic constant matrix, z the phase vector, the control parameters u and v belong to compact, convex sets P and Q, are chosen respectively by the first and second player.

The main feature of the control problem in a game consist in that the first player /from the part of whom we shall investigate the control process/ does not know at the moments $t > 0$ the values $v(s)$, $s > t$, that is the control of his opponent in the future. He shall construct his control $u(t)$ by what we call a behaviour, using only the knowledge of the matrix C, sets P and Q and the values $z(s)$, /perhaps $v(s)$/, for $s \leq t$. For each measurable function $v(t)$ the constructed $u(t)$ should also be measurable. Following the idea of L.S.Pontrjagin we use behaviours which are build up in discrete steps in time. We consider cases /Theorem 1/, in which it is sufficient to know only the values $z(t_o n)$, $n = 1, 2, \ldots$, in the periods $[n\, t_o, (n+1)\, t_o]$.

Let us mention briefly that to search for optimal behaviour /without "memory"/ of the type $u(t) = U(z(t))$ with a continous /possibly multivalued/ function $U(z)$, would lead, even in the simple game (1), not only to a strongly nonlinear /and discontinous!/ partial differential equation of first order, of Bellman-Isaacs type, but would not give a better result that the open loop control $u(t) = \bar{u}(t)$, see [3].

The methods we propose make use from linearity more fully.

<u>Definition</u>: A closed set $K \in R^n$ is said to be invariant /from the point of view of the first player/, iff for each $z_o \in K$ there is an appropriate behaviour for the first player which quarantees $z(t) \in K$, for all $t > 0$.
A set K is said to be strongly stable, if for each $\varepsilon > 0$ there exist $\delta(\varepsilon) > 0$, $\delta(\varepsilon) \to 0$ for $\varepsilon \to 0$, such that for each z_o $d(z_o, K) \leq \delta(\varepsilon)$, the first player can quarantee $z(\varepsilon) \in K$; here $d(.,K)$ is the distance function from the set K.

Let us remark, that in a pursuit game the optimal pursuit time $T(z_o)$ is continous iff the terminal manifold is strongly stable.

The optimisation problem we deal with may be formulated as the problem of finding that invariant set K, for which a functional, $f_o(K)$, defined on sets in R^n assumes its minimal value.

As in the evasion problem, the above problem relates to the whole interval $[o, \infty)$, so might be called stationary. Let M be a linear subspace, called terminal manifold, π the operator of orthogonal projection to the orthogonal complement of M, L, and take, with $\|\pi z\| = d(z,M)$, $\|z\|^2 = \langle z,z\rangle$,

$$f_0(K) = -\min\{d(z,M) \mid z \in K\}. \tag{3}$$

So we have another formulation of the evasion problem.

In many engeneering, economic and biological applications we have the following problem of "stabilisation": hold $\|\pi z(t)\|$ as small as possible:

$$f_0(K) = \max\{d(z,M) \mid z \in K\}. \tag{4}$$

As an example let us consider the following game: $x,y \in R^k$

$$\begin{gathered} x^{(n)} + a_{n-1}x^{(n-1)} + \dots + a_1\dot{x} = \varrho u, \quad y^{(m)} + b_{m-1}y^{(m-1)} + \dots + b_1\dot{y} = \sigma v, \\ \|u\| \le 1, \quad \|v\| \le 1, \quad M = \{z \mid z^1 = x - y = 0\}, \quad \varrho, \sigma > 0 \end{gathered} \tag{5}$$

where for $d_i = a_i$ or b_i the polinomals $\sum d_i p^{i-1}$ are stable; i.e. their roots have negativ real parts, therefore $d_i > 0$. A further interesting example, for which we shall investigate the problems (3), (4), is

$$\begin{gathered} \dot{x}^1 = \lambda x^2 + u, \quad \dot{x}^2 = -\lambda x^1 + v, \quad |u| \le \varrho, \quad |v| \le \sigma, \quad z \in R^2, \quad M = 0 \\ \lambda, \varrho, \sigma > 0. \end{gathered} \tag{6}$$

3. The idea of periodisation. Sufficient conditions of existence of compact invariant sets

Here we are interested to give some simple sufficient conditions for the existence of an invariant set for which (4) is finite and /possibly/ small. The idea of periodisation in a simple form may be expressed so: find a set K_0 and a positive number t_0 such, that for each $z_0 \in K_0$ there is a behaviour of the first player /from a given class of behaviours/, which quarantees $z(t_0) \in K_0$. Then the set of all points $z(s)$, $0 \le s \le t_0$ which are realized when the second player selects arbitrary controls $v(\tau)$, is an invariant set $K(t_0, K_0)$.

In the paper [1] L.S.Pontrjagin has given a procedure to compute the set $W(\vartheta, M)$ of all points $z(0)$ for which $z(\vartheta) \in M$ can be quaranted for a fixed time, ϑ, and closed, not necessarily convex, set M, by introducing the operations $(+)$ and $(\overset{*}{-})$ for convex sets and notion of alternating integral. For brevity we do not repeat here this definitions, only $A \overset{*}{-} B = \{w \mid A \supseteq B + w, w \in R^n\}$.

Taking $M=K_o$, $\vartheta = t_o$, the condition of periodizability is

$$-e^{t_oC} K_o \subseteq \int_{-K_o,0}^{t_o} (e^{\tau C} P \overset{*}{-} e^{\tau C} Q)\, d\tau = -e^{t_oC} W(t_o, K_o) \tag{7}$$

and $K(t_o, K_o) = \bigcup_{t \in [0,t_o]} W(t, K_o)$.

The computation of the alternating integral is relatively simple if we have, for all $0 \leq t \leq t_o$, exact "sweeping" /M.S.Nikolskii, B.N. Pseniitsmii/, see [4] . For this reason we shall search for the solutions (K_o, t_o) of the form

$$K_o = -\int_0^{t_o} e^{\tau C} Q\, d\tau - K^o(t_o). \tag{8}$$

This leads to the condition /for simplicity we write $K^o = K^o(t_o)$)

$$e^{t_oC}(K^o + Q(t_o)) \subseteq K^o + P(t_o) \tag{9}$$

where $Q(t_o) = \int_0^{t_o} e^{\tau C} Q\, d\tau$, $P(t_o) = \int_0^{t_o} e^{\tau C} P\, d\tau$.

<u>Theorem 1</u>: Let us suppose that there exist a positive number t_o, for which the inclusion (9) has a /compact/ solution K^o, then there exist a /compact/ invariant set $K=K(t_o,K_o)$. To construct the corresponding behaviour of the first player it is enough for him to know /remember/ only the values $z(nt_o)$, $n=1,2,\ldots$, on the time intervals $[nt_o, (n+1)t_o]$

The proof of this theorem uses only the well known integral representation of the solutions of (2) and the basic properties of the operation $(+)$. /Using the theory of the alternating integral one can show that the same periodisation can be attained by another behaviour which uses only arbitrarily small "memory" on $z(t_i)$, without knowing or computing, the values of $v(s)$./ This is proved in showing that, then, in the construction of the alternating integral the order of the operations $(\overset{*}{-})$ and + can be reversed. A third way to prove the first part of the Theorem 1 is to introduce the variables x, y, $\dot{x}=Cx+u$, $\dot{y}=Cy+v$ $x(o)=z$, $y(o)=0$ and to show that if $[x(nt_o) - e^{t_oC} \cdot y((n-1)t_o)] \in -K^o$, then it is possible to quarantee: $[x((n+1)t_o) - e^{t_oC} y(nt_o)] \in -K^o$.

The set K^o in (9) /or K_o in (7)/ might be interpreted as a "generalized" fixed-point of a /not everywhere defined/ mapping in the space, Ω, of convex, compact sets: $K^o \subseteq e^{-t_oC}[K_o + P(t_o) \overset{*}{-} Q(t_o)]$.

Remark: (9) is fulfilled with $K^o=0$, if for some t_o :

$$e^{t_o C} Q(t_o) \subseteq P(t_o), \qquad \text{specially if} \quad e^{t_o C} Q \subseteq P. \tag{10}$$

In the case $\pi \neq E$: identity, the choice $K^o=(E-\pi)Q(2t_o)$ is natural is some cases. Then (9) is satisfied if

$$e^{t_o C}\pi Q(t_o) + \pi e^{t_o C}(E-\pi)e^{t_o C}Q(t_o) \subseteq P(t_o) \tag{11}$$

and

$$(E-\pi)e^{t_o C}(E-\pi)e^{t_o C}Q(t_o) \subseteq (E-\pi)Q(2t_o). \tag{12}$$

We used $Q(2t_o)=Q(t_o)+e^{t_o C}Q(t_o)$, and the inclusion $A \subseteq (E-\pi)A+\pi A$, for an arbitrary set A.
The equation (12) is satisfied for all t_o if the mapping $(E-\pi)e^{t_o C}$ is a contraction on the subspace $(E-\pi)R^n = M$, see example (5).

Technically, the following lemma is often useful in proving (9) or (11).

Lemma 1: Let $A(s)=\left\{\begin{pmatrix} a_1(s)u \\ a_2(s)u \end{pmatrix} \Big| \|u\| \leq 1\right\} \in \Omega$, $B(s)=\left\{\begin{pmatrix} b(s)u \\ 0 \end{pmatrix} \Big| \|u\| \leq 1\right\} \in \Omega$

then $\int_{t_1}^{t_2} A(s)\,ds \supseteq \int_{t_1}^{t_2} B(s)\,ds$ holds if the equations

$$\int_{t_3}^{t_2} a_1(s)\,ds - \int_{t_1}^{t_3} a_1(s)\,ds \geq \int_{t_1}^{t_2} b(s)\,ds, \qquad \int_{t_3}^{t_2} a_2(s) - \int_{t_1}^{t_3} a_2(s) = 0$$

have a solution for some $t_3 \in (t_1, t_2)$.

Applications. The condition (1o) is fulfilled in the example (6) (4), if $\rho > \sigma$ and $t_o = \pi \kappa^{-1}$, $\pi = 3.14 \ldots$. In the game (1), (4) it will be fulfilled with $t_o = 4\sigma\rho^{-1}$. In the example (5) the inclusion (9) will be satisfied if we chose $K^o=(E-\pi)\ Q(2t_o)$, where E: identity, if $\rho a_1^{-1} > \sigma b_1^{-1}$ and t_o is large enough, see [5] . Another reasonable choice is: $K_o = \{z \mid x=y,\ \dot{x}=\dot{y}\}$ which however, requires a stronger superiority, see [4] .

A sufficient condition for the existence of a set $K^o \neq R^n$, satisfying (9) , which is more special, is established in the following theorem.

Theorem 2: Let us suppose that

$$\tilde{K}=[\tilde{K}^o+\int_0^\infty e^{-\tau C}P\,d\tau] \stackrel{x}{-} \int_{-t_o}^{\infty} e^{-\tau C}Q\,d\tau \left\{\tilde{K}^o+\sum_{n=1}^{\infty} e^{-nt_o C}P(t_o)\right\} \stackrel{x}{-} \sum_{n=1}^{\infty} e^{-nt_o C}e^{t_o C}Q(t_o) \neq \emptyset, \tag{13}$$

$e^{t_o C}\tilde{K}^o \subseteq \tilde{K}^o$, /$\tilde{K}^o$ need not to be finite/, e.g. $e^{t_o C}M=M=\tilde{K}^o$) then $K^o=\tilde{K}$ satisfies (9), so it is possible to periodize the process with the set:

$$K_o^x = -[\{\tilde{K}^o+\int_0^\infty e^{-\tau C}P\,d\tau\} \stackrel{x}{-} \int_0^\infty e^{-\tau C}Q\,d\tau].$$

This condition is necessary for the existence of a nontrivial set K^o, for which (9) is satisfied, if $\int_0^\infty e^{\tau C} P d\tau$ is finite, which is true if the matrix C has roots only which positiv real parts. If, on the contrary, $P(\infty)$, see (9), is finite, e.g. C is stable, then a necessary and sufficient condition for the existence of an invariant set, or for the existence of a solution of (9) is equally: $P(\infty) \overset{*}{-} Q(\infty) \neq \emptyset$.

Taking in (13) not difference of sums, but the "alternating" sum with an initial set $\bar{K}^o$, for example $\bar{K}^o=0$, and going to the limit $t_o \to 0$, we get that if the alternating integral:

$$\bar{K}_o^o \subseteq K^* = -\lim_{T\to\infty} e^{-TC}_{\bar{K}^o,} \int_0^T (e^{+\tau C}P \overset{*}{-} e^{+\tau C} Q)\, d\tau \tag{14}$$

exists in the limit, then K^* is an invariant set. If C has roots only with positive real parts, then taking $\bar{K}^o=0$, we get in (14) an invariant set which is extremal in the sense, that is it is not contained in any other invariant set, if the sets K^* and K_o^* are equal.

In order to diminish $f_o(K(t_o,K_o))$, the behavioral prescription for the first player might be changed to: remain in K_o as long as possible, return to K_o as soon as possible /this is because the smallest, fixible time of return is not necessarily t for all boundary points of K_o/. If we have other, for example integral type, restrictions on the functions $u(\tau)$, $v(\tau)$: $\int_0^{t_o} u(\tau)\, d\tau \le p_o$, $(\int_0^{t_o} v^2(\tau)\, d\tau \le q_o)$, then only the sets $P(t_o)$, $Q(t_o)$ should be changed.

The set of all solutions K^o, in Ω, of the inclusion (9), or (7), is convex. Therefore the set of all so constructed invariant sets $K=K(t_o,K_o)$, is also convex, and closed /in the topology, generated by Hausdorff metric $d(K_1,K_2)$ yet an individual $K(t_o,K_o)$ is, generally, not convex!

The functional $f_o(K)$ is convex, continuous. Therefore our optimisation problem fits in the framework of convex optimisation theory. Let us remind that Ω can be embedded in a linear space. $f_o(.)$ is Gateaux--differentiable. For approximation the Galerkin method might be used treating the problem as one, leading to a variational inequality, for the monotone operator $f_o'(K)$.

To represent the elements of Ω, a very useful device is introducing support functions of the sets K^o, $Q(t_o)$, $P(t_o)$,...,

$m(K,\varphi)=\max\{\langle\varphi,z\rangle | z\in K\}$, $\|\varphi\|=1$ then $f_o(K)=\max\{m(\varphi,K) | \varphi\in L\}$.

(9) means, that: $m(\varphi,K^o)-\mu m(\varphi e^{t_oC},K^o) \ge m(\varphi,e^{t_oC}Q(t_o)) - m(\varphi,P(t_o))$, for all φ, $\|\varphi\|=1$, $\mu=\|\varphi e^{t_oC}\|^{-1}$.

4. Application of a special representation of convex sets

In analyzing the inclusions (7), (9) another representation of the sets $S=K^\circ$, $P(t_\circ)$, $Q(t_\circ)$,...., in the form

$$\left\{S=\int_0^{\bar{s}t_\circ} s(\tau)e^{\tau C}W d\tau,\ K^\circ=\int_0^{k^\circ t_\circ} k^\circ(\tau)e^{\tau C}W d\tau,\dots\right\}=\Omega(C,W), \qquad (15)$$

is useful, where W is a fixed compact, convex set, s a natural number, $s(\tau)$, $(k^\circ(\tau)$,....,), a measurable and integrable scalar function /or more generally taken from a suitable class of "generalized" functions, distributions, containing the $\delta(\tau-\tau_i)$ functions, which we do not specify here/. The reason to introduce such representations lies in the simplicity of operations expressing, in terms of the scalar functions $s(\tau)$, the operations $S\to e^{\tau C}s$, S_1+S_2, $\int S(\tau)d\tau$. If we ask for solutions, K°, of the inclusion (9), which admit a representation (15), then $k^\circ(\tau)$ should satisfy the inequality, /supposing the uniqueness of (15)

$$k^\circ(\tau)+p(\tau)\geq k^\circ(\tau-t_\circ)+q(\tau-t_\circ), \qquad \text{for all } \tau>0, \text{ if } W=-W \qquad (16)$$

For an arbitrary initial, positive function $\underline{k}(\tau)$, $0\leq\tau\leq t_\circ$ it is possible to compute from (16) the prolongation $K^\circ(\tau)$, $\underline{k}=k^\circ$ for $0\leq\tau\leq t$, to the intervalls $[t_\circ,2t_\circ]$, $[2t_\circ,3t_\circ]$,...., and so on, as the smallest positiv solution of (11). A sufficient condition for the existence of $k^\circ(\tau)\geq 0$, with compact support, is given in the following

<u>Lemma 2</u>: Let $r(t)=q(t-t_\circ)-p(t)$, $t>0$, $(q(s)\equiv 0$ for $s<0)$, where the functions $p(.)$, $q(.)$ are taken from (15), /see (9)/. If there exist an integrable function $\underline{k}(\tau)$, $0\leq\tau\leq t$, for which $h(\tau,n)\equiv 0$ for all large enough $n\geq N(\tau)$, where

$$h(\tau,0)=\underline{k}(\tau),\quad h(\tau,n)=\operatorname{pos}h(\tau,n-1)+r(\tau+nt_\circ),\quad \operatorname{pos}y=\begin{cases}y & \text{if } y>0\\ 0 & \text{if } y<0\end{cases}$$

then the inequality has a measurable solution $k(\tau)\geq 0$, which has finite support, if $N(\tau)\leq N$ for all $0\leq\tau\leq t$.

<u>Corollary</u>: if for each τ, $0\leq\tau\leq t$ the function $\ell(\tau)=\sum_{n=1}^{\infty}r(\tau+nt_\circ)<\infty$ and is integrable, then there exist a compact set K° satisfying (9). The special case $P\supseteq\exp(t_\circ C).Q$ was mentioned before in (10).

In general, the above representation (15) do not exist for all compact convex sets S, neither is he unique, here one can apply the theory of G.Choquet. It will be unique if we require $\max_\tau|S(\tau)|$, or $\int S^2 d\tau$ to be minimal /for a fixed interval/. A natural choice, for which (15) is unique,:

$W=\{sw_o | w_o \in R^n, \quad s\in[-1,1]\}, \lambda e^{\tau C} w_o \neq w_o$ for $\lambda>0, \; \tau\in[0,t_o s]$,

then from $S=S_1+S_2$ and $S\in \mathfrak{R}(C,W)$ follows, that $S_1,S_2\in\mathfrak{R}(C,W)$. For greater range of $\mathfrak{R}(C,W)$ representable sets, one wishes the control system $\dot{z}=Cz+w_o$ to be controllable that is: rank $(w_o, Cw_o, C^{n-1}w_o)=n$. In the game (6), then, every centralsymmetrical convex compact set can be represented by (15), for arbitrary $\lambda, W=Q$ and $\bar{s}t_o=\pi\lambda^{-1}$. In dimensions greater then 2 we get a special class of representable sets, which might be useful in other control problems too.

5. Construction of invariant sets in the evasion problem

First let us remark that it is possible to apply the same method, as in theorem 1, for treating the "evasion" problem (2), (3). However then one should be sure that the transient states $z(s)$, $0\le s\le t$, do not belong to M, and one wishes that $z(t)\in K_o$ can be quaranteed for all $z_o\in R^n$ for a finite time $T=T(z_o)$.

Theorem 3: Suppose that there exist a positive number t_o such that $\mathrm{int}[\exp(t_oC)Q(t_o)]\supset P(t_o)$, then with the choice of the periodizing set: $K_o=R^n\setminus Q(t_o)$, one can construct an invariant set which has positive distance from the point z=0.

Application: in the game (6), (3) if $\sigma>\rho$, taking $t_o=\pi\lambda^{-1}$.

As another application of the idea of periodisation we are able to get a slight improvement of the construction of the evading strategy given in [2].

Theorem 4: Assume that the conditions of evasion in [2] are fulfilled. Then, there exist positive numbers ε, Θ such that for each initial position z_o, the evading player can quarantee $\|\pi z(t)\|\ge\varepsilon$ for all $t\ge T(z_o)$, where $T(z_o)\le\Theta$. The numbers depend only on the game, that is on (C, P, Q, M).

To prove it we need to modify the behaviour, described in [2], briefly saying, in that we define the condition of switching to the evading menoeuvre, at moment $\underline{t}$, being the violation of the condition $z(\underline{t})\in \mathrm{int}\,K_o$, where

$$K_o=\{z \,|\, \|\pi e^{\tau C}z\|\ge\varepsilon_o, \quad \text{for all} \quad 0\le\tau\le\Theta_o\}=K_o(\varepsilon_o,\Theta_o) \qquad (17)$$

We use a different evading manoeuvre only for great "drifts" that is, if $\|\pi Ce^{\bar{\tau}C}z(\underline{t})\|\ge\mu_o$, here $\mu_o,\varepsilon_o,\Theta_o$ are constans depending only on the game. In the latter case the curve $\pi e^{\tau C} z(\underline{t})$ is almost a straight line /near the origine, $\pi z=0$/, therefore it is easy to Construct, using the "two dimensional superiority" a simple evading manoeuvre which shall quarantee $z(t+\theta_0) \in K(\varepsilon_0,\theta_0)$, see the game (1) for example.

To "connect" the two behaviours we need the

Lemma 3: Suppose we have $z(t) \notin K_0(\varepsilon_0,0)$ for $t=t_1+k\theta_0$, $k=1,2,\ldots,N$, then $\|\pi C z(s)\| \le \mu_0 = \mu_0(\varepsilon_0)$ for all $t_1 \le s \le t_1+N\theta_0$ if θ_0, N^{-1} are small enough, and depending only on the game. In the evading game (1) it is possible to construct the evading manoeuvre, which quarantees $\|x(t)-y(t)\| \geq \varepsilon$ for all $t>\theta_0$ using only the values $z(n\theta_0) \in K_0$, $n=1,2,\ldots$, in the intervalls $[n\theta_0,(n+1)\theta_0]$, for some sufficiently small θ_0.

References

1. L.S.Pontrjagin: Linear differential games. II.Dokl.Akad.Nauk SSSR 175 No.4,(1967), 764-766.
2. L.S.Pontrjagin: Linear differential games of evasion, Trudy Mat. Inst.Steklova, t CXII.(1971), 30-63.
3. N.N.Krasovskii: Extremal strategies in a differential game, Actesdu Cong.Intern. des Math., Nice,(1970). t.3 177-181.
4. G.Sonnevend: On a type of player superiority, D.A.N.SSSR, 208. No 3.(1973)
5. L.S.Pontrjagin: On the evading process, Applied Mathematics and Optimisation, Vol.1.No.1.(1974).
6. G.Sonnevend: On a class of nonlinear control processes, Dokl. Akad.Nauk T.213, No 2,(1974), 274-277.

A NON COOPERATIVE GAME IN A DISTRIBUTED PARAMETER SYSTEM

J.P. YVON

IRIA - LABORIA

78 - ROCQUENCOURT (FRANCE)

ABSTRACT :

This paper is devoted to study a class of differential games for distributed parameter systems. Essentially we study a Nash equilibrium point for a system governed by a parabolic equation. A method based on the SCARF-HANSEN [1] algorithm for solution of non-cooperative games is given.

I - INTRODUCTION

The general formulation of non-cooperative games is the following .

(1.1) $\begin{cases} \text{Let } \mathcal{U}_i \quad i=1,2,\ldots,n \text{ be a family of Banach spaces} \\ \text{and } J_i \quad i=1,2,\ldots,n \text{ a family of functionals, each of} \\ \text{them being defined on } \mathcal{U} = \prod_{i=1}^{n} \mathcal{U}_i \text{. Then we have the following :} \end{cases}$

Definition

Let $\mathcal{U}_{ad}$ be a subset of $\mathcal{U}$, a point $u \in \mathcal{U}_{ad}$ is a Nash equilibrium point if it satisfies :

(1.2) $\begin{cases} J_i(u_1,\ldots,u_{i-1},u_i,u_{i+1},\ldots,u_n) \leq J_i(u_1,\ldots,u_{i-1},v_i,u_{i+1},\ldots,u_n) \\ \text{for all } v_i \text{ such that} \\ (u_1,\ldots,v_i,\ldots,u_n) \in \mathcal{U}_{ad} \end{cases}$

This definition follows the paper of Nash [1]. The main result is then,

Theorem 1.1 ROSEN [1]

We assume that

(1.3) $\mathcal{U}_{ad}$ is a convex compact subset of $\mathcal{U}$

(1.4) $\begin{cases} \text{the mapping } v \to J_i(v) \text{ is continuous on } \mathcal{U} \text{ and the mapping} \\ v_i \to J_i(v_1,\ldots,v_i,\ldots,v_n) \\ \text{is convex and continuous on } \mathcal{U}, \text{ for } i=1,2,\ldots,n. \end{cases}$

Then there exists at least an equilibrium point on $\mathcal{U}_{ad}$.

In order to be able to treat the infinite dimensionnal case we have the following extension of the result of ROSEN.

Theorem 1.2

(1.5) If $\mathcal{U}_i$ is a reflexive Banach space for i=1,2,...,n and $\mathcal{U}_{ad}$ is a closed bounded convex subset of $\mathcal{U}$ then Theorem 1.1 still holds with assumption (1.4) with continuity in the weak topology of $\mathcal{U}$. ∎

Remark 1.1

Uniqueness of the solution requires more specific (and unfortunately restrictive) hypothesis. The following assumption is as extension of monotonicity property :

If J_i is Gateaux-differentiable with respect to its ith component and verifies the following property :

$$(1.6)\begin{cases}\text{There exists a } \lambda_1,\dots,\lambda_n \quad \lambda_i \geq 0 \quad \text{such that} \\ \sum_{i=1}^{n} \lambda_i \left(\frac{\partial J_i}{\partial u_i}(u) - \frac{\partial J_i}{\partial u_i}(v),\ u_i - v_i\right)_i \geq 0\end{cases}$$

where $(,)_i$ denote the duality pairing between V_i and V_i', then there exists a unique equil.point. ∎

II - A CLASS OF DIFFERENTIAL GAMES

Now we can state our game in the frame of optimal control theory for linear parabolic systems (cf. LIONS [1]).

Let V and H be two Hilbert spaces with V dense in H with continuous injection so that we have

$$V \subset H = H' \subset V' .$$

In addition we consider a family of operator

$$A(t) \in \mathcal{L}(V;V') \qquad t \in [0,T]$$

and

$$\{B_i \in \mathcal{L}(\mathcal{U}_i; L^2(0,T;V'))\}_{i=1,2,\dots,n}.$$

If f is given in $L^2(0,T;V')$ et $y_0 \in H$ we may consider the parabolic system

$$(2.1)\begin{cases}\frac{dy}{dt} + A(t)y = f + \sum_{i=1}^{n} B_i v_i \quad \text{sur} \quad]0,T[\\ y(0) = y_0 .\end{cases}$$

The equation admits a unique solution $y(v)$ depending continuously on v under assumption of "measurability" of $A(t)$ and coercivity of $A(t)$:

$$(A(t)\varphi,\varphi)_{V'V} + \lambda|\varphi|_H^2 \geq \alpha \|\varphi\|_V^2 \qquad \forall\, \varphi \in V.$$

Observation is given by a family of criteria

$$(2.2)\begin{cases}J_i(v) = \psi_i(y(v)) + \theta_i(v) \quad \text{where } \psi_i \text{ is convex continuous in y and} \\ \theta_i \text{ is convex continuous with respect to } v_i, \text{ for } i=1,2,\dots,n \\ \text{continuous with aspect to v.}\end{cases}$$

If we assume that $\mathcal{U}_{ad}$ is convex compact we can apply directly Theorem 1.1. If $\mathcal{U}_{ad}$ is compact only for the weak topology of $\mathcal{U}$ (eg. $\mathcal{U}_{ad}$ is a bounded closed convex subset of $\mathcal{U}$) then we need a further assumption :

$$(2.3)\quad \begin{cases} v \to \psi(y(v)) + \theta_i(v) \\ \text{is continuous for the weak topology of } \mathcal{U} \, ; \end{cases}$$

An example of this situation is the following :

<u>The linear-quadratic problem with distributed observation</u>

We know (cf. LIONS [1]) that the solution $y(v)$ of (2.1) belongs to $W(0,T)$ where

$$(2.4)\qquad W(0,T) = \{\varphi \in L^2(0,T;V) \mid \frac{d\varphi}{dt} \in L^2(0,T;V')\}$$

which is an Hilbert space with the norm

$$(2.5)\qquad \|\varphi\|_W^2 = \int_0^T \{\|\varphi(t)\|_V^2 + \|\frac{d\varphi}{dt}(t)\|_{V'}^2\}dt \; .$$

(2.6) { Furthermore if the injection of V into H is compact then "the injection of W in $\mathcal{H} = L^2(0,T;H)$ is compact".

Hence if we consider the family of functional

$$(2.7)\qquad J_i(v) = \int_0^T \|y(v) - z_{d_i}\|_H^2 \, dt$$

The assumption (1.6) is verified ; then we have :

<u>Theorem</u> 2.1

<u>The linear-quadratic game given by</u> (2.1) <u>and</u> (2.2) <u>admits at least an equilibrium point in the bounded closed convex subset</u> $\mathcal{U}_{ad}$ <u>of</u> $\mathcal{U}$, <u>under assumption</u> (2.6).

<u>Remark</u> 2.1

Another approach for existence of a Nash point is, in the case of uncoupled contraints

$$(2.8)\qquad \mathcal{U}_{ad} = \prod_{i=1}^{n} \mathcal{U}_{ad,i} \qquad \mathcal{U}_{ad,i} \subset \mathcal{U}_i$$

is the theory of variational inequalities (cf. BENSOUSSAN [1], BENSOUSSAN-LIONS-TEMAM [1]) and the compactness argument of Theorem 2.1 corresponds to a pseudo-monotonicity property (in terminology of LIONS [2]) of the operator $\mathcal{A}$ defined by

$$(2.9)\qquad (\mathcal{A}(u),v) = \Sigma \left(\frac{\partial J_i}{\partial v_i}(u_i), v_i\right). \qquad \blacksquare$$

III - <u>AN ALGORITHM FOR SOLUTION OF THE GAME</u>

In this section we give an example of a feasible algorithm (for solution of the game) based upon a method of SCARF [1].

If we consider the function

$$(3.1) \qquad Z(u,v) = \sum_{i=1}^{n} J_i(u_1,\dots,u_{i-1},v_i,u_{i+1},\dots,u_n)$$

it is easy to show that the equilibrium point [(1)] is a <u>fixed point</u> for the mapping

$$(3.2) \qquad u \to \Gamma(u) = \{ v \in \mathcal{U}_{ad} \mid Z(u,v) \leq Z(u,w) \quad \forall \; w \in \mathcal{U}_{ad} \} .$$

Let us note that the mapping (3.2) is a point-to-point mapping in the case of strict convexity of J_i for $i=1,2,\dots,n$.

It follows from this remark that the search of a Nash point for the game is equivalent to the search of a fixed point for (3.2).

The original algorithm of SCARF consists in finding a fixed point for a continuous mapping Γ of a simplex of R^N in itself. Let S be the simplex of R^N :

$$S = \{v \in R^N \; / \; v_i \geq 0 \; \sum_{i=1}^{N} v_i = 1\} .$$

and let us consider a partition T_k of S in sub-simplices $S_1,\dots,S_k$. The principle of the method is to give a systematic way for walking over the vertices of the triangulation up to a vertex u "which is not far from its image $\Gamma(u)$". More precisely the result is

<u>Theorem</u> 3.1 (SCARF [1]).

<u>Let</u> $\varepsilon > 0$ <u>be given</u>. <u>Then there exists a partition</u> T_k <u>of</u> S (<u>in sub-simplices</u>) <u>such that the final vertex</u> u <u>of the algorithm satisfies</u>.

$$\|\Gamma(u)-u\|_{R^N} \leq \varepsilon . \quad \blacksquare$$

<u>Remark</u> 3.1

There is an other version of this algorithm which finds a fixed point for a continuous mapping of a convex polyedron in itself but the principle of the method is the same. ■

<u>Remark</u> 3.2

There are many others algorithm to solve a fixed point problem (to say nothing of contraction mappings !), but there are all based upon the SPERNER's lemma. Hence they possess a combinatorial aspect very sensitive to the dimensionality cf. EAVES [1], KUHN [1]. ■

An implementable version of the SCARF's algorithm is due to SCARF-HANSEN [2]. ■

IV - <u>AN EXAMPLE</u>

Let Ω be an open subset of R^p, Γ its boundary, $T > 0$ given and

$$Q = \Omega \times \left]0,T\right[, \quad \Sigma = \Gamma \times \left]0,T\right[.$$

We consider the following equation

---

(1) In the case of uncoupled constraints (2.8).

$$(4.1)\quad \begin{cases} \dfrac{dy}{dt} - \sum\limits_{i=1}^{p} \dfrac{\partial}{\partial x_i} \left(a_i(x) \dfrac{\partial y}{\partial x_i}\right) = f + \sum\limits_{j=1}^{n} v_j(t)\, \delta(x-c_j) \\ y|_{\Sigma} = 0 \\ y(x,0) = y_0(x) \end{cases}$$

where c_j are given points in Ω.

If we denote by $y(v)$ the unique solution of (4.1) with $v=(v_1,\dots,v_n)$ then we introduce the family of functionals :

$$(4.2)\quad J_i(v) = \int_0^T \int_{\Omega_i} |y(x,t;v) - z_{d_i}(x,t)|^2 dx dt + \nu_i \int_0^T |v_i(t)|^2\, dt \quad (\nu_i \geq 0)$$

Ω_i being a subset of Ω with meas. $(\Omega_i) \neq 0$ for $i=1,2,\dots n$. Then if we introduce

$$\mathcal{U}_{ad,i} = \{v_i(t) \,|\, \alpha_i \leq v_i(t) \leq \beta_i\} .$$

We can define our game by :

(4.3) "To find a Nash point for the family of functionals (4.2)". ∎

Remark 4.1

In spite of the (relative) simplicity of this problem, we cannot apply directly the theory of section 2 because of the Dirac's measures in (4.1). But this difficulty can be skipped easily by using an approximation of the δ-function.

Remark 4.2

This game is an idealization of an economic problem arising in allocation of water resources between regions.

The first equation of (4.1) represents the pressure of water in a phreatic. This phreatic supplies regions, each of them represented by its withdrawal :

$$v_j(t) \quad \delta(x-c_j)$$

where c_j is the location of the well number j and $v_j(t)$ is the flow-rate of water.

The cost function given by (4.2) has no direct interpretation. Actually it is necessary to modelize the economical structure of the region to exhibit the dependance of the profit with respect to the supply of water.

For a general approach of this problem cf. BREDEHOEFT-YOUNG [1].

Remark 4.3

The SCARF's algorithm is not directly applicable to problem (4.3) but it is possible to use a method analogous to the relaxation method combined with SCARF's algorithm cf. YVON [1]. ∎

REFERENCES

A. BENSOUSSAN [1] "Point de Nash pour des jeux différentiels à n personnes". To appear in SIAM J. of Control.

A. BENSOUSSAN, J.L. LIONS, R. TEMAM [1] Cahier de l'IRIA n°11. June 1972.

J.D. BREDEHOEFT, R. YOUNG [1] "The temporal allocation of a ground-water a simulation approach". Water Resource Research. Vol. 6 n°1. (1970)

B.C. EAVES [1] "Computing Kakutani fixed points" Siam J. of Appl. Math. Vol. 21 n° 2 (1971).

T. HANSEN, H. SCARF [1] "On the approximation of a Nash equil.point" Cowles Foundation. Discussion paper n°272.

H.W. KUHN [1] "Simplicial approximation of fixed points". Proc. N.A.S. n°6 (1968).

J.L. LIONS [1] Contrôle optimal des systèmes distribués. DUNOD (1968).

[2] Quelques méthodes de résolution des problèmes non linéaires. DUNOD Paris (1969).

J. NASH [1] "Equilibrium points in N person game". Proc. of N.A.A. Vol. 36 (1950).

J.B. ROSEN [1] "Existence and uniqueness of Equilibrium point for concave n-person game". Econometrica Vol. 33 n°3 (1965).

H. SCARF [1] I6 "The approx. of fixed points...". SIAM J. of App. Math. Vol. 15 n°5 (1967).

J.P. YVON [1] To appear.